NETWORKS AND TELECOMMUNICATIONS

PREFACE

The networks I have personally had to develop and operate have bridged not only the classical subdivisions of 'voice' and 'data', but also national boundaries, technological standards and regulations. As a result, I have found it necessary to become familiar with a broad range of technology, telecommunications practices and worldwide regulations. I have needed to know how things work and interwork, and the practical problems to be overcome. I have needed to understand the basic electrical engineering detail underlying telecommunications, but experience has also given me two equally valuable resources—an armoury of practical techniques and a knowledge of both technical and regulatory constraints.

In my time I have used plenty of books, each addressing one of the many individual technologies I have had to deal with. But to date I have not discovered a book aimed at technical managers who need a broad range of knowledge about how to develop and operate a number of different types of networks in a pragmatic way. This seems perverse, since many of today's professional 'telecommunications managers' are expected to know about all the various types of network making up their trade. It is also perverse given the fact that, in my experience, network problems cannot be neatly categorized by technology. I have rarely needed to know all the technical details of a particular technology and cannot remember having been able to use any network to its full capability. Seldom have I had the opportunity, since seldom have I had a problem where one technology alone will suffice, and never have I had the option anyway to scrap the established network with its contraints and start entirely afresh.

Surprisingly perhaps, my major problems have arisen not from technical incompatibility of different international networks—such problems are soluble with time and money. Rather the difficulty has been gaining a basic understanding of how different types of network operate, and therefore how they could be made to co-exist, interwork or co-operate to serve a given user's need. It is from this background that I decided to write this book—a day-to-day encyclopaedia for practical people like me—people faced with 'voice', 'data' and a whole compendium of other networks to be responsible for. Here in 33 chapters we have a picture of an evolving jungle of techniques and administrative controls, all about telecommunications today and how they got that way.

The book provides an insight for all of telecommunications affected parties: end users, engineers, private companies, public telecommunications operators, regulators and governments. The sort of people I hope will read it and use it—the interested parties—are students and academics; telecommunications and computer practitioners; and technical managers. Through our greater common awareness and understanding,

4

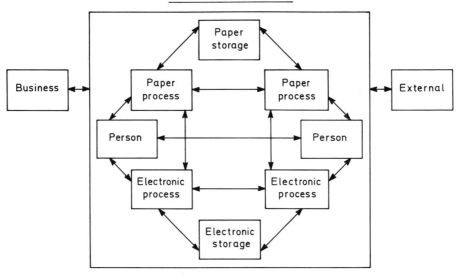

Figure 1.1
Categorizing information flows.

Mode	Examples	
	One - to - one	Broadly aimed
Paper	Sending a letter Leaving a note	Advertising board
Person - to - person	Talking Winking	Acting Television Radio
Electronic	← ———— Computer networking ———— →	

Figure 1.2
Categorizing simple communication methods.

an 'information environment', across which information flows in one of a number of different forms.

The simplest form of information flow (illustrated by Figure 1.1) might be directly from one person to another—by word of mouth or by a visual signal. Alternatively the information could have been conveyed on paper or electrically. The advantage of either of the latter two methods is that the information in paper or electronic form may also be readily stored for future reference.

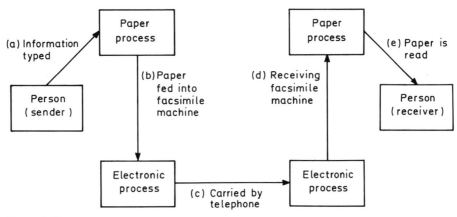

Figure 1.3
Information conveyance by facsimile.

In this book we shall use the model of Figure 1.1 twice. Here we use it to illustrate how different methods of communication may be categorized into one of the three broad types, and as a basis for explaining the prerequisite components of a telecommunications system. In Chapter 30 it is used to illustrate the analysis, simplification and planning of business information flows. This double theme runs throughout the book—understanding telecommunications technology, and explaining its exploitation in a pragmatic business-oriented manner. Well-known examples of communications methods falling into the three categories of paper, person-to-person and electronic are given in Figure 1.2.

Some types of communication are hybrids of the three basic methods. Modern facsimile machines, for example, are capable of relaying images of paper documents over the telephone network and recreating them at a distant location. This would appear as quite a complex information path on our model of Figure 1.1, as Figure 1.3 shows.

First a person must record the relevant information on paper (shown as (a) on Figure 1.3), then he must feed it into the facsimile machine which converts it into electronic format (b). Next the telephone network conveys the electronic information (c), before the distant facsimile machine reconverts the information to paper (d) and the receiver reads it (e).

The example we have chosen is rather convoluted, involving several successive conversions to take place, changing the format of the information from 'personal' to 'paper' to 'electronic' and back again. All these conversions make the process inefficient, and as we shall find out later, companies who have recognized this fact have already set about converting all their key business information into electronic (computer) format—not only for conveyance, but also for storage and processing purposes.

1.1 TYPES OF INFORMATION

Put specifically in the context of telecommunications, information might be a page of written text, a conversation or a television picture. The 'information' usually requires conversion into an electrical signal in order to be conveyed by 'telecommunication' means. But there are many different types of information—so can they all be treated identically? The answer to this is 'no', because each type of information makes slightly different demands on the system.

Information conveyed over telecommunications systems is usually classed as either 'analogue signal' information or as 'data'. An 'analogue signal' is an electrical wave-form which has a shape directly analogous to the information it represents (e.g. speech or a television picture). 'Data', on the other hand, is the word given to describe infor-mation in the form of text and numbers—in the jargon called 'alphanumeric character form'.

Different forms of data and analogue signal information require different treatment. For example, when conversing with someone we expect their reply to follow shortly after our own speech, but when we send a letter we do not expect a reply for some days. The analogy runs directly into telecommunications. Thus, an electrical represen-tation of conversation must allow the listener to respond instantly. But in the case of data communication, slightly more leeway exists, since a computer is prepared to accept 1–2 second response times. A human would find this length of delay intolerable in everyday speech. Another difference between electrical representations designed for different applications will be the speed with which information can be transferred. This is normally referred to as the 'information rate' or the 'bandwidth'. A speech circuit requires more bandwidth to carry the different voice tones than a telegraph wire needs simply to carry the same text. Later chapters in the book discuss the various methods of electrical representation and the technical standards used.

1.2 TELECOMMUNICATIONS SYSTEMS

There are four essentials for effective information transfer between two points, all of which are provided in well-designed telecommunications systems:

(i)   a transmitting device

(ii)   a transport mechanism

(iii)   a receiving device

(iv)   the fourth requirement is that only information which is compatible and compre-hensible to the receiver is transmitted by the transmitter

All the four components together form a telecommunications system.

In the example of a communications system consisting of two people talking to one another, the transmitting device is the mouth, the transport mechanism is the sound through the air, and the receiving device is the other person's ear. Provided that both people talk the same language, then the fourth requirement has also been met, and conversation can continue. But if the talker speaks English, and the listener only understands French, then, despite the availability of the 'physical components' of the system (i.e. mouth, ear and air), communication is ineffective owing to the incompat-ibility of the information. The coding and method of transfer of the information over the transport mechanism is to said to be the 'protocol'. In our example the protocol would be either the English or the French language: the fact that talker is English and listener French is an example of protocol incompatibility. Protocol also defines the procedure to be used. An example of the procedural part of protocol is the use of the word 'over' to signify the end of radio messages (for example 'Come in Foxtrot, Over'). The protocol in this case prompts a reply and prevents both parties speaking at once. The hardest part of telecommunications system design is often the need to ensure the compatibility of the protocol. In some cases, this necessitates the provision of 'interworking' devices. In our example, the interworking device might be a human English/French interpreter.

1.3   A BASIC TELECOMMUNICATIONS SYSTEM

Figure 1.4 illustrates the physical elements of a simple telecommunications system including the transmitter, the receiver and the transport mechanism. As already discussed, the physical element shown in Figure 1.1 must be complemented by the use of a compatible protocol between transmitter and receiver. Together with such a protocol, we have all the means for communication from point A to point B in Figure 1.4. We do not however, have the wherewithal for communication in reverse (i.e. from B-to-A). Such single direction communication, or 'simplex operation' as it is called, may suffice for some purposes. For many more examples of communications, two way, or 'duplex operation' is normally required. For duplex operation, a transmitter and a receiver must be provided at both ends of the connection as shown in Figure 1.5. A telephone handset, for example, contains both a microphone and an earphone.

The transport mechanism can be one of a range of different media, ranging from sound waves passing through air to light pulses passing down the latest technology optical fibre. Furthermore the transport mechanism may or may not comprise an element of switching, as we describe later in the chapter.

Most transport mechanisms demand an encoding of the information or data into a signal form suitable for conveyance over electrical transmission media. Chapters 2 to 5 describe how many of the common forms of information (e.g. speech, TV, telex, computer data, facsimile etc.) are converted into a transmittable signal—carried in either 'analogue' or 'digital' form. In Chapters 6 and 9 we discuss various methods of switching and in Chapter 8 we discuss a range of different transmission media (cables,

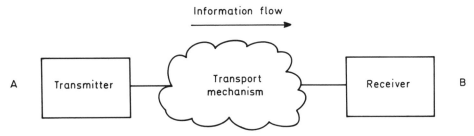

Figure 1.4
Basic physical elements of a telecommunications system (*simplex operation*).

radio systems etc.), describing how different ones provide the optimum balance of low cost and good transmission performance for individual cases of application.

1.4 COMMON TYPES OF TELECOMMUNICATIONS SYSTEMS

In order to meet differing communications needs, a number of different types of telecommunications equipment have been developed over time. These include in chronological order:

- telegraph
- telephone
- telex
- data networks—using either 'circuit-switched' or 'packet-switched' conveyance,
- computer 'local' and 'wide area networks' (LANs and WANs),
- integrated voice and data networks.

This book covers the principles involved in each of the above telecommunications types, it also aims to give adequate background to enable the reader to tackle basic network planning of any of these types. The book covers networking from simple interconnection of two computers right up to complex, globally spread, telecommunications networks.

In the main, the book describes the general principles applying to all types of telecommunications networks, though specific chapters give coverage on particular types of network where this is necessary. Telephone and computer network practice is covered extensively.

An illustration of the similarity of the underlying principles is afforded in current development trends in telecommunications. Today's emerging standard is for integrated voice and data networks, carried by the so-called 'integrated services digital network' (ISDN). The ISDN is covered in Chapter 24.

1.5 NETWORKS

Let us now consider the ideal properties of the various components of the telecommunications system illustrated in Figure 1.5. This system is ideal provided the two 'stations', A and B, wish to communicate with one another but never with anyone else. If A and B are both provided with telephones the transport mechanism need be no more than a single transmission line, as illustrated in Figure 1.6.

If a third station C, wishes to be interconnected for private interconnection with either or both of the other two stations (A and B) then this could be achieved by duplication of the simple layout. In this way a triangular network between A, B and C is created as shown in Figure 1.7.

The configuration of Figure 1.7 is used today by some companies in their private

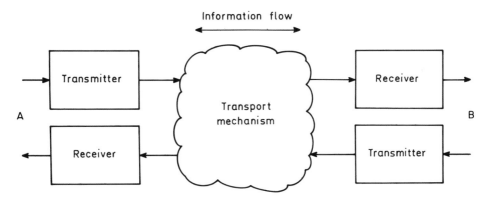

Figure 1.5
A basic duplex telecommunications system.

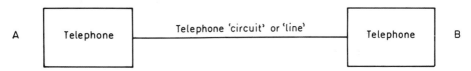

Figure 1.6
A simple two station telephone system.

networks, where a dedicated 'private line telephone' may operate over a special telephone line, leased from a telecommunications administration, to connect different premises. The configuration is uneconomical in equipment, for a network of numerous 'stations'. In the three station (A, B, C) case illustrated, six telephones and three lines are needed to interconnect the stations, but only two telephones and one line can ever be used at any one time (unless one of the people is superhuman and can talk and listen on more than one telephone at a time).

As even more stations are introduced to the configuration, the relative inefficiency grows. In a system of N stations in which each has a direct link to each other, a total of $1/2\{N(N-1)\}$ telephone lines will be needed, together with $N(N-1)$ telephone sets. If configured in the manner shown in Figure 1.7, the linking of 100 stations would need 5000 links (and 10 000 telephones) and a 10 000 station system would need 50 million lines and 100 million telephones. We need to find a more efficient configuration.

Let us limit each station in Figure 1.7 to one telephone only. To make this possible we install a switching device at each station to enable appropriate line selection, in order that connection to the desired destination may be achieved on demand. This is now a simple switched network, as Figure 1.8 shows. Now the transport mechanism (stylized in Figure 1.5) is no longer just a single 'line', but is a more complex switch and line arrangement.

Let us develop Figure 1.8 further, by permitting more stations (telephones in this case) to be connected to each of the three switches. Three more stations, A', B', and

C′ are shown in Figure 1.9. The new configuration allows the idle lines of Figure 1.8 (A–C and B–C) to be put to use.

Figure 1.9 illustrates simultaneous calls involving A and B; B′ and C′; A′ and C. In this example each of the switches (which are now labelled instead as 'exchanges') is shared by a number of stations, each of which is switched and connected to the exchange by a 'local line'. Our example now resembles a public-switched telephone

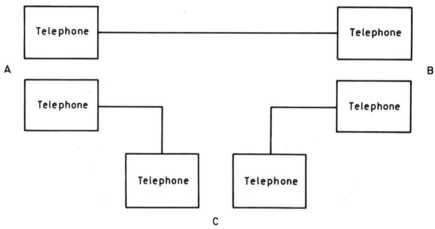

Figure 1.7
Three stations interconnected by independent telephone lines.

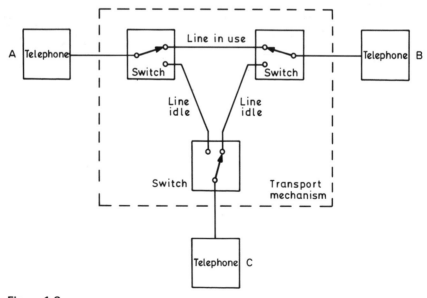

Figure 1.8
A simple three station switched network.

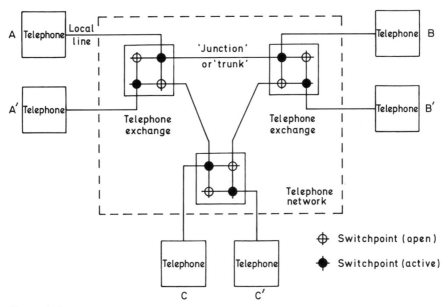

Figure 1.9
A simple telephone network.

network (PSTN). And since the lines between exchanges are correctly referred to as 'junctions' or 'trunks', they have been labelled accordingly.

In a real telephone network the number of exchanges and their locations are governed by the overall number and geographical density of stations requiring interconnection. Similarly, the number of 'junctions' or 'trunks' provided between the various exchanges will be made sufficient to cater for the normal telephone call demand. In this way, far fewer junctions than stations need to be provided. This affords a significant cost saving over the configuration of Figure 1.6.

Before leaving Figure 1.9, notice how each of the exchanges has been drawn as an array of individual switch points. This allows either of the telephones connected to the exchange to access either of the junctions, and is a so-called 'full availability' switch since any one of the incoming lines may be connected to any one of the available junctions. We shall come back to circuit theory of switching and availability in Chapter 6.

In our example we have justified on economic grounds alone, the use of 'switched' as opposed to 'transmission line only' networks. The particular case we have developed is an example of a '*circuit-switched network*'. Other important switched network types, especially used for data transmission, are those of *packet switching* and *message switching*. The following sections describe, compare, and contrast circuit- and packet-switched networks. Message switching networks may employ either the circuit switching or packet switching technique. The term is simply used to describe the practice of ensuring the final delivery of the message (e.g. letter or document) in one go. Communication over 'message-switched networks' is much like sending and receiving letters. We discuss the subject further in Chapter 25.

1.6 CIRCUIT- AND PACKET-SWITCHED NETWORKS

The distinguishing property of a *circuit-switched connection* is the existence, throughout the communication phase of the call, of an unbroken physical and electrical path between origin and destination points. The path is established at 'call set-up' and 'cleared-down' after the call. The path may offer either one direction (simplex) or two direction (duplex) use.

Conversely, in *packet-switched networks*, an entire physical path from origin to destination will not generally be established at any time during communication. Instead, the total information to be transmitted is broken down into a number of elemental 'packets' each of which are sent in turn. An analogy might be sending the text of a large book through the post in a large number of envelopes. In each individual envelope might be just a single page. The envelopes can either be sent sequentially (say one each day), or numbered so that the receiver can reassemble the pages in order. Figure 1.10 illustrates this simple packet-switched communication system. Should the receiver in Figure 1.10 not receive any given numbered envelope, he may write back and re-request it. In this way, very accurate and reliable communication may be established. Circuit-switched networks are generally necessary when very rapid or instantaneous interaction is required (as is the case with speech). Conversely, packet-switched networks are more efficient when instantaneous reaction is not required, but when very low 'corruption' (either through distortion or loss of information) is paramount. Message-switched networks have thus become the backbone of many data communications systems.

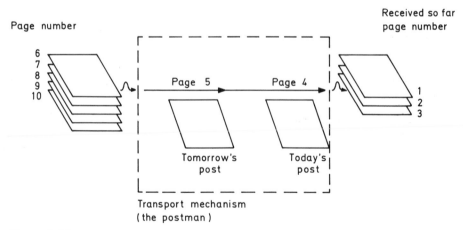

Figure 1.10
A simple 'message switched' communication system.

1.7 OTHER CONSIDERATIONS FOR NETWORK PLANNERS

We have established the economic value of switched networks (as opposed to direct wire systems), and briefly addressed the needs for basic communication (the physical, switching and protocol needs). But what other factors have to be provided in order to enable networks to function in a manner fit for purpose? The following list is a brief summary of some of the factors which require consideration.

● *Transmission plan*. A plan laying out guidelines according to which appropriate transmission media may be arranged in order that adequate end-to-end conveyance of information is achieved. The plan will include safeguards to ensure that the received signal is loud enough, clear enough and free of noise and interference.

● *Numbering and routing plans*. The numbering plan is crucial to the ability of the network to deliver communications to the appropriate destination. Much like reading an address on an envelope, it is the inspection of the destination number that permits the determination of destination and appropriate route within the network.

● *Usage monitoring plan*. A usage monitoring plan is needed in order to ensure continuing and future suitability of the network. Adequate measurements of network performance are required in order that early steps may be taken, as necessary, to adjust the network design, or expand its overall throughput capacity to meet demand.

● *Charging and accounting plan*. Public telecommunication operators (PTOs) need reimbursement for their services. Specific equipment may be required to monitor individuals' use of the network, in order that bills can be generated.

● *Maintenance plan*. A maintenance plan is needed to guarantee continuation of service without a break.

Any of these items may be appropriate to any type of telecommunications network regardless of what type of information it is carrying (e.g. telex, telephone). Each is discussed more fully in later chapters.

1.8 TECHNICAL STANDARDS FOR TELECOMMUNICATIONS SYSTEMS

As a way of promoting greater compatibility between various telecommunications systems in different parts of the world, a number of technical standards defining bodies are active in various parts of the world. These are discussed in more detail in Chapter 21. However, two of the most significant and influential bodies are worth mentioning now. These are:

● the International Organization for Standardization.

● the International Telecommunications Union (ITU), and specifically its sub-entities, the International Telegraph and Telephone Consultative Committee (CCITT) and the International Radiocommunication Consultative Committee (CCIR).

The 'standards' and 'recommendations' of these two bodies are commonly used throughout telecommunications, and are often referred to in this book.

BIBLIOGRAPHY

Aries, S. J., *Dictionary of Telecommunications*, Butterworths, 1981.

Davies, E., *Telecommunications—A Technology for Change*. London, HMSO, London Science Museum, 1973.

From Semaphore to Satellite—The Centenary of the International Telecommunication Union. ITU Geneva, 1965.

Murphy, R. J., *Telecommunications Networks—A Technical Introduction*. Harold W Sams, 1987.

Overman, M., *Understanding Telecommunications*. Lutterworth Press, 1974.

Smale, P. H., *Telecommunications Systems (TEC Level 1)*. 2nd edn. Pitman, 1981.

INTRODUCTION TO SIGNAL TRANSMISSION AND THE BASIC LINE CIRCUIT

2.1 INFORMATION ENCODING AND ELECTRICAL SIGNALS

To make it suitable for carriage over most telecommunications networks, information must first be encoded in an electrical manner, as an electrical signal. Only such signals can be conveyed over the wires and exchanges that comprise the 'transport mechanism' of telecommunications networks, as we discussed in Chapter 1.

Over the years, a variety of different methods have been developed for encoding different types of information. This chapter briefly reviews these methods. It starts by describing the most basic element of all; the electrical transmission link itself.

The chapter categorizes all methods of information into one of two basic types, 'analogue transmission' and 'digital transmission', describing the principles of both types but concentrating on the technique of analogue transmission. Chapter 3 continues the subject of analogue transmission, explaining its use for information conveyance over long distances, while Chapters 4 and 5 expand the details of digital transmission.

2.2 ANALOGUE AND DIGITAL TRANSMISSION

The simplest type of transport mechanism used in telecommunications networks is a pair of electrical wires. These allow the conveyance of an electrical current of 'signal' from a transmitter at one end of the wires to the receiver at the other end.

There are two basic methods for information encoding which may be applied by the transmitter and decoded by the receiver. These are analogue signal encoding and transmission, and digital encoding and transmission.

As the name suggests, analogue signal encoding involves the creation of an electrical signal waveform which is analogous to the waveform of the original information (e.g.

(a) Typical analogue signal

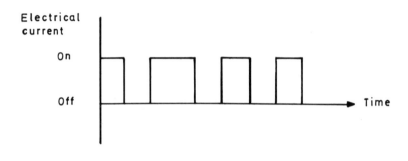

(b) Typical digital signal

Figure 2.1
Typical analogue and digital signals.

a speech wave pattern). In this way the electrical signal is similar in shape to the speech or other information waveform which it represents. Analogue transmission lines are used to convey analogue encoded signals and have historically predominated as the links of the world's telecommunications networks. 'Analogue exchanges' have provided the nodal interconnections between these links. Digital encoding and transmission of information (a technique in which information is conveyed as a string of 'digits', or more precisely, pulses of 'on' and 'off' electrical current) is rapidly taking over as the main method of telecommunications transmission. This is because of the significant benefits that digital transmission offers, both in terms of performance and cost. Figure 2.1 illustrates typical analogue and digital signal waveforms.

As for the transmission line systems themselves, Table 2.1 presents a short summary of some of the physical media which may be used, and compares their relative merits. In general any physical medium may be designed to work in either an analogue or a digital mode, and each of the different transmission media listed in Table 2.1 have been used as the basis of both analogue and digital transmission systems. Most transmission system manufacturers offer equipment capable of the different permutations of transmission medium and encoding type. Choosing which medium and encoding system should be used is a decision for the network planner, who needs to consider

Table 2.1
Comparison of common transmission media.

Transmission medium used as basis of line system	Comments
Copper wire (or aluminium wire)	Ordinary wire. The gauge (or thickness) will be carefully selected for the particular use. The pairs of wires may be either 'parallel' or 'twisted' within the 'cable', which usually comprises many pairs. Copper and aluminium wire is the predominant system used for the 'local loop network'.
Coaxial cable	Centre conductor plus cylindrical conducting sheath. Used on long-haul or 'trunk' systems. Particularly used to carry many tens of hundreds of communications channels simultaneously; in 'multiplexed mode' (described later).
Microwave radio	Small dish, 'line-of-sight' radio system. Used for long-haul 'trunks', particularly where terrain is difficult for cable laying.
Tropospheric scatter	Radio system using high power to achieve 'over the horizon' communication. 100–600 km but susceptible to fading.
High-frequency radio (HF, VHF and UHF)	Used historically for ship-to-shore communications. Now increasingly used by the flourishing mobile cellular radio networks of portable telephone handsets.
Satellite	Satellites are generally intended to be 'geostationary' (permanently above a particular point on the equator). They give single hop access to approximately one third of the globe. The disadvantage is the comparatively long half-second transmission time.
Optical fibre	Glass fibre, of a thickness comparable with a human hair. Used in conjunction with lasers, optical fibres are capable of the carriage of thousands of simultaneous calls. Generally only suitable as 'digital' systems.

the relative costs and the ease with which any new type of line system could be incorporated into the existing network. We shall discuss the relative benefits of different transmission media in more detail in Chapter 8.

Switching equipment and exchanges are also classified into 'analogue' and 'digital' types. A switch, is after all, only a specialized and very flexible form of transmission equipment.

To develop our understanding of information encoding, we now discuss the early encoding techniques, concentrating in particular on the principles of analogue transmission.

2.3 TELEGRAPHY

Electric telegraphy was the earliest common method of digital transmission. It began in the early part of the nineteenth century, and became the forerunner of modern telecommunications. Samuel Morse, inventor of the Morse code, developed a working telegraph system in 1835. By 1837 a practical system was established from Euston railway station in London to Camden, a mile away. The system came about through the desire of railway signalmen to ensure the safety of trains, the number of which was booming as a result of the industrial revolution. In 1865 Morse code was accepted

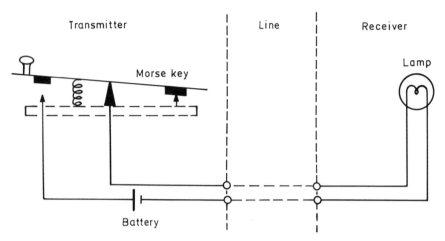

Figure 2.2
A simple telegraph circuit.

as a world standard during the first International Telegraph convention, and this gave birth to the International Telecommunications Union (ITU). Public interest in telegraphy grew rapidly until about 1880, when telephones became more widely available.

Telegraphy formed the basis of the early telegram service. It provided the means of sending short textual messages between post offices so that immediate delivery could then be arranged by the 'telegram boy' carrying it by cycle to the recipient's premises.

A simple telegraph circuit is illustrated in Figure 2.2. The system works by transmitting simple on/off electrical pulses from the transmitter (the key) to the receiver (the lamp). A single pair of wires provides the transmission medium for the electrical current between origin and destination.

In the early days all telegraphy was carried out by manual keying (switching the circuit 'on' and 'off') according to the Morse code. In this code the message is split up into its component alphabetic characters, each of which is represented by a unique sequence of short or long electrical pulses (so-called 'dots' and 'dashes'). For those readers not familiar with the code it is described more fully in Chapter 4.

Dots and dashes are sent by pressing the 'key' of Figure 2.2 for an appropriate length of time. Consecutive dots or dashes are separated by a 'space' (a period of no current—key not pressed) equal in length to one 'dot'. Three successive spaces are used to separate the sequences of dots and dashes which together represent a single character. Words are separated by five spaces.

As telegraph networks were gradually replaced by telex networks, which essentially are no more than 'automatic telegraph' networks, so the Morse code was superseded by the Murray code, another digital code, but one which was especially designed for automatic working. It differs from the Morse code in that every alphabetic and numeric character is represented by a fixed number (five) of electrical pulses, so-called 'marks', or 'spaces'. Unlike the dots and dashes of the Morse code, the marks and spaces of the Murray code are all of equal length. The code is more suitable for automatic working since it is easier to design mechanical equipment to respond to pulses (either marks or spaces) which have a predetermined duration. For example, a telex receiver or 'teleprinter' can more easily select and print characters at a fixed speed than it can at a variable one. The evolution from the Morse to the Murray code took place in a number of stages, first to a manual and then an automatic teleprinter system, finally to the widespread telex system. The output of the early teleprinter devices was a narrow strip of paper which could be stuck on to a receiving telegram form. The telex system by contrast prints direct on to a roll of ordinary paper.

In both telegraph and telex networks, two-way sequential transmission over a single pair of wires is possible simply by providing a device capable of both the transmitting and the receiving functions at both ends of the line.

2.4 TELEPHONY

Telephony, invented by Alexander Graham Bell and patented in the USA in 1875–7 is perhaps today's most important means of long-distance communication. Certainly in terms of overall volume of information carried, it is far and away the most important of the long-distance telecommunications methods.

Telephony is the transmission of sound, in particular, speech, to a distant place. The literal translation of 'tele–phone' from the Greek means 'long distance–voice'.

Telephony is achieved by conversion of the sound waves of a speaker's voice into an equivalent electrical signal wave. Let us consider the characteristics of sound, in order that we may understand how this conversion is performed. Sound is really only vibrations in the air around us, caused by localized pockets of high and low air pressure which have been generated by some form of mechanical vibration, for example an object hitting another, or the vibration of the human vocal chords during speech.

Sound waves in the air around us cause objects in the vicinity to vibrate in sympathy. The human ear detects sound by the use of a very sensitive diaphragm, which vibrates in synchronism with the sound hitting it.

The pitch of a sound (how 'high' or 'low' it sounds) depends upon the 'frequency' of the sound vibration. A typical range of frequencies which are audible to the human ear are 20–20 000 Hertz (or vibrations per second). Low frequencies are heard as low notes, high frequencies as high notes, and most sound is a very complex mixture of frequencies as Figure 2.3 illustrates. Figure 2.3(a) is a 'pure' tone of a single frequency. To a listener this is the rather dull and insipid note produced by a tuning fork. By contrast, Figure 2.3(b) shows the waveform of the vowel 'o', as in 'hope'. Notice how the waveform is not as smooth as before. In reality the waveform will be even more complicated, preceded and followed by other complicated patterns making up the sounds of the rest of the word. Real sounds have an irregular and spiky waveform.

Concert grade musical performances use a wide range (or 'bandwidth') of frequencies within the human aural range. However, for the purpose of intelligible speech, only the frequencies from 300–3400 Hz (Hertz) are necessary. If only this bandwidth of frequencies is received the quality is slightly downgraded but is still entirely comprehensible and acceptable to most listeners.

So how can the sound waveform be converted into an electrical signal? Figure 2.4

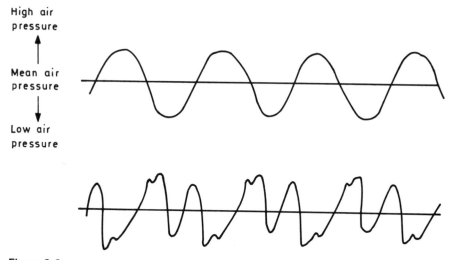

Figure 2.3
(Top) Waveform of 'pure' (single frequency) tone. (Bottom) Typical speech waveform.

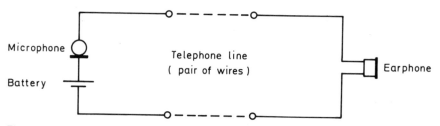

Figure 2.4
A simple telephone circuit.

illustrates a simple telephone circuit capable of the conversion. It consists of a microphone, a battery, a telephone line and an earphone.

When no one speaks into the microphone, then both the microphone and earphone have a steady electrical resistance, and so a constant electrical current flows around the circuit according to Ohm's law (current = voltage divided by resistance). If someone speaks into the microphone, then the sound waves hitting the microphone diaphragm cause the electrical resistance of the device to alter slightly. The change in resistance results in a corresponding change in the circuit current. To illustrate this, Figure 2.5 shows the circuit components in more detail. On the left of the diagram is a 'carbon microphone', common in many countries. The diaphragm of the microphone compacts or rarefacts the carbon granules, and this results in the change of resistance. The change in resistance causes a change in the current around the circuit. The current varies about the steady state value in the same way as the air pressure in the sound wave fluctuates about its average pressure. The effect is to create an electrical signal fluctuating in almost the same way as the original sound wave.

Conversely, in the earphone the electrical signal is reconverted to sound. The simple earphone shown in Figure 2.5 is constructed from an iron diaphragm and an electro-magnet. An electromagnet is an electrical device which acts like a magnet when current passes through its winding and tends to attract the iron diaphragm accordingly. The strength of the attraction depends upon the magnitude of the current flowing in the coil of the electromagnet (i.e. upon the telephone circuit current). The varying electrical signal created by the microphone results in a varying force of attraction on the diaphragm. This vibrates the surrounding air and recreates the original soundwave.

The conversion process described above is an 'analogue' one, since the electrical waveform carried through the 'analogue network' is analogous to the original sound waveform.

2.5 RECEIVED SIGNAL STRENGTH, SIDETONE AND ECHO

In an analogue network, it is important that the received electrical signal strength is sufficient to reproduce sound waves loud enough for the listener to hear. Herein lies a problem with the circuit of Figure 2.5. The circuit works adequately at short line lengths, but if too long a line were to be used, then because the electrical resistance of the microphone would be substantially smaller than the resistance of the line itself, the microphone would not be able to stimulate much fluctuation in the circuit current. The result is faint sound reproduction. To get around this problem, the circuit is improved by the use of transformers, as shown in Figure 2.6. The greater number of turns on the secondary winding of each transformer (as compared with the primary winding) reduces the effect of the line resistance.

You will also notice in Figure 2.6 that the circuit connects the equivalent of two complete telephone handsets (i.e. it contains two microphones and two earphones). The circuit is therefore capable of two-way communication.

A real telephone circuit normally uses a more complicated circuit. The added complication is needed to control the effect of 'sidetone' and reduce the chance of 'echo'. Consider the following example in explanation of these effects. As party A speaks, the voice will be heard in earphone B, but also in earphone A. The speaker will therefore hear 'sidetone' of his or her own voice. If the sidetone is too loud, the speaker may be distracted. Conversely, if there is no sidetone, the speaker sometimes gets the impression that the phone is 'dead'. An anti-sidetone circuit is normally incorporated

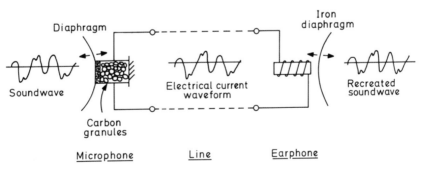

Figure 2.5
The carriage of a soundwave over a simple telephone circuit.

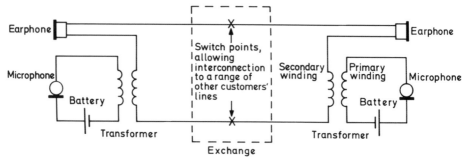

Figure 2.6
Basic two-way telephone circuit.

into the telephone circuitry to control the level of sidetone. It does so by restricting the volume of speech and other noises picked up by the microphone from being heard in the speaker's own earphone. The anti-sidetone circuit also eases the difficulty of hearing weak incoming signals in noisy places (e.g. a public telephone in a café). The circuit has the further benefit that it controls the undesirable effect of echo. Consider again speaker A's position. The voice is emitted by earphone B, and if too loud is picked up by microphone B and returned to earphone A. The slight delay introduced by the signal transit time (along the line and back) means that 'A' hears a slightly delayed echo of his or her own voice. The anti-sidetone circuit in instrument B helps to control this effect, by reducing the signal 'feedback' from the earphone to the microphone of B's handset. The 'echo path' from microphone A—via earphone B, microphone B and earphone A—is therefore largely eliminated. We will learn in Chapter 13, however, that this is not the only possible echo path in practical networks, and that even more precautions need to be taken.

2.6 AUTOMATIC SYSTEMS—CENTRAL BATTERY AND EXCHANGE CALLING

A practical problem encountered by the circuit in Figure 2.6, when it was employed in early telephone networks was that it required an individual battery to be provided in each telephone customer's line. Thus a 'common (or central) battery' (CB) system was developed. One large battery, or power system, is positioned in each exchange and provides the necessary power for all customers' lines terminated at that exchange. The use of a CB system (in preference to individual line batteries) has significant benefit in simplifying the job of maintenance.

Figure 2.7 illustrates a central battery configuration. Note that a high impedance coil has been added to the circuit. This forms part of another normal circuit feature called the 'transmission bridge'. Without the transmission bridge coil the speech currents from one telephone would be shorted out by the battery and not heard in the receiving telephone. An inductive 'coil' prevents this 'shorting'. A high resistance can be used but this has the disadvantage of causing conversation 'overhearing' in all the other telephones connected to the same central battery.

Returning now to the circuit of Figure 2.6 there are three more deficiencies that

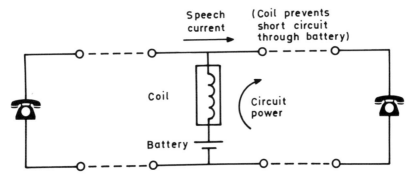

Figure 2.7
A simple transmission bridge.

make it unsuitable for direct use in a real telephone network. These are:

(i)   the circuit is always drawing electrical power,

(ii) the customer has no means of calling the exchange, in order to set-up calls to other customers,

(iii) the exchange has no means of 'alerting' the called customer (for example, ringing the telephone).

The first two problems above can be solved by providing a switch in the telephone handset which is activated when the handset is lifted off the 'cradle'. In this way the telephone circuit is 'open circuit' when the handset is 'down' and does not draw power when not in use. When required for use, lifting the handset completes the circuit, an action which is sometimes called 'looping' the circuit. On detection of the 'loop' the exchange readies itself to receive the telephone number of the 'called party', and indicates this readiness by returning 'dialling tone'. The 'calling customer' may then dial the number on the telephone handset and the number is relayed to the exchange by one of the following two signalling methods:

(i)   'loop' and 'disconnect' pulsing,

(ii) multi-frequency (MF) tone signalling.

'Loop' and 'disconnect' pulsing or 'LD signalling' is the older of the two methods and the most prevalent. The line is 'disconnected' and re-looped in a very rapid sequence. The effect is to cause a short period, or 'pulse' of disconnection. The pulses are used to indicate the digits of the destination number, one pulse for a digit of value 'one', two pulses for value 'two' etc., up to 10 pulses for digit 'zero'. The pulses are repeated at a frequency of ten per second. Thus the digit '1' will take one tenth of a second to send, while digit '0' will take a whole second. Between each digit, a longer gap of half a second is left, in order that the exchange may differentiate between the sequences of pulses representing consecutive digits of the overall numbers. The gap is called the 'inter-digit pause (IDP)'. A telephone number is typically ten digits in length and takes

Table 2.2
Multi-frequency (MF) tone signalling (customer to exchange).

Low frequency group	High frequency group			
	1209 Hz	1336 Hz	1477 Hz	1633 Hz (not used)
697 Hz	1	2	3	spare
770 Hz	4	5	6	spare
852 Hz	7	8	9	spare
941 Hz	*	0	#	spare

6–15 seconds to dial from the telephone, depending upon the values of the actual digits.

Multi-frequency (MF), or 'tone' signalling is a more recent invention. Its use is growing because the digits may be relayed much faster (in 1–2 seconds) and this reduces the call set-up time or 'post dialling delay' (PDD) if the exchanges are fast in their switching action. The exchange still needs to be alerted initially by looping the line, but following the dial tone the digits are dialled by the customer as a number of short bursts of tones. Rather than detecting the disconnect pulses, the exchange must instead detect the frequency of the tones in order to determine the digit values of the destination number. Each digit is represented by a combination of two pure frequency tones. The use of two tones in combination reduces the risk of mis-operation if other interfering noises are present on the line. Table 2.2 gives the frequencies used.

Following receipt of the number, the exchange network sets up a connection to the desired destination and 'alerts' the 'called customer'. 'Ringing' the telephone is the normal means of 'alerting', or attracting the called customer's attention. To achieve ringing, a large alternating current is applied to the called customer's line to activate the bell. Simultaneously, a 'ring tone' is returned to the caller. The bell does not ring at other times because the normal line current is insufficient to cause the bellhammer vibration.

In response to the ringing telephone, the called customer picks up the handset, and therefore 'loops' the line. On detection of the 'loop', the exchange 'trips' (ceases) the ringing, and conversation may then take place. At the end of the conversation the caller replaces the handset thereby removing the loop, so that the line is disconnected. The exchange responds by 'clearing' the rest of the connection.

2.7 REAL COMMUNICATIONS NETWORKS

We have now developed all the basic principles of telephone communication. Similar principles apply in nearly all communication network types, and the remaining chapters of this book develop these principles, discussing in detail the solutions to the complications that exist in practice. Before concluding the chapter, however, it is valuable to point out another reality of practical networks. While similar principles might form the foundation of nearly all telecommunications networks, it is not true to say that all

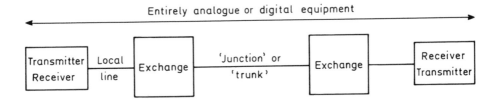

(a) A simple network of entirely analogue or digital components

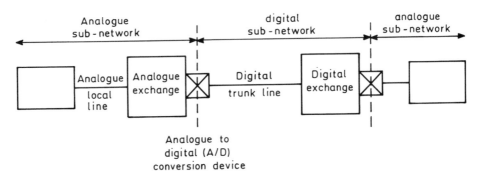

(b) A network comprising both analogue and digital components

Figure 2.8
Interworking dissimilar networks.

networks are fully compatible, allowing direct interconnection. Indeed, direct inter-connection of dissimilar networks (e.g. telephone and telex) is almost impossible, with little exception. Where interconnection of dissimilar networks is possible, it is usually by the use of special 'interworking' or conversion equipment at the interface of the two dissimilar networks.

Even two telephone networks, one operating using the analogue encoding tech-nique, and the other using digital, cannot be directly interconnected without the use of conversion equipment. The equipment is needed to convert the information from its digitally encoded form into the equivalent analogue signal, and vice versa. The final diagram in this chapter, Figure 2.8 shows an example of interworking dissimilar net-works. It illustrates the positioning of analogue to digital (and vice versa) conversion equipment at the interface of interconnected analogue and digital sub-networks.

BIBLIOGRAPHY

Lowe, R. and Nave, D., *The Electrical Principles of Telecommunications*. Macmillan, 1973.
Rhodes, F. L., *Beginnings of Telephony*. Harper & Brothers Publishers, 1929.

Roddy, D., *Radio and Line Transmission* (Vols 1 and 2). Pergamon Press, 1967.

Seshadri, S. R., *Fundamentals of Transmission Lines and Electromagnetic Fields*. Addison-Wesley, 1971.

Sinnema, W., *Digital, Analog and Data Communication*. Reston Publishing Company, 1982.

Transmission Systems for Communications, Fourth edn. Bell Telephone Laboratories, 1970.

3

LONG-HAUL COMMUNICATION

None of the circuits described in the previous chapter are suitable as they stand for long-haul communication, not even with a range-boosting transformer. To get us anywhere with long haul we have to address ourselves to the following pertinent topics:

● Attenuation (loss of signal strength over distance)

● Line loading (a way of reducing attenuation on medium length links)

● Amplification (how to boost signals on long-haul links)

● Equalization (how to correct tonal distortion)

● Multiplexing (how to increase the number of 'circuits' that may be obtained from one physical cable)

In this chapter we discuss predominantly how these line issues affect analogue transmission systems and how they may be countered. The effects on digital transmission are discussed in later chapters.

3.1 ATTENUATION AND REPEATERS

Sound waves diminish the further they travel and electrical signals also get weaker as they pass along electromagnetic transmission lines. With electrical signals the attenuation (as this type of loss in signal strength is called) is caused by the various electrical properties of the line itself. These properties are known as the 'resistance', the 'capacitance', the 'leakance' and the 'inductance'. The attenuation becomes more severe as the line gets longer. On very long-haul links the received signals become so weak as to be imperceptible, and something has to be done about it. Usually the attenuation in analogue transmission lines is countered by devices called 'repeaters', which are located at intervals along the line, their function being to restore the signal to its original wave shape and strength.

Signal attenuation occurs in simple wireline systems, in radio, and in optical fibre systems. The effect of attenuation upon a line follows the function shown in Figure

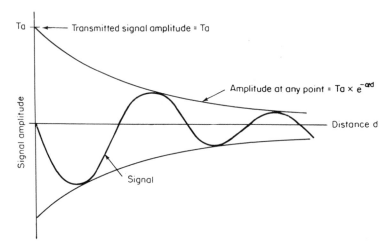

Figure 3.1
The effect of distance on attenuation. α = attenuation constant.

3.1, where the signal 'amplitude' (the technical term for signal strength) can be seen to fade with distance travelled according to a negative exponential function, the rate of decay of the exponential function along the length of the line being governed by the attenuation constant, alpha (α).

Complex mathematics, which we will not go into here, reveals that the value of the attenuation constant for any particular signal frequency on an electrical wire line transmission system is given by the following formula:

$$\alpha = \sqrt{(\tfrac{1}{2}\{\sqrt{[(R^2 + 4\pi^2 f^2 L^2)(G^2 + 4\pi^2 f^2 C^2)]} + (RG - 4\pi^2 f^2 LC)\})}$$

The larger the value of α the greater the attenuation, the exact value depending upon the following line characteristics:

R: the electrical resistance per kilometre of the line, in ohms

G: the electrical leakance per kilometre of the line, in ohms

L: the inductance per kilometre of the line, in henries

C: the capacitance per kilometre of the line, in farads

f: the 'frequency' of the particular component of the signal.

The *resistance* of the line causes direct power loss, by impeding the onward passage of the signal. The *leakance* is the power lost, by conduction through the insulation of the line. The *inductance* and *capacitance* are more complex current-impeding phenomena caused by the magnetic effects of alternating electric currents.

It is important to realize that because the amount of attenuation depends upon the frequency of the signal, then the attenuation may differ for different frequency components of the signal. For example, it is common for high frequencies (treble tones) to be disproportionately attenuated, leaving the low frequency (bass tones) to dominate. This leads to 'distortion', also called 'frequency attenuation distortion' or simply 'attenuation distortion'.

3.2 LINE LOADING

Since the inductance and capacitance work against one another, a simple means of minimizing the effect of unwanted attenuation and distortion is to increase the inductance of the line in order to counteract some or all of the line capacitance.

Taking a closer look at the equation given in the last section we find that attenuation is zero when both resistance and leakance are zero, but will have a real value when either resistance or leakance is non-zero. The lesson is that both the resistance and the leakance of the line should be designed to be as low as is practically and economically possible. This is done by using large 'gauge' (or diameter) wire and good quality insulating sheath.

A second conclusion from the formula for the attenuation constant is that its value is minimized when the inductance has a value given by the expression:

$$L = CR/G \quad \text{henries/km}$$

This in effect, reduces the electrical properties of the line to a simple resistance—minimizing both the attenuation and the distortion simultaneously. In practice, the

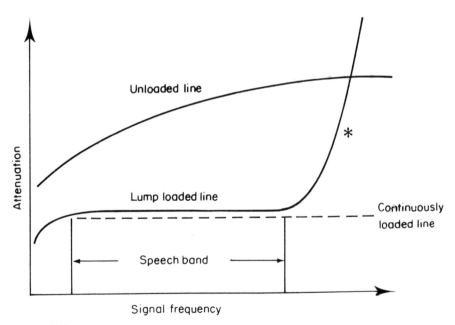

Figure 3.2
Attenuation characteristics of loaded line.

attenuation and distortion of a line can be reduced artificially by increasing its inductance, L—ideally in a continuous manner along the line's length. The technique is called 'line loading'. It can be achieved by winding iron tape or some other magnetic material directly around the conductor, but it is cheaper and easier to provide a 'lumped loading coil' at intervals (say 1–2 km) along the line. The attenuation characteristics of unloaded and loaded lines is shown in Figure 3.2.

The use of line loading bears the disadvantage that it acts as a high frequency filter—tending to suppress the high frequencies. It is therefore important when lump loading is used, to make sure that wanted speech band frequencies suffer only minimal attenuation. If the steep part of lump loaded line curve (marked by an asterisk in Figure 3.2) were to occur in the middle of the speech band, then the higher frequencies in the conversation would be disproportionately attenuated, resulting in heavy and unacceptable distortion of the signal at the receiving end.

3.3 AMPLIFICATION

Although line loading reduces speech-band attenuation, there is still a loss of signal which accumulates with distance, and at some stage it becomes necessary to 'boost' the signal strength. This is done by the use of an electrical amplifier. The reader may well ask what precise distance line loading is good for. There is unfortunately, no simple answer since it depends on the gauge of the wire and on the transmission bandwidth required by the user. In general, the higher the bandwidth, the shorter the length limit of loaded lines (10–15 km is a practical limit).

Devices called 'repeaters' are spaced equally along the length of a long transmission line, radio system, or other transmission medium. Repeaters consist of amplifiers and other equipment, the purpose of which is to boost the basic signal strength.

Normally a repeater comprises two amplifiers, one for each direction of signal transmission. A splitting device is also required in order to separate 'transmit' and 'receive' signals. This is so that each signal can be fed to a relevant 'transmit' or 'receive'

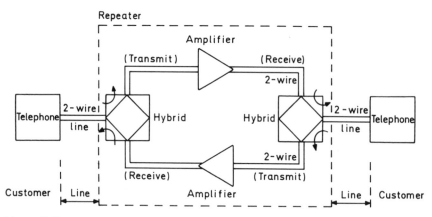

Figure 3.3
A simple telephone repeater system.

amplifier, as Figure 3.3 illustrates. The splitting device is called a 'hybrid' or 'hybrid transformer'. Essentially it converts a two-way communication over 2-wires into two one-way, 2-wire connections, and it is then usually referred to as a '4-wire communication' (one direction of signal transmission on each pair of a two-pair set). So, while a single 2-wire line is adequate for two-way telephone communication over a short distance, as soon as the distance is great enough to require amplification then a conversion to 4-wire communication is called for. Figure 3.3 shows a line between two telephones with a simple telephone repeater in the line. Each repeater consists of two hybrid transformers and two amplifiers (one for each signal direction).

The amplification introduced at each repeater has to be carefully controlled so as to overcome the effects of attenuation, without adversely affecting what is called the 'stability' of the circuit, and without interfering with other circuits in the same cable. Repeaters too far apart, or with too little amplification would allow the signal current to fade to such an extent as to be subsumed in the electrical 'noise' present on the line. Conversely, repeaters that are too close together or have too much amplification, can lead to circuit instability, and to yet another problem known as 'crosstalk'.

A circuit is said to be unstable when the signal it is carrying gets over-amplified, causing feedback and even more amplification. This in turn leads to even greater feedback, and so on and on, until the signal is so strong that it reaches the maximum power that the circuit can carry. The signal is now distorted beyond cure and all the listener hears is a very loud singing noise. For the causes of this unfortunate situation let us look at the simple circuit of Figure 3.4.

The diagram of Figure 3.4 shows a poorly engineered circuit which is electrically unstable. At first sight, the diagram is identical to Figure 3.3. The only difference is that various signal attenuation values (indicated as negative) and amplification values (indicated as positive) have been marked using the standard unit of measurement, the decibel (dB). The problem is that the net gain around the loop is greater than the net loss. Let us look more closely. The hybrid transformer H1 receives the incoming signal from telephone Q and transmits it to telephone P, separating this signal from the one that will be transmitted on the outgoing pair of wires towards Q. Both signals suffer

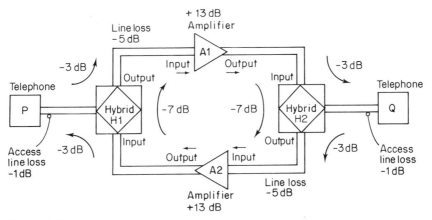

Figure 3.4
An unstable circuit.

a 3 dB attenuation during this 'line-splitting' process. Adding the attenuation of 1 dB which is suffered on the local access line by the outgoing signal coming from telephone P, the total attenuation of the signal by the time it reaches the output of hybrid H1 is therefore 4 dB. The signal is further attenuated by 5 dB as a result of line loss. Thus the input to amplifier A1 is 9 dB below the strength of the original signal. Amplifier A1 is set to more than make up for this attenuation by boosting the signal by 13 dB, so that at its output the signal is actually 4 dB louder than it was at the outset. However, by the time the second hybrid loss (in H2) and Q's access line attenuation have been taken into account, the signal is back to its original volume. This might suggest a happy ending, but unfortunately the circuit is unstable. Its instability arises from the fact that neither of the hybrid transformers can actually carry out their line-splitting function to perfection; a certain amount of the signal received by hybrid H2 from the output of amplifier A1 is finding its way back onto the return circuit (Q-to-P).

A well designed and installed hybrid would give at least 30 dB separation of receive and transmit channels. However, in our example, either the hybrid has been poorly installed or has become faulty, and the unwanted retransmitted signal (originated by P but returned by hybrid H2) is only 7 dB weaker at hybrid H2's point of output than it was at the output from A1, and is then attenuated by 5 dB and amplified 13 dB before finding its way back to hybrid H1, where it goes though another undesired retransmission, albeit at a cost of 7 dB in signal strength. The strength of this signal, which has now entirely 'lapped' the 4-wire section of the circuit, is 2 dB less than that of the original signal emanating from telephone P. But on its first 'lap' it had a strength 4 dB lower than the original. In other words, the recirculated signal is actually louder than it was on the first lap! What is more, if it goes around again it will gain 2 dB in strength for each lap, quickly getting louder and louder—and out of control. This phenonomen is called instability. The primary cause is the feedback path available across both hybrids, which is allowing incoming signals to be retransmitted (or 'fedback') on their output. The path results from the non-ideal performance of the hybrid. Particularly in practice it is impossible to exactly balance the hybrid's resistance to that of the end telephone handset.

One way to correct circuit instability is to change the hybrids for more efficient (and probably more expensive) ones. This solution also requires careful balancing of each telephone handset and corresponding hybrid—and is not possible if the customer lines and the hybrids are connected via the switch matrix (as would be the case for a 2-wire customer local line connected via a local exchange to a 4-wire trunk or junction). A cheaper and quicker alternative to the instability problem is simply to reduce the amplification in the feedback loop. This is done simply by adding an attenuating device. Indeed, most variable amplifiers are in fact fixed gain amplifiers (around 30 dB) followed immediately by variable attenuators or 'pads'. For example, in the case illustrated in Figure 3.4 a reduction in the gain of both amplifiers A1 and A2 to 12 dB will mean that the feedback signal is exactly equal in amplitude to the original. In this state the circuit may just be stable, but it is normal to design circuits with a much greater margin of stability, typically at least 10 dB. For the Figure 3.4 example this would restrict amplifier gain to no more than 7 dB. Under these conditions the volume of the signal heard in telephone Q will be 6 dB quieter than that transmitted by telephone P, but this is unlikely to trouble the listener.

'Crosstalk' is the name given to an overheard signal on an adjacent circuit. It

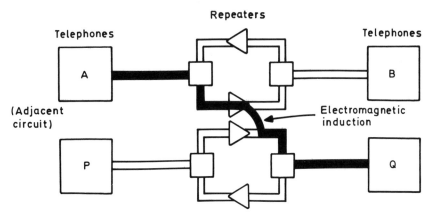

Figure 3.5
Crosstalk. Q hears A.

is brought about by electromagnetic induction of an over-amplified signal from one circuit on to its neighbour, and Figure 3.5 gives a simplified diagram of how it happens.

Because the transit amplifier on the circuit from telephone A to telephone B in Figure 3.5 is over-amplifying the signal, it is creating a strong electromagnetic field around the circuit and the same signal is induced into the circuit from P to Q. As a result the user of telephone Q annoyingly overhears the user of telephone A as well as the conversation from P. It is resolved either by turning down the amplifier or by increasing the separation of the circuits. If neither of these solutions is possible then a third, more expensive option is available involving the use of specially 'screened' or 'transverse-screened' cable. In such cable a foil 'screen' wrapped around the individual pairs of wires makes it relatively immune to electromagnetic interference.

3.4 2- AND 4-WIRE CIRCUITS

The diagram in Figure 3.3 illustrated a single repeater, used for boosting signals on a 2-wire line system. If the wire is a long one, a number of individual repeaters may be required. Figure 3.6 shows an example of a long line in which three amplifiers have been deployed.

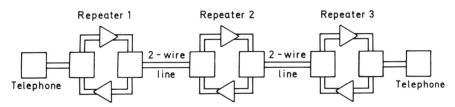

Figure 3.6
A repeatered 2-wire circuit.

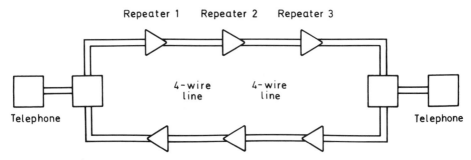

Figure 3.7
A repeatered 4-wire circuit.

Such a system may work quite well but it has a number of drawbacks, the most important of which is the difficulty of maintaining circuit stability and acceptable received signal volume simultaneously; this difficulty arises from the interaction of the various repeaters. An economic consideration is the high cost of the many hybrid transformers that have to be provided. All but the first and last hybrids could be dispensed with if the circuit were wired instead as a 4-wire system along the total length of its repeatered section, as shown in Figure 3.7. This arrangement reduces the problem of achieving circuit stability when a large number of repeaters are needed, and it eases the maintenance burden. This is why most amplified long-haul (i.e. 'trunk') circuits are set up on 4-wire transmission lines. Shorter circuits, typically requiring only 1–2 repeaters (i.e. junction circuits) can, however, make do with 2-wire systems.

3.5 EQUALIZATION

We have mentioned the need for equalization on long-haul circuits, in order to minimize the signal distortion. Speech and data signals comprise a complex mixture of pure, single frequency components, each of which is affected differently by transmission lines. The result of different attenuation of the various frequencies is tonal degradation of the received signal; at the worst the entire high or low-frequency range could be lost. Figure 3.8 shows the relative amplitudes of individual signal frequencies of a distorted and an undistorted signal.

In Figure 3.8 the amplitude of the frequencies in the undistorted (received) signal is the same across the entire speech bandwidth, but the high and low-frequency signals (high and low notes) have been disproportionately attenuated (lost) in the distorted signal. The effect is known as 'attenuation distortion' or 'frequency attenuation distortion'.

To counteract this effect equalizers are used which are circuits designed to amplify or attenuate different frequencies by different amounts. The aim is to 'flatten' the frequency response diagram—to bring it in line with the undistorted frequency response diagram. In our example above, the equalizer would have to amplify the low and high frequencies more than it would amplify the intermediate frequencies.

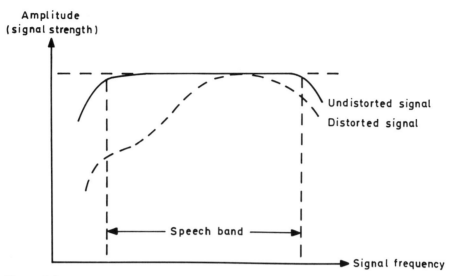

Figure 3.8
Amplitude spectrum of distorted and undistorted signals (attenuation distortion).

Equalizers are normally included in repeaters, so that the effects of distortion can be corrected all the way along the line, in the same way as amplifiers counteract the effect of attenuation.

3.6 FREQUENCY DIVISION MULTIPLEXING (FDM)

When a large number of individual communication channels are required between two points a long distance apart, providing a large number of individual physical wire circuits, one for each channel, can be a very expensive business. For this reason, what is known as multiplexing was developed as a way of making better use of lineplant. Multiplexing allows many transmission channels to share the same physical pair of wires or other transmission medium. It requires sophisticated and expensive transmission line-terminating equipment (LTE), but has the potential for overall saving in cost because the number of wire pairs required between the end points can be reduced.

The method of multiplexing used in analogue networks is called frequency division multiplex (FDM). FDM calls for a single, high grade, 4-wire transmission line (or equivalent), and both pairs must be capable of supporting a very large bandwidth. Some FDM cables have a bandwidth as high as 12 MHz (million cycles per second), or even 60 MHz. This compares with the modest 3.1 kHz (thousand cycles per second) required for a single telephone channel. The large bandwidth is the key to the technique, since it sub-divides readily into a much larger number of individual small bandwidth channels.

The lowest constituent bandwidth going to make up an FDM system is a single channel bandwidth of 4 kHz. This comprises the 3.1 kHz needed for a normal speech channel, together with some spare bandwidth to create separation between channels on the system as a whole. Various other standard bandwidths are then integral multiples of a single channel. Table 3.1 illustrates this and gives the names of these standard bandwidths.

The overall bandwidth of the FDM transmission line is equal to one of the standard bandwidths (e.g. 'supergroup', 'group') as named in Table 3.1, and is then broken down into a number of sub-bandwidths, called 'tributaries' in the manner shown in Figure 3.9. The equipment which performs this segregation of bandwidth is called 'translating equipment'. Thus 'supergroup translating equipment' (STE) subdivides a supergroup into its five component 'groups', and 'channel translating equipment' (CTE) subdivides a group into twelve individual channels.

Not all the available bandwidth need be broken down into individual 4 kHz channels. If required, some of it can be used directly for large bandwidth applications such as concert grade music or television transmission. In Figure 3.9, two 48 kHz circuits are derived from the supergroup, together with 36 individual telephone channels. The same principle can be applied at other levels in the hierarchy. Thus, for example, all the supergroups of a hypergroup could be broken down into their component groups using HTE and STE. Alternatively, some of the supergroups could be used directly for bandwidth applications of 240 kHz.

Before multiplexing, the audio signals which are to be multiplexed-up are first converted to 4-wire transmission, if they are not so already. The signals on each transmit pair are then accurately filtered so that stray signals outside the allocated bandwidth are suppressed. In fact, a telephone channel is filtered to be only 3.1 kHz in bandwidth (of the available 4 kHz). The remaining 0.9 kHz 'separation' prevents speech interference between adjacent channels. Groups are filtered to 48 kHz, supergroups to 240 kHz etc. Each filtered signal is then modulated by a carrier frequency, which has the effect of 'frequency shifting' the original signal into another part of the frequency spectrum. For example, a telephone channel starting out in the bandwidth range

Table 3.1
Frequency Division Multiplex (FDM) hierarchy.

Bandwidth name	Consists of	Bandwidth	Usual baseband	Number of channels
Channel (1 telephone channel)	24 telegraph subchannels 120 Hz spacing	4 kHz	0–4 kHz	1
Group	12 channels	48 kHz	60–108 kHz	12
Supergroup	5 groups	240 kHz	312–552 kHz	60
Basic hypergroup (also called a 'super mastergroup')	15 supergroups (3 mastergroups)	3.7 MHz (3.6 MHz used) (240 kHz per supergroup with 8 kHz spacing normally between each)	312–4028 kHz (4 MHz line)	900
Basic hypergroup (alternative)	16 supergroups	4 MHz	60–4028 kHz	960
Mastergroup	5 supergroups	1.2 MHz	312–1548 kHz	300
Hypergroup (12 MHz)	9 mastergroups	12 MHz	312–12 336 kHz	2700
Hypergroup (60 MHz)	36 mastergroups	60 MHz	4404–59 580 kHz	10 800

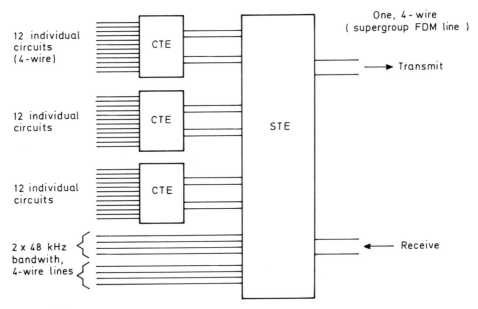

Figure 3.9
Breaking-up bandwidth in FDM.

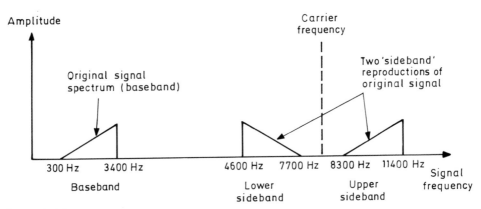

Figure 3.10
Frequency shifting by carrier modulation.

300–3400 Hz might end up in the range 4600–7700 Hz. Another could be shifted to the range 8300–11 400 Hz, and so on.

Each component channel of an FDM group is frequency shifted by the CTE to a different bandwidth 'slot' within the 48 kHz available; so that in total, 12 individual channels may be carried. Likewise, in a supergroup, five already made-up groups of 48 kHz bandwidth are slotted-in to the 240 kHz bandwidth by the STE.

The frequency shift is achieved by 'modulation' of the component bandwidths with different 'carrier signal' frequencies. The frequency of the carrier signal which is used to modulate the original signal (or 'baseband') will be equal to the value of the frequency shift required. Each carrier signal must be produced by the translating equipment.

The modulation of a signal in the frequency band 300–3400 Hz using a carrier frequency of 8000 Hz produces a signal of bandwidth from 4600 Hz (8000 − 3400) to 11 400 Hz (8000 + 3400). The original frequency spectrum of 300–3400 Hz is reproduced in two mirror image forms, called 'sidebands'. One sideband is in the range 4600–7700 Hz and the other is in the range 8300–11 400 Hz. Both sidebands are shown in Figure 3.10.

Since all the information is duplicated in both sidebands, only one of the sidebands need be transmitted. For economy in the electrical power needed to be transmitted to line, it is normal for FDM systems to operate in a 'single sideband' (SSB) and 'suppressed carrier' mode. The original signal is reconstructed at the receiving end by modulating (mixing) a locally generated frequency. (*Note*: single sideband operation

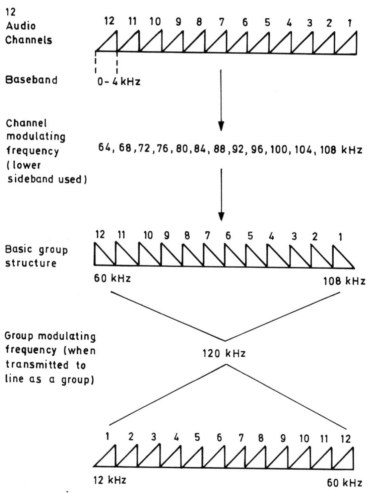

Figure 3.11
The structure of an FDM group.

may also be used in radio systems, but the carrier is not suppressed in this case since it is often inconvenient to make a carrier signal generator available at the receiver. When the carrier is not suppressed a much simpler and cheaper detector can be used.)

Each baseband signal to be included in an FDM system is modulated with a different carrier frequency, the lower sideband is extracted for conveyance. It may seem inefficient not to double up the use of carrier frequencies—adopting alternating upper and lower sidebands of different channels, but by always using the lower sideband we get a better overall structure, allowing easier extraction of single channels from higher order FDM systems. The carrier frequencies needed to produce a standard group are thus 64 kHz, 68 kHz, 72 kHz, ..., 108 kHz and the overall structure is as shown in Figure 3.11.

Supergroups and hypergroups may be modulated in a similar fashion, using appropriate carrier frequencies and single sideband operation.

3.7 CROSSTALK AND ATTENUATION ON FDM SYSTEMS

There can be as much crosstalk and attenuation in FDM systems as in the single channel or 'audio' circuits which we discussed in the first part of the chapter. They require just as much, if not more planning, since FDM systems are generally more sensitive and complex.

BIBLIOGRAPHY

Green, D. C., *Transmission Systems (TEC Level 2)*. Pitman Publishing, 1978.
Haykin, S., *Communication Systems*, 2nd edn. John Wiley & Sons, 1983.
Hill, M. T. and Evans, B. G., *Transmission Systems*. George Allen & Unwin, 1973.
Sinnema, W., *Digital, Analog and Data Communication*. Reston Publishing Company, 1982.

4

DATA AND THE BINARY CODE SYSTEM

'Data', a plural noun, is the term used to describe information which is stored in and processed by computers. It is essential to know how such data is represented electronically before we can begin to understand how it can be communicated between computers, communication devices (e.g. facsimile machines) or other data storage devices. As a necessary introduction to the concept of 'digital' transmission, this chapter is devoted to a description of that method of representing textual and numeric information which is called the 'binary code'.

4.1 THE BINARY CODE

'Binary code' is a means of representing numbers. Normally, numbers are quoted in decimal (or 'ten-state') code. A single digit in decimal code may represent any of ten different 'unit' values, from nought to nine, and is written as one of the figures 0, 1, 2, 3, 4, 5, 6, 7, 8, 9. Numbers greater than nine are represented by two or more digits: twenty for example is represented by two digits, 20, the first '2' indicating the number of 'tens', so that 'twice times ten' must be added to '0' units, making twenty in all. In a three digit decimal number, such as 235, the first digit indicates the number of 'hundreds' (or 'ten times tens') the second digit the number of 'tens' and the third digit, the number of 'units'. The principle extends to numbers of greater value, comprising four or indeed many more digits.

Consider now another means of representing numbers using only a 'two-state' or 'binary' code system. In such a system a single digit is restricted to one of two values, either zero or one. How then are values of 2 or more to be represented? The answer, as in the decimal case, is to use more digits. 'Two' itself is represented as the two digits one–zero, or 10. In the binary code scheme therefore, 10 does not mean 'ten', but 'two'. The rationale for this is similar to the rationale of the decimal number system with which we are all familiar.

In decimal the number one thousand three hundred and forty-five is written '1345'. The rationale is:

$$(1 \times 10^3) + (3 \times 10^2) + (4 \times 10) + 5$$

the same number in binary requires many more digits, as follows:

1345 (decimal) = 10101000001 (binary)

$$
\begin{array}{rlr}
 & \text{(binary)} & \text{(decimal)} \\
= & 1 \times 2^{10} & 1024 \\
+ & 0 \times 2^{9} & + \quad 0 \\
+ & 1 \times 2^{8} & + \quad 256 \\
+ & 0 \times 2^{7} & + \quad 0 \\
+ & 1 \times 2^{6} & + \quad 64 \\
+ & 0 \times 2^{5} & + \quad 0 \\
+ & 0 \times 2^{4} & + \quad 0 \\
+ & 0 \times 2^{3} & + \quad 0 \\
+ & 0 \times 2^{2} & + \quad 0 \\
+ & 0 \times 2 & + \quad 0 \\
+ & 1 & + \quad 1 \\
\hline
 & & 1345
\end{array}
$$

Any number may be represented in the binary code system, just as any number can be represented in decimal.

All numbers, when expressed in binary consist only of 0s and 1s, arranged as a series of Binary digITS, a term which is usually shortened to the jargon 'BITs'. The 'string' of 'bits' of a binary number are usually suffixed with a 'B', to denote a binary number. This prevents any confusion that the number might be a decimal one. Thus 41 is written '101001B'.

4.2 ELECTRICAL REPRESENTATION AND STORAGE OF BINARY CODE NUMBERS

The advantage of the binary code system is the ease with which binary numbers can be represented electrically. Since each digit, or 'bit', of a binary number may only be either 0 or 1, the entire number can easily be transmitted as a series of 'off' or 'on' (sometimes also called 'space' and 'mark') pulses of electricity. Thus forty-one (101001B) could be represented as on-off-on-off-off-on, or mark-space-mark-space-space-mark. The number could be conveyed between two people on opposite sides of a valley, by flashing a torch, either on or off, say every half second. Figure 4.1 illustrates this simple binary communication system in which two binary digits (or bits) are conveyed every second. The speed at which the binary code number, or other information can be conveyed is called the 'information conveyance rate' (or more briefly the 'information rate'). In this example the rate is two bits per second, which can be expressed also as 2 bit/s.

Figure 4.1 illustrates a means of transmitting numbers, or other binary coded data by a series of 'on' or 'off' electrical states. Transmission of data, however, is not in itself sufficient to permit proper exchange of information between the computers or other equipment located at either end of the line; some means of data storage is needed as well. At the sending end the data has to be stored prior to transmission, and at the receiving end a storage medium is needed not only for the incoming data, but also for the computer programs required to interpret it.

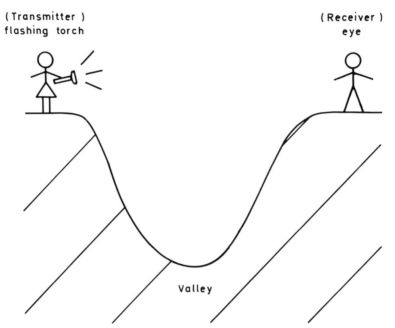

Figure 4.1
A simple binary communication system.

4.3 USING THE BINARY CODE TO REPRESENT TEXTUAL INFORMATION

The letters of the alphabet can be stored and transmitted over binary coded communication systems in the same way as numbers provided they have first been 'binary-encoded'. There are four notable binary coding systems for alphabetic text. In chronological order these are the 'Morse code', the 'Baudot code' (used in Telex, and also known as international alphabet number 2 IA2), EBCDIC (Extended Binary Coded Decimal Interchange Code), and ASCII (American (National) Standard Code for Information Interchange, also known as 'international alphabet IA5'). These four coding schemes are now described briefly.

4.4 MORSE CODE

The Morse code system of dots and dashes was for use over key and lamp telegraph systems. It was also used for signalling by heliograph and by flag. Its two binary elements are 'dit' and 'da' ('dot' and 'dash'). Thirty-nine characters were coded as shown in Figure 4.2. When transmitting, a short pause is inserted to mark the beginning and end of each character; and between words there is a longer pause. As an example of Morse code we see from the figure that the word 'Morse' is transmitted as 'da da' (pause) 'da da da' (pause) 'dit da dit' (pause) 'dit dit dit' (pause) 'dit' (which would be written as − − / − − − / · − · / · · · / ·).

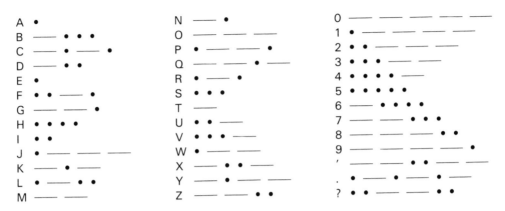

Figure 4.2
The Morse code.

4.5 BAUDOT CODE (ALPHABET IA2)

When the telex system was introduced, the Baudot code (now called the international alphabet IA2) was developed, with significant advantages over the Morse code for 'automatic' use. Each character is represented by five binary elements (usually called 'mark' and 'space'), but seven elements are transmitted in total, since start (space) and stop (mark) bits are also used. Fixing the number of elements cuts out the need for gaps or pauses between alphabetic characters, and separate words are delimited without a break by introducing the space (SP) character (00100). The regular flow of these signals suits automatic transmitting and receiving devices, and makes them easier to design. Figure 4.3 illustrates the Baudot code. Thus the sequence of seven bits sent to represent the letter A are

'space(start)-mark-mark-space-space-space-mark(stop)'.

The word 'Baudot' would thus be transmitted in Baudot code as:

Order of	B	A	U	D	O	T
transmit	10011	11000	11100	10010	00011	00001

Note in passing that the term 'Baud' is commonly used in data communications as the unit of rate of signal change on the line transmission medium (the so-called 'Baud rate'). Telex networks usually operate at a rate of 50 Baud (50 signal changes per second) and they use the Baudot code. Since 5 line state *changes* (from mark-to-space, space-to-mark, space-to-space or mark-to-mark) are required to convey each character, this produces an 'information rate' of 50 divided by 5, that is to say 10 alphabetic characters per second, which incidentally corresponds roughly to ordinary human speech, when we are speaking or reading deliberately.

Case	Character (letters)	(figures)	Pattern 5 4 3 2 1		Case (letters)	(figures)	Pattern 5 4 3 2 1
	A	-	0 0 0 1 1		Q	1	1 0 1 1 1
	B	?	1 1 0 0 1		R	4	0 1 0 1 0
	C	:	0 1 1 1 0		S	,	0 0 1 0 1
	D	£	0 1 0 0 1		T	5	1 0 0 0 0
	E	3	0 0 0 0 1		U	7	0 0 1 1 1
	F	!	0 1 1 0 1		V	;	1 1 1 1 0
	G	&	1 1 0 1 0		W	2	1 0 0 1 1
	H		1 0 1 0 0		X	/	1 1 1 0 1
	I	8	0 0 1 1 0		Y	6	1 0 1 0 1
	J	(Bell)	0 1 0 1 1		Z	"	1 0 0 0 1
	K	(	0 1 1 1 1		Shift (figures to letters)		1 1 1 1 1
	L	)	1 0 0 1 0		Shift (letters to figures)		1 1 0 1 1
	M	.	1 1 1 0 0		Space (SP)		0 0 1 0 0
	N	,	0 1 1 0 0		Carriage Return‹		0 1 0 0 0
	O	9	1 1 0 0 0		Line Feed		0 0 0 1 0
	P	0	1 0 1 1 0		Blank		0 0 0 0 0

1 = Mark (Punch hole on paper tape)
0 = Space (No hole)

Figure 4.3
Baudot code (International Alphabet IA2).

4.6 ASCII

With the advent of semiconductors and the first computers, 1963 saw the development of a new 7-bit binary code for computer characters. This code encompassed a wider character range, including not only the alphabetic and numeric characters but also a range of new 'control' characters which are needed to govern the flow of data in and around the computers. The code, named ASCII (pronounced 'Askey') is now common in computer systems. The letters stand for American (National) Standard Code for Information Interchange. It is also known as International Alphabet number 5 (IA5). Figure 4.4 illustrates it.

Note that the bit numbers 1 through 7 (top left-hand corner of table) represent the 'least' through 'most significant bits' respectively. Each letter, however, is usually written most significant bit (i.e. bit number 7) first. Thus the letter C is written '1000011'. But to confuse matters further, the least significant bit is transmitted first. Thus the order of transmission for the word 'ASCII' is:

(I)	(I)	(C)	(S)	(A)	Order
1001001	1001001	1000011	1010011	1000001	of
Last				First	transmit

The characters need not be transmitted directly in the form of the 35 bits shown, but are usually separated by other 'control' characters. In particular delimiting bits (so called 'start' and 'stop' bits) may be used to separate the strings representing individual characters. We shall return to this subject in Chapter 9 when discussing 'asynchronous' and 'synchronous' transmission methods.

			7	0	0	0	0	1	1	1	1
	Bits		6	0	0	1	1	0	0	1	1
4	3	2	1 〳 5	0	1	0	1	0	1	0	1
0	0	0	0	NUL	DLE	SP	0	@	P	\	p
0	0	0	1	SOH	DC1	!	1	A	Q	a	q
0	0	1	0	STX	DC2	"	2	B	R	b	r
0	0	1	1	EXT	DC3	#	3	C	S	c	s
0	1	0	0	EOT	DC4	$	4	D	T	d	t
0	1	0	1	ENQ	DAK	%	5	E	U	e	u
0	1	1	0	ACK	SYN	&	6	F	V	f	v
0	1	1	1	BEL	ETB	'	7	G	W	g	w
1	0	0	0	BS	CAN	L	8	H	X	h	x
1	0	0	1	HT	EM	)	9	I	Y	i	y
1	0	1	0	LF	SUB	*	:	J	Z	j	z
1	0	1	1	VT	ESC	+	;	K	[	k	{
1	1	0	0	FF	FS	'	<	L	\	l	:
1	1	0	1	CR	GS	−	=	M	]	m	}
1	1	1	0	SO	RS	.	>	N	^	n	~
1	1	1	1	SI	US	/	?	O	-	o	DEL

Figure 4.4
The ASCII code (International Alphabet IA5).

EBCDIC code chart. Columns are selected by bits 5,6,7,8 (high-order group) and rows by bits 4,3,2,1.

Bits 4321 \ Bits 5678	0000	0001	0010	0011	0100	0101	0110	0111	1000	1001	1010	1011	1100	1101	1110	1111
1111	SI	IUS	BEL	SUB	|	¬	?	"								□
1110	SO	IRS	ACK		+	;	>	=								
1101	CR	IGS	ENQ	NAK	(	)	_	'								
1100	FF	IFS		DC4	<	*	%	@								
1011	VT				.	$	,	#								
1010	SMM	CC	SM		¢	!	¦	:								
1001		EM							i	r	z		I	R	Z	9
1000		CAN							h	q	y		H	Q	Y	8
0111	DEL	IL	PRE	EOT					g	p	x		G	P	X	7
0110	LC	BS	EDB	UC					f	o	w		F	O	W	6
0101	HT	NL	LF	RS					e	n	v		E	N	V	5
0100	PF	RES	BYP	PN					d	m	u		D	M	U	4
0011	ETX	DC3							c	l	t		C	L	T	3
0010	STX	DC2	FS	SYN					b	k	s		B	K	S	2
0001	SOH	DC1	SOS				/		a	j			A	J		1
0000	NUL	DLE	DS		SP	&	-									0

Figure 4.5
EBCDIC code.

4.7 EBCDIC

EBCDIC, or 'Extended Binary Coded Decimal Interchange Code' is an extension of ASCII, given even more control characters. It uses an 8-bit representation for each character, as shown in Figure 4.5 and is widely used in IBM computers and compatible machines.

4.8 USE OF THE BINARY CODE TO CONVEY GRAPHICAL IMAGES

Besides representing numerical and alphabetical (or textual) characters, the binary code can also be used to transmit pictorial and graphical images. One of the devices capable of converting graphical images into binary data is the 'facsimile' machine. These machines work in pairs, separated by some form of transmission link. At the 'transmitting' end of the link, one facsimile machine 'scans' a piece of paper, and converts the black-and-white image which it sees into a binary-coded stream of data. This data is then transmitted to the receiving facsimile machine, where it is used to produce a black-and-white facsimile reproduction of the original paper image. The working principle of these machines is simple enough, as we may now see.

The image on the original is assumed to be composed of a very large number of tiny dots, arranged in a grid pattern on the paper. Figure 4.6, for example shows how one word on the paper may be broken down into a grid of dots.

The image is reproduced by making a copy of that same grid pattern of dots at the receiving end. The procedure is as follows. Starting at the top left-hand corner, the transmitting facsimile machine 'scans' the original paper document from left to right, following each line of the grid in turn. At the end of each line, the machine returns to the left-hand side of the grid, and moves down to the line below. Each line scanned is transposed by the machine into a string of binary coded data, comprising a series of variable length 'code words'. Each 'code word' represents a number of consecutive squares, or 'runs', along the horizontal row of the grid, either an 'all-black run' or 'all-white' one.

White 'runs' and black 'runs' necessarily alternate, since these are the only two colours distinguishable by the scanning device. A small section of the grid was shown in Figure 4.6. For A4 paper, 1728 small picture elements represent one scan of a horizontal row of the grid, some 215 mm in length. (In other words, there are around 64 dots, termed 'picture elements', per square millimetre.) The data sent to represent each

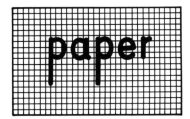

Figure 4.6
Facsimile scanning grid.

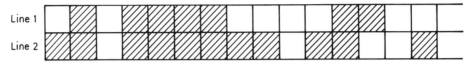

Figure 4.7
Facsimile scanning and coding.

Figure 4.8
Facsimile terminal. A group 3 facsimile terminal receiving an incoming document. Typically around 25–60 seconds is required to transmit one page, though darker documents may take longer (*Courtesy of British Telecom.*)

line of the grid is thus in the form 'two white, three black, ten white, two black etc.', describing the colours of each consecutive picture element along the row. The end of the row is indicated in the data stream by a 'terminating code word'. Each string of data, corresponding to one horizontal scan of the grid, starts on the assumption that the first colour on the left-hand side is going to be 'white' by indicating the 'white run length'. This allows the receiver always to be in the correct colour synchronization at the beginning of the line. If, as frequently, the new line starts with a black picture element, then the initial signal will be 'white run length of zero elements'. Figure 4.7 shows a small section of two consecutive runs, as a way of explaining the coding method.

Starting on line 1 of Figure 4.7, the scanning and transmitting facsimile machine sends a string of data saying 'white-run length, one; black-run length, one; white, one; black, four; white, four; black, two; white ... end of line'. For the second line, the transmitting facsimile machine carries on 'white-run length, zero; black, two; white, one; black, six etc.' At the receiving end, the second facsimile machine slavishly prints out a corresponding series of black and white picture elements, which reproduce the original image. Returning to Figure 4.6, we see how the image of the word 'paper' has been coded for transmission and subsequent reproduction with the aid of the scanning grid.

Not surprisingly, facsimile machines actually use slightly more sophisticated techniques than those described but the principles are the same. The purpose of these enhancements to basic technique is to improve the accuracy and overall speed of transmission and so reduce the time required for conveying each paper sheet.

Any type of image can be conveyed using facsimile machines: typed text, manuscript, pictures and diagrams. The scanning and image reproducing machinery works in the same way for all of them.

Since 1968, when recommendations for CCITT's first, 'Group 1' standard apparatus were published, various generations of facsimile machines have been developed. The latest 'Group 4' facsimile machines produce extremely high quality pictures, and can transmit a page of A4 in a few seconds, as compared with the six minutes that group 1 apparatus took over the same job.

Facsimile machines are not the only way of transmitting graphic information in binary form. New equipment appears on the market almost daily, with all kinds of applications from slow motion security surveillance television, to high quality and high definition colour graphics or moving video images.

4.9 DIGITAL TRANSMISSION

Nowadays most data and much other information is communicated in one or other of the binary coded forms which we have discussed. Since binary coded data is transmitted as a sequence of 'on' or 'off' states, with each 'on' or 'off' representing the value '1' or '0' of consecutive binary digits or 'bits', all information is conveyed essentially as a string of *digits*, and so the process has acquired the name 'digital transmission'. We go on now to assess its considerable advantages over the analogue technique, and how it can be extended to speech and other analogue signals.

BIBLIOGRAPHY

CCITT, T-Series Recommendations.
CCITT Recommendation T4, 'Standardisation of Group 3 Facsimile Apparatus for Document Transmission'.
CCITT Recommendation T50, 'International Alphabet Number 5'.
Fitzgerald, J. and Eason, T. S., *Fundamentals of Data Communications*. John Wiley & Sons, 1978.
Housley, T., *Data Communications and Teleprocessing Systems*. Prentice-Hall, 1979.
Maynard, J., *Computer and Telecommunications Handbook*. Granada, 1984.

5

DIGITAL TRANSMISSION AND PULSE CODE MODULATION

By the 1960s experimental digital line systems were on trial in a number of countries, and by 1970 the systems which set today's standards were in operation. At first digital transmission was only cost-effective in junction networks, where it was introduced to alleviate cable congestion, but as a result of the miniaturization and large-scale integration of electronic components and the digitalization of telephone exchanges, it rapidly became the obvious choice for all new transmission systems. Now, in the networks of many countries and with wider world international satellite and submarine networks, digital transmission has no rivals. It has already become the standard for conveying information and other signals over telecommunications networks, and as the world's networks are modernized it is taking over so rapidly from the old analogue technology that within a few years nearly all telecommunications equipment will be digital. So what exactly is it, and what can we gain by it?

5.1 DIGITAL TRANSMISSION

In investigating analogue transmission we found a relationship between bandwidth and overall information carrying capacity, and we described 'frequency division multiplexing' (FDM). This was a method of reducing the number of physical wires needed to carry a multitude of individual channels between two points, and it worked by sharing out the overall bandwidth of a single set of four-wires (transmit and receive pairs) between all the channels to be carried. We now discuss digital transmission in detail, how it works, and the equivalents of analogue 'bandwidth' and channel multiplexing. In contrast with analogue networks, digital networks are ideal for the direct carriage of data, since as the name suggests, a digital transmission medium carries information in the form of individual 'digits'. And not just any type of digits, but 'binary digits' in particular.

The medium used in digital transmission systems is usually designed so that it is only electrically stable in one of two states, equivalent to 'on' (binary value '1') or 'off' (binary value '0'). Thus a simple form of digital line system might use an electrical current as the conveying medium, and control the current to fluctuate between two values, 'current on' and 'current off'. A more recent digital transmission medium using 'optical fibre' (which consists of a hair-thin (50 μm in diameter) strand of glass) conveys the digital signal in the form of 'on' and 'off' light signals, usually generated by some sort of semiconductor electronic device such as a 'laser' or a 'light emitting diode (LED)'. Any other medium capable of displaying distinct 'on/off' states could also be used.

For simultaneous two-way (or 'duplex') digital transmission a 4-wire (or equivalent) transmission medium is always required. Duplex digital transmission cannot be achieved on a 2-wire medium. Just as with 4-wire analogue transmission (as discussed in Chapter 3) one 'pair' of wires (or its equivalent, for example an optical fibre) is used for the 'transmit' direction while the other pair is used for the 'receive' direction. This allows the digital pulses to pass in both directions simultaneously without interference. Since simultaneous two-way transmission is nearly always required, digital transmission links are nearly always 4-wire. Thus they differ from simple local analogue telephone systems which can be made to work adequately in a two-way mode over only 2-wires. (Recall Figure 2.6 of Chapter 2.)

The principal advantage of digital over analogue transmission is the improved quality of connection. With only two 'allowed' states on the line ('off' and 'on') it is not all that easy to confuse them even when the signal is being distorted slightly along the line by 'electrical noise interference' or some other cause. Digital line systems are thus relatively immune to interference. As Figure 5.1 demonstrates, the receiving end only needs to detect whether the received signal is above or below a given threshold value. If the pulse shape is not a 'clean' square shape, it does not matter. Allow the same electrical disturbance to interfere with an analogue signal, and the result would be a low volume crackling noise at the receiving end, which could well make the signal incomprehensible.

To make digital transmission still more immune to noise, it is normal practice to 'regenerate' the signals at intervals along the line. A regenerator reduces the risk of misinterpreting the received bit stream at the distant end of a long-haul line, by counteracting the effects of attenuation and distortion, which show up in digital signals as

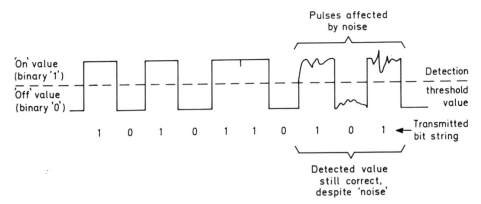

Figure 5.1
Digital signal and immunity to noise.

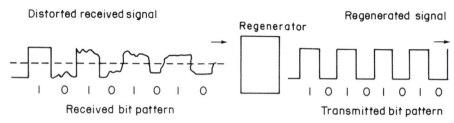

Figure 5.2
The principle of regeneration.

pulse shape distortions. In this corrective function a digital regenerator may be regarded as the equivalent of an analogue repeater.

The process of regeneration involves detecting the received signal and recreating a new, clean square wave for onward transmission. The principle is shown in Figure 5.2. The 'regeneration' of digital signals is all that is needed to restore the signal to its original form; there is no need to amplify, equalize or process it in any other way. The fact that the signal can be regenerated exactly is the reason why digital transmission produces signals of such high quality.

Errors at the detection stage can be caused by noise, giving the impression of a pulse when there is none. Their likelihood can, however, be reduced by stepping up the electrical power (which effectively increases the overall pulse size or height), and a probability equivalent to one error in several hours or even days of transmission can be obtained. This is good enough for speech, but if the circuit is to be used for data transmission it will not be adequate; the error rate may need to be reduced still further, and that requires a special technique using an error checking code.

A digital line system may be designed to run at almost any bit speed, but on a single digital circuit it is usually 64 kbit/s. This is equivalent to a 4 kHz analogue telephone channel, as we shall see shortly. The bit speed of a digital line system is roughly equivalent to the bandwidth of an analogue line system; the more information there is to be carried, the greater the required bit speed. Later in the chapter we also discuss how individual 64 kbit/s digital channels can be multiplexed together on a single physical circuit, by a method known as 'time division multiplex' (or TDM). TDM has the same multiplying effect on the circuit-carrying capacity of digital line systems as FDM has for analogue systems.

5.2 PULSE CODE MODULATION

You may well ask, how is a speech signal, a TV signal or any other analogue signal to be converted into a form that can be conveyed digitally? The answer lies in a method of analogue to digital signal conversion known as 'pulse code modulation'.

Pulse code modulation (PCM) functions by converting analogue signals into a format compatible with digital transmission, and it consists of four stages:

(1)   There is the translation of analogue electrical signals into digital pulses.

(2)   These pulses are coded into a sequence suitable for transmission.

(3)   They are transmitted over the digital medium.

(4)   They are translated back into the analogue signal (or an approximation of it) at the receiving end.

PCM was invented as early as 1939, but it was only in the 1960s that it began to be widely applied. This was mainly because, before the day of solid state electronics, we did not have the technology to apply the known principles of PCM effectively.

Speech or any other analogue signals are converted into a sequence of binary digits by sampling the signal waveform at regular intervals. At each sampling instant the waveform amplitude is determined and, according to its magnitude, is assigned a numerical value, which is then coded into its binary form and transmitted over the transport medium. At the receiving end, the original electrical signal is reconstructed by translating it back from the incoming digitalized signal. The technique is illustrated in Figure 5.3, which shows a typical speech signal, with amplitude plotted against time. Sampling is predetermined to occur at intervals of time 't' (usually measured in microseconds). The numerical values of the sampled amplitudes, and their 8-bit binary translations, are shown in Table 5.1.

Because the use of decimal points would make the business more complex and increase the bandwidth required for transmission, amplitude is represented by integer values only. When the waveform amplitude does not correspond to an exact integer value, as occurs at time $4t$ in Figure 5.3, an approximation is made. Hence at $4t$, value -2 is used instead of the exact value of -2.4. This reduces the total number of digits that have to be sent. The signal is reconstituted at the receiving end by generating a stepped waveform, each step of duration 't', with amplitude according to the digit

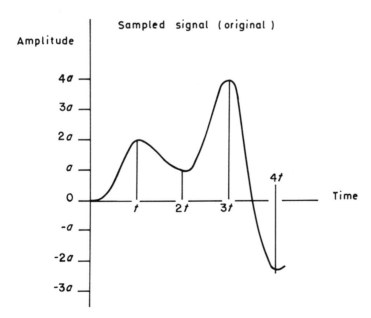

Figure 5.3
Sampling a waveform.

Table 5.1
Waveform samples from Figure 4.1.

Time	Amplitude	Decimal numeric value	8-bit binary translation
0	0	0	00000000
t	$2a$	2	00000010
$2t$	a	1	00000001
$3t$	$4a$	4	00000100

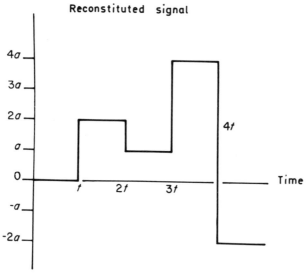

Figure 5.4
Reconstruction of the waveform of Figure 4.1 from transmitted samples.

value received. The signal of Figure 5.3 is therefore reconstituted as shown in Figure 5.4.

In the example, the reconstituted signal has a square waveform rather than the smooth continuous form of the original signal. This approximation affects the listener's comprehension to an extent which depends on the amount of inaccuracy involved. The similarity of the reconstituted signal to the original may be improved by:

(i) Increasing the sampling rate (i.e. reducing the time separation of samples) so as to increase the number of points on the horizontal axis of Figure 5.3 at which samples are taken.

(ii) Increasing the number of 'quantum' levels (i.e. wave amplitude levels). The quantum 'levels' are the points on the vertical scale of Figure 5.3.

However, without an infinite sampling rate and an infinite range of quantum values, it is impossible to match an original analogue signal precisely. Consequently an irrecoverable element of 'quantization noise' is introduced in the course of translating the original analogue signal into its digital equivalent. The sampling rate and the number of quantization levels have to be carefully chosen to keep this 'noise' down to levels at which the received signal is comprehensible to the listener. The snag is that the greater the sampling rate and the greater the number of quantization levels, the greater is the digital bit rate required to carry the signal. Here again a parallel can be found with the bandwidth of an analogue transmission medium, where the greater the required fidelity of an analogue signal, the greater is the bandwidth required.

The minimum acceptable sampling rate for carrying a given analogue signal using digital transmission is calculated according to a scientific principle known as the 'Nyquist criterion' (after the man who discovered it). The criterion states that the sample rate must be at least double the frequency of the analogue signal being sampled. For a standard speech channel this equals $2 \times 4\,\text{kHz} = 8000$ samples per second, the normal bandwidth of a speech channel being 4 kHz.

The number of quantization levels found (by subjective tests) to be appropriate for good speech comprehension is 256. In Binary digIT (bit) terms this equates to an eight-bit number, so that the quantum value of each sample is represented by eight 'bits'. The required transmission rate of a digital speech channel is therefore 8000 samples per second times 8 bits, or 64 kbit/s. In other words a digital channel of 64 kbit/s capacity is equivalent to an analogue telephone channel bandwidth of 4 kHz. This is the reason why the basic digital channel is designed to run at 64 kbit/s.

5.3 QUANTIZATION

When the amplitude level at a sample point does not exactly match one of the quantization levels, an approximation is made which introduces what is called 'quantization noise' (also 'quantizing noise'). Now, if the 256 quantization levels were equally spaced over the amplitude range of the analogue signal, then the low amplitude signals would incur far greater percentage quantization errors (and thus distortion) than higher amplitude signals. For this reason, the quantization levels are not linearly spaced, but instead are more densely packed around the zero amplitude level, as shown in Figure 5.5. This gives better signal quality in the low amplitude range and a more consistently clear signal across the whole amplitude range. Two particular sets of quantization levels are in common use for speech signal quantization. They are called the 'A-law code' and the 'Mu-law code'. Both have a higher density of quantization levels around the zero amplitude level, and both use an eight-bit (256 level) coding technique. They only differ in the actual amplitude values chosen as their respective quantum levels. The A-law code is the European standard for speech quantization while the Mu-law code is used in North America. Unfortunately, since the codes have non-corresponding quantization levels, conversion equipment is required for interworking and this adds to the quantization noise of a connection comprising both A-law and Mu-law digital transmission plant.

Conversion from A-law to Mu-law code (or vice versa) amounts to a compromise between the different quantization levels. An 8-bit binary number in one of the codes corresponds to a particular quantum value at a particular sample instant. This 8-bit

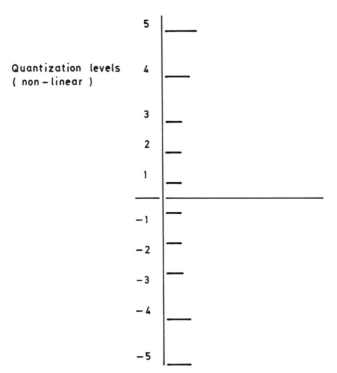

Figure 5.5
Non-linear quantization levels.

number is converted into the 8-bit number corresponding to the nearest quantum value in the other quantization code. The conversion is therefore a relatively simple matter of mapping (i.e. converting) between one 8-bit value and another.

5.4 QUANTIZATION NOISE

Most 'noise' heard by the listener on a digital speech circuit is the noise introduced during quantization rather than the result of interfering electromagnetic noise added along the line, and it is minimized by applying the special A-law and Mu-law codes as already discussed. The total amount of quantization noise (quantizing noise) on a received signal is usually quoted in terms of the number of quantization levels by which the signal differs from the original.

This value is quoted as a number of 'quantization distortion units' (or 'qdus'). Typically the acceptable maximum number of qdus allowed on a complete end-to-end connection is less than 10 (taking into account any A-to-Mu law code conversion or other signal processing undertaken on the connection). Another possible type of speech processing is the technique of speech compression, and we shall see in Chapter 20 the overall bit rate can be reduced by speech compression, at the cost of some increase in quantization noise.

Quantization noise only occurs in the presence of a signal. Thus, the quiet periods during a conversation are indeed quiet. This 'quietness' gives an improved subjective view of the quality of digital transmission.

5.5 TIME-DIVISION MULTIPLEXING

Since digital transmission is by discrete pulses and not continuous signals, it is possible for the information of more than one 64 kbit/s channel to be transmitted on the same path, as long as the transmission rate (i.e. bit rate) is high enough to carry the bits from a number of channels. In practice this is done by 'interleaving' the pulses from the various channels in such a way that a sequence of eight pulses (called a 'byte' or an 'octet') from the first channel is followed by a sequence of eight from the second channel, and so on. The principle is illustrated in Figure 5.6, in which the TDM equipment could be imagined to be a rotating switch, picking up in turn 8 bits (or 1 'byte') from each of the input channels A, B and C in turn. Thus the output bit stream of the TDM equipment is seen to comprise, in turn, byte A1 (from channel A), byte B1 (from channel B), byte C1 (from channel C), then, cycling again, byte A2, byte B2, byte C2 and so on. Notice that a higher bit rate is required on the output channel, in order to ensure that all the incoming data from all three channels can be transmitted onward. Since $3 \times 2 = 6$ bytes of data are received on the incoming side during a time period of 250 μs (1 byte on each channel every 125 μs), all of them have to be transmitted on the outgoing circuit in an equal amount of time. Since only a single channel is used for output, this implies a rate of $6 \times 8 = 48$ bits in 250 μs, i.e. 192 kbit/s. (Unsurprisingly, the result is equal to 3×64 kbit/s.) Thus in effect the various channels 'time-share' the outgoing transmission path. The technique is known as time-division multiplexing, or TDM.

TDM can either be carried out by interleaving a byte (i.e. 8 bits) from each 'tributary' channel in turn, or it can be done by single bit interleaving. Figure 5.6 shows the more common method of byte interleaving. The use of the TDM technique is so common on digital line systems that physical circuits carrying only 64 kbit/s are extremely rare, so that digital line terminating equipment usually includes a multiplexing function. Figure 5.7 shows a typical digital line terminating equipment, used to convert between a number of individual analogue channels (carried on a number of individual physical circuits) and a single digital bit stream, carried on a single physical circuit. The equipment shown is called a primary multiplexor. A primary multiplexor (or PMUX) contains an analogue to digital conversion facility for individual telephone channel conversion to 64 kbit/s, and additionally a time division multiplex facility. In Figure 5.7, a PMUX of European origin is illustrated, converting 30 analogue channels into A-law encoded 64 kbit/s digital format, and then time division multiplexing all of these 64 kbit/s 'channels' into a single 2.048 Mbit/s digit line system. (2.048 Mbit/s is actually equal to 32×64 kbit/s, but channels '0' and '16' of the European system are generally used for purposes other than carriage of information.)

We could equally well have illustrated a North American version PMUX. The difference would have been the use of Mu-law encoding and the multiplexing of 24 channels into a 1.544 Mbit/s transmission format also called a 'T span', a 'T1'

58

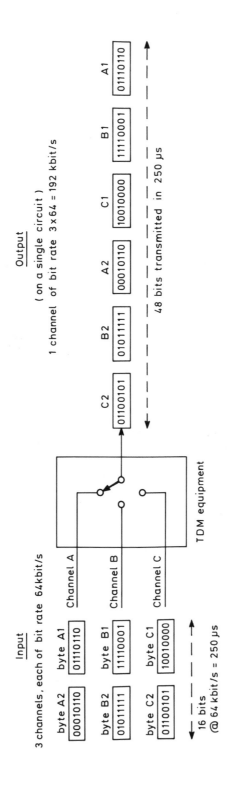

Figure 5.6
The principle of time division multiplexing (TDM).

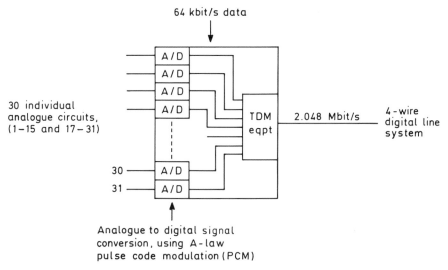

64 kbit/s data

A/D
A/D
A/D
A/D

30 individual
analogue circuits,
(1–15 and 17–31)

TDM
eqpt

2.048 Mbit/s

4-wire
digital line
system

30 —— A/D
31 —— A/D

Analogue to digital signal
conversion, using A-law
pulse code modulation (PCM)

Figure 5.7
European primary multiplexor.

line or a 'DS1' (T = Transmission, DS = Digital Line system). (1.544 Mbit/s = 24 × 64 kbit/s plus 8 kbit/s.)

The transmitting equipment of a digital line system has the job of multiplexing the bytes from all the constituent channels. Conversely, the receiving equipment must disassemble these bytes in precisely the correct order. This requires synchronous operation of transmitter and receiver, and to this end particular patterns of pulses are transmitted at set intervals, so that alignment and synchronism can be maintained. These extra pulses are sent in channel 0 of the European 2 Mbit/s digital system, and in the extra 8 kbit/s of the North American 1.5 Mbit/s system.

5.6 HIGHER BIT RATES OF DIGITAL LINE SYSTEMS

The number of channels multiplexed on a carrier depends on the overall rate of bit transmission on the line. Given that each channel must be transmitted at 64 kbits/s, the overall bit speed is usually related to an integer multiple of 64 kbit/s. As we have already seen there are two basic hierarchies of transmission rates which have been standardized for international use, but these extend to higher bit rates than the 2.048 MBIT/s and 1.544 Mbit/s versions so far discussed:

(i) The CCITT (International Telegraph and Telephone Consultative Committee), and also CEPT (European Conference for Posts and Telecommunications) have standardized 2.048 Mbit/s as the primary digital bandwidth. This has 32 channels, 30 for speech and two for alignment synchronization and signalling, more of which we shall discuss later in the chapter.

Higher transmission rates in the European digital hierarchy are attained by

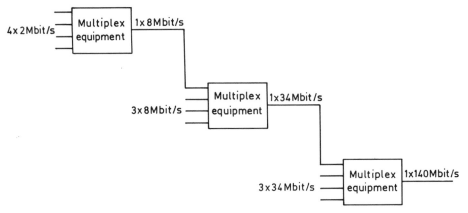

Figure 5.8
European digital multiplex hierarchy.

interleaving a number of 2 Mbit/s systems as illustrated in Figure 5.8. The standardized rates are:

2.048 Mbit/s, referred to as	2 Mbit/s
8.448 Mbit/s, referred to as	8 Mbit/s (4 × 2 Mbit/s)
34.368 Mbit/s, referred to as	34 Mbit/s (4 × 8 Mbit/s)
139.264 Mbit/s, referred to as	140 Mbit/s (4 × 34 Mbit/s)

Multiplexing equipment is available for any of the rate conversions as Figure 5.8 shows.

(ii) In the second CCITT standard (which currently predominates in North America and some Far Eastern Countries), a different multiplex hierarchy is recommended and is shown below. The principles of multiplexing, however, are largely the same, and similar diagrams to Figure 5.8 could have been drawn.

DS1 =	1.544 Mbit/s or	1.544 Mbit/s
	(this is called the T-span, T1 or DS1 system)	
DS2 =	6.312 Mbit/s (4 × 1.5 Mbit/s)	6.312 Mbit/s (4 × 1.5 Mbit/s)
DS3 =	44.736 Mbit/s (7 × 6 Mbit/s)	32.064 Mbit/s (5 × 6 Mbit/s)
DS4 =	139.264 Mbit/s (3 × 45 Mbit/s)	97.728 Mbit/s (3 × 32 Mbit/s)

The North American and European hierarchies are incompatible at all levels (including the basic speech channel level), on account of the different quantum coding used by the A- and Mu-law PCM algorithms. Interworking equipment is therefore required for international links between administrations employing the different hierarchies. In general, this interworking is undertaken in the country which uses the 1.544 Mbit/s standard.

5.7 DIGITAL FRAME FORMATTING

As we noted earlier in the chapter, it is common in a 2.048 Mbit/s system to use only thirty 64 Mbit/s channels (representing only 1.920 Mbit/s) for actual carriage of information. This leaves an additional 128 kbit/s bit rate available. Similarly, in the 1.544 Mbit/s system, the bit rate required to carry twenty-four 64 kbit/s channels is only 1.536 kbit/s, and 8 Mbit/s are left over. The burning question: what becomes of this spare capacity? The answer: it is used for synchronization and signalling functions.

Consider a 2.048 Mbit/s bit stream, and in particular the bits carried during a single time interval of 125 μs. During a period of 125 μs a single sample of 8 bits will have been taken from each of the 30 constituent or 'tributary' channels making up the 2.048 Mbit/s bit stream. These are structured into an imaginary 'frame', each frame consisting of 32 consecutive 'timeslots', one 'timeslot' of eight bits for each tributary channel. Overall the 'frame' represents a snapshot image, one sample of eight bits taken from each of the 30 channels, at a frequency of one frame every 125 s. Each frame is structured in the same way, so that the first timeslot of 8-bits holds the eight-bit sample from tributary channel 1, the second timeslot the sample from channel 2, and so on. The principle was shown in Figure 5.6. It is very like a single frame of a movie film; the only thing missing is the equivalent of the film perforations which allow a movie projector to move each 'freeze-frame' precisely. This 'film perforations' function is in fact performed by the first timeslot in the frame. It is given the name timeslot '0'. It carries so-called 'framing' and 'synchronization' information, providing a clear mark to indicate the start of each 'frame'. The principle is shown in Figure 5.9, which illustrates a single frame of 32 'timeslots'.

Timeslot 0 then provides a mark for framing. Timeslots 1–15 and 17–31 are used to carry the tributary channels. That leaves timeslot 16 which, as we shall see, is used for 'signalling'.

We cannot leave timeslot 0, without briefly discussing its 'synchronization' function which serves to keep the line system bit rate running at precisely the right speed. Consider a wholly digital network consisting of three digital exchanges A, B and C interconnected by 2.048 Mbit/s digital transmission links, as shown in Figure 5.10, with end-users connected to exchanges A and C.

Each of the exchanges A, B and C in Figure 5.10 will be designed to input and receive data from the digital transmission links A–B and B–C at 2.048 Mbit/s. But what happens if link A–B actually runs at 2 048 000 bit/s, while link B–C runs at 2 048 001 bit/s? This, or something even worse could quite easily happen in practice if we did not take 'synchronization' steps to prevent it. In the circumstances shown, the bit stream received by exchange B from exchange A is not fast enough to fill the outgoing timeslots on the link from B to C correctly, and a 'slip' of 1 'wasted' bit will occur once per second. Conversely, in the direction from C to A via B, unsent bits will gradually be stored up by exchange B at a rate of one extra unsent bit per second, since the exchange is unable to transmit the bits to A as fast as it is receiving them from exchange C. Ultimately bits are lost when the store in exchange B overflows. Neither 'slip' nor 'overflow' of bits is desirable, so networks are normally designed to be synchronous at the 2.048 Mbit/s level, in other words are controlled to run at exactly the same speed. Some of the bits in timeslot zero of a 2.048 Mbit/s line system are used

62

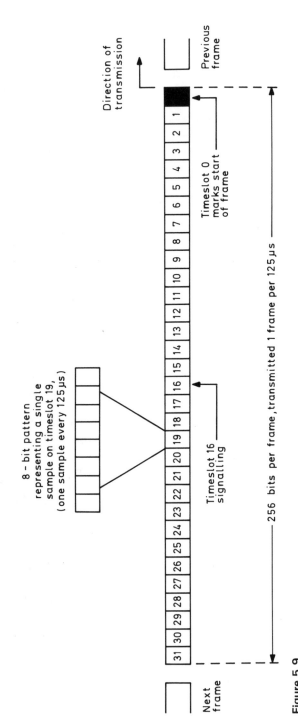

Figure 5.9
2.048 Mbit/s digital frame format.

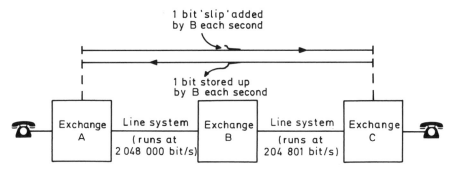

Figure 5.10
The need for synchronization.

to achieve this synchronization. The same functions of 'framing' and synchronization'
are carried out by the surplus 8 kbit/s capacity of the 1.544 Mbit/s digital line system.

Now let us consider the function of timeslot 16 in a 2.048 Mbit/s digital line system.
This 'timeslot' is usually reserved for carrying the signalling information needed to set
up the calls on the 30 'user channels'. The function of signalling information is to
convey the intended destination of a call on a particular channel between one exchange
and the next.

From the above, we see that the maximum usable bit rate of a 2.048 Mbit/s system
is 30×64 kbit/s or 1.920 Mbit/s. In occasional circumstances however, this can be
increased to 1.984 Mbit/s when the signalling channel (timeslot 16) is not needed.

When required on a 1.544 Mbit/s line system, a signalling channel can either be
made available by 'stealing' a small number of 'bits' (equivalent to 4 kbit/s) from one
of the tributary channels (thereby reducing the capacity of that particular channel to
60 kbit/s), or alternatively one whole 64 kbit/s channel may be dedicated for signalling use. The method of stealing bits to create a 4 kbit/s signalling channel is known
in North America as 'robbed bit signalling'. It is only permissible to 'rob' the bits from
a voice channel and not from a data channel. Robbing a small number of bits from
a voice channel is permissible since the quality lost thereby is almost imperceptible to
a human telephone listener. Robbing bits from a channel which is carrying data, however, will result in quite unacceptable data corruption. Where a signalling channel is
required on a 1.544 Mbit/s digital line system carrying only data circuits, a whole
channel should be dedicated for signalling. Such a dedicated signalling channel is
necessary to create CCITT 7 signalling links.

To return to the two different bit rate hierarchies, observant readers may have
noticed that the higher bit rates of both hierarchies are not exact integer multiples of
the basic 2.048 Mbit/s and 1.544 Mbit/s tributaries. Instead, some extra 'framing' bits
have been added once again at each hierarchical level. These are provided for the same
framing reasons as have already been described in connection with the 2.048 Mbit/s
line system, and illustrated in Figure 5.9. However, unlike their 2 Mbit/s or
1.5 Mbit/s tributaries, synchronization of higher bit-rate line systems is not usually
undertaken. Instead, higher order systems are generally allowed to 'free run'. The
extra bits allow free running, since a slightly higher bit rate is available than the tributaries can feed. The higher bit rate ensures that there is no possibility of bits building

up between the tributaries and the higher bit rate line system itself. Instead there will always be a few bits to spare. The benefit is that the need for synchronization at the higher bit rate is avoided, but the 'penalty' is the complicated frame structure needed at the higher rates of the hierarchy. The problem arises from the fact that although each frame has a very strict format, it is not possible to say for sure how many spare bits will separate each frame. The start of each frame must therefore be clearly marked by the framing bits. The variable number of spare bits between frames means that it is virtually impossible to extract a 2 Mbit/s tributary from a 140 Mbit/s line system without virtually separating out each individual 2 Mbit/s tributary, using a large number of multiplex equipments, as already shown in Figure 5.8.

The same problem faces users of the 1.544 Mbit/s hierarchy, but may disappear in a few years time with the gradual emergence of a new digital line system hierarchy known as the 'synchronous digital hierarchy' (SDH). Developed from the SONET (synchronous optical network) standards of the American National Standards Institute, SDH demands synchronous operation of the line system with no spare bits between frames, but in return it offers a far simpler and more regular frame structure of 2 Mbit/s and 1.544 Mbit/s tributaries in high bit rates. The removal of the padding bits between 2 Mbit/s frames is what gives the regular frame spacing, in turn giving much greater scope for direct 'drop and insert' of individual tributaries within line systems, since consecutive samples of each tributary occur at a fixed bit spacing within the overall higher bit rate stream. SDH will work at speeds in excess of 150 Mbit/s.

5.8 INTERWORKING THE 2 MBIT/S AND 1.5 MBIT/S HIERARCHIES

Interworking of digital line systems running in the 2 Mbit/s hierarchy and 1.5 Mbit/s hierarchy is relatively straightforward, given the availability of proprietary equipment for the conversion. At its simplest, the 24 channels of a 1.5 Mbit/s system can be carried within a 2 Mbit/s system, effectively wasting the remaining capacity of the 2 Mbit/s system. Alternatively, a 2 Mbit/s system can be entirely carried on two 1.5 Mbit/s systems, wasting 16 channels of the second 1.5 Mbit/s system. More efficiently, however, four 1.5 Mbit/s systems fit almost exactly into three 2 Mbit/s systems or vice versa. (They appear to fit exactly, but usually some bits are taken for separating the different frames—so that the efficiency is reduced slightly.)

The interworking of one digital hierarchy into the other need only involve mapping the individual 8-bit timeslots from one hierarchy into corresponding timeslots in the other. The technique is called timeslot interchange. The only complication is when the 8-bit patterns in the timeslots are not simple data patterns (data patterns should be mapped across unchanged) but when they are sample patterns corresponding to A- or Mu-law pulse code modulated speech. In this instance, an A- to Mu-law speech conversion is also required at the 2 Mbit/s-to-1.5 Mbit/s interworking point.

Figure 5.11 illustrates a typical timeslot interchange between four 1.5 Mbit/s and three 2 Mbit/s digital line systems. Note that the timeslot interchange equipment in Figure 5.11 is also capable of Mu- to A-law conversion (and vice versa). This has to be available on each of the channels, but is only employed when the channel is carrying a speech call. When there are consecutive speech and data calls on the same channel, the Mu- to A-law conversion equipment will have to be switched on for the

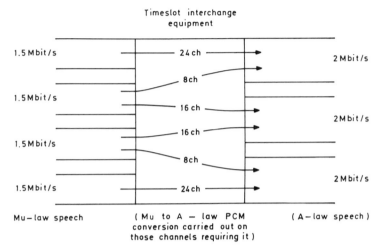

Figure 5.11
Timeslot interchange between 1.5 Mbit/s and 2 Mbit/s.

first call and off for the second. Some means is needed therefore of indicating to the timeslot interchange equipment whether, at any particular time it is carrying a speech or a data call. Alternatively, particular channels could be preassigned either to speech or data use. In this case the Mu- to A-law conversion equipment will be permanently on and permanently off, respectively.

5.9 LINE CODING

The basic information to be transported over any digital line system, irrespective of its hierarchical level, is a sequence of ones and zeros, also referred to as 'marks' and 'spaces'. The sequence is not usually sent directly to line, but is first arranged according to a 'line code'. This aids intermediate regenerator timing and distant end receiver timing, maximizing the possible regenerator separation and generally optimizing the operation of the line system. The potential problem is that if either a long string of 0's or 1's were sent to line consecutively then the line would appear to be either permanently 'on' or permanently 'off'—effectively a direct current condition is transmitted to line. This is not advisable for two reasons. First the power requirement is increased and the attenuation is greater for direct as opposed to alternating current. Second, any subsequent devices in the line cannot distinguish the beginning and end of each individual bit. They cannot indeed tell if the line is actually still alive. The problem gets worse as the number of consecutive 0's or 1's increases. Line codes therefore seek to ensure that a minimum frequency of line state changes is maintained.

Examples of common line codes recommended by CCITT are AMI (alternate mark inversion) and HDB3 (high density bipolar). Both AMI and HDB3 are actually three-state, rather than simple two-state (on/off) codes. In these codes, the two extreme states are used to represent 'marks', and the mid state is used to represent 'space'. The three states could be positive and negative values, with a mid value of 0. Or in the

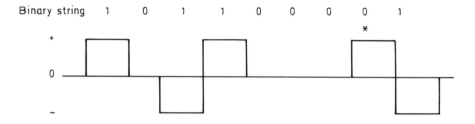

Binary string 1 0 1 1 0 0 0 0 1

✱ Fourth consecutive zero in 'violation'

Figure 5.12
HDB3 code—alternate mark inversion with fourth zeros marked in 'violation'.

case of optical fibres, where light is used, the three states could be 'off', 'low intensity' and 'high intensity'. In both AMI and HDB3 line codes, alternative 'marks' are sent as positive and negative pulses. Alternating the polarity of the pulses helps to prevent direct current being transmitted to line. In a two-state code, a string of marks would have the effect of sending a steady 'on' value to line.

The HDB3 code (used widely in Europe and on international transmission systems) is an extended form of AMI in which the number of consecutive zeros that may be sent to line is limited to 3. Limiting the number of consecutive zeros brings two benefits: first a null signal is avoided, and second a minimum 'mark density' can be maintained (even during idle conditions such as pauses in speech). A high mark density aids the regenerator timing and synchronization. In HDB3, the fourth zero in a string of four is 'marked' (i.e. forcibly set to 1) but this is done in such a way that the 'zero' value of the original signal may be recovered at the receiving end. The recovery is achieved by marking fourth zeros in 'violation', that is to say in the same polarity as the previous 'mark', rather than in opposite polarity mark (opposite polarity of consecutive marks being the normal procedure). Figure 5.12 illustrates the HDB3 line code.

5.10 OTHER LINE CODES AND THEIR LIMITATIONS

A number of different line codes are in use around the world. Many of them seek to maintain a minimum frequency of line state changes, in order to ease the job of regenerator timing and synchronization, and are therefore similar in seeking to eliminate long spells of consecutive 'marks' (1's) or consecutive spaces (0's). One of the line codes commonly used in North America in association with the 1.5 Mbit/s line system is called 'zero code suppression'. This technique seeks to eliminate patterns of eight or more consecutive zeros, but it does so in an irreversible manner—by forcibly changing the value of the eighth consecutive bit of value 0, so that instead of transmitting 00000000, 00000001 is transmitted instead. Unfortunately, since it is only a two-state code, the receiving end device—unlike an HDB3 receiver—is unable to tell that the eighth bit value has been altered. An error results. The error is not perceptible to speech users, but would cause unacceptable corruption of data carried on a 64 kbit/s channel.

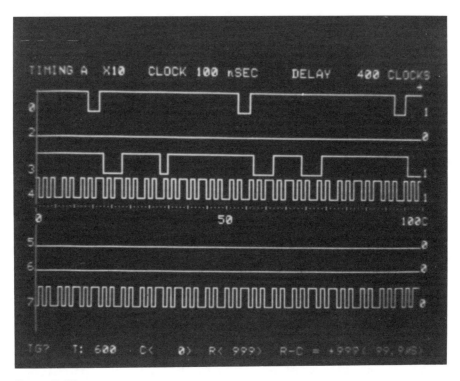

Figure 5.13
Digital signal pattern. These oscilloscope patterns result from testing of circuits using a standard line format for the Bell Systems digital network. (*Courtesy of AT&T.*)

Once 'eighth bit encoding' using the zero code suppression technique had begun, it became acceptable to 'rob' the eighth bit for other internal network uses. A 'Robbed bit signalling' channel, equivalent to the Europeans' 'timeslot 16' signalling channel, was created as already discussed.

Both of the above uses of 'eighth bit encoding' reduced the usable portion of the 64 kbit/s channel. For this reason, it is common for data terminals in North America to use only seven of the eight available bits in each byte. This has the effect of reducing the usable bit rate to 56 kbit/s (8000 samples of 7 bits per second) even though 64 kbit/s is carried on the line. North American readers may be familiar with the 56 kbit/s user rate.

In connections from Europe to North America where the 56 kbit/s user data rate is employed, it is necessary to employ a rate adapter at the European end to accommodate the lower rate. In essence the rate adapter is programmed to waste the eighth bit of each byte, giving a 56 kbit/s user rate even at the European end. Alternatively a rate adapter may be used at both ends to employ an even lower bit rate, such as the CCITT standard bit rate of 48 kbit/s. In this case 2 bits of each byte are ignored. Stimulated by worldwide customer pressure for 64 kbit/s services (including ISDN—see Chapter 24), the restriction to 56 kbit/s channel capacity in North America looks set to disappear with the adoption of various new line codes. These include B8ZS (bipolar 8-zero substitution) and ZBTSI (Zero Byte Time Slot Interchange). Like HDB3, both eliminate long strings of zeros—but in a recoverable way.

Some US readers may additionally have come across the '64 kbit/s RESTRICTED' bandwidth. As its name suggests this provides a bit rate close to 64 kbit/s. It derives from the use of a zero code suppressed channel during the time prior to the availability of either B8ZS or ZBTSI. Any bit pattern can be sent by the user at the normal 64 kbit/s rate—as long as the '00000000' pattern is never used.

A number of other line codes are also in use around the world. Without going into their methods of working, they include Manchester code, WAL2 and Miller code.

5.11 THE FUTURE OF DIGITAL TRANSMISSION

As for the future of digital transmission, equipment is already being installed for rates of 280 Mbit/s and 565 Mbit/s. These will allow many thousand conversations to be carried on a single transmission medium. Furthermore, systems with rates as high as 1.76 Gbit/s have already been proven on the laboratory bench. Meanwhile other researchers have demonstrated that speech and other signals can be transmitted quite adequately at very low bit speeds using techniques such as ADPCM (discussed in Chapter 20). Rates even as low as 8 kbit/s have undergone promising tests on speech transmission. To quantify the significance of these two independent trends in technology: if 8 kbit/s channels were multiplexed to form a 565 Mbit/s system, more than 70 000 simultaneous telephone calls could be carried on a single cable (containing transmit and return pairs or channels)! With the flexibility of the synchronous digital hierarchy (SDH) also looming, analogue transmission looks set for an early retirement!

BIBLIOGRAPHY

Clark, A. P., *Principles of Digital Data Transmission*, 2nd edn. Pentech Press, 1983.

Green, D. C., *Digital Techniques and Systems*, 2nd edn. Pitman, 1980.

Inose, H., *An Introduction to Digital Integrated Communications Systems*. Peter Peregrinus (for IEE), 1979.

Sinnema, W., *Digital, Analog and Data Communication*, Reston Publishing Company, 1982.

Smith, D. R., *Digital Transmission Systems*. Van Nostrand Reinhold/Lifetime Learning Publications, 1985.

Uiet, P. G. F., *Telecommunications Systems*. Artech House, 1986.

Wade, J. G., *Signal Coding and Processing—an Introduction Based on Video Systems* Ellis Horwood, 1987.

THE PRINCIPLES OF CIRCUIT SWITCHING

In this chapter we shall deal with the mechanics of the exchange itself, and describe the necessary sequence of its functions. We shall cover the principles of circuit switching (as would be used in voice or circuit data networks) and describe in outline some of the better known types of circuit-switched exchange.

6.1 CIRCUIT-SWITCHED EXCHANGES

In circuit-switched networks a physical path, or 'circuit', must exist for the duration of a call, between its point of origin and its destination, and three particular attributes are needed in all circuit-switched exchanges:

- The ability not only to establish and maintain (or 'hold') a physical connection between the 'caller' and the 'called party' for the duration of the call but also to disconnect (CLEAR) it afterwards.

- The ability to connect any circuit carrying an incoming call (a so-called 'incoming circuit') to one of a multitude of other (so-called 'outgoing') circuits. Particularly important is the ability to select different outgoing circuits when subsequent calls are made from the same incoming circuit. During the set-up period of each call the exchange must determine which outgoing circuit is required usually by extracting it from the dialled number. This makes it possible to put through calls to a number of other network users.

- The ability to prevent new calls intruding into circuits which are already in use. To avoid this the new call must either be diverted to an alternative circuit, or it must temporarily be denied access, in which case the caller will hear a 'busy' or 'engaged' tone.

Exchanges are usually designed as an 'array' or 'matrix' of 'switched crosspoints' as illustrated in Figure 6.1.

The switch matrix illustrated in Figure 6.1 has five incoming circuits, five outgoing circuits and 25 switch crosspoints which may either be 'made' or 'idle' at any one time.

Any of the incoming circuits, A to E, may therefore be interconnected to any of the outgoing circuits, 1 to 5, but at any particular instant no incoming circuit should be connected to more than one outgoing circuit, since each caller can only speak with one party at a time. In Figure 6.1, incoming circuit A is shown as connected to outgoing circuit 2, and simultaneously C is connected to 1, D to 4, and E to 3. Meanwhile, circuits B and 5 are idle. Therefore, at the moment illustrated, four calls are in progress. Any number up to five calls may be in progress, depending on demand at that time, and on whether called customers are 'free' or not. Let us assume, for example, that only a few moments before the moment illustrated, customer B had attempted to make a call to customer 3 (by dialling the appropriate number) only to find 3's line engaged. Any telephone user will recognize B's circumstance. But only a moment later, customer E might cease conversation with customer 3, and instantly make a call to customer 5. If B then chances to pick up the phone again and redial customer 3's telephone number, the call will complete since the line to C is no longer busy. Figure 6.2 shows

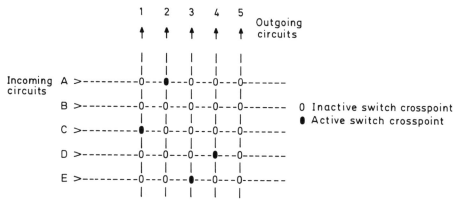

Figure 6.1
A basic switch matrix at a typical instant in time.

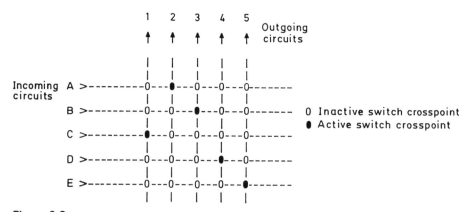

Figure 6.2
The same switch matrix a few moments later.

the new switchpoint configuration at this subsequent point in time, when five calls (the maximum for this switch) are simultaneously in progress.

At any one time, between nought and five of the 25 crosspoints may be in use, but an incoming circuit can never be connected to more than one outgoing circuit, nor any outgoing circuit be connected to more than one incoming circuit.

What exactly do we mean when we say a 'connection' is made? In previous chapters on line transmission methods we concluded that a basic circuit needs at least two wires, and that a long distance one is best configured with four wires (a 'transit' and a 'receive pair'). How are all these wires connected by the switch? And how exactly is intrusion prevented?

The answer to the first question is that each of the two (or four) wires of the connection is switched separately, but in an array similar to that in Figure 6.2. Thus a number of switch array 'layers' could be conceived, all switching in unison as shown in Figure 6.3, which illustrates the general form of a 4-wire switch.

Each layer of the switch shown in Figure 6.3 is switching one wire of the four. Each of the four layers switches at the same time, using corresponding crosspoints in order that all of the four wires comprising any given incoming circuit are connected to all the corresponding wires of the selected outgoing circuit (note that in Figure 6.3, circuit A is connected to circuit 5).

Additionally in Figure 6.3 you will see that the fifth or so-called 'P-wire' has also been switched through. The use of such an 'extra' wire is one way of designing switches to prevent call intrusion. Usually this method is used in an electro-mechanical exchange and it works as follows.

When any of circuits A to E are idle, there is an electrical voltage on their corresponding P-wires. When any of the callers A to E initiates a new call, the voltage on the P-wire is dropped to earth (zero volts). When the call is switched through the matrix to any of the outgoing circuits 1 to 5, the P-wire of that circuit will also be earthed as a result of being connected to the P-wire of the incoming circuit. When the call is over, the caller replaces the handset. This causes a voltage to be re-applied to

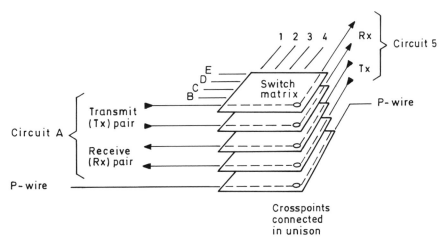

Figure 6.3
Switching a 4-wire connection.

the P-wire, to which the switch matrix responds by clearing the connection. Intrusion is prevented by prohibiting connection of circuits to others for which the P-wire is already in an earthed (i.e. busy) condition. In this way, the 'earth' on the P-wire is used as a marker to distinguish lines in use.

The P-wire also provides a useful method of circuit holding, and for initiation of circuit 'clearing'. To this end the switch is designed to maintain (or 'hold') the connection as long as the P-wire is earthed. As soon as the caller replaces the handset, the P-wire is reset to a non-zero electrical voltage, and the switch responds by clearing the connection (i.e. releasing the switch point).

In the latest stored program control (SPC) exchanges call intrusion is prevented, and the job of 'holding' the circuit is carried out by means of the exchange processor's electronic 'knowledge' of the circuits in use. 'P-wires' are thus becoming obsolete.

To return to the example in Figures 6.1 and 6.2, what if party A wishes to call party E? Our diagrams show A and E as only able to make outgoing calls through the switch, so how can they be connected together? The answer lies in providing either or both circuits with access to both 'incoming' and 'outgoing' sides of the exchange, and it is done by 'commoning' (i.e. wiring together) circuits 1 and A, 2 and B, 3 and C, 4 and D, 5 and E; as shown in Figure 6.4. (Incidentally, in the example shown in Figure 6.4 as well as in other diagrams in the remainder of the chapter, it is necessary to duplicate the layers for each of the two or four wires of the connection, as already

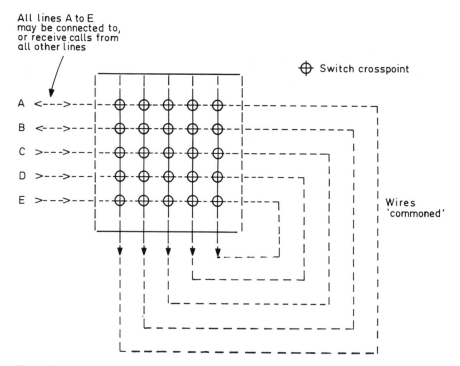

Figure 6.4
A simple small local exchange.

explained in Figure 6.3. For simplicity however the diagrams only illustrate one of the layers.)

In Figure 6.4, any of the customers A to E may either make calls to, or receive calls from, any of the other four customers. The maximum number of simultaneous calls now possible across the matrix is only two as compared with the five possible in Figure 6.2. This is because two calls are sufficient to engage four of the five available lines. One line must therefore always be idle.

A switch matrix designed in the manner illustrated in Figure 6.4 would serve well in a small exchange with only a few customers, such as an office 'private branch exchange' (PBX) or a small 'local exchange' in a public telephone network. The exchange illustrated in Figure 6.4 is actually a 'full availability' and 'non-blocking' system. Fully available means that any line may be connected to any other; non-blocking indicates that as long as the destination line is free a connection path can be established across the switch matrix regardless of what other connections are already established. These terms will be more fully explained later in the chapter and we shall also discover some of the economies that may be made in larger exchanges, by introducing 'limited availability'.

But first, let us consider how the isolated system of Figure 6.4 can be connected to other similar systems in order to give more widespread access to other local exchanges say, or to trunk and international exchanges. Figure 6.5 illustrates how this is done. It shows how some circuits, which are designated as incoming or outgoing 'junctions' connect the exchange to other exchanges in the network. Such inter-exchange circuits allow connections to be made to customers on other exchanges.

Networks are built up by ensuring that each exchange has at least some 'junction', 'tandem' or 'trunk' circuits to other exchanges. The exchanges need not be fully interconnected however; connections can also be made between remote exchanges by the use of 'transit' (also called 'tandem') routes via third exchanges, as shown in Figure 6.6. 'Junction' and 'trunk' circuits are always provided in multiple numbers. This means that if one particular circuit is already in use between two exchanges, a number of equally suitable alternative circuits (interconnecting the same exchanges) could be used instead. Figure 6.6 shows four circuits interconnecting exchanges P and Q; two each for incoming and outgoing directions of traffic. The consequence is that when circuit 1 from exchange P to exchange Q is already busy, circuit 2 may be used to

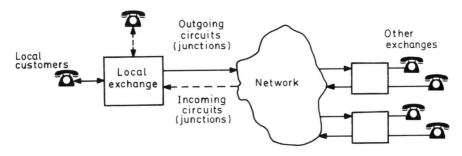

Figure 6.5
Junction connection to other exchanges.

establish another call. Only when both circuits are busy need calls be failed, and callers given the 'busy tone'.

Each of the exchanges shown in the networks of Figures 6.5 and 6.6 must have its switch matrix and circuit-to-exchange connections configured as illustrated in Figure 6.7. Figure 6.7 shows how the local customers' lines are connected to both incoming and outgoing sides of the switch matrix, and how in addition a number of 'uni-directional' (i.e. single direction of traffic) incoming and outgoing junctions are connected.

The junctions of Figure 6.7 could have been designed to be 'bothway' junctions. In this case, like the customers' lines illustrated in Figure 6.4, they would need access to both incoming and outgoing sides of the switch matrix. In some circumstances this can be an inefficient use of the available switch ports, because it may reduce the number of calls that the switch can carry at any given time. (Remember that the matrix in Figure 6.4 may only carry a maximum of two calls at any time, while the same size matrix in Figure 6.1 could carry five calls.)

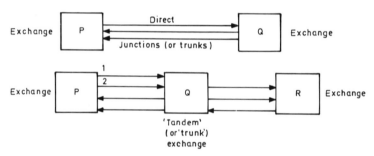

Figure 6.6
Typical networking arrangements.

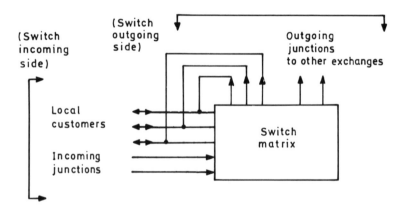

Figure 6.7
Local exchange configuration.

6.2 CALL BLOCKING WITHIN THE SWITCH MATRIX

Telecommunications networks which are required to have very low call blocking probabilities, have to be designed with excess equipment capacity over and above that needed to carry the average call load. Indeed, to achieve zero call blocking, we would need to provide a network of an infinite size. This would guarantee enough capacity even in the unlikely event of everyone wanting to use the network at once. But because an infinitely sized network is impractical, telecommunications systems are normally designed to be incapable of carrying the last very small fraction of traffic. Switching matrices are similarly designed to lose a small fraction of calls as the result of internal switchpoint congestion.

In the case of switch matrices we refer to the designed lost fraction of calls as the 'switch blocking coefficient'. This coefficient exactly equates to the 'grade-of-service' that we shall define in Chapter 10, and the dimensioning method is exactly the same. Thus a switch matrix with a blocking coefficient of 0.001 is *designed* to be incapable of completing one call in 1000. That one call will be lost as a direct consequence of 'switch matrix congestion'. By comparison, a 'non-blocking switch' is designed in such a way that no calls fail due to internal congestion.

How does switch blocking come about anyway? And how can costs be cut by designing switches with relatively large switch blocking coefficients? There are two methods of economizing hardware, both of which inflict some degree of call blocking due to switch matrix congestion. They rely either on:

- limiting circuit 'availability', or on

- employing 'fan-in fan-out' switch architecture.

and they are described separately below.

6.3 FULL AND LIMITED AVAILABILITY

All switches fall into one of two classes:

- 'full availability' switches, or

- 'limited' (or 'partial') availability switches

The difference between the two lies in the internal architecture of the switch. The term 'availability', in this context, is used to describe the number of the circuits in a given outgoing *route* which are 'available' to any individual incoming circuit. As an example, Figure 6.8 illustrates a simple network in which five customers, A to E, are connected to an exchange P, which, in turn, has five junction circuits to exchange Q. But Figure 6.8 is not drawn in sufficient detail to show the 'availability' of circuits within the group of junction circuits joining P and Q, since the architecture of the switch matrix itself is not shown.

Figures 6.9 and 6.10 illustrate two of many possible switch matrix architectures for

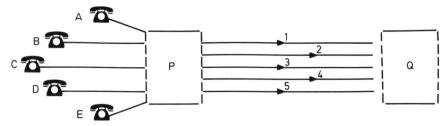

Figure 6.8
Customers A to E on exchange P, which has five circuits to exchange Q.

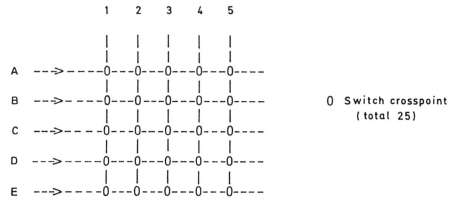

Figure 6.9
Switch matrix of exchange P configured as 'fully available'.

the exchange P which was shown topologically in Figure 6.8. In Figure 6.9, all the outgoing trunk circuits from P (numbered 1 to 5) may be accessed by any of the customers' lines, A to E. This is the 'fully available' configuration, since all outgoing circuits are 'available' to all the incoming circuits.

By contrast, in Figure 6.10 each of the customers may only access four of the five outgoing circuits. Not all of them get through to the same four though. Customer A may access circuits 1, 2, 3, 4, customer B circuits 1, 2, 3, 5, customer C circuits 1, 2, 4, 5, customer D circuits 1, 3, 4, 5 and customer E circuits 2, 3, 4, 5. Figure 6.10 shows only one of a number of possible permutations (called 'gradings') in which the outgoing circuits could be made 'available' to the incoming ones. Figure 6.10 therefore illustrates one particular 'limited availability' grading. The 'availability' of the grading shown is 4, since only a maximum of four outgoing circuits (within the outgoing route PQ) are 'available' to any individual incoming circuit. This despite the fact that more than four circuits exist within the route as a whole. In this example the total route size PQ is five circuits.

Notice in Figure 6.10 how the total number of switch crosspoints is only 20 compared with the 25 that were required in Figure 6.9. This may give the advantage of reducing the cost of the exchange, particularly if the switch matrix hardware is expensive. On the other hand it may be an unfortunate limitation of the hardware design that only four outgoing ports are possible per incoming 'circuit'. As we will find later in this chapter electromechanical switches are often not configurable as full availability switches because of the way they are made.

The disadvantage of limited availability switches is that more calls are likely to fail through internal congestion than with an equivalent full availability switch. The difference between the two is plain in Figures 6.9 and 6.10. In Figure 6.10, when circuits 1 to 4 are busy but circuit 5 is not, call attempts made on line A are 'failed', whereas the same attempt made on the configuration in Figure 6.9 will succeed, hence the lower 'switch blocking' of the latter. Similarly, B cannot reach circuit 4, nor C circuit 3, D ditto 2 and E ditto 1.

In the past a whole statistical science grew up in order to minimize the grade of service impairments encountered in limited availability systems. It was based on the study of 'grading' which involves determining the 'slip-pattern' of wiring (for example see Figure 6.10) in which optimum use of the limited available circuits is achieved. The resultant 'grading chart' is often diagrammatically represented in a form similar to that shown in Figure 6.11.

Figure 6.11 illustrates the grading chart of a number of selectors (switching mechanisms) sharing a common group of outgoing circuits to the same destination (e.g. a distant exchange). In total, 65 outgoing circuits are available in the grading, but of these, each individual incoming circuit (and its corresponding selector) can only be connected to 20 of the 65. In other words, each selector (and therefore its corresponding incoming circuit) has a limited availability of 20. Thus the top horizontal row of 20 circuits shown on the grading chart of Figure 6.11 are the outlets available to a particular incoming circuit. The second row, meanwhile, represents the 20 outlets available to a different incoming circuit.

The action of a particular incoming circuit's selector mechanism is to scan across its own part of the grading (i.e. its horizontal row) from left to right—and to select the first available free circuit nearest the left-hand side. Other selectors similarly scan their rows of the grading to select free outgoing circuits. This means that outgoing circuits towards the left-hand side of the grading chart are generally more heavily used

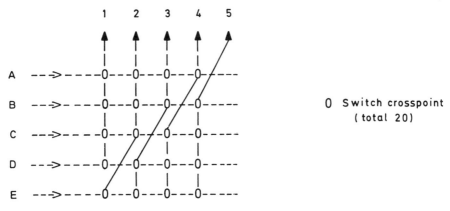

Figure 6.10
Switch matrix of exchange P configured as 'limited availability' (4).

than those on the right. To counteract this effect, the right-hand outlets of the grading (i.e. the later choices) are combined as 'doubles', 'trebles', 'quads', 'fives' and 'tens' etc. This means that some outgoing circuits are accessible from more than one selector. This helps to boost the average traffic carried by outgoing circuits in the 'later part' of the grading and so create even loading. A whole statistical science of 'grading' grew up during the period of electromechanical exchange predominance, and different types of grading are named after their inventors (e.g. the O'Dell grading).

In the diagram of Figure 6.11 you will see that apparently 10 incoming circuits (the number of horizontal rows) have access to a far greater number (65) of outgoing trunks circuits—all to the same destination exchange. Absurd you might think—and you would be right. There is no point having more outlets to the same destination than the incoming demand could ever need. The explanation is that in practice the grading horizontal reflects identical wiring of ten or twenty selectors making up a whole shelf. In our case, then, 100 (10 × 10) or 200 (20 × 10) incoming circuits are vying for 65 outgoing trunks. You will agree this is much more plausible. The reason the grading is simplified in this way is that it is much easier to design and wire a 10 × 20 grading and duplicate it than it is to create a 100 × 20 or 200 × 20 grading.

In our example, if 20 or less circuits would have sufficed to meet the traffic demand to the destination then the grading work is much easier. In this case, all the selector outlets of Figure 6.11 may be 'commoned' and full availability of outgoing circuits is possible—each incoming circuit capable of accessing each outlet.

Limited availability switches are now becoming less common, as technology increasingly enhances the sophistication and reduces the cost of modern exchanges, thereby removing many of the hardware constraints and extra costs associated with limited availability switch design. Amongst older exchange technologies, limited availability was a common hardware constraint. Typically, 'Strowger' type exchanges (described

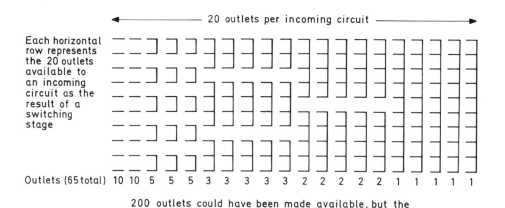

Figure 6.11
A 20-availability grading.

later in this chapter) were either 10- 20- or 24-availability. In other words each incoming circuit had access only to a maximum of either 10, 20 or 24 circuits.

6.4 FAN IN–FAN OUT SWITCH ARCHITECTURE

In Figure 6.4 we developed a simple exchange, suitable on a small scale to be a fully available and non-blocking switching mechanism for full interconnectivity between five customers. It was achieved with a switch matrix of 25 crosspoints. Earlier in the chapter we suggested that economies of scale could be made within larger switches. The main technique for achieving these economies is the adoption of a 'fan in–fan out' switch 'architecture'.

Consider a much larger equivalent of the exchange illustrated in Figure 6.4. A typical local exchange, for example, might have 10 000 customer lines plus a number of junction circuits, so that using a configuration like Figure 6.4, a matrix of around 10 000 × 10 000 switch crosspoints would be required. Bearing in mind that a typical residential customer might only contribute on average 0.5 calls to the busy hour traffic, and each call has an average duration of 0.1 hour (6 minutes) then the likely maximum number of these switchpoints that will be in use at a given time is only around 10 000 × 0.05, or 500. The conclusion is that a similar arrangement to Figure 6.4 is rather inefficient on this much larger scale.

Let us instead set a target maximum switch blocking of 0.0005. In other words, we intend that a small fraction (0.05 per cent) of calls be lost as the result of internal switch congestion. This is a typical design value and such target switch blocking values can be met by using the 'fan in–fan out' architecture illustrated in Figure 6.12.

Notice how the total number of switchpoints needed has been reduced by breaking

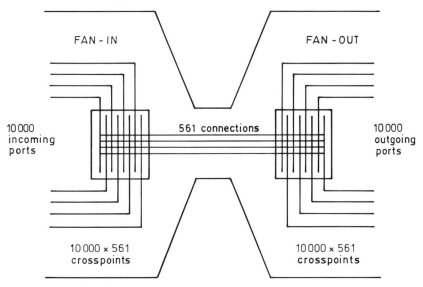

Figure 6.12
The principle of fan-in/fan-out.

the switch matrix into 'fan-in and fan-out' parts. A total of 561 connections join the two parts, the significance of the value 561 being that this is the number of circuits theoretically predicted to carry 500 simultaneous calls, with a blocking probability of 0.05 per cent. The switch is now limited to a maximum carrying capacity of 561 simultaneous calls, but the benefit is that the total number of switchpoints required is only $2 \times 561 \times 10\ 000$ or around 10 million (a tenth of the number required by the configuration like that of Figure 6.4). This provides the potential for savings in the cost of switch hardware.

In our example it is intended that the internal blocking should never exceed 0.05 per cent of calls failed due to internal switch congestion. In practice the actual switch blocking depends on the actual offered traffic, and it may be slightly higher or lower than this nominal value.

6.5 SWITCH HARDWARE TYPES

We have dealt with the general principles of exchange switching and the need for a number of incoming lines to be able to be connected to a range of outgoing lines, using a matrix of switched crosspoints. We now go on to discuss the different ways in which the matrix can be achieved in practice, and describe four individual switch types. In chronological order these are:

Strowger (or 'step-by-step') switching.
Crossbar switching.
Reed relay switching.
Digital switching.

A number of other types, namely, 'Rotary', '500-point', 'panel' and 'X–Y' switching systems have been developed through the years. They are not discussed in detail here.

Strowger switching

Strowger switches were the first widely used type of automatic exchange systems. They were developed by and named after an American undertaker who was keen to prevent operators transferring calls to his competitors. His patent was filed on 12 March 1889.

Strowger exchanges are a marvel of engineering ingenuity, using precisely controlled mechanical motion to make electrical connections. The combination of electrical and mechanical components leads to the much used expression 'electro-mechanical switching'.

The switching components of Strowger exchanges are usually referred to as 'selectors', and they work in a manner which is marvellously easy to understand. In its simplest form, a selector consists of a moving set of contacting arms (known as a 'wiper assembly') which moves over another fixed set of switch contacts known as the 'contact bank'. The act of switching consists of moving (or 'stepping') the contactor arm over each contact in turn until the desired contact is reached. Two main types of

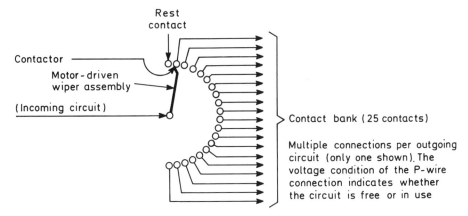

Figure 6.13
A simple uniselector.

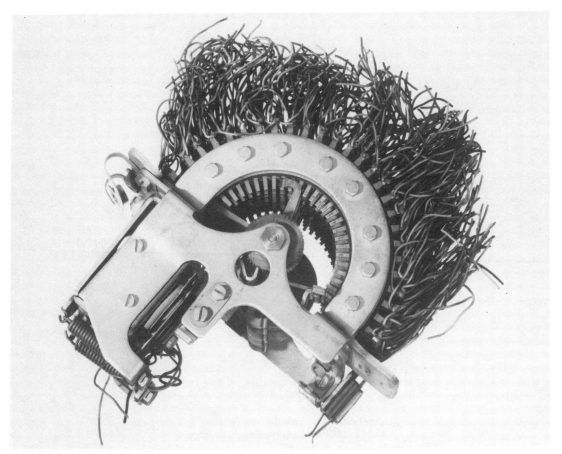

Figure 6.14
Strowger uniselector. The wipers move in the arc of a circle around 25 outlet contacts, stopping at a free outlet. Uniselectors are typically used on the customer side of an exchange, helping to find free exchange equipment to handle the call.

Strowger (or 'step-by-step') selectors are used in most exchanges of this type. They are called 'uniselectors' and 'two-motion selectors'.

A uniselector is a type of selector in which the wiper assembly rotates in one plane only, about a central axis. The contactors move along the arc of a circle, on which the fixed contact bank is arranged. Figure 6.13 illustrates the principle. A single incoming circuit is connected to the uniselector's contactor on the wiper assembly, and

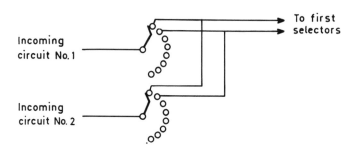

(a) Uniselectors graded to share 'first' selectors

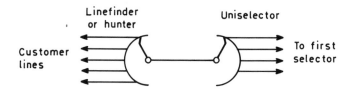

(b) Linefinder used to achieve uniselector economy

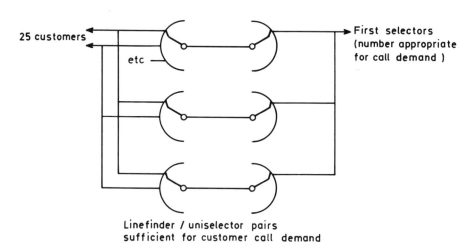

(c) Grading of linefinders and uniselectors

Figure 6.15
Using uniselectors.

25 possibly outgoing circuits are connected, one to each of the individual contacts making up the contact bank. The first contact in the bank is not connected to any outgoing circuit, but serves as the 'rest' position for the wiper arm during the idle period between calls, thus in practice only 24 outlets are available.

When a call comes in on the incoming circuit—indicated by a loop (say because the

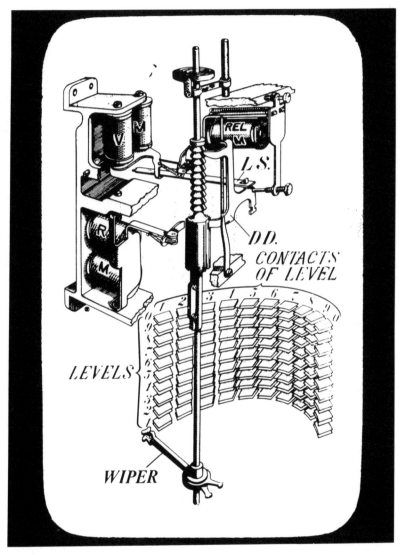

Figure 6.16
The principle of the Strowger two-motion selector. An illustration from an old manuscript, showing the basic action of the two-motion selector. Pulsing the VM relays initially steps the wiper in the vertical plane. RM relay pulses then move the wiper an appropriate number of contacts in the horizontal plane. (*Courtesy of Telecom Technology Showcase, London.*)

customer has picked his phone up), the uniselector automatically selects a free outgoing connection to a 'first selector'. The 'first selector' is a Strowger two-motion selector which initially returns dial tone to the caller and subsequently responds to the first dialled digit. We shall discuss the mechanism of two-motion selectors shortly.

Uniselectors consist of a number of rows (or planes) of bank contacts, as the photograph in Figure 6.14 illustrates. The use of a number of planes allows all the contacts necessary for 2, 3 or 4-wire and P-wire switching to be carried out simultaneously. But how do we cope with more than one incoming circuit? One answer is to provide a uniselector corresponding to each individual caller's line. The outlets of these uniselectors can then be 'graded' as we have seen to provide access to a suitable number of first selectors—sufficient to meet traffic demand. A simple arrangement of this type is shown in Figure 6.15(a). But unless the traffic on the incoming circuit is quite heavy then this arrangement is relatively inefficient and uneconomic—requiring a large number of uniselectors which see little use. For this reason it is normal to provide also a 'hunter' or 'linefinder'. This is a second uniselector, wired back-to-back with the first as shown in Figure 6.15(b). The linefinder (or hunter) is used to enable the first uniselector to be shared between a number of incoming lines. A number of linefinders (sufficient to meet customers' traffic demand) are graded together—giving each individual line a number to choose from (Figure 6.15(c)).

Different selectors in the same grading are prevented from simultaneously choosing the same outgoing circuit by the action of the P-wire, as we saw earlier in the chapter. It is also the P-wire that invokes the 'release' of the selectors at the end of a call, whereupon a spring or other mechanical action returns the wiper assembly to the rest or home position.

The most common type of selector found in Strowger exchanges is the 'two-motion selector'. These are the type capable of responding to dialled digits. The wipers of a two-motion selector can, as the name implies, be moved in two planes. The first motion is linear, up-and-down between the ten planes (or 'levels') of bank contacts under dialled digit control. This is followed by a circular rotation into the bank itself. The second motion can be an automatic motion—scanning across the 'grading' to find a free connection to a subsequent two-motion selector to analyze the next digit. Alternatively, if the two-motion selector is a 'final selector', then the selector analyzes both the final two digits of the called customer's number. In this case the rotary motion of the selector is controlled by a dialled digit.

Each contact bank plane in a two-motion selector appears as an arch-shaped layer or 'level' with a number of sets of equally spaced contacts in each bank. The appearance of the selector contact bank is thus somewhat akin to a portion of a cylinder, as Figure 6.16 shows.

In practice the cylindrical bank of a two-motion selector needs to be duplicated to allow each of the wires of a connection (2-wire plus P-wire etc.) to be connected simultaneously, so that an actual two-motion selector looks a little more complicated, as the photograph of Figure 6.17 reveals.

Having heard dial tone returned from the first selector (when it is ready), the caller dials the 'called number'. This is indicated to the exchange by a train of electrical loop-disconnect ('off–on') pulses on the incoming line itself. These are the pulses which activate 'first', 'second' or subsequent two-motion selectors accordingly. The number of 'off–on' pulses used to represent a particular digit corresponds to the value of the digit

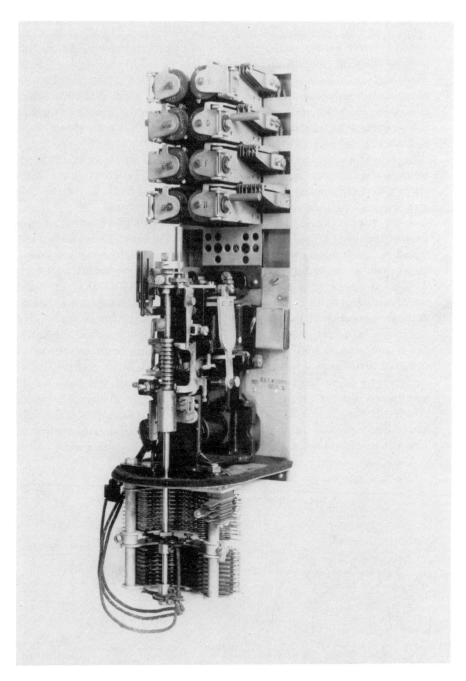

Figure 6.17
Strowger two-motion selector. This one is actually a final selector, used in the British Post Office network. (*Courtesy of Telecom Technology Showcase, London.*)

dialled. Thus one pulse equals digit value '1', two pulses equals value '2' etc. The digit value '0' is represented by ten pulses. This simple form of signalling is called 'loop disconnect' or 'LD' signalling; it was described in Chapter 2. Each pulse steps the selector upwards in the vertical plane by one 'level'. The gap between dialled trains of pulses indicates the end of the train and so marks the end of the 'stepping' sequence. When a two-motion selector detects the gap between dialled digits then the rotary action of the selector commences automatically, moving the wiper into the bank to find a free connection. Alternatively, in the case of a final selector, the 'inter-digit pause' is used by the selector to ready itself for receiving a second dialled digit to control its rotary motion.

Digressing for a moment, it is worth noting that a Strowger two-motion selector has a limited availability of 10 (or sometimes 20). The constraint arises because the selector may only automatically scan the horizontal levels of the bank—and on each a maximum of 10 (or 20) outlets may be accommodated.

Apart from being ideal for the direct stepping of two-motion Strowger selectors, another benefit of loop disconnect signalling is the ease with which pulses may be generated by a dial telephone. Figure 6.18 shows a dial telephone and how the pulses of loop disconnect signalling are created by a rotating cam attached to the telephone dial. As the dial is turned, the cam operates a set of contacts which connect and disconnect the circuit. Depending on how far the dial is turned, a number of pulses are generated. A longer period of electrical current 'on' separates the bursts corresponding to consecutive digits of the number. This longer 'on-period' is called the 'inter-digit pause' and is generated by an initial wide tooth on the rotating cam, as shown in Figure 6.18.

Actual Strowger exchanges comprise both uniselector and two-motion selector types. Figure 6.19 shows an example of a permutation of uniselectors and two-motion selectors to support up to 1000 customers on a three-digit numbering scheme. The uniselector finds a free 'first selector', which analyzes the first dialled digit and then finds a free 'final selector' in the correct range (corresponding to the first dialled digit). The final selector provides the final connection to the customer.

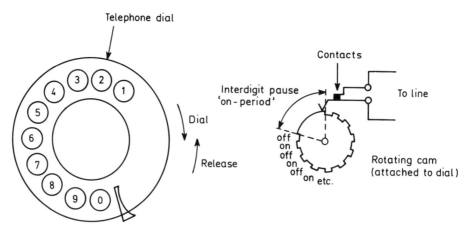

Figure 6.18
Generating 'loop-disconnect' signals.

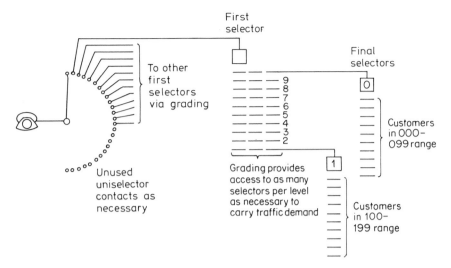

Figure 6.19
Selectors for a 1000-customer Strowger exchange.

For networks consisting of more than one exchange it is clear that the same destination customer will not always be reached using the same route from the calling customer, since the starting place is not always the same. All routes, however, will comprise a number of switching stages, located in the various exchanges. These need to be fed with trains of digit impulses in order to operate. But since the path is not common, then each caller will either have to dial a different string of digits, or else each exchange will have to be capable of 'translating' a common dialled string into the string necessary to activate the selectors on the route needed from that particular exchange. In the latter case, a so-called 'common' or 'linked numbering scheme', a 'register/translator' is needed to store the dialled digits and 'translate' (or convert) them into the string of 'routing digits' which are needed to step the selectors as previously described. The function of the register and the technique of number translation are both described more fully in the next chapter.

Finally, let us close our discussion of Strowger switching by explaining briefly why the phenomenon of limited availability arises. It comes about because each selector is limited in the number of bank contacts. On a uniselector this is typically 24. On a two-motion selector it is usually only 10 or 20. Thus no matter how many circuits make up the route in total, each incoming circuit may have access only to a limited number of them (24, 10 or 20 in the examples given above).

The major drawback of Strowger switching is the relatively large amount of space that it takes up, the relatively high electrical power needed for busy-hour operation, and the labour intensive demands of maintaining it. The mechanical parts are highly prone to wear, and the electrical contacts are very sensitive to damage and dirt. As a result Strowger switching has now been largely superseded, but many Strowger exchanges still operate in some countries and may well remain in those lacking the capital to replace them.

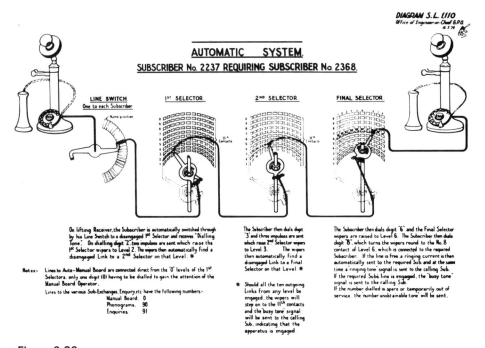

Figure 6.20
The principle of Strowger automatic switching. An extract from an early British General Post Office (GPO) article, explaining the principle of Strowger automatic switching. (*Courtesy of Telecom Technology Showcase, London.*)

Crossbar switching

Crossbar technology emerged in the 1940s and was partly though not wholly responsible for a change in equipment usage away from Strowger and other similar step-by-step exchange types. Crossbar switching offered the advantage of reduced maintenance and accommodation requirements, but was sometimes more expensive than Strowger for large and complex exchange applications.

As early as 1926 the first public crossbar exchange opened in Sweden. It worked in a step-by-step, digit-by-digit, manner which was too unwieldy for exchanges exceeding 2000 lines. It was not until the 1940s that crossbar systems became more common, following development in America of a system employing 'marker control'. This is by far the most common type of crossbar exchange now used and is the type described here.

The technique of crossbar switching is relatively easy to understand, since the switch matrix is immediately apparent from the physical structure of the components. As the photograph in Figure 6.21 shows, the switch looks like a matrix, consisting of ordered vertical and horizontal members.

The operation of the crossbar switch is most readily understood by considering the action of two adjacent crosspoints. Figure 6.22 shows a single crosspoint of a crossbar assembly. The adjacent crosspoint of interest is a mirror image of the one shown, immediately below it.

Figure 6.21
Crossbar switching matrix. A typical basic crossbar matrix. This one comprises ten armatures and bridge magnets, six select bars and twelve select magnets. (*Courtesy of British Telecom.*)

The crosspoint shown in Figure 6.22 is capable of simultaneous switching of three inlet wires to three outlet wires. The three inlet wires (a 'pair' plus a P-wire) are connected to three fixed 'bridge common' contacts and the three outlet wires are connected to three movable outlet spring contacts. The three outlet spring contacts are joined together by a piece of insulating material, which also connects the outlet springs to the 'lifting spring'. Thus when the lifting spring is moved to the right of the diagram the outlet springs are all simultaneously pushed into contact with the bridge common, so completing the connection. (As an aside, more wires can also be switched simultaneously, by adding more outlet springs and bridge commons.)

The lifting spring is actuated by the action of the 'select finger' and the 'bridge armature'. The 'select finger' is really a thin piece of wire, connected to the 'select bar' by a small spring, and this gives the 'select finger' a small amount of flexibility. The select magnets are electromagnets, activated by passing an electric current through their coils. Not more than one of the select magnets is used at any one time. To activate the crosspoint shown in Figure 6.22, select magnet 2 must be operated. This has the effect of tilting the select bar, and so moving the select finger into the position marked by the dashed line on our diagram. At this point the bridge magnet is activated. This in turn pivots the 'bridge armature', with the effect of 'trapping' the select finger on to the lifting spring, so moving the lifting spring towards the right of the diagram, and making contact between the outlet springs and the bridge commons. The connection is now complete. The bridge magnet must remain held as long as the connection is required so trapping the select finger for the entire call. The select magnet, however, may be released once the connection has been made. The slight flexibility in the select finger permits it to remain trapped under the 'bridge armature', even when the select magnet is released.

At the end of the call, the bridge magnet is released, and the select finger springs back to its normal position.

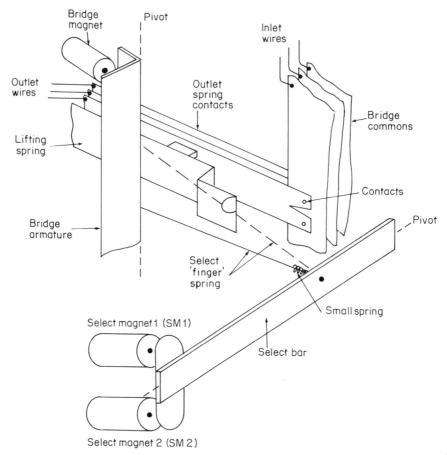

Figure 6.22
Operation of a crossbar switch.

By a similar mechanism, another set of outlet spring contacts (beneath the set illustrated) can be connected to the same inlet bridge commons (inlet circuit). In fact, this is the function of select magnet 1. The contacts are arranged in a mirror image beneath those of our illustration, and work in precisely the same way except that they use select magnet 1 instead of select magnet 2. Thus select magnet 1 selects one outlet circuit while select magnet 2 selects the other. It does not matter that there is only one select finger available, since we would not wish to connect the inlet to both outlets at the same time anyway.

Figure 6.23 illustrates a larger portion of a crossbar switch, showing three bridge magnets and armatures and two select bars. The form of the overall switch matrix is now apparent. To activate any particular crosspoint of the matrix one of the select magnets must be activated (according to the desired outlet circuit required) and then the appropriate bridge magnet is also activated, to connect the desired inlet circuit.

Crossbar switch assemblies usually consist of ten bridge armatures and six select bars, requiring ten bridge magnets and twelve select magnets. (Note: Some manufacturers of crossbar switches use more armatures and bridge magnets than discussed

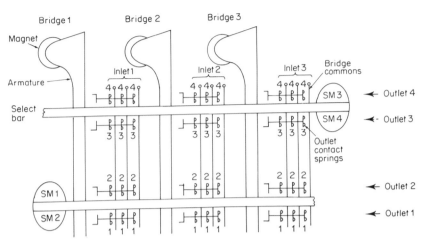

Figure 6.23
A crossbar matrix.

here. The principle, however, is the same.) Used in its most straightforward manner therefore, a single crossbar switch assembly can be used for ten inlets and twelve outlets. By combining a number of these assemblies a much bigger matrix can be built up.

The 'limited availability' of twelve outlets from the basic assembly can raise problems, as we saw earlier in the chapter. So let us consider how we could improve things so as to allow 20 outlets. A single bridge of a 20-outlet switch is shown in Figure 6.24. Notice that there are now six bridge commons and corresponding sets of outlet springs instead of only three. Notice also that the top two magnets have been labelled as 'auxiliary magnets' instead of 'select magnets'.

In the assembly shown in Figure 6.24 each crosspoint is made by activating three magnets: one of the auxiliary magnets, one of the select magnets, and the appropriate bridge magnet. The auxiliary magnet essentially has the effect of connecting the inlet circuit either to the bridge commons corresponding to the even-numbered outlet circuits, or to the bridge commons corresponding to the odd-numbered outlet circuits. (The three right-hand bridge commons correspond to even-numbered outlet circuits and to auxiliary magnet 2. The three left-hand bridge commons correspond to the odd-numbered outlet circuits and auxiliary magnet 1.) The overall assembly of ten bridge armatures and six select bars is thus increased to a capacity of ten inlets and 20 outlets. With even more sophisticated wiring and the use of nine bridge commons, up to 28 outlet circuits can be made available to the ten inlets.

The individual assemblies of crossbar switches are usually arranged in a fan-in/fan-out manner as described earlier in the chapter. In order to pick an appropriate path between inlet and final outlet circuit, the exchange must first examine the dialled digit train. This is done by the register, the action of which is described more fully in the next chapter. Having chosen the outlet circuit required, a path is usually selected by one of two methods. The easiest to explain is the method used in 'stored program

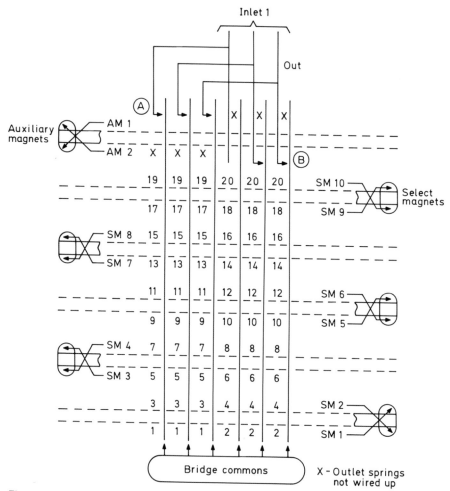

Figure 6.24
Twenty-outlet crossbar switch.

control' (SPC) crossbar exchanges. In SPC exchanges (i.e. relatively recent ones), the exact path is determined by a special control computer from its knowledge of the instantaneous state of the exchange. The necessary cascade of crosspoints across the exchange is then activated. As an alternative we have the original method of 'marker control', which was electro-mechanical. In marker control, the final destination outlet is 'marked', and a large number of trial paths are then set up in a backwards direction across the switch. However, only the path which happens to find its way right across the chain of individual switches to reach the desired inlet is actually connected, while all the other trial paths are released.

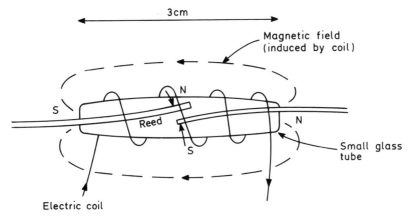

Figure 6.25
A reed relay crosspoint.

Reed relay switching

In reed relay switched exchanges the individual crosspoints consist of two nickel-iron 'reeds' sealed in a glass tube, approximately 3 cm long, containing dry nitrogen. The free ends of the reeds are plated with gold to give a low resistance contact area. When the contact is not made, the reeds rest in their idle position with their contacts apart. Figure 6.25 shows this assembly.

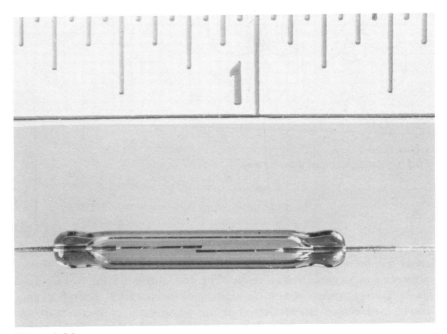

Figure 6.26
A reed relay, showing the glass encasement and contacts. This reed would be inserted into a bobbin magnet, which when activated will induce a magnetic field in the contacts, making them close. The size is around 1 inch (3 cm). (*Courtesy of British Telecom.*)

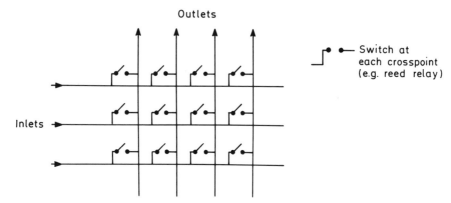

Figure 6.27
Crosspoint matrix of reed relays.

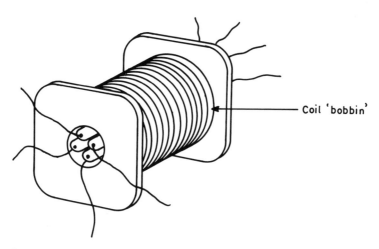

Figure 6.28
Four contact cross-point reed relay.

The whole glass tube is located along the axis of an electromagnetic coil, and when the current is switched on in this coil it induces a magnetic field in the tube. The two reeds take on opposite polarity in the overlapping free ends, so that they are attracted together and a connection is made. When the current in the coil is switched off, the reeds return to their idle rest position, and the contact is again broken. This action can be used as the basis for activating the crosspoints of a switch matrix, as discussed in the early part of the chapter, and re-illustrated in Figure 6.27.

Multiple wire connections are achieved by packing the electromagnetic coil or 'bobbin' (as it is called), with more than one reed. Figure 6.28 illustrates a bobbin containing four reeds and is therefore capable of simultaneous 4-wire connection.

Reed relay exchanges invariably have stored program control. The exchange processor analyzes the dialled number and activates the appropriate switchpoints of the switch matrix by applying current to the coil bobbins.

Digital switching

Digital switching differs considerably from the other three types of switching previously discussed in this chapter. All of the three types discussed earlier are what are called 'space switching' techniques. A space switch connects and disconnects physical contacts using a matrix of switchpoints. When a connection has been established through a space switch, a permanent electrical path exists throughout the duration of the call. This is not so with a digital switch.

Most digital switches use instead a time–space–time switch architecture; they are not simple space switches. The need for 'time switching' as well as 'space switching' arises from the fact that the line systems connected at the periphery of the switch are not individual circuits; they are usually either 2 Mbit/s (32-channel) or 1.5 Mbit/s (24-channel) digital line systems.

As we saw in Chapter 5, a digital line system carries either 24 or 32 channels in a multiplexed form, which means that the total bit-stream carried on the system has been created by interleaving 8-bit samples of each of the constituent channels. Thus the incoming bit-stream to a digital exchange appears as shown in Figure 6.29. The first eight bits are channel 0, or 'timeslot 0'; the next eight bits are channel 1, or 'timeslot 1'; right up to channel 23 or channel 31 as appropriate. Figure 6.29(a) shows the 2 Mbit/s line system frame structure, and Figure 6.29(b) shows the frame structure of 1.5 Mbit/s line systems.

It is no good merely to space switch the digital line systems, since this would have the effect of switching all the individual 24 or 32 channels through on to the same outgoing line system. We need instead to be able to connect any channel (or 'timeslot') of one incoming digital line system on to any channel (or 'timeslot') of any of the other digital line systems which are connected to the same exchange. To do this we need both a 'time-switch' capability to shift channels between timeslots, and a 'space-switching' capability to enable different physical outgoing line systems to be selected.

Figure 6.30 shows a digital switch using a time–space–time architecture to switch the timeslots of three 2 Mbit/s digital line systems. On the left of the diagram the 'incoming' or 'receive' timeslots are shown. On the right-hand side, the corresponding 'outgoing' or 'transmit' timeslots of the same 2 Mbit/s line systems are shown. (Recall that any 4-wire transmission system, including all digital line systems, comprises the equivalent of two pairs of wires, one pair each for the receive and the transmit directions of transmission.)

The incoming (receive) line pairs are fed directly into a time switch, the output of which feeds the space switch (shown in the middle). The output of the space switch feeds another time switch to which the outgoing (transmit) line pairs are connected.

Imagine now that we wish to switch timeslot 2 in line system A to timeslot 31 in line system B. The space switch allows us to connect line system A to line system B. Point (2) of the space switch allows the receive pair from line system A to be connected to the transmit pair of line system B, while point (6) caters for the other direction of

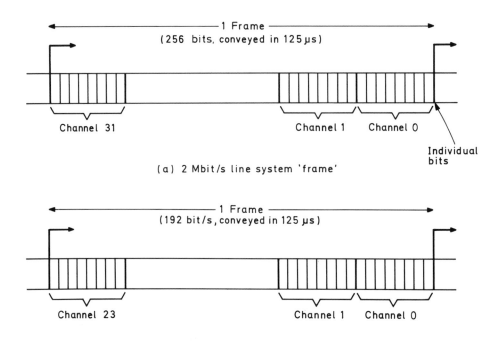

Figure 6.29
Frame structure of digital line systems.

transmission (receive from line system B, and transmit on line system A). But this by itself is not enough since it would mean that timeslot 2 in line system A ended up in timeslot 2 of line system B, and not in the desired timeslot 31 of line system B. The two time switches allow this conveyance of bits between timeslots. A time switch works by storing the received pattern of 8-bits from an incoming 'timeslot', and by waiting for anything up to a whole frame (i.e. anything up to 125 μs) in order to feed the pattern out into any desired outgoing timeslot. For example, delaying an 8-bit pattern by $125/32$ μs will move the bit pattern from one timeslot of a 2 Mbit/s system to the next, for example from timeslot 2 into timeslot 3. Similarly, a delay of $2 \times 125/32$ μs will move timeslot 2 to timeslot 4 and so on. Finally, a delay of $31 \times 125/32$ μs will move timeslot 2 into timeslot 1 of the next frame (see Figure 6.29).

But why the need for two stages of timeswitching in Figure 6.30? For an answer, let us suppose we want not only to switch timeslot 2 of line system A into timeslot 31 of line system B, but also to switch timeslot 3 of line'system A into timeslot 31 of line system C. Then, if we only had one time switch (say the one connected to the incoming line pairs), we would end up trying to superimpose both timeslot 2 and timeslot 3 on the receive side of the time switch on to timeslot 31 before switching the space switch across to its output side. To overcome this problem, the space switch uses its own internal timeslots.

When asked for a connection between incoming and outgoing timeslots, the switch control system carries out a search for free internal timeslots on both the receive and

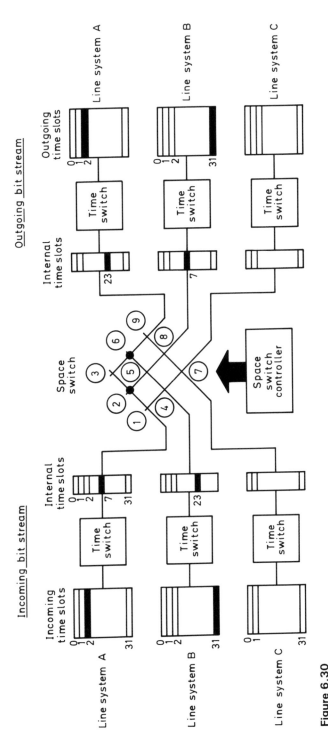

Figure 6.30
A simple time–space–time digital switch.

transmit sides of the space switch matrix, and when an idle timeslot (number 7 in the case illustrated) has been found, switching can begin. The incoming timeslot, number 2 of line system A, is time switched into internal timeslot 7. Thereafter whenever internal timeslot 7 comes up, contact (2) of the space switch is activated momentarily and switches the timeslot content towards the transmit pair of line system B. Finally the contents of internal timeslot 7 are time-switched again, this time into timeslot 31 on the transmit pair of line system B. Meanwhile the configuration of the space switch is being changed by the switch control system every 125 μs. Necessarily so, since not all incoming timeslots of line system A are to be connected to the same outgoing line system (B in this case). The space switch must be ready to send each incoming timeslot of a single line system into various timeslots of up to 32 outgoing line systems.

At the same time, internal timeslot 23 (the antiphase timeslot of timeslot 7 i.e. the timeslot $\pm$ 16 timeslots) is being used in conjunction with space switch contact (6) to convey the incoming bit pattern on timeslot 31 of line system B to timeslot 2 on the transmit pair of line system A.

The antiphase timeslot is the timeslot half of the frame after the first allocated internal timeslot. By always using the antiphase timeslot, we guarantee that the return path will never suffer internal timeslot congestion, and furthermore we save the switch control system the effort of searching for a second free internal timeslot. The same method of time–space–time switching can be used in conjunction with both 1.5 Mbit/s and 2 Mbit/s line systems. In practice the time switches are rarely designed to operate at such low rates, but instead work simultaneously on a number of such line systems which have been previously multiplexed together. Thus rather than running at 1.5 Mbit/s or 2 Mbit/s (24- or 32-channel rates), time switches are often designed to run at 512-channel or an even higher rate.

Finally, before we leave the subject of digital switching, it is worth observing that the whole switch operation can in fact be carried out with only a time switch, and no space switch at all. This is done by multiplexing all the incoming line systems together into a very high bit stream of interleaved 8-bit channel patterns; input to the switch is then from one source only and there is no need for a space switch. For small exchanges this single time-switch configuration (without space switch) is practicable, and indeed some digital PBXs do function on a time switch alone. Large public exchanges must be designed as time–space–time (TST) switches, since the high bit rates required to multiplex all the incoming channels together are unattainable with today's electronic technology.

BIBLIOGRAPHY

Brieley, B. E., *Introduction to Telephone Switching*. Addison-Wesley, 1973.

Grinsec, *Electronic Switching*. North-Holland Studies in Telecommunications, 1983.

Joel, A. E. Jr., *A History of Engineering (1925–1975)*. Bell Telephone Laboratories, 1982.

Joel, A. E. Jr., *Electronic Switching: Digital Central Office Systems of the World*. IEEE Press, 1982.

Noll, A. M., *Introduction to Telephones and Telephone Systems*. Artech House, 1986.

One Hundred Years of Telephone Switching (1878–1978). North-Holland in Telecommunication, 1982.

Pearce, J. G., *Telecommunications Switching*. Plenum Press, 1981.

Renton, R. N., *The International Telex Service*. Pitman, 1974.

Ronayne, J., *Digital Communications Switching*. Pitman, 1986.

Smith, S. F., *Telephony and Telegraphy—An Introduction to Instruments and Switching Systems (Electromechanical and Reed—Electronic)*, Third edn. Oxford University Press, 1978.

Talley, D., *Basic Electronic Switching for Telephone Systems*. Hayden Book Company, 1982.

7

SETTING UP AND CLEARING CALLS

The establishment of a physical connection across a circuit-switched network relies not only upon the availability of an appropriate topology of exchanges and transmission links between the two end-points but also upon the correct functioning of a logical 'call set-up' and 'cleardown' procedure. This is the logical sequence of events for establishing calls. It includes the means by which the caller may indicate the desired destination, the means for establishment of the path, and the procedure for subsequent 'cleardown'. In this chapter we discuss these 'call control' capabilities of circuit-switched networks, and we shall describe the related principles of inter-exchange signalling. The chapter ends with a review of various standard signalling systems.

7.1 ALERTING THE CALLED CUSTOMER

Figure 7.1 shows the very simple kind of communication system which we have considered earlier in this book—two telephones are directly connected by a single pair of wires, without any intervening exchange.

The users of the system in Figure 7.1, A and B, are able to talk at will to one another without fear of interruption. The problem with the equipment illustrated is the difficulty of 'alerting' the other party in the first place, so as to bring him or her to the phone. One easy solution would be to connect a bell at both ends in parallel with each telephone set. If the bell is designed to respond to a relatively high alternating current, whenever such a current is applied from the calling end, the bell at the *called* end rings. This was the earliest form of signalling used on telephone networks. The alternating current (properly called 'ringing current') was applied at the calling end by a manually cranked magneto-electric generator and the technique is known as 'generator', 'bothway generator' or 'ringdown' signalling. The term 'ringdown' originates from the fact that call clearing in manual exchanges was done by means of the operators ringing one another a second time at the end of the call, hence 'ringdown'. Figure 7.2 illustrates a possible though crude adaptation of our network to include generator signalling. Actually, the circuit would normally include an inbuilt contact to disconnect the 'own' telephone from the circuit when the handle was turned.

Magneto-generator signalling was the only form of signalling used in early manual

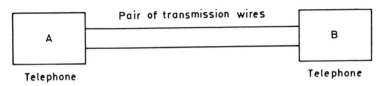

Figure 7.1
A simple communication system.

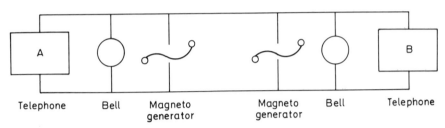

Figure 7.2
A crude 'generator signalling' circuit.

networks. The operator was alerted by use of the generator and was then told by the caller who it was he or she wished to call. The operator then alerted the destination customer (again using the generator) or alternatively referred the call to another operator (if the called customer was on another exchange). Finally the connection was made.

7.2   AUTOMATIC NETWORKS

Automatic networks are required to undertake quite a complicated logical sequence of events, first to set up calls, then to make sure they are maintained during conversation (or data transfer), and finally to cleardown the connections after use. In support of this sequence of events, 'call control' functions are carried out by the exchanges of automatic networks. These functions monitor the state of the call and initiate whatever actions are indicated, such as switching of the connection, applying 'dialtone', performing 'number analysis' (to determine the destination), and 'signalling' the desired number to a subsequent exchange. Information about the state of the call is communicated from one exchange to another by 'signalling systems', so that automatic connections can be established across a whole string of exchanges. We shall now go on to discuss the sequence of call control functions which makes these things possible.

7.3   SET UP

Our first step when making a call is to tell the exchange that we want to do so, by lifting the handset from the telephone cradle or 'hook'. (The word 'hook' dates from the earliest days when the handset was hung on a hook at the side of the telephone.

In those days, a horizontal cradle was no good since the early carbon microphones needed gravity in order to remain compacted.) This sends an *off-hook* signal to the exchange. The signal itself is usually generated by 'looping' the telephone-to-exchange access line, thereby completing the circuit as shown in Figure 7.3.

On receipt of the off-hook signal the exchange has to establish what is called the *calling line identity*, i.e. which particular telephone of the many connected to the exchange has generated the signal. The exchange's control system has to know this in order to identify which access line 'termination' requires onward cross-connection. This information also serves to monitor customers' network usage, and shows how much to charge them.

One way to identify the calling line is to use its so-called 'directory number', some-times abbreviated as 'DN'. This is the number which is dialled by a customer when calling the line. In early exchanges, particularly the Strowger type, line terminations were arranged in consecutive directory-number order; the functioning of the exchange did not allow otherwise. But with the advent of computer-controlled exchanges, it is no longer necessary to use physically-adjacent line terminations for consecutive direc-tory numbers; directory numbers can now be allocated to line terminations almost at random.

For instance it may be convenient that directory numbers 25796 and 36924 should be connected to adjacent line terminations in the exchange. (Such a situation might arise when a customer moves house within the same exchange area, and wishes to retain the same telephone number.) For convenience the line terminations in the exchange can still be numbered consecutively by using an internal numbering scheme of so-called 'exchange numbers' (or ENs). The extra flexibility that is required for random directory number allocation is achieved by having some form of 'mapping' mechanism, of DNs to ENs, as shown in Figure 7.4.

Customers' line terminations are not alone in being given 'exchange numbers'; trunks to other exchanges, and even digit sending and receiving equipment, can also be allocated an exchange number, depending on the design of the exchange. Alloca-tion of exchange numbers to all the equipment allows the exchange to 'recognize' all

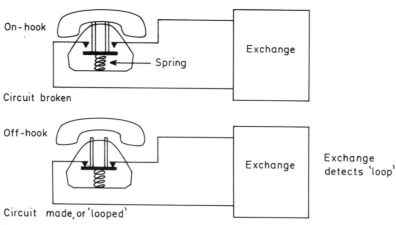

Figure 7.3
The 'off-hook' signal.

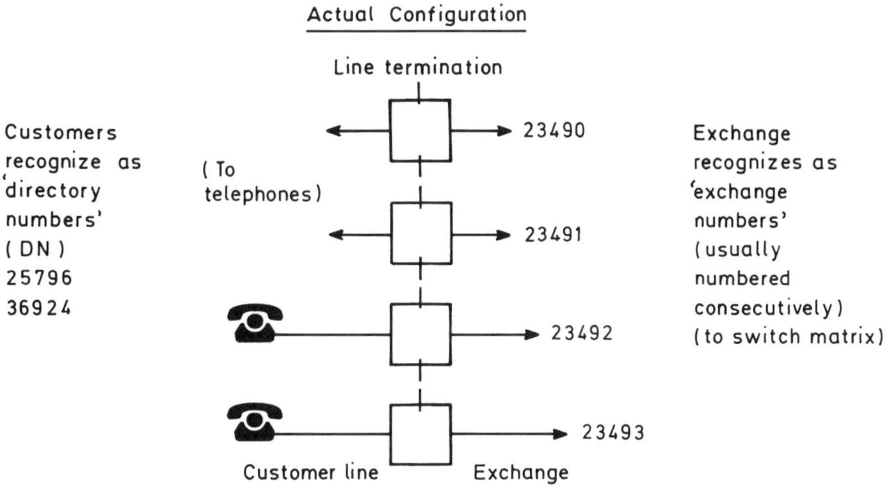

Mapping table (Held in exchange)

EN	DN
23490	
23491	
23492	25796
23493	36924

EN	Device
25932	Digit receiver
27631	Outgoing trunk

Figure 7.4
'Directory numbers' (DNs) and 'exchange numbers' (ENs).

the items which may need to be connected together by the switching matrix. Commands to the switching matrix may thus take the form 'connect EN23492 to EN23493'.

In the example of Figure 7.4 the command 'connect EN23492 to EN23493' would connect the directory numbers 25796 and 36924 together. Another command might connect a line to a digit receiving device. This command would be issued just prior to receiving dialled digits from the customer—at the same time that the dialtone is applied.

Having identified the calling line, the exchange's next job is to allocate and connect equipment, ready for the receipt of *dialled digits* from the customer. This equipment normally consists of two main parts, the 'code receiver' which recognizes the values of the digits dialled, and the 'register' which stores the received digit values, ready for analysis.

Once the exchange has prepared code receivers and a register, it announces its readiness to receive digits, and prompts the customer to dial the directory number of the desired destination. This it does by applying *dial tone*, which is the familiar noise heard by customers on lifting the handset to their ear. Because the whole sequence of events (the 'off-hook' signal, the preparation of code-receiver and register, and the return of dial tone) is normally almost instantaneous, dial tone is usually heard by the customer before the earphone reaches the ear. Noticeable delays occur in exchanges where there are insufficient code receivers and registers to meet the call demand. The remedy lies in providing more of them.

On hearing dial tone, the customer dials the directory number of the desired destination. There are two prevalent 'signalling systems' by which the digit values of the number may be indicated to the exchange: 'loop disconnect' signalling, and 'multifrequency (MF)' signalling. (Multifrequency signalling is also sometimes called 'dual tone multifrequency' (DTMF).)

In loop disconnect (or LD) signalling, as described in earlier chapters, the digits are indicated by connecting and disconnecting the local exchange access line, or 'loop'.

LD signalling was first mentioned in Chapter 2. In Chapter 6 we saw how well it worked with step-by-step electromechanical exchange systems such as Strowger, and we discovered that the pulses themselves could easily be generated by telephones with rotary dials. Both these characteristics have contributed to the widespread and continuing use of LD signalling in customers' telephones.

Modern telephone exchanges also permit the use of an alternative access signalling system (often the customer may even change the signalling type of his telephone without informing the telephone company). The alternative to LD uses multifrequency tones (i.e. DTMF). This system has the potential for much faster dialling and call set-up (if the exchange can respond fast enough). It too was discussed briefly in Chapter 2.

DTMF telephones, almost invariably have 12 push-buttons, labelled 1–9, 0, '∗' and '#'. The two extra buttons '∗' and '#' (called 'star' and 'hash') are not, however, always used, and their function may vary between one network and another. Where they *are* used, they often indicate a request for some sort of special service; for example, '∗9-58765' might be given the meaning 'divert incoming calls to another number—58765'.

When any of the buttons of an MF4 (DTMF) telephone are pressed, two audible tones are simultaneously transmitted on to the line. The frequencies of the two tones depend on the actual digit value 'dialled'. The relationship was shown in Figure 2.7 of Chapter 2.

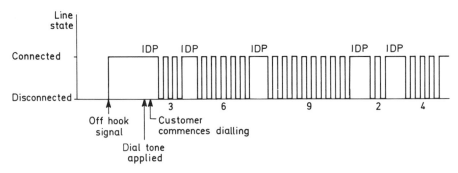

Figure 7.5
Loop disconnect signalling train.

While 'dialling' in MF4 the customer hears the tones in the earpiece. Only a very short period of tone (much less than half a second) is needed to indicate each digit value of the dialled number.

Once the exchange has started to receive digits, it can set to work on *digit analysis*. This is the process by which the exchange is able to determine the appropriate onward routing for the call, and the charge per minute to be levied for the call. During digit analysis, the exchange compares the 'dialled number' held by the register with its own list of permitted numbers. The permitted numbers are held permanently in 'routing tables' within the exchange. The routing tables give the 'exchange number' identity of the outgoing route required to reach the ultimate destination.

Routing tables are normally constructed in a tree-like structure allowing a 'cascade' analysis of the digit string. This is shown in Figure 7.6 where customer a on exchange A wishes to call customer b on exchange B. The number customer a must dial is 222 6129. The first three digits are the 'area code', identifying exchange B, and the last four digits, 6129, identify customer b in particular. The routing table tree held by exchange A is also illustrated. As each digit dialled by customer a is received by exchange A, a further stage of the analysis is made possible until, when '222' has been analysed, the command 'route via exchange B' is encountered. At this stage the switch path through exchange A may be completed to exchange B, and subsequent dialled digits may be passed on directly to exchange B for digit analysis. The register, code receivers, and other common equipment in use at exchange A are released at this point, and are made available for setting up calls on behalf of other customers. Exchange B is made aware of the incoming call by a 'seizure signal' (equivalent to the off-hook signal on a calling customer's local line). The signal is sent by means of an inter-exchange signalling system which will be discussed in more detail later in the chapter.

Exchange B prepares itself to receive digits by allocating 'common equipment' including code receivers and register. Then, having received the digits, digit analysis is undertaken in exchange B, again using a routing table. This time, however, the analysis is concluded only after analysing the number '6129', at which point the

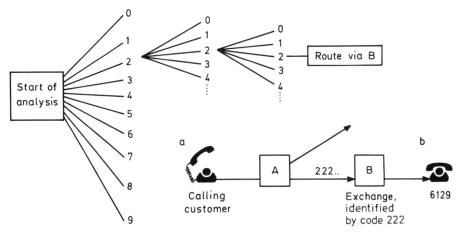

Figure 7.6
Routing table tree.

command 'connect to customer's line' is issued. In principle the method of analysis of the area code '222' and the customer's number '6129' is the same. The difference is that one or more different routes may be available to get from one exchange to another, but for a customer's line of course only one choice is possible. The former analysis may thus require production or translation of routing digits while the latter only involves a line selection.

Once the connection from calling to called customer has been completed, *ringing current* will be sent to ring the destination telephone, and simultaneously ringing tone will be sent back to the calling customer (a).

As soon as customer b (the called customer) 'answers' the telephone (by lifting the handset), then an *answer signal* is transmitted back along the connection. This has the effect of 'tripping' the ringing current and the ringing tone, and commencing the process of charging the customer for his call.

Conversation (or the equivalent phase of communication) may continue for as long as required until the calling customer replaces the telephone handset to signal the end of the call. This is called the *clear signal*, and it acts in the reverse manner to the 'off-hook' signal, breaking the access line 'loop'. The signal is passed to each exchange along the connection, releasing all the equipment and terminating the call-charging.

Depending upon the network and the type of switching equipment, it may be possible for both the calling and the called customers to initiate the 'cleardown' sequence. The ability of called customers to initiate cleardown was not prevalent in all early automatic exchanges (for example, UK Strowger exchanges). However, in many modern switch types either party may clear the call.

7.4 NUMBER TRANSLATION

Modern exchanges use stored program control (SPC—actually a computer processor) for the purpose of digit analysis and route determination, often using a routing data tree, as figure 7.6 showed. The administration needed to support such exchanges and their routing tables is fairly straightforward. In the past, however, particularly in the days of electromechanical exchanges, digit analysis and call routing mechanisms were often very complex. Having to be formed out of hard-wired and mechanical components, their efficient operation often demanded slightly different call routing techniques. An important tool in effective call routing was, and still is, the process known as number 'translation'.

Number translation is a means of reducing the number of times that digit analysis has to be undertaken during a call connection. Digit analysis still takes place at the first exchange in the connection, and may have to be repeated at another exchange later in the connection (typically a trunk exchange), but any further digit analysis (at other exchanges) can be minimized by the use of number translation, thereby enabling subsequent exchanges to respond to the received digit string without analyzing more than one digit at a time. As we learned in chapter 6, the direct response of selectors to each digit in turn is crucial for the correct operation of some types of switching equipment (e.g. Strowger). Number translation also remains in use even in digital networks—for reason of flexibility to change route, signalling system or number length (e.g. abbreviated dialling).

Number translation involves detecting the actual number dialled by the customer and replacing it with any convenient string of digits which will make the operation of the network and the connection to the called customer easier. The translated number may thus be entirely unrelated to the dialled number. Figure 7.7 gives a typical example of digit translation. The example illustrates the use of number translation in the United Kingdom Public Switched Telephone Network (PSTN) in the 1950s, when automatic long-distance calling was introduced to the existing Strowger network. In the example, the number dialled by the customer is composed of three parts:

Trunk code '0' + Area code '703' + Customer Number '62314'.

For the same destination area, the same area code is dialled by any 'calling customer' in the network, but the route taken by the call will obviously have to take account of the different starting points. Different intermediate exchanges will have to be crossed, depending upon whether a particular call has originated from the north or the south of the UK. Different number translations are therefore used in each case. The sequence of events is as explained below.

On receiving the digit string from the calling customer, exchange A recognizes the first digit '0' as signifying a trunk call, and so routes the call to the nearest trunk exchange, passing on all other digits '703 62314', but deleting the '0', which has served its purpose. Trunk exchange C then analyzes the next three digits (the area code) '703'. This is sufficient to establish that exchange E is the destination trunk exchange and that the route to be taken is via exchange D. Two options are now available for onward routing, either:

(a)   the call may be routed to exchange D, and the original digit string (70362314) transmitted with it, in which case exchange D will have to re-analyze the area code 703, or:

(b)   the call may be routed to exchange D, together with a translated number string, thereby easing the digit analysis at D.

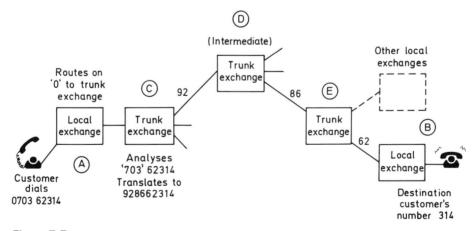

Figure 7.7
Number translation in Strowger networks.

Method (a) is more commonly used with stored program control (SPC) exchanges, since it affords greater flexibility of network administration. This is because routing changes can be made at individual exchanges without altering the 'translated number' which the previous exchange must send. The disadvantage of this method is that it requires more digit analysis.

In different cases, when exchange D is of Strowger type say, or when digit analysis resources at exchange D are short, it may be important to use translated number method (b) so as to minimize digit analysis after exchange C. This method is described in detail below. Either method (a) or (b) may be used, but whichever is chosen one method usually prevails throughout the network; it is rare to find both methods simultaneously in use in the same network.

Method (b) above is the translation method. In the Figure 7.7 example the digits '92' are required by the Strowger selectors in exchange C, in order to select the route to exchange D. Similarly the digits '86' select the route to exchange E from exchange D. Because exchange C translates the area code '703' into the 'routing digits' (9286) required by exchanges C and D, no further digit analysis will be required at either the intermediate trunk exchange D or the destination trunk exchange E. Instead the Strowger selectors absorb and respond directly to the digits. Thus even exchange C will 'step' its Strowger selectors by using and absorbing the digits '92', and exchange D will do the same with the digits '86'. By the time the call reaches exchange E, only digits 62314 remain. Exchange E uses digits 62 to select the appropriate local exchange and routes the call to exchange B, the destination local exchange, where digits 314 identify the actual customer's line which will then be rung.

Translation is also used in more modern networks as a way of providing new and special services. Chapter 26, on 'intelligent networks', will describe the 'freephone', or '0800-service', where the calling customer dials a specially allocated '800' number rather than the actual directory number of the destination. Calls made to an 800 number (e.g. 0800 800800) are charged to the account of the destination number (i.e. to the account of the person who rents the '800' number) and are therefore free to calling customers. Although each '800' is unique to a particular destination customer, it is not recognized by the network itself for the purpose of routing and must therefore be translated into the actual directory number of the destination. Let us imagine, in Figure 7.7 that the number 0800 80800 has also been allocated to the destination customer shown, so that customers can dial either '0800 80800' or '0703 62314'. Calls to the first of these numbers are free to the caller; calls to the second will be charged. (In the former case, it is the renter of the '800' number who will be charged.) In both cases, however, the call needs to be routed to the real directory number, 0703 62314. In the former case this is done by number translation, setting up a bill for the '800' account on the way.

7.5 UNSUCCESSFUL CALLS

We have run through the sequence of events leading up to 'successful call', when the caller gets through and a conversation follows. But as we all find out, our calls do not always succeed, and when a call fails, perhaps because of network congestion, or because the called party is busy or fails to answer, the network has to tell the caller what has happened, and then it has to clear the connection so as to free the network for more fruitful use.

When it is a case of network congestion or called customer busy, the caller usually hears either a standard 'advisory tone', or a recorded announcement. Telephone users will be miserably familiar with 'busy tone' and recorded announcements of the form 'all lines to the town you have dialled are busy—please try later', to say nothing of the 'number unobtainable' tone which tells us we have dialled an invalid number.

A caller who hears one of the call unsuccessful advisory announcements, or a prolonged ringing tone, usually gives up and clears the connection by replacing the handset, to try again later. When, however, the caller fails to do this, the network has to 'force' the release of the connection. 'Forced release', if needed, is put in hand between 1 and 3 minutes after the call has been dialled, and it is initiated if there has been no reply from the called party during this period, whatever the reason. Forced release once initiated—normally by the originating exchange even though the handset of the calling telephone is left off-hook—forces the calling telephone into a 'number obtainable' or 'park' condition. To normalize the condition of a telephone, the handset must be returned to the cradle.

Release is also forced in a number of other 'abnormal' circumstances, as when a calling party fails to respond to dial tone (by dialling digits), or when too few digits are dialled to make up a valid number. In such cases the A-party (i.e. the calling) telephone may be entirely disconnected from the local exchange, silencing even its dial tone. This has the advantage of freeing code receivers, registers, and other common equipment for more worthwhile use on other customers' calls. The calling customer who is a victim of forced release may hear either silence or 'number unobtainable' tone. To restore dial tone, a new 'off-hook' signal must be generated by replacing the handset and then lifting it off again.

It is clearly important for exchanges to be capable of 'forced release' so that their common equipment is not unnecessarily 'locked up' by a backlog of unsuccessful calls.

7.6 INTER-EXCHANGE AND INTERNATIONAL SIGNALLING

Inter-exchange signalling is the process by which the destination number and other call control information is passed between exchanges with the object of establishing a connection. Inter-exchange signalling systems have come a long way since the early days of automatic telephone switching, when ten pulse-per-second, loop disconnect (LD) and similar, relatively simple, signalling systems were the fashion. Today, many different types of inter-exchange signalling are available, and which type a particular network will use depends on the nature of the services it provides, the equipment it uses, its historical circumstances, and the length and type of the transmission medium. For example, small local networks may use relatively slow and cheap signalling systems. By contrast, in extensive public international networks, or in private networks connected to public international networks, consideration is needed for the greater demands—for faster signalling, longer numbers and greater sophistication. We can now review the signalling systems defined in CCITT recommendations (see Table 7.1), with particular attention to the system called CCITT R2. This is a 'multifrequency code' (MFC) inter-exchange signalling system, in some ways similar to the DTMF signalling system used for customer dialling, but far more sophisticated.

All the systems listed in Table 7.1 are for inter-exchange use. In fact they are all CCITT-standard systems designed for international use. Each of them can be used by one exchange to establish calls to another exchange, but they vary considerably in sophistication. CCITT 1, a simple system enabling operators in different manual exchanges to call one another up, has been described previously. In the course of time CCITT 2 to CCITT 7, plus the R1 and R2 signalling systems were developed. Each tends to be slightly more sophisticated than its predecessor and therefore better attuned to the developing technologies of switching and transmission. The most advanced of the CCITT signalling systems developed to date is CCITT 7. A common channel signalling system, CCITT 7 has a number of powerful network control features and is capable of supporting a wide range of advanced services.

The CCITT standard signalling systems are only a small subset of the total range available, but they are the most suitable for international interconnection of public telephone networks because they are widely available. Other systems have evolved either as national standards or have been developed specially for particular applications. Signalling systems in general can be classified into one of four different classes according to how the signalling information is conveyed over the transmission medium, as follows:

(i) *Direct current (DC) signalling systems*, which use on/off current pulse or vary the magnitude and polarity of the circuit current to represent the different signals; 'loop disconnect (LD)' is an example. It has the disadvantage that it will only work when there is a distinct set of wires for each channel (i.e. on 'audio' lineplant), and it is therefore only suitable for short ranges. DC signalling is not possible on either FDM or TDM lineplant, though the on/off states can be mimicked using speechband or voice frequency (VF) tones over FDM (or TDM) (example pulse on/off = tone on/off) or alternatively over TDM by crudely converting the pulse on/offs into strings of binary 1's and 0's. The advantage of DC signalling (when possible) is its cheapness.

(ii) *Voice frequency (VF) signalling systems*, VF signalling is the name given to single or two-tone signalling systems (otherwise '1VF' and '2VF'). As stated above, these are similar to DC signalling systems, merely mimicking pulse 'on' and 'off' (and varying lengths and combinations of same) with tone on/tone off conditions. The principal engineering problems associated with VF signalling systems arise from the difficulty in keeping speech frequencies and signalling tones logically separate from one another—while sharing the same circuit. CCITT 4 is an example of a 2VF system.

(iii) *Multifrequency code (MFC) signalling systems*, which use tones within or close to the frequencies heard in normal speech to represent the signalling information. The advantage of MFC signalling is that it can easily be carried over FDM or TDM lineplant, the tones being processed through the multiplexor in exactly the same way as the speech frequencies. Owing to their compatibility with FDM and TDM lineplant this type of signalling system has become very

Table 7.1
CCITT signalling systems.

Signalling type	Description
CCITT 1	Now obsolete signalling system intended for manual use on international circuits. A 500 Hz tone is interrupted at 20 Hz for 2 seconds.
CCITT 2	Never implemented. CCITT 2 was a 2 Voice Frequency (VF) tone system, using 600 Hz and 750 Hz tones for 'line signalling'[*] and dialling pulses respectively. It was the first system for international automatic working.
CCITT 3	Now obsolete, system designed for manual and automatic operation using a 1 VF tone at 2280 Hz for both line signalling[*] and inter-register[‡] signalling. Inter-register signalling using a binary code at 20 baud.
CCITT 4	Intra-European signalling system still used for automatic and semi-automatic use. Line signalling[*] using 2040/2400 Hz (2 VF) code. Inter-register[‡] signalling using the same 2 VF tones; each digit comprising four elements, transmitted at 28 baud. (2040 Hz = binary 0; 2400 = binary 1).
CCITT 5	System designed and still used for intercontinental operation via satellite and using circuit multiplication equipment (Chapter 20 refers). Line signalling[*] using 2400/2600 Hz (2 VF) code. Multifrequency (MF) inter-register[‡] signalling, each digit represented by a permutation of two of six available tones.
CCITT 5 (bis)	Never used; a compelled version of CCITT 5.
CCITT 6	Common channel signalling system intended for international use between analogue SPC (stored program control exchanges. Signalling link speed typically 2.4 kbit/s.
CCITT 7	Common channel signalling system intended for widespread use between digital SPC exchanges. Multifunctional with various different 'user parts' for different applications (see Chapter 28). Signalling link speed 64 kbit/s.
CCITT R1	Regional signalling system somewhat akin to CCITT 5 and formerly used particularly for trunk network signalling in North America.
CCITT R2	Regional signalling system used widely within Europe. Described fully later in this chapter.
CCITT R2D	Digital version of R2. Adapted particularly for use after the European Communications Satellite (ECS, or 'Eutelsat').

[*] Line signalling and [‡] inter-register signalling are described more fully later in this chapter.

common in trunk and international networks. CCITT 5 and CCITT R2 are examples of MFC signalling systems.

(iv) *Digital signalling systems*, which code their signalling information in an efficient binary code format, each byte of information having a particular meaning. This type of signalling is therefore ideal for carriage over TDM lineplant. Typically, timeslot 16 in the European 2 Mbit/s digital transmission system is reserved for digital signalling, while in the 1.5 Mbit/s system either an entire 64 kbit/s channel or a 'robbed bit' channel is used. Alternatively, the signals can be encoded using a 'modem' (as described in Chapter 9) to make them suitable for carriage over FDM lineplant. Examples of digital signalling systems are CCITT 6, CCITT 7 and (in part) CCITT R2D.

CCITT R2 is typical of signalling systems in analogue network usage today, and as it provides a useful introduction to the principles of interexchange call control and multifrequency signalling we shall discuss it next, returning to CCITT 7 in Chapter 28.

7.7 THE CCITT R2 SIGNALLING SYSTEM

The CCITT R2 signalling is typical of the many multifrequency code (MFC) systems used in the networks of the world. R2 is one of CCITT's two 'regional' systems, and is used extensively within and between the countries of Europe. It comprises two functional parts, an 'outband' 'line signalling' system, together with an 'inband' and 'compelled sequence' MFC 'inter-register signalling' system. (The new terms are explained later in this section).

CCITT R2 may be used on international as well as national connections, but since there are significant differences between these two applications, it is normal to refer to the variants as if they were two separate systems, 'International R2' and 'National R2'.

R2 is a 'channel-associated' signalling system. By this, we mean that all the signals pertinent to a particular channel (or circuit) are passed down the circuit itself (in other words are 'associated' with it). By contrast, the more modern 'common channel signalling' systems use a dedicated 'signalling link' to carry the signalling information for a large number of traffic carrying circuits. The traffic carrying (i.e. speech carrying) circuits take a separate route, so that speech and signalling do not travel together. Figure 7.8 illustrates the difference between 'channel-associated' and 'common channel' signalling systems.

In channel associated signalling systems (exchanges A and B of Figure 7.8), a large

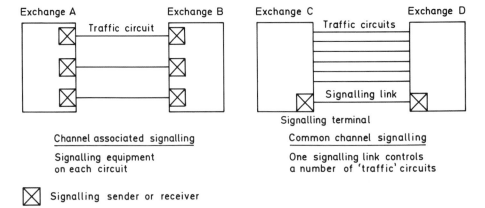

Figure 7.8
'Channel-associated' and 'common channel' signalling method.

number of code senders and receivers are required—one for each circuit. By contrast, in common channel systems (exchanges C and D of Figure 7.8), a smaller number of so-called 'signalling terminals' are required. Examples of channel associated signalling systems are 'loop disconnect' (LD) and R2 (CCITT R2). Common channel signalling systems include CCITT 6 and CCITT 7.

7.8 R2 LINE SIGNALLING

Multifrequency, channel-associated signalling systems nearly always have two parts, the line signalling part and the inter-register signalling part, each with its own distinct function. The line signalling part controls the line and the common equipment; it also sends line 'seizures' (described earlier), and other 'supervisory' signals such as the 'cleardown' signal. The inter-register signalling part carries the 'information', such as 'number dialled', between exchange registers. Splitting channel-associated signalling systems into two parts in this way helps to minimize the overall number of signalling code senders and receivers required in an exchange, as we shall see.

The line signalling part is usually only a single or two-frequency system. In R2, it is a single tone, out-of-band system. Out-of-band means that the frequency used is OUTside the (3.1 kHz) BANDwidth which is made available for conversation; the frequency, none the less, lies within the overall 4 kHz bandwidth of the circuit. Figure 7.9 illustrates the *inband* and *out-of-band* ranges of a normal 4 kHz telephone channel.

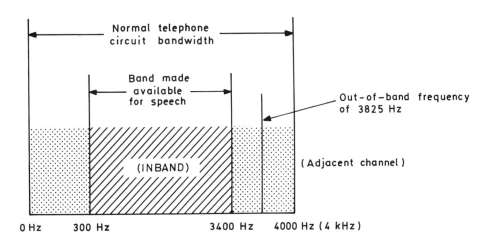

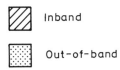

Figure 7.9
'Inband' and 'out-of-band' signals.

We learned about telephone circuit bandwidth in Chapter 3, and how 4 kHz of bandwidth is allocated for each individual channel on an FDM system, but that only the central 3.1 kHz bandwidth is used for conversation. The unused bandwidth provides for separation of channels as a means of reducing the likelihood of adjacent channel interference. Figure 7.9 illustrates the relationship, showing the total 4 kHz bandwidth actually allocated for the telephone circuit, and the 3.1 kHz range from 300 Hz to 3400 Hz which is available for speech. The ranges from 0–300 Hz and 3400–4000 Hz are normally filtered out from the original conversation (with little customer-perceived disadvantage) and give the bandwith separation between channels. In short, 'out-of-band' means a frequency in the range 0–300 Hz or 3400–4000 Hz, and 'inband' frequencies are those in the range 300–3400 Hz.

The use of an out-of-band signal (rather than an inband one) for R2 line signalling has two advantages: first, it does not disturb the conversation (without affecting the channel separation); second, line signals cannot be sent fraudulently by a telephone customer since the out-of-band region is not accessible to the end user. Despite this advantage, some other signalling systems do use inband line signalling. The advantage of using inband tones lies in simplifying the circuit configuration through FDM multiplexors.

The line signalling part of the signalling system is the only part of the signalling which is always active. It is the line signalling that actually controls the circuit. From the circuit idle state, it is the line signalling part that 'seizes' the circuit (alerting the distant exchange for action), and in so doing will activate the inter-register signalling part. The 'seizure' involves making a register ready at the distant (incoming) exchange, as well as activating appropriate inter-register signalling 'code senders' and 'receivers'. 'Seizure' will be followed by a phase of inter-register signalling, to convey dialled number and other call set-up information between the exchanges; at the end of this phase the inter-register signalling equipment and register will be released for use on other circuits, but the line signalling will remain active. The line signalling has more

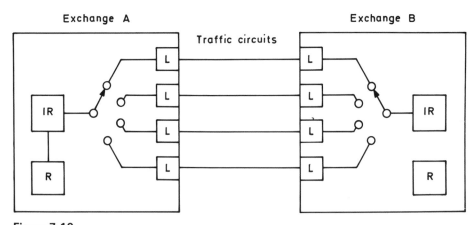

Figure 7.10
Channel associated signalling: line and inter-register signalling parts. L, line signalling equipment (always active on each circuit); IR, inter-register signalling code senders and receivers (shared); R, register (for storing and analysing call set-up information) (shared).

work to do in detecting the answer condition (in order to meter the customer), and at the end of the call it must carry the signals necessary to clear the connection and stop the metering. Even when the call has been cleared, both exchanges must continue to monitor the line signalling in order to detect any subsequent call seizures.

It is because the line signalling part is always active that it is normally designed as a single or two-tone system. This reduces the complexity and cost of signalling equipment that has to be permanently active on each and every circuit. The inter-register signalling part is only used for a comparatively short period on each call, during call set up. It is necessarily more complicated because of the range of information that it must convey, but a small number of *common* devices (code senders, code receivers and registers) may be shared between several circuits. This is cheaper than employing an inter-register sending and receiver with every circuit. The equipment is switched to an active circuit in response to a line signalling seizure, as already outlined. Figure 7.10 illustrates the interrelationship of line and inter-register signalling parts.

The line signalling part of R2 operates by changing the state of the single frequency (3825 Hz) tone, from on-to-off and vice versa. When the circuit is not in use, a tone of 3825 Hz can be detected in both 'transmit' and 'receive' channels of the circuit.

When R2 signalling is used directly on audio circuits (i.e. unmultiplexed analogue circuits) the line signalling tone sender and receiver is located in the exchange termination. When FDM (frequency division multiplex) is used on the circuits, the tones themselves are usually generated in the FDM channel translating equipment (CTE), otherwise the tones would be filtered out together with other signals outside the normal speech.

Two further wires connect the exchange to the CTE, and enable the exchange to control and monitor the state of tones on incoming (receive) and outgoing (transmit) channels, whether on or off. These extra leads are called the E&M leads (hence the common term 'E&M signalling'). In total therefore, each circuit between the exchange

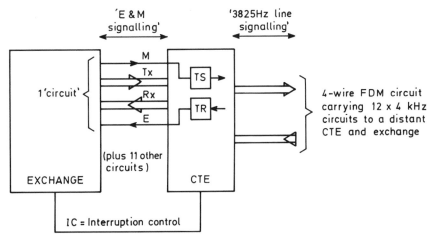

Figure 7.11
Typical configuration for R2 signalling. Tx, Transmit pair; Rx, Receive pair; IC, Interruption control wire; TS, 3825 Hz tone sender; TR, 3825 Hz tone receiver; CTE, channel translating equipment.

and the CTE comprises six wires: a transmit pair, a receive pair, plus E-wire and M-wire. The M-wire controls the tone state on the transmit channel, activating the CTE to send a tone when the M-wire is at high voltage, and not sending a tone when the M-wire is earthed. Similarly, the CTE conveys information concerning the state of the tone on the receive channel by using the E-wire; high voltage means there is a tone on the receive channel, earth (zero voltage) means there is not. A useful way of remembering the job of the E and M wires is to make the association E = Ear; M = Mouth. In addition to the 72 wires needed for the 12 channels, an extra wire is provided for interruption control (IC), so that if the FDM carrier fails the CTE does not immediately seize all 12 circuits. Figure 7.11 shows a typical arrangement of exchange and CTE where R2 signalling is being used in conjunction with FDM lineplant.

From the tone-on-idle state (i.e. circuit not in use, with 3825 Hz tones passing in both forward and backward directions), the tones may be turned 'off' and 'on' in sequence to indicate the different stages of a call: set up, conversation, and cleardown. We next discuss the signalling sequence and describe the terms 'incoming' and 'out-

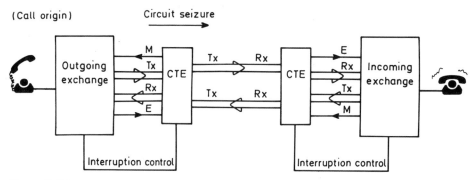

Figure 7.12
'Incoming' and 'outgoing' exchanges of an R2 controlled circuit.

Table 7.2
R2 line signalling sequence. (*Courtesy of CCITT—derived from Table 1/Q411.*)

State no:	Forward signal tone	Backward signal tone	Line signal meaning	Moves to state
1	Tone-on	Tone-on	Circuit idle	2 or 6
2	Tone-off	Tone-on	Seized (by outgoing end)	3 or 5
3	Tone-off	Tone-off	Answered-conversation	4 or 5
4	Tone-off	Tone-on	Clear-back	3 or 5
5	Tone-on	Tone-on or Tone-off	Release (forward)	1
6	Tone-on	Tone-off	Blocked (at incoming end)	1 (when unblocked)

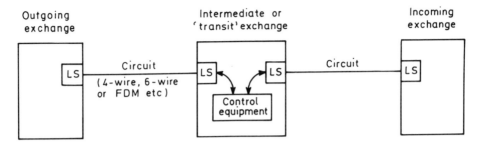

LS = Line Signalling Equipment (Sender and Receiver)

Figure 7.13
'Link-by-link' operation of R2 line signalling.

going' exchange. The 'outgoing' exchange is that originating the call. It effects the 'seizure' of an idle circuit by turning 'off' the forward signal tone as shown in Figure 7.12. The 'incoming' exchange (the one which did not generate the 'seizure') responds by allocating common equipment, including a register and inter-register signalling equipment.

In signalling systems other than R2 the readiness to receive digits is indicated at this stage by returning a 'proceed-to-send' (PTS) signal. On receipt of the PTS the outgoing exchange sends the dialled number digits by changing over to 'inter-register signalling'. Since R2 has no PTS signal the change over to inter-register signalling is unprompted and the digit is sent anyway. It continues to be sent until it is acknowledged by the incoming end exchange; a technique known as 'compelled signalling'. The call set up continues with similar inter-register and line signal interchanges. Table 7.2 shows the entire line signalling sequence.

When the inter-register signalling has conveyed the necessary number of dialled digits to allow the incoming exchange to decide its action, then the connection can be assumed to be made through the incoming exchange. If a further onward link is necessary (e.g. to another trunk or international transit exchange) then either the 'incoming exchange' changes its role to that of an 'outgoing exchange' (for signalling purposes) or else the 'outgoing exchange' retains its role and signals transparently through the previous 'incoming exchange' (now switched through) to the new incoming exchange (next link). The former method is called 'link-by-link' signalling, the latter 'end-to-end' signalling.

The line signalling part of R2 works in a link-by-link mode. In other words, on a connection comprising a number of links in tandem the line signalling on each link of the connection works in an independent and sequential mode. The 'cleardown' signal for instance does not pass straight through from one end of the connection to the other; it is passed one link at a time (link-by-link), and is interpreted by the control equipment of each exchange and passed on as necessary. Independent line signalling equipment is therefore required on every link of a tandem connection, as shown in Figure 7.13.

The link-by-link operation of the line signalling part is in contrast to the end-to-end operation of the inter-register signalling part, as we shall shortly see.

7.9 'COMPELLED' OR 'ACKNOWLEDGED' SIGNALLING

R2 line signalling is said to be a 'compelled' system, meaning that each signal is sent and continues to be sent, whether forward or backward, until a signal is received from the opposite end—the return signal acting as an acknowledgement and a prompt for the next action. This ensures receipt of the signal and readiness for the next. Thus the acknowledgement of the first digit sent in the forward direction prompts the outgoing exchange to send the next. Until the signal is acknowledged the outgoing exchange merely continues to send it. Compelled signalling is more reliable than 'non-compelled'. But the advantage that non-compelled signalling systems have, is the ability to send a string of signals all in one go, so potentially reducing the time needed for call set up. This can be particularly valuable if the circuit propagation time is quite long. For if each signal has to be acknowledged over a satellite circuit, then the 'loop-delay' (there-and-back time) for signal transmission and acknowledgement over the circuit, means that a maximum rate of around one signal (or digit) per second is all that can be achieved.

'*Acknowledged*' signalling is slightly different from '*compelled* signalling'—it is slightly less burdensome. While compelled signalling systems are always also acknowledged systems, the reverse is not necessarily so. In an acknowledged (but not *compelled*) signalling system, short sequences of signals may be sent, and may be repeated if not acknowledged, but are not sent continuously. They may be acknowledged by a single signal. CCITT 4 is a good example of acknowledge signalling—each forward digit is pulsed and acknowledged by a pulsed acknowledgement.

7.10 R2 INTER-REGISTER, MULTI-FREQUENCY CODE (MFC) SIGNALLING

In the previous section we saw how the line signalling part of R2 controls the circuit itself, conveying circuit seizure, answer, clearing, and other 'circuit supervision' signals. It cannot convey the crucial information for actual call set up, including the dialled number etc. This is the function of the inter-register signalling part, which is activated following the 'seizure signal' of the line signalling as we saw above.

The inter-register signalling part of R2 (R2-MFC, or 'multifrequency code') works between R2 registers in an end-to-end fashion. At the outgoing exchange, an access loop signalling system (such as LD or MF4) will have stored the information required for call set up, in a register. Analysis of the digits, as we saw earlier in the chapter, then allows selection of an outgoing route to the destination. If this route is via another exchange, then inter-register signalling will be needed to relay the information to the register located in the subsequent (incoming or transit) exchange. Figure 7.10 showed how inter-register signalling equipment is configured so as to convey call information from the register in exchange A to that in exchange B, during call set up. If a further link, via exchange C, were added to the connection (as now shown in Figure 7.14) then the information required by the register in exchange C could be derived direct from the register in the outgoing exchange A. That being so, the register in exchange B can be released after the link B–C has been seized and the switch path through exchange B has been established. This method of signalling is described as

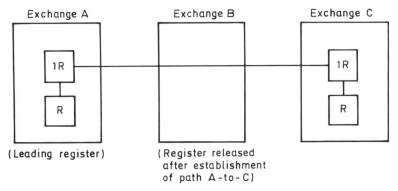

Figure 7.14
End-to-end inter-register signalling. IR, inter-register signalling code senders and receivers;
R, register.

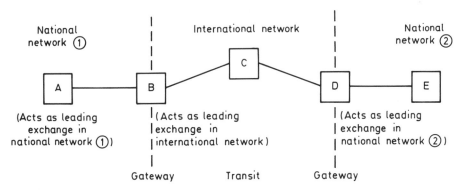

Figure 7.15
Regeneration of R2 MFC.

'end-to-end' signalling. The originating register of an 'end-to-end' signalling configuration (in our example, that in exchange A) is often termed the 'leading' register.

End-to-end operation of inter-register signalling is common within national or international networks, but at the boundary between such networks (at the international gateway exchange for example) it is normal to undertake signal regeneration. By this we mean that a new 'leading register' function is assumed by the gateway exchange. Thus a trunk exchange in one country is not expected to work on an end-to-end basis as an 'outgoing exchange' with an 'incoming' trunk exchange in some other country. Instead of that, signal regeneration is carried out at the outgoing international gateways and maybe the incoming one as well. In fact R2 is designed to work end-to-end from an outgoing R2 international register through to the distant local exchange. Incoming calls to Denmark, Switzerland and the Netherlands work in this way. Regeneration at the incoming international gateway is necessary when the national R2 is incompatible—or in cases where some other inland signalling system is used. The principle of regeneration is illustrated in Figure 7.15. The network designer has at least two reasons for choosing to regenerate R2 signalling:

- Either he will want to reduce the likelihood of excessively long holding times of 'leading' registers, which can cause congestion;

- Or he may be anxious to accommodate the differences between national and international variants of R2.

Since international as compared with national connections usually require a greater number of transmission links and exchanges to be connected in tandem, it follows that international connections generally take longer to set up. The 'holding time' per call of 'leading registers' which control the set up of international connections will therefore be longer than that of their solely national counterparts, and an accordingly greater number of registers will be required. In order, therefore, to prevent the congestion of national network registers which might result from a large throughput of international calls, it is normal to provide a separate set of international R2 registers at international gateways (such as exchange B of Figure 7.15). The number of international R2 registers provided at exchange B can account for the longer holding time. Furthermore, these registers can accommodate and 'interwork' the slight differences between the different national and international versions of R2. Similar interworking needs call for further regeneration of national version R2 at exchange D.

Like the line signalling, the inter-register signalling in R2 MFC is carried out by means of 'compelled' forward and backward signals. Subsequent signals waiting to be sent in the same direction (forward or backward) are only sent in the relevant direction as the result of requests and acknowledgements from the other end—and the previous signal removed at this time. The inter-register MFC signals themselves consist of 'two-out-of-six' 'in-band' frequencies, sent together. The use of two frequencies makes the signalling more reliable and less subject to line interference, since before reacting to a signal a receiver must detect two, and only two, frequencies. In this way single stray frequencies do not cause mis-operation of the connection of 'wrong numbers'.

Fifteen 'forward' and fifteen 'backward' signals are available, constituted in the manner shown in Table 7.3. When R2 is used over 4-wire or equivalent circuits, the use of separate frequency groups in the two directions gives no particular benefit, but over 2-wire circuits it allows simultaneous signalling in both directions without interference.

Table 7.3
Frequencies used in R2 MFC. (*Courtesy of CCITT—derived from Table 5/Q441.*)

Signal frequencies			Signal number													
Forward	Backward	Signal value	1	2	3	4	5	6	7	8	9	10	11	12	13	14	15
1380	1140	0	*	*		*			*				*				
1500	1020	1	*		*		*			*				*			
1620	900	2		*	*			*			*				*		
1740	780	4				*	*	*				*				*	
1860	660	7							*	*	*	*					*
1980	540	11											*	*	*	*	*

Tables 7.4 and 7.5 illustrate the signals used in international R2 MFC. The first inter-register signal sent is a forward signal, usually the first dialled digit. This is acknowledged by signal A-1 requesting the next digit. Signals are sent in turn in the forward and backward directions, the incoming register controlling the sending of digits either by using the 'send next digit' signal, A-1, or by requesting the re-sending of certain digits using any of signals A2, A7 or A8.

Once sufficient information has been received by the incoming register, digit analysis, route selection, and the connection path switch-through may be carried out in the normal manner.

International R2 MFC is more complicated than national R2 MFC, in that more signals are generally required to carry the information pertaining to international

Table 7.4
International R2 MFC forward signals. (*Courtesy of CCITT—derived from Tables 6 and 7/Q441.*)

Signal number	Group I: Address information etc.		Group II: Category of the calling party
	Ia: First signal on an international circuit	Ib: Subsequent signals on an international circuit	
1	Language digit: French	Digit: 1	Signals assigned for national use. If received by an international gateway they are converted to one of signals II-7 to II-10.
2	English	2	
3	German	3	
4	Russian	4	
5	Spanish	5	
6	Spare (language digit)	6	
7	Spare (language digit)	7	Subscriber (or operator without forward transfer facility)
8	Spare (language digit)	8	Data transmission call
9	Spare (discriminating digit)	9	Subscriber with priority
10	Discriminating digit	0	Operator with forward transfer facility
11	Country code indicator; outgoing half-echo suppressor required	Code 11: call to incoming operator	Spare signals for national use
12	Country code indicator; no echo suppressor required	(a) Code 12: call to operator for delay service	
		(b) Request not accepted	
13	Code 13: call by automatic test equipment	Code 13: call to automatic test equipment	
14	Country code indicator; incoming half-echo suppressor required	Incoming half-echo suppressor required	
15	Spare	Code 15: end of pulsing	

Table 7.5

International R2 MFC backward signals. (*Courtesy of CCITT—derived from Tables 8 and 9/Q441.*)

Signal number	Group A: Control signals	Group B: Condition of the called subscriber's line
1	Send next digit $(n+1)$	Spare signal for national use
2	Send last but one digit $(n-1)$	Subscriber transferred
3	Send category of calling party and changeover to reception of B-signals	Subscriber's line busy
4	Congestion in the national network	Congestion
5	Send category of calling party	Number not in use
6	Set up speech conditions	Subscriber's line free, charge
7	Send last but two digits $(n-2)$	Subscriber's line free, no charge
8	Send last but three digits $(n-3)$	Subscriber's line out of order
9	Spare signals for national use	Spare signals for national use
10		
11	Send country code indicator	
12	Send language or discriminating digit	
13	Send identity code (location) of outgoing international R2 register	Spare signals for international use
14	Is an incoming half-echo suppressor required?	
15	Congestion in an international exchange or at its output	

networks, including signals which:

● Indicate the country of destination;

● Control the selection of the type of transmission system (e.g. satellite or cable);

● Request overseas operator assistance (if required), and the preferred language to be used for such assistance;

● Control specific items of equipment, such as echo suppressors.

Later chapters, on routing, numbering and transmission plans, as well as on operator services, describe the reason for and use of these signals.

The signals used in international signalling systems such as R2 are standardized and detailed in CCITT recommendations. However, when used within a single operator's network some scope exists for using either a more sophisticated range of signals or a simpler one.

A more sophisticated range of signals might be invaluable in enabling more advanced call routing mechanisms to be adopted, as will be described in Chapter 12. Another reason for adopting a sophisticated signalling repertoire might be the network operator's desire to introduce some of the 'network management' techniques to be described in Chapter 19. Not all these techniques, however, can be supported by the limited facilities of R2 signalling, and some more advanced signalling system such as

CCITT 7 may be needed to give network operators what they want. On the other hand a national network operator can usually reduce his overall costs by simplifying the signalling systems; in particular, R2 MFC can successfully be operated in national networks without either the forward or backward signals numbered 11 to 15. This means that some of the tone sending and receiving equipment is not needed. Specifically both the forward tone of 1980 Hz and the backward tone of 540 Hz can be dispensed with.

Signalling system R2 is described more fully in CCITT Recommendations Q400–Q490.

7.11 DIGITAL LINE SYSTEMS AND CHANNEL ASSOCIATED SIGNALLING

In the earlier part of this chapter we learned how, in channel associated signalling systems (such as CCITT 1, 2, 3, 4, 5, R1 and R2), the call set up and control information is carried over the channel itself. This is the case where the circuits are carried over analogue transmission equipment (i.e. using either audio 'pairs' or FDM lineplant), but in cases where the circuits are carried by digital line systems, it is usual to encode even 'channel associated' line signalling into a single channel. In 2.048 Mbit/s (European) digital transmission practice, the channel normally used for this purpose is timeslot 16 (as we discovered in chapter 5), and all the line signals for the 30 usable circuits (channels 1–15 and 17–31) are encoded on this single channel. Similarly, in 1.544 Mbit/s digital transmission practice, either an entire 64 kbit/s channel can be used for signalling, or more commonly, the robbed-bit-signalling channel 9 (also described in chapter 5) can be used.

Even though the line signalling may be concentrated on a single digital channel, it is always correct to refer to CCITT 1, CCITT 2, CCITT 3, CCITT 4, CCITT 5, CCITT R1 and CCITT R2 signalling systems as 'channel associated'. In all of them the multifrequency inter-register signalling is still conveyed within the channel itself. Concentrating the line signalling on a single channel, however, enables a small economy to be made in the number of line signalling senders and receivers.

A special type of R2 line signalling was developed for use over digital line systems. The system is called R2D (standing for R2 Digital). The line signalling of R2D is defined by CCITT Recommendation Q421. The MFC part of R2D is identical to the analogue version. Similarly, special digital variants of the line signalling parts are to be found in other signalling systems.

7.12 SIGNALLING INTERWORKING

The choice of signalling system for a network is determined by a number of factors, including the network topology, the length and propagation delay of the transmission links, the services to be provided, the transmission media to be used, and of course

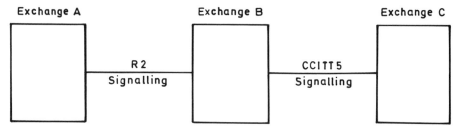

Figure 7.16
Signalling interworking.

the nature of the established network. Investments in obsolescent signalling systems often mean that old and modern equipment practices and signalling systems have to co-exist. The job of designing and operating cheap and effective networks against such an evolving background is challenging and will remain so.

Figure 7.16 illustrates a typical situation where R2 signalling is at work on the link between exchanges A and B, with CCITT 5 signalling on link B–C of the same connection. The configuration demands that similar call set up and cleardown information can be carried by both signalling systems, and even transferred from one to the other through some form of signalling interworking. Thus exchange B must comprehend both the R2 and CCITT 5 signalling repertoires and be capable of interworking between the two.

The provision of signalling interworking demands careful appreciation by exchange designers of the logical sequences and information flows of the two signalling systems. The exchange itself, or the signalling conversion device, needs to be able to interpret correctly the signals of both the systems and convey the same meaning when converting between the two. The complexities of interworking reach their peak when the two signalling systems that are to be interworked have different signal vocabularies (e.g. one system may only have a 'network busy' signal while the other may have a range of signals to determine whether the exchange or the called customer is busy). In these circumstances some information may well be lost.

The loss of information may, or may not, cause the network to operate in an unpredictable manner. For example in CCITT 5 there is only one backward signal to indicate that the call cannot be completed for any reason, including far-end customer congestion, invalid number etc. The signal is called the 'busy flash' signal. However, in R2 signalling a number of different backward signals enable the cause of the unsuccessful call completion also to be returned, e.g. A4—national network congestion; A15—international network congestion; B3—subscriber line busy; and B4—congestion. So in Figure 7.16 if exchange B receives a 'busy flash' signal in CCITT 5 signalling from exchange C, which unsuccessful message signal should be conveyed in R2 signalling back to exchange A? There is no perfect answer: there is however a 'standard' answer, as defined by the set of CCITT signalling interworking recommendations in the Q600 series, to the effect that either the A4 (national network congestion) signal should be returned or the speech channel should be connected to 'busy tone'.

7.13 ADVANCED SIGNALLING APPLICATIONS

Having described the basic principles of signalling for setting up connections on circuit-switched networks, we can now approach some of the sophisticated network control methods made possible by more advanced signalling systems. These include controlling network traffic, call routing, and overall transmission quality which will be dealt with in Chapters 12–14. Later chapters cover the advanced signalling requirements of what is called network management, and the most advanced of today's signalling systems, CCITT 7.

7.14 SIGNALLING SEQUENCE DIAGRAMS

It is helpful, in the design, operation and maintenance of signalling systems, to understand the correct sequences of signals. This simplifies the task of understanding and tracking exchange responses to signals and gives an insight into the interaction of different signalling systems, so making easier the diagnosis of faults. The simplest means of illustrating signalling sequences and the one commonly employed in specification documents is the signalling sequence diagram, an example of which is Figure 7.17.

Figure 7.17 illustrates a hypothetical connection between two telephones, A and B, interconnected by means of exchanges C, D and E. Signalling from customer A to exchange C is by means of 'off-hook' and DTMF signalling. From exchange C to exchange D, R2 signalling is in use, and from D to E, CCITT 5 signalling.

In reality, it is highly unlikely that customers would be directly connected to international exchanges in the manner illustrated in Figure 7.17; none the less it is theoretically possible. I have chosen to illustrate international signalling sequences since they are slightly more complex. International signalling systems generally need to convey more information, and here we show the conveyance of information about 'country code' of destination, 'language digit' (discussed in more detail in Chapter 15), and echo suppressor needs (Chapter 13). National interexchange signalling systems do not always need these complications and are generally simpler, so if you can understand the signalling sequence described here, you should be well set up for assimilating other national signalling sequence diagrams.

Ringing current and 'loop' signalling provide the final means of alerting customer B and him signalling his answer.

The sequence diagram in Figure 7.17 shows the signalling interactions which must occur in order to establish the call and clear it down afterwards. When illustrated like this we see just how many interactions the exchanges are having to cope with!

Sidelined on the diagram are annotated the various stages of call set-up—as described earlier in the chapter—thus 'off-hook', 'dial-tone', 'dialled digits' etc.

The reason for illustrating the use of different signalling systems on links CD and DE is to contrast their different methods of operation, and the 'interworking' between them that must be conducted by exchange D. You will note two major differences. First, some of the signals do not have identical meanings in the two signalling systems, so a 'best fit' approach occasionally has to be used. Second, the sequencing of the signals is quite different in the two systems. R2 is operating in an 'overlap' signalling mode. By this we mean that exchange C is retransmitting the early dialled digits on

126

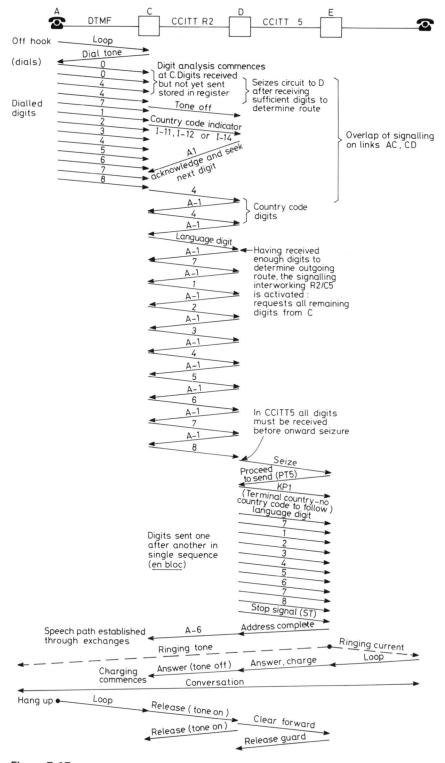

Figure 7.17
An example signalling sequence diagram.

to exchange D even before the final digit has been received from the caller, A. Thus signalling on links AC and CD is 'overlapping'. But in contrast, exchange D does not commence signalling on link DE until all the digits have been received from C. This is because CCITT 5 signalling is required to operate in an *en-bloc* signalling mode, in which all the digits are sent together.

Historically, overlap mode signalling tended to be employed with early electromechanical exchanges, since it allowed exchanges late in the connection to commence their relatively slow switching actions as early in time as possible—so minimizing the overall time required for call set-up. The much more rapid digit analysis and switching action of SPC and digital exchanges favours the use of *en-bloc* signalling, because the exchange processor's time is not wasted simply waiting to receive digits from a previous exchange.

The end user will not be concerned about whether the connection is established using either 'overlap' or *en-bloc* mode signalling, but there is one word of warning for the network operator. Since the two modes do not interact efficiently (see Figure 7.17), a connection made up of links alternating in signalling types employing both 'overlap' and *en-bloc* modes will tend to be relatively slow in establishing calls—5 to 10 seconds can be added to the call set-up time. If callers are used to being connected in around 15–20 seconds and it takes 25–30 seconds there is a risk that callers will clear calls prematurely—thinking that the set-up phase has failed.

BIBLIOGRAPHY

CCITT Recommendations. Q Series.
CCITT Recommendations Q400–Q490. Specification of Signalling System R2.
Flood, J. E., *Telecommunication Networks*. Peter Peregrinus (for IEE), 1975.
Pearce, J. G., *Telecommunications Switching*. Plenum Press, 1981.
Voiceband Signalling for Delivery or Network Data (i.e. Call Routing Information) on Local Exchange Lines. Bellcore. TR-820-23125-84-02. April 1984.
Welch, S., *Signalling in Telecommunications Networks*. Peter Peregrinus (for IEE), 1979.

8

TRANSMISSION SYSTEMS

The basic line transmission theory we have considered so far, together with the principles of analogue and digital signal transmission over pairs of electrical wires is quite suitable for short-range conveyance of a small number of circuits. As we have seen, it is not always practical or economic to use many multiple numbers of physical 'pairs' between exchanges, so we have recourse to frequency division and time division multiplexing in order to reduce the number of physical pairs of wires needed to convey a large number of long-haul circuits between common end-points. There are of course other transmission media to be considered, some of them much better suited to particular applications than plain electric wire. They include coaxial cable, radio, satellites and optical fibre. Once we are familiar with the main features of each of these transmission types, we can decide which one is best suited to any particular application.

8.1 AUDIO CIRCUITS

The 2-wire and 4-wire transmission systems described in the first chapters of this book, comprising respectively, one or two physical 'pairs' of copper (or aluminium) wires together with amplifiers and equalizers as appropriate, are correctly called audio circuits. (In actual use the wires themselves are often wound or 'twisted' around each other inside the cable, hence the terms 'twisted pairs' and 'twisted pair cables'.) The title 'audio' circuit is given because the frequencies of the electrical signals carried by the circuit are a direct match with those of the audio signal they represent. In other words, the electrical signal frequencies are in the 'audio' range; the signal has not been processed in any way, neither frequency shifted nor multiplexed. As a reminder of a typical audio circuit configuration, Figure 8.1 shows a 4-wire repeatered circuit as already discussed in Chapter 3 on 'long-haul communication'.

Audio circuits may be used to carry any of a wide range of bandwidths, although in general the greater the bandwidth to be carried, the shorter the maximum range of the circuit, since the electrical properties of the line (its impedance, reactance and capacitance) tend to interfere. It is known that high-frequency signals are carried only by the outer surface of wire conductors (the 'skin effect'), and this makes high-

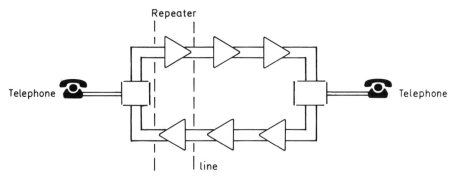

Figure 8.1
A 4-wire repeatered 'audio circuit'.

frequency signals prone to interference from electrical and magnetic fields emanating from adjacent wires.

Unamplified audio circuits are commonly used to provide relatively short-range telephone 'junction' circuits of 4 kHz bandwidth between exchanges over a range up to about 20 km, and amplification can be used to extend this range. A typical use of an audio circuit is to connect an analogue 'local' (or 'end office') exchange to the nearest 'trunk' (or 'toll') exchange. Beyond a distance of about 15 km, even if amplifiers are used, economics tend to favour other transmission means. (As an example an unamplified pair of wires using 0.63 mm diameter conductors has an insertion loss (i.e. a line loss which must be added to other circuit losses) of 3 dB at a range of 7 km. The transmission range for speech is increased to 33 km by the use of an amplifier but this range is usually precluded by the line quality needs of the signalling system.)

As also explained in Chapter 3, the rate of signal degradation (i.e. attenuation and distortion) in twisted pair transmission media can be reduced by 'loading' the cable (see Chapter 3) and more simply, by increasing the diameter ('gauge') of the wires themselves. Bigger conductors cause less signal attenuation because they have less resistance. However, the relative cheapness of alternative transmission media makes it uneconomic to employ very heavy gauge (larger diameter) twisted pairs over more than about 20 km. At this distance the conductors of an unamplified circuit need to be around 1 mm thick, and since the amount of conducting material required (copper or aluminium) is proportional to the square of the diameter the cost escalates rapidly as the cable gauge is increased. On a more cheerful view, the converse also applies, so that at shorter distances narrower cables are economically favourable. In one way and another a network operator has extensive information available to help him choose the right conductor (copper, aluminium etc.) and the appropriate gauge of wire, thereby cutting costs and still giving the transmission performances required.

Because they are susceptible to external electromagnetic interference, audio circuits are not normally suitable for use on very wide-bandwidth systems, including FDM (frequency division multiplex) systems, although early systems used screened copper wires. Audio circuits are most commonly found in small scale networks, and within the 'local' area of a telephone exchange. A number of twisted pairs are normally packed together into large cables. A 'cable' may vary in size from about 3 mm diameter to about 80 mm, carrying anything between one and several thousand individual pairs. Thus running out from a 10 000 line local exchange, a small number of (say 10–20) main backbone cables of several hundred or thousand pairs may run out along the street conduits to streetside cabinets where a simple cross-connect frame is used to extend the pairs out in a star fashion on smaller cables (say 25 or 50 pair). Finally, at the 'distribution point' individual pairs of wires may be crimped within a cable 'joint' (at the top of a telegraph pole) and run out from here to individual customers' premises.

8.2 TRANSVERSE SCREEN AND COAXIAL CABLE TRANSMISSION

A considerable problem with 'twisted pairs' is signal distortion caused by the 'skin effect'. The interference may be manifested as noise or 'hum' on the line, or in extreme cases a stronger voice or other signal may be induced from adjacent wire pairs. The likelihood of such interference is increased when higher bandwidth signals (including FDM) are carried. Two options are available to reduce the interference. The first is to screen the cable with a light metal foil, usually wound into the cable sheath. This is called 'screened' or 'shielded cable'. In addition, transverse screen cable also includes screens between wires within the cable—isolating 'transmit' from receive pairs. A second type of cable not prone to electromagnetic interference of this kind is coaxial cable. This is constructed so that most of the signal is carried by an electro-magnetic field near the centre of the cable. Coaxial cable was the workhorse of frequency division multiplex (FDM), high bandwidth analogue and early digital transmission systems. Lately, optical fibre is supplanting it for new digital lines.

Instead of a 'twisted pair' of wires, a coaxial cable consists of two concentric conductors, made usually of aluminium or copper. The central wire, or pole is separated from the cylindrical outer conductor by a cylindrical spacing layer of insulation, as shown in Figure 8.2.

The outer conductor is a metal foil or mesh, wound spirally around the insulation layer, and outside this is a layer of sheathing to provide external insulation and physical protection for the cable.

A single coaxial cable is equivalent to a single pair of twisted wires. To achieve the

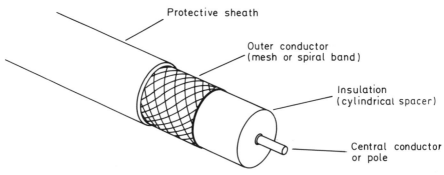

Figure 8.2
Coaxial cable.

equivalent of 4-wire transmission, two coaxial cables must be used, one to carry the 'transmit' signal and one to carry the 'receive' signal.

Coaxial cables function in much the same way as twisted pair circuits, but there are some important differences in performance. Unlike 'twisted pair' audio circuits, coaxial cable circuits are said to be 'unbalanced'. By this we mean that the two conductors in a coaxial cable do not act equally in conveying the signals. In a 'balanced' circuit such as a twisted pair, where the signal is carried by the electrical currents in the two wires passing in opposite directions, the currents generate equal but opposite electromagnetic fields around them which tend to cancel each other out. By contrast, the signal in a coaxial cable is carried largely by the electromagnetic field surrounding the inner and outer conductors, which is largely induced by the central conductor. The outer conductor, sometimes called the 'shield' is usually operated at 'earth' or 'ground' voltage, and prevents the electromagnetic field from radiating outside the cable sheath. The signal is thereby protected from interference to some extent.

Unfortunately the overall rate of signal attenuation is greater on unbalanced circuits like coaxial cables than on balanced ones such as twisted pairs, and this has to be counteracted by the use of amplifiers. On the other hand, unlike twisted pair circuits which are liable to lose balance as the result of damp or damage, coaxial cables are more stable and suffer less from the variable attenuation and extraneous signal reflections which sometimes affect twisted pairs. Coaxial cables are more reliable in service, easier to install, and they require less maintenance. Further, because of their insensitivity to electromagnetic interference they perform better than twisted pairs when the cable route passes near metal objects or other electrical cables.

8.3 RADIO

In 1887, Heinrich Hertz produced the first man-made radio waves. Radio waves, like electricity and light, are forms of electromagnetic radiation—the energy is conveyed by 'waves' of magnetic and electrical fields. In a wire, these waves are induced and guided by an electrical current passing along an electrical conductor, but that is not the only way of propagating electromagnetic (EM) waves. By using a very strong electrical signal as a transmitting source, an electromagnetic wave can be made to spread far and wide through thin air. That is the principle of radio. The radio waves are produced by transmitters, which consist of a radio wave source connected to some form of antenna, examples of which are:

- low frequency antenna (aerial) or radio mast,

- HF (high frequency), VHF (very HF), or UHF (ultra-HF) antenna or mast,

- microwave dish antenna,

- tropospheric scatter antenna,

- satellite antenna.

Radio is a particularly effective means of communication between remote locations and over difficult country, where cable laying and maintenance is not possible or is prohibitively expensive.

One way of communicating information by radio waves is by encoding (or more correctly 'modulating') a high-frequency 'carrier' signal prior to transmission. The technique of modulation in radio is very similar to that used in FDM, and single side-band (SSB) operation is employed though the carrier is rarely suppressed (as explained in Chapter 13). A distinctive feature of a radio carrier signal is its high frequency relative to the frequency bandwidth of the information signal. The frequency of the carrier has to be high in order for it to propagate as radio waves.

The modulation of the radio carrier signal may follow either an analogue or a digital regime. Analogue modulation is carried out in a similar way to FDM. Digital modulation can be by 'on/off' carrier signal modulation (i.e. by switching the carrier on and off), or by other methods such as 'frequency shift keying' (FSK), 'phase shift keying' (PSK) or 'quadrature amplitude modulation' (QAM), details of which are given in the next chapter.

After modulation of the carrier frequency, the combined signal is amplified and applied to a radio antenna. Amplification boosts the signal strength sufficiently for the antenna to convert the electrical current energy into a radio wave.

Figure 8.3 illustrates a simple radio transmitter: an audio signal is being filtered (to cut out extraneous signals outside the desired bandwidth) and amplified. Next it is modulated on to a radio-frequency (RF) carrier signal which is produced by a high quality oscillator; then the modulated signal is filtered again to prevent possible interference with other radio waves of adjacent frequency. Finally the signal is amplified by a high-power amplifier and sent to the antenna where it is converted into radio waves.

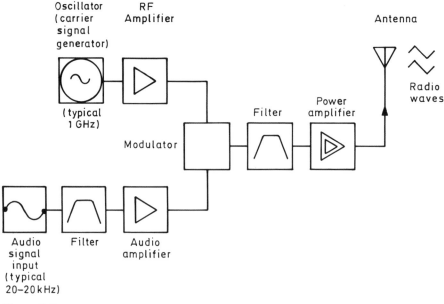

Figure 8.3
A simplified radio transmitter.

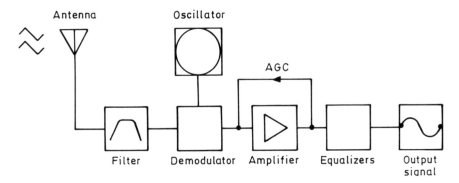

Figure 8.4
A simplified radio receiver.

Figure 8.4 illustrates a simplified radio receiver, the components of which are similar to those of the transmitter shown in Figure 8.3. The radio waves are received by the antenna and are reconverted into electrical signals. A filter removes extraneous and interfering signals before demodulation. As an alternative to a filter, a 'tuned circuit' could have been used with the antenna. A tuned circuit allows the antenna to select which radio wave frequencies will be transmitted or received.

Then comes demodulation. In the receiver shown in Figure 8.4 (typical of a microwave receiver), demodulation is carried out by 'subtracting' a signal equivalent to the original carrier frequency, leaving only the original audio or 'information' signal. A cruder and cheaper method of demodulation, not requiring an oscillator, employs a 'rectifying circuit'. This serves to have the same subtracting effect on the carrier signal. After demodulation, the information signal is processed so as to recreate the original audio signal as closely as possible. The signal is next amplified using an amplifier with 'automatic gain control' (AGC) to ensure that the output signal volume is constant, even if the received radio wave signal has been subject to intermittent 'fading'. Finally, the output is adjusted to remove various signal distortions called 'group delay' and 'frequency distortion', by devices called equalizers. We will consider the cause and cure of these distortions in Chapter 13.

The simplified transmitter and receiver shown in Figures 8.3 and 8.4 could be used together for 'simplex' or one-way radio transmission. Such components in use are thus equivalent to one 'pair' of a 4-wire transmission line discussed in Chapter 3. For full 'duplex' or two-way transmission—equivalent to a 4-wire system—the equipment must be duplicated, with two radio transmitters and two receivers, one of each at each end of the radio link. Two slightly different carrier frequencies are normally used for the two directions of transmission, to prevent the two channels interfering with one another.

8.4 RADIO WAVE PROPAGATION

When radio waves are transmitted from a point, they spread and propagate as spherical wavefronts. The wavefronts travel in a direction perpendicular to the wavefront, as shown in Figure 8.5.

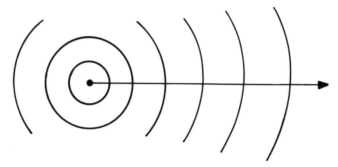

Figure 8.5
Radio wave propagation.

Radio waves and light waves are both forms of electromagnetic radiation, and they display similar properties. Just as a beam of light may be reflected, refracted (i.e. slightly bent), and diffracted (slightly swayed around obstacles), so may a radio wave, and Figure 8.6 gives examples of various different wave paths between transmitter and receiver, resulting from these different wave phenomena.

Four particular modes of radio wave propagation are shown in Figure 8.6. A radio transmission system is normally designed to take advantage of one of these modes. The four modes are:

● 'line-of-sight' propagation,

● surface wave (diffracted) propagation,

● trophospheric scatter (reflected and refracted) propagation,

● skywave (refracted) propagation.

A 'line-of-sight' radio system, such as microwave, relies on the fact that waves normally travel in a straight line. This is perhaps the simplest type of system; provided the receiver is 'in-sight' of the transmitter, then line-of-sight propagation is possible. The range of a line-of-sight system is limited by the effect of the earth's curvature, as Figure 8.6 shows. Line-of-sight systems therefore can reach beyond the horizon only when they have tall masts. Radio systems can also be used beyond the horizon by utilizing one of the other three radio propagation effects also shown in Figure 8.6.

Surface waves have a good range, depending on their frequency. They propagate by diffraction using the ground as a 'waveguide'. Low-frequency radio signals are the best suited to surface wave propagation, because the amount of bending (the effect properly called 'diffraction') is related to the radio wavelength. The longer the wavelength, the greater the effect of diffraction. Therefore the lower the frequency, the greater the bending. A second means of over-the-horizon radio transmission is by 'tropospheric scatter'. This is a form of radio wave reflection. It occurs in a layer of the earth's atmosphere called the 'troposphere' and works best on ultra high frequency (UHF) radio waves. The final last example of over-the-horizon propagation given in Figure 8.6 is known as skywave propagation. This comes about through refraction

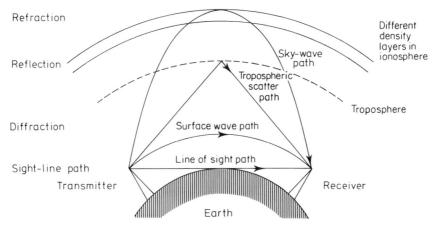

Figure 8.6
Different modes of radio wave propagation.

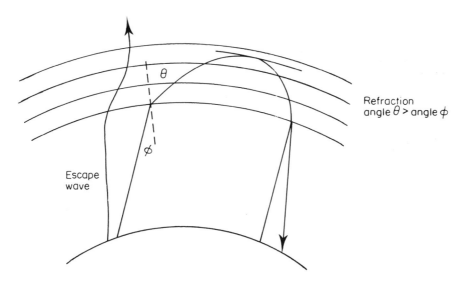

Figure 8.7
Refraction causing skywaves.

(deflection) of radio waves by the earth's atmosphere, and it occurs because the different layers of the earth's upper atmosphere (the 'ionosphere') have different densities, with the result that radio waves propagate more quickly in some of the layers than in others, giving rise to wave deflection between the layers, as Figure 8.7 shows in more detail. But as Figure 8.7 also demonstrates, not all waves are refracted back to the ground; some escape the atmosphere entirely, particularly if their initial direction of propagation relative to the vertical is too low (angle ϕ, phi).

8.5 RADIO ANTENNAS

Each individual radio system is required to perform slightly different tasks. The distinguishing qualities of a radio system are:

● its range,

● its received signal power,

● its directionality (i.e. whether the transmitted signal is concentrated in a particular direction or not).

These qualities are determined by the mode of propagation for which the system is designed (e.g. surface wave, tropospheric scatter, line-of-sight, skywave etc.), by the frequency of the radio carrier frequency, by the power of the system, and most particularly, by the suitability of the transmitting and receiving antennas. Antenna design has a significant effect on radio systems' range, directionality and signal power.

The simplest type of radio antenna is called a dipole. A dipole antenna consists of a straight metal conductor. Examples can be found fitted on many household domestic radios, to receive 'VHF' (or 'FM') radio signals. Another, larger, example of a dipole aerial is a radio mast. Radio masts may be up to several hundred feet in height, consisting of one enormous dipole.

The qualities of dipole antennas make them suitable for broadcast or 'omnidirectional' transmission and receipt of radio signals. A dipole antenna sends out a signal of equal strength in all (or 'omni') directions, and a receiving antenna can receive signals equally well from most directions. Dipole aerials are therefore much used for 'broadcast' transmission from one transmitting station to many receiving stations. Dipole antennas are also used for point-to-point transmission when the location of the receiver or transmitter is unknown, as in the case of mobile radio or ships at sea.

The disadvantage of omnidirectional receiving antennas is their susceptibility to radio interference from undesired sources. Omnidirectional transmitting antennas also have a disadvantage: they expend a lot of power transmitting their radio waves in all conceivable directions, but only a small number of points within the overall coverage area pick up a very small fraction of the signal power; the remainder is wasted. For this reason, some antennas are specially designed to work directionally. One example of a directional antenna is the dish-shaped type used for microwave, satellite, and tropospheric scatter systems—see Figure 8.8(a). Dish-shaped antennas work by focusing the transmitted radio waves in a particular direction. Another type called an 'array antenna' is illustrated in Figure 8.8(b).

Figure 8.8
(a) Microwave radio repeater station. A 68-foot tower and associated microwave radio repeater station, at the brow of a hill. The antenna dishes on opposite sides of the tower are slightly offset in direction to prevent the possibility of signal overshoot. (*Courtesy of Telecom Technology Showcase, London.*) (b) Array antenna. Sixteen metre sterba array antenna. The elements of the antenna are the barely visible wires, strung from the main pylon, and arranged in a regular and rectangular formation. The pattern of repeated elements makes the antenna highly directional—cutting out all signals except that from the desired direction. (*Courtesy of Telecom Technology Showcase, London.*)

(a)

(b)

Omnidirectional
lobe pattern

Highly directional
lobe pattern

Figure 8.9
Aerial lobe diagrams.

By using a directional rather than an 'omnidirectional' antenna, the range of the radio system and the signal power transmitted in, or received from a given direction can be deliberately controlled.

Directional antennas are ideal for use on point-to-point radio links; the overall power needed for transmission is reduced, and the received signal suffers less from interference.

The directionality of an antenna is best illustrated by its 'lobe' diagram, and two examples are given in Figure 8.9.

The lobes of an antenna lobe diagram show the transmitting and receiving efficiency of the antenna in the various directional orientations. The length of a lobe on a lobe diagram is proportional to the relative signal strength of the radiowave transmitted by the antenna in that particular direction. Hence, the circle in the omnidirectional antenna lobe diagram represents an equal strength of signal transmitted in all directions. Compare this with the directional aerial which has one very strong lobe in one particular direction, and a number of much smaller ones. This sort of lobe pattern is typical of many directional aerial types.

Any antenna can be used either to transmit or to receive, and it has the same lobe pattern when used for either purpose. The lobe pattern shows an antenna's directionality. On the transmission side we have seen that the lobe pattern shows the relative signal strengths transmitted in various directions. In 'receive mode', the lobe pattern shows the antenna's relative effectiveness in picking up radio waves from sources in the various directions. As we have said, a directional receive antenna also helps to reduce the likelihood of interference from other radio sources; the lobe pattern shown in Figure 8.10 shows a directional antenna being used to eliminate interference from a second radio source.

The receiving antenna, R, in Figure 8.10 is so oriented that its principal lobe is pointing at the transmitting source T1. Meanwhile, source T2 aligns with a 'null' in the lobe pattern. The receiving antenna will thus be much more effective in reproducing the signal from T1 than in reproducing the signal from T2, so that the signal from T2 is suppressed relative to that from T1.

An antenna is required at each end of any radio link, but the two antennas need not be similar. For example, the transmitting antenna may be a high power, omnidirectional, broadcast radio mast, while receiving antennas may be highly directional, and perhaps relatively small and cheap, as would be the case with television broadcasting, whereas with ships' radio it is difficult to keep directional antennas correctly oriented, and so an omnidirectional antenna will be a cheaper solution.

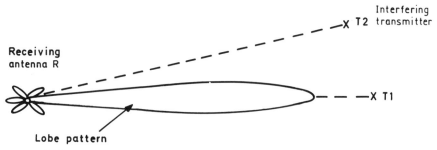

Figure 8.10
Excluding interfering radio transmitters.

In the next few sections, some practical radio systems are considered in more detail. The characteristics of the antennas, including their directionality, are described and an explanation is given of how the various antennas induce the desired mode of radio wave propagation (e.g. surface wave, skywave etc.).

8.6 SURFACE-WAVE RADIO SYSTEMS

Surface-wave radio systems have a good range when using relatively low frequency radio waves. (In this context low frequency is the range 50 kHz–2 MHz.) Surface-wave radio is typically used for broadcast radio transmissions, particularly for public radio stations, maritime radio, and navigation systems. Broadcast, surface-wave radio transmitters are usually very tall radio masts, several hundred feet high as illustrated in Figure 8.11.

Broadcast radio transmitting antennas usually combine high-power transmitters with an omnidirectional lobe pattern. The high power enables domestic radio receivers to be relatively unsophisticated, and therefore cheap. The omnidirectional pattern allows them to transmit over a wide area.

8.7 HIGH-FREQUENCY (HF) RADIO

Radio waves in the frequency band from 3–30 MHz propagate over greater distances in a 'skywave' form than they do as a surface wave. For skywave propagation, a directional antenna is set up to transmit a radio wave at a specific angle to the horizon. The angle must not be so close to the vertical as to allow an escape wave such as we saw in Figure 8.7. The signal is transmitted via the earth's ionosphere (a region about 200–350 km above the earth's surface). Layers of the ionosphere which have different physical properties cause the signal to be refracted and eventually reflected back to the earth's surface, an effect shown in Figure 8.7. Unfortunately, high-frequency skywave radio is very sensitive to weather conditions in the ionosphere. Time-of-day and sun spot cycles can break up reliable service, giving rise to fading and interference. None the less, high-frequency radio using skywaves was one of the earliest transmission methods employed for international telephone service. As early as 1927, the first commercial radio telephone service was in operation between Britain and the United States, many years before the first transatlantic telephone cable was laid in 1956.

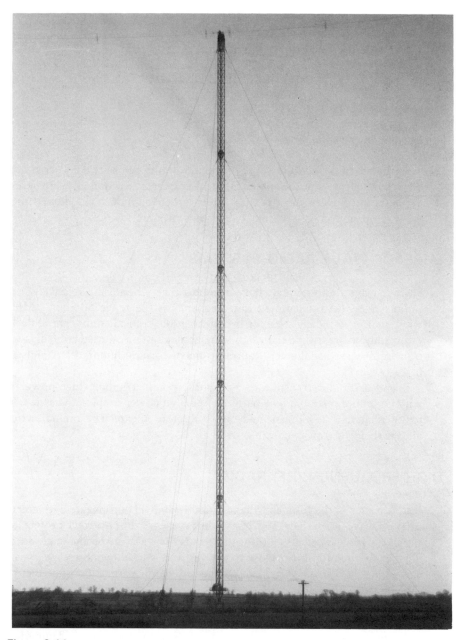

Figure 8.11
Radio mast. 250-metre high radio mast, weighing 200 tonnes. The visible wires are probably steel, provided merely to support the mast. The active element is mounted within the main pylon. (*Courtesy of Telecom Technology Showcase, London.*)

8.8 VERY HIGH-FREQUENCY (VHF) AND ULTRA HIGH-FREQUENCY (UHF) RADIO

At frequencies above about 30 MHz, neither skywave propagation nor surface-wave over-the-horizon propagation is possible. For this reason the VHF and UHF radio spectrum is used mainly for line-of-sight radio transmission systems. 'Just-over-the-horizon' transmission is also possible, but only if the antennas are elevated clear of the electrical 'ground effects' of the earth's surface.

VHF and UHF radio systems are becoming increasingly common as the basis of a large number of applications:

- local radio stations,

- citizens band (CB) radio,

- cellular radio (mobile telephones),

- radiopagers.

The advantage of VHF and UHF radio is that much smaller antennas can be used. It has made possible the wide range mobile communications handsets, some of which have antennas only a few inches long. A drawback is the restricted range of VHF and UHF systems, although this has been turned to advantage in cellular radio telephone systems, where a number of transmitters are used, each covering only a small 'cell' area. Each cell has a given bandwidth of the radio spectrum to establish telephone calls to mobile stations within the cell. Adjacent cells use different radio bandwidths, thereby preventing radio interference between the signals at the cell boundaries. The fact that the radio range is short also means that a non-adjacent cell can 'reuse' the same radio

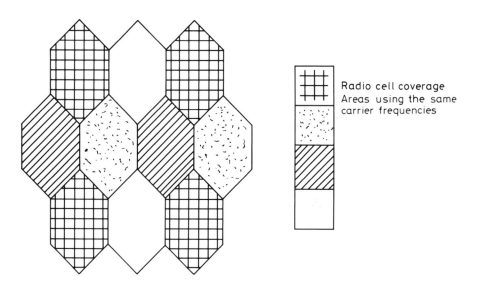

Figure 8.12
Radio spectrum reuse in cellular radio networks.

spectrum without chance of interference, so that very high radio spectrum utilization can be achieved. Figure 8.12 illustrates the reuse of radio spectra, as achieved in 'cellular radio' networks for portable telephone sets. Cells marked with the same shading are using identical radio bandwidth.

We shall return to the subject of cellular radio in Chapter 27.

8.9 MICROWAVE RADIO

Microwave is the name given to radio waves in the frequency above 1000 MHz (1 GHz). The prefix 'micro' is in recognition of the very short wavelength (only a few centimetres). Microwave systems are commonly used as high capacity, point-to-point transmission systems in telecommunications networks, such as high-capacity trunk telephone network connections between major cities, or on a smaller scale (using smaller antennas) between company offices. The high frequency and short wavelength of microwave radio allows high capacity radio systems to be built using relatively small but highly directional antennas. The small scale yields benefits in terms of cost, installation, and maintenance.

Microwave antennas are operated in a line-of-sight mode, usually spaced 40–50 kilometres apart as governed by the amount of radio signal 'fade' and also, practically, by the availability of suitable locations for radio towers. A simple empirical formula recommended by CCIR Report 338 (Vol. V, Geneva, 1978) to estimate the probability of a given amount of fade is as follows:

The probability of incurring a given radio fade F (where $F > 15$ dB) and clear line-of-sight paths with negligible earth reflection is given by P, in the expression:

$$P = KQ10^{F10}f^B d^c$$

where K is a factor for climatic conditions (1.4×10^{-8} for Europe), Q is a factor for terrain (0.4 for mountains, 1.0 for plain), F is the fade in decibels, f is the frequency in GHz, B is a regional terrain factor ($B = 1$ for Europe, 1.2 for Japan), d is path length in kilometres, c is another terrain factor, value 3.5.

Distances beyond single hop links of 40–50 km are achieved using a multilink path, passing through a number of intermediate radio repeater stations. The path itself is usually arranged in a slight zigzag formation of 'repeater stations', as shown in Figure 8.13(a). At each repeater station there is a tower or radio mast, covered in a large number of microwave antennas. For each transmission path passing through the repeater stations there will be at least one dish antenna facing one 'hop' of the path and one facing the next. (Two dishes may face each link, depending upon whether the same antenna reflector is used for both transmit and receive purposes.)

The radio signal is received by one antenna and retransmitted in a slightly different direction on the next hop or link of the path by another antenna. As Figure 8.13(a) shows, the use of a zigzag path allows the use of the same radio carrier frequency on all the links of the overall path, without any risk of radio interference which might result from 'overshoot' on a straight line path between repeater stations (Figure 8.13(b)).

Slight refraction, caused by the earth's atmosphere causes radio waves to propagate along a curved path as shown in Figure 8.14, and this has to be taken into account when setting up the antennas. Figure 8.14 also shows some of the path interference problems associated with trees and other obstacles which can be experienced with microwave radio links and have to be taken into account during installation.

Like other forms of radio transmission, microwave radio is an effective way of spanning difficult terrain, although providing power for the intermediate 'repeater stations' on long-haul, multiple 'hop' links can be a problem. Like other forms of radio propagation, microwave systems are prone to fading owing to the prevailing weather conditions. Disturbed weather can cause radio path interference as shown in Figure 8.14. Owing to the short wavelength of microwaves, reflections off buildings and other obstacles near to the radio path, cause greater interference than that experienced by longer wavelength signals. This is called 'multipath interference', and it too is

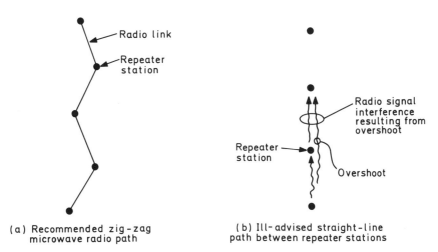

(a) Recommended zig-zag microwave radio path

(b) Ill-advised straight-line path between repeater stations

Figure 8.13
Microwave radio route planning.

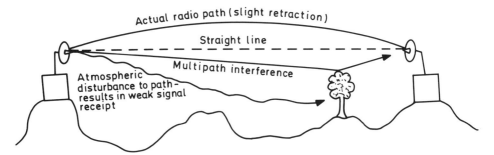

Figure 8.14
Microwave radio path and disturbances.

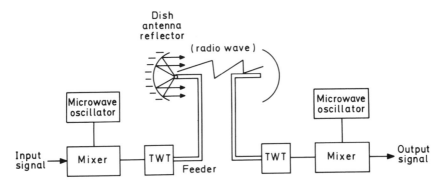

Figure 8.15
A microwave radio system.

shown in Figure 8.14. Careful path planning and choice of site for the antenna towers helps to minimize this problem.

Microwave systems are commonly operated at 2, 4, 6, 11, 12, 14, 16 and 30 GHz. The microwave carrier is produced by a microwave oscillator and modulated in a similar way to high-frequency radio signals, using modulating equipment called a mixer. The radio wave is then amplified immediately prior to transmission with the aid of a sophisticated and very high power gain amplifier, called a 'travelling wave tube (TWT)'. The radio wave is then fed into the antenna itself using a hollow metal tube called a 'feeder'. The signal is emitted from the 'focus' of a parabolic reflector antenna to produce a highly directional single beam. At the receiving end a similar aerial and set of electronics works in reverse order to reproduce the original signal. Figure 8.15 illustrates a simple microwave radio system.

8.10 SATELLITE SYSTEMS

Satellite transmission provides an excellent means of long-distance communication, either around the globe or across difficult terrain. It also provides an effective means of 'broadcasting' the same signal to a large number of receiving stations.

The type of satellites most commonly used in telecommunications networks, called 'geostationary' satellites, orbit the earth directly above the equator at such a height that they travel once around the earth's axis every 24 hours. Because both the satellite and the earth move at the same speed, the satellite appears to be geographically (or geo)stationary above a particular location on the equator.

When used for telecommunications purposes, a geostationary satellite is equipped with microwave antennas, which allow 'line-of-sight' radio contacts between the satellite and other microwave antennas at 'earth stations' on the ground. Communication between two earth stations can then be established by a tandem connection consisting of an 'uplink' from the transmitting station to the satellite, and a 'downlink' from the satellite to the receiving station. On board the satellite the uplink is connected to the downlink by a 'responder' (receiver) for each uplink and a transmitter for each downlink, and because the two normally work in pairs they are often designed as a single piece of equipment usually referred to as a 'transponder'.

The first international telecommunications satellite, called 'Telstar', was launched in 1962. This was a low *orbiting* satellite rather than a *geostationary* one, and it orbited the earth's axis faster than the earth's own spin rate. Consequently the satellite was mutually visible to American and European earth stations for three or four periods a day, but each period was of 30–40 minutes duration. Sophisticated 'tracking' mechanisms were therefore required on the early earth stations in order to keep the reflector of the antenna pointing directly at the satellite as it made its path from one horizon to the other in about 30 minutes.

A major breakthrough in international satellite telecommunications came in 1964 with the establishment of the International Telecommunications SATellite consortium, or INTELSAT for short. The first generation of INTELSAT's series of Satellites, INTELSAT 1—'Early Bird' to give it its popular name—was launched in 1965. 'Early Bird' was a geostationary satellite, and the consortium of 11 countries (Australia, Canada, Denmark, France, Italy, Japan, Netherlands, Spain, UK, USA and Vatican City) which set INTELSAT up were able to take advantage of its services.

The INTELSAT satellites brought on a boom in international telecommunications

Figure 8.16
Telstar satellite. The first international telecommunications satellite, Telstar, launched in 1962. It was a low orbiting satellite, about 0.9 metre in diameter, and supported 12 telephone channels. (*Courtesy of British Telecom.*)

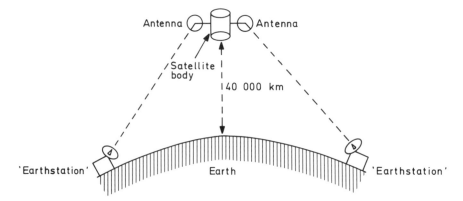

Figure 8.17
A satellite transmission system.

Figure 8.18
Satellite earth station. Goonhilly satellite earth station on the Lizard peninsular, Cornwall, England. One of the world's first established telecommunications satellite earth stations. Five main dishes and a number of smaller ones are visible. Note also the microwave tower, providing onward connection into the UK terrestrial network. (*Courtesy of British Telecom.*)

Figure 8.19
INTELSAT VI satellite. Artist's impression of the enormous INTELSAT VI satellite, the workhorse for the early 1990s. Twelve metres in height, 4 metres in diameter, 2 tonnes and 110 000 telephone channels. (*Courtesy of British Telecom.*)

throughout the late 1960s and 1970s, owing to the high capacity of satellite transmission and the relatively low cost of the extremely long connections which it provided. (Satellite transmission costs are not determined by the distance between earth
stations.) The satellite's one drawback is the significant signal delay resulting from the
time it takes a microwave signal to travel up to and back from the satellite. For geostationary satellites, the delay is quite significant — being 40 000 km from the Earth it
adds a period of silence between a talker ending speaking and hearing the reply of
between a half to one second. For lower, orbiting satellites, the delay is much less. The
historical problem of satellite tracking returns (as with Telstar), but the user is much
happier with the shorter delay. For this reason, current technological research is
returning to develop low orbit systems, using sophisticated Earth station technology
for tracking. Figure 8.17 illustrates a simple satellite transmission system.

Today's satellites are a far cry from their predecessors. Compared with the Telstar
satellite, which was 0.9 m in diameter and 77 kg in mass, INTELSAT's sixth generation satellites, INTELSAT VI which will serve the world's telecommunication needs
in the early 1990s are truly 'spacecraft'. Towering edifices, 12 m in height, 4 m in
diameter and 2 tonnes in mass, each satellite supports up to 110 000 telephone
channels. This compares with the meagre 12 circuits supported by the first Telstar.

8.11 TROPOSPHERIC SCATTER

It takes 'tropospheric scatter' to make over-the-horizon communication possible.
Radio waves appear to be reflected by the earth's atmosphere. It is not a well

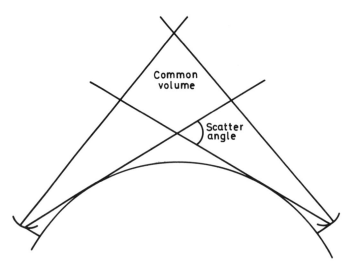

Figure 8.20
Tropospheric scatter.

understood phenomenon and various explanations have been offered. Perhaps the easiest to grasp is the notion of radio wave reflection (commonly called 'scatter'), caused by irregularities in the tropospheric region of the earth's atmosphere. The scatter occurs in a 'common volume' of the earth's atmosphere, corresponding to the region 'visible' to both the receiving and transmitting tropospheric scatter antennas (Figure 8.20). The scatter angle is the angle of path deviation (or the 'reflection angle') caused by the scatter effect. Figure 8.20 illustrates this model of tropospheric scatter.

Large dish-shaped antennas are used for tropospheric scatter systems, together with microwave radio frequencies in the range of 800 MHz–5 GHz. Communication distances achievable with tropospheric scatter radio systems are 100–300 km. Their main drawback is the bad and continually varying signal fading that is experienced.

In the United Kingdom, the tropospheric scatter method of 'trans-horizon radio' transmission has become a standard method of communicating with offshore oil drilling platforms in the North Sea.

Figure 8.21
Land radio station for tropospheric scatter radio system. This one is located on the north-east coast of England, providing telecommunications services to oil rigs in the North Sea, about 100–300 km offshore. (*Courtesy of British Telecom.*)

Figure 8.22
Tropospheric scatter radio station. The oil rig receiving antennas for a tropospheric scatter radio system. (*Courtesy of Telecom Technology Showcase, London.*)

8.12 OPTICAL FIBRES

Most transmission types can be used to support either analogue or digital signal transmission: twisted pair circuits, coaxial cables, and radio systems can all be used as the basis of either analogue or digital line transmission systems. But the fact that digital transmission requires only two distinct line states (on/off; mark/space) has opened up a new wider range of digital propagation methods; and the most important new digital transmission medium is optical fibre.

The growing importance of optical fibres stems from their extremely high bit-rate capacity and their low cost. Optical fibres are a hair's breadth in diameter and are made from a cheap raw material—glass. They are easy to install because they are small, and because they allow the repeaters to be relatively far apart (nowadays over 100 km spacing is possible) they make for easy maintenance.

An optical fibre conveys the bits of a digital bit pattern as either an 'on' or an 'off' state of light. The light (of wavelength 1.3 or 1.5 μm) is generated at the transmitting end of the fibre, either by a laser, or by a cheaper junction diode device called a 'light emitting diode' (LED for short). A diode is used for detection. The light stays within the fibre (in other words is guided by the fibre) due to the reflective and refractive properties of the outer skin of the fibre, which is fabricated in a 'tunnel fashion'.

Fibre-optic technology is already into its third generation. In the first generation, fibres had two cylindrical layers of glass, called the core and the cladding. In addition a 'cover' or 'sheath' of a different material (say plastic) provides protection. The core and the cladding were both glass, but of different refractive index. A 'step' in the refractive index existed at the boundary of core and cladding, causing reflection of the light rays at this interface, so guiding the rays along the fibre. Unfortunately, however,

Figure 8.23
An optical fibre cable, showing clearly the overall construction. At the centre, a steel twisted wire provides strength, enabling the cable to be pulled through ducts during installation. Around it are ten plastic tubes, each one containing and protecting a hair's breadth fibre, each capable of carrying many megabits of information per second. (*Courtesy of British Telecom.*)

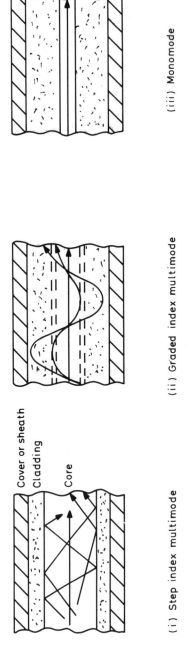

Cover or sheath
Cladding
Core

(i) Step index multimode

(ii) Graded index multimode

(iii) Monomode

Figure 8.24
Types of optical fibre.

because of the relatively large diameter of the 'core', a number of different light-ray paths could be produced, all reflecting off the cladding at different angles, an effect known as 'dispersion'. The different overall ray paths would be of different lengths and therefore take different amounts of time to reach the receiver. The received signal is therefore not as sharp—it is dispersed. It is most problematic when the user intends very high bit-rate operation. The different wave paths (or modes) explain the name *step index multimode* fibre which has been given to these fibres. *Step index multimode* fibre is illustrated in Figure 8.24(i).

A refinement of step index multimode fibre is achieved by using a fibre with a more gradual grading of the refractive index from core to cladding. A *graded index multimode* fibre is shown in Figure 8.24(ii). This has a slightly improved high bit-rate performance over step–index multimode fibres. Finally, monomode fibres (illustrated in Figure 8.24(iii) are the third generation of optical fibre development, and have the best performance. In monomode fibre, more advanced fibre production techniques have produced a very narrow core area in the fibre, surrounded by a cladding area—there is a step change in the refractive index of the glass at the boundary of the core and the cladding. The narrow core of a monomode fibre allows only one of the ray paths (or 'modes') to exist. As a result there is very little light pulse dispersion in monomode fibre, and much higher bit rates can be carried.

The light travelling down optical fibres suffers attenuation just as electrical signals attenuate in copper conductors but the rate of attenuation is much lower, which means that 'repeaters' or 'regenerators' can be spaced much further apart. Optical fibres have to be jointed carefully at every section, since joints can be a major source of loss in light strength.

Development work abounds in the optical fibre field. An exciting development discussed in Chapter 20 is that of wavelength division multiplex (WDM). This is a technique allowing second and third high bit-rate signals down the same fibre—merely by using second and third lasers and diode detectors operating at different light wavelengths.

8.13 SUMMARY

This chapter has covered the common methods of signal transmission: twisted pairs, coaxial cables, radio systems, and optical fibres. All these media can be used for digital transmission, and all but optical fibres for analogue transmission. In practice most networks combine a variety of transmission media, and to each of their varied operations the most economic and effective technique is applied.

BIBLIOGRAPHY

Dunlop, J. and Smith, D. G., *Telecommunications Engineering*. Van Nostrand Reinhold (UK), 1984.

Franco, G. L., *World Telecommunications—Ways and Means to Global Integration*. Le Monde economique, 1987.

Freeman, R. L., *Radio System Design for Telecommunications* (1–100 GHz). Wiley, June 1987.

Fthenakis, E., *Manual of Satellite Communications*. McGraw-Hill, 1984.

Inglis, A. F., *Electronic Communications Handbook*. McGraw-Hill, 1988.

Maslin, N., *HF Communications—A Systems Approach*. Pitman, 1987.

Morgan, W. L. and Gordon, G. D., *Communications Satellite Handbook*. Wiley, February 1989.

Picquenard, A., *Radio Wave Propagation*. Macmillan, 1974.

Prentiss, S., *Satellite Communications*. 2nd edn., Tab Books, 1987.

Radio Communication Handbook, Radio Society of Great Britain (RSGB), Fifth edn., 1987.

Uiet, P-G. F., *Telecommunications Systems*. Artech House, 1986.

9

DATA NETWORKS AND PROTOCOLS

We have described the binary form in which data are held by computer systems, and how such data are conveyed over digital line transmission systems, but we shall need to know more than this before we can design the sort of devices which can communicate sensibly with one another in something equivalent to human conversation. In this chapter we shall discuss in detail the conveyance of data between computer systems, the networks required, and the so-called 'protocols' they will need to ensure that they are communicating properly. In the second half of the chapter some practical data and computer network topologies, and the equipment needed to support them are described.

9.1 COMPUTER NETWORKS

Between 1950 and 1970 computers were large and unwieldy, severely limited in their power and capabilities, and rather unreliable. Only the larger companies could afford them, and they were used for 'batch-processing' scientific, business or financial data on a large scale. Data storage in those days was laborious, limited in capacity, long in preparation, not at all easy to manage. Many storage mechanisms (e.g. paper tape and punched cards) were very labour intensive; they were difficult to store, and prone to damage. Even magnetic tape, when it appeared, had its drawbacks; digging out some trivial information 'buried' in the middle of a long tape was a tedious business, and the tapes themselves had to be painstakingly protected against data corruption and loss caused by mechanical damage or nearby electrical and magnetic fields.

Computing was for specialists. Computer centre staff looked after the 'hardware', while 'software' experts spent long hours improving their computer programmes, squeezing every last drop of 'power' out of the computers' relatively restricted capacity.

By the mid 1970s all this began to change, and very rapidly. Cheap semiconductors heralded the appearance of the microcomputer, which when packaged with the newly developed floppy diskette systems, opened a new era of cheap and widespread computing activity. 'Personal computers' (PCs) began to appear on almost every manager's desk, and many even invaded peoples' homes. Suddenly, computing was within reach of the masses, and the creation and storage of computer data was easy, cheap and fast.

All sorts of individuals began to prepare their own isolated databases and to write computer programmes for small-scale applications, but these individuals soon recognized the need to share information, and to pass data between different computers. This could be done by transferring floppy diskettes from one machine to another, but as time went on that method on its own proved inadequate. There was a growing demand for more geographically widespread, rapid, voluminous data transfer. Advanced 'distributed processing' computer networks have since emerged.

A very simple computer network consists of a computer linked to a piece of peripheral equipment, such as a printer. The link is necessary so that the data in the computer's memory can be reproduced on paper. The problem is that a 'wires-only' direct connection of this nature is only suitable for very short connections—typically up to about 20 metres. Beyond this range, some sort of 'line driving' telecommunications technique must be used. A number of techniques are discussed here. A long-distance point-to-point connection may be made using 'modems'. A slightly more complex computer network might connect a number of computer terminals in outlying buildings back to a 'host' (mainframe computer) in a specialized data centre. Another network might be a 'local area network' (or LAN), used in an office to interconnect a number of desktop computing devices, laser printers, data storage devices (e.g. file servers) etc. More complex computer networks might interconnect a number of large mainframe computers in the major financial centres of the world, and provide dealers with 'up-to-the-minute' market information.

The basic principles of transmission, as set out in the early chapters of this book, apply equally to data which is communicated around computer networks. So circuit-switched networks or simple point-to-point lines may also be used for data communication. Data communication, however, makes more demands on its underlying network than a voice or analogue signal service, and additional measures are needed for coding the data in preparation for transmission, and in controlling the flow of data during transmission. Computers do not have the same inherent 'discipline' to prevent them talking two at a time. For this reason special 'protocols' are used in data communications to make quite sure that information passing between computers is correct, complete and properly understood.

9.2 BASIC DATA CONVEYANCE

As we learned in Chapter 4, data is normally held in a computer or computer storage medium in a binary code format, as a string of digits having either value '0' or value '1'. A series of such binary digits can be used to represent alphanumeric characters (e.g. ASCII code), or graphical images, such as those transmitted by facsimile machines.

In Chapter 5 we went on to discuss the principles of digital transmission, and found that it was ideal for the conveyance of binary data. Digital transmission is now fast becoming the backbone of both private and public networks. But despite the increasing availability and the ideal suitability of digital transmission for data communication, it is unfortunately not always available. In circumstances where it is not, digitally-oriented computer information must instead be translated into a form suitable for transmission across an analogue network. This translation is carried out by a piece of equipment called a MODulator/DEModulator, or MODEM for short. Modems transmit data by imposing the binary (or 'digital') data stream on to an audio-frequency carrier signal. The process is very similar to that used in the frequency division multiplexing of voice channels described in Chapter 3.

Figure 9.1 illustrates two possible configurations for data communication between two computers using either a digital or an analogue transmission link. The configurations look very similar, comprising the computers themselves (these are specific examples of data terminal equipment (DTE)); sandwiched between them in each case is a line and a pair of 'Data Circuit terminating Equipments' (DCE).

A digital DCE (Figure 9.1(a)) connects the customer's digital DTE to a digital transmission line, perhaps provided by the public telecommunications operator (PTO). The DCE provides several network functions. In the transmit direction, it regenerates the digital signal provided by the DTE and converts it into a format, level and line code suitable for transmission on the digital line. In the receive direction, it establishes a reference voltage for use of the DTE and reconverts the received line signal into a form

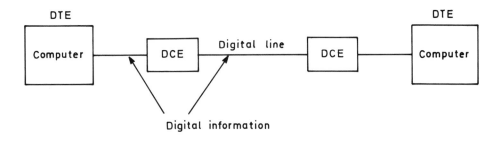

(a) Digital line connection

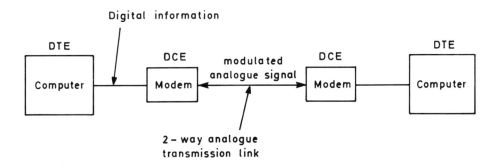

(b) Analogue line connection using modems

Figure 9.1
Point-to-point connection of computers. DTE, Data Terminal Equipment; DCE, Data Circuit Terminating Equipment.

suitable for passing to the DTE. In addition, it uses the clocking signal (i.e. the exact bit rate of the received signal) as the basis for its transmitting bit rate. The received clocking signal is used since this derives from the highly accurate 'master clock' in the PTO's network.

Digital DCEs can be used to provide various digital bit speeds, from as low as 2.4 kbit/s, through the standard channel of 64 kbit/s, right up to higher order systems such as 1.544 Mbit/s (called T1 or DS1), 2.048 Mbit/s (called E1), or 45 Mbit/s (called T3 or DS3). When bit speeds below the basic channel bit rate of 64 kbit/s are required by the DTE, then the DCE has an additional function to carry out, breaking down the line bandwidth of 64 kbit/s into smaller units in a process called sub-rate multiplexing. The process derives a number of lower bit-rate channels, such as 2.4 kbit/s, 4.8 kbit/s, 9.6 kbit/s etc., some or all of which may be used by a number of different DTEs.

In their various guises, digital DCEs go by different names. In the United Kingdom, British Telecom's 'Kilostream' digital private line service uses a DCE called a NTU (network terminating unit). Meanwhile, in the United States, AT&T's digital data system (DDS) and similar digital line services are provided by means of channel service units (CSU) or data services units (DSU). Sometimes the term LTU (line terminating unit) is used.

In the case of analogue transmission lines, the DCE (Figure 9.1(b)) cannot rely on the network to provide accurate clocking information, since the bit rate of the received signal is not accurately set by the PTO's analogue network. For this reason, an internal clock is needed within the DCE in order to maintain an accurate transmitting bit rate. The other major difference between analogue and digital DCEs is that analogue DCEs (modems) need to convert and reconvert digital signals received from the DTE into an analogue signal suitable for line transmission.

Three basic data modulation techniques are used in modems but there are more sophisticated versions of each type and even hybrid versions, combining the various techniques. The three techniques are illustrated in Figure 9.2 and described below.

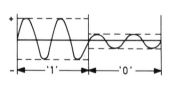

(a) Amplitude modulation

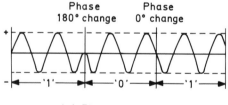

(c) Phase modulation

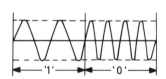

(b) Frequency modulation

Figure 9.2
Datamodulation techniques.

Amplitude modulation Modems employing amplitude modulation alter the amplitude of the carrier signal between a set value and zero (effectively 'on' and 'off') according to the respective value '1' or '0' of the modulating bit stream. Alternatively two different, non-zero values of amplitude may be used to represent '1' and '0'.

Frequency modulation In frequency modulation (Figure 9.2(b)), it is the frequency of the carrier signal that is altered to reflect the value '1' or '0' of the modulating bit stream. The amplitude and phase of the carrier signal is otherwise unaffected by the modulation process. If the number of bits transmitted per second is low, then the signal emitted by a frequency modulated modem is heard as a 'warbling' sound, alternating between two frequencies of tone. Modems using frequency modulation are more commonly called FSK, or 'frequency shift key' modems. A common form of FSK modem uses four different frequencies (or tones), two for the transmit direction and two for the receive. This allows simultaneous sending and receiving (i.e. 'duplex' transmission) of data by the modem, using only a 2-wire circuit.

Phase modulation In phase modulation (Figure 9.1(c)), the carrier signal is advanced or retarded in its 'phase' cycle by the modulating bit stream. At the beginning of each new bit, the signal will either retain its phase or change its phase. Thus in the example shown the initial signal phase represents value '1'. The *change* of phase by $180°$ represents next bit '0'. In the third bit period the phase does not *change*, so the value transmitted is '1'. Phase modulation (often called 'phase shift keying' or 'PSK') is conducted by comparing the signal phase in one time period to that in the previous period, thus it is not the absolute value of the signal phase that is important, rather the phase change that occurs at the beginning of each time period.

9.3 HIGH BIT-RATE MODEMS

The transmission of high bit rates can be achieved by modems in one of two ways. One is to modulate the carrier signal at a rate equal to the high-bit-rate of change of the modulating signal. Now the rate (or frequency) at which we fluctuate the carrier signal is called the 'baud rate', and the disadvantage of this first method of high-bit-rate carriage is the equally high 'baud rate' that it requires. The difficulty lies in designing a modem capable of responding to the line signal changes fast enough. Fortunately an alternative method is available in which the baud rate is lower than the bit rate of the modulating bit stream. The lower baud rate is achieved by encoding a number of consecutive bits from the modulating stream to be represented by a single line signal state. The method is called 'multilevel transmission', and is most easily explained using a diagram. Figure 9.3 illustrates a bit stream of 2 bits per second (2 bit/s being carried by a modem which uses four different line signal states. The modem is able to carry the bit stream at a 'baud rate' of only 1 per second (1 Baud)).

The modem used in Figure 9.3 achieves a lower baud rate than the bit rate of the data transmitted by using each of the line signal frequencies f_1, f_2, f_3 and f_4 to represent two consecutive bits rather than just one. It means that the line signal always has to be slightly in delay over the actual signal (by at least 1 bit as shown), but the benefit is that the receiving modem will have twice as much time to detect and interpret the

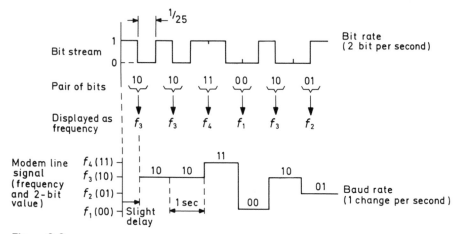

Figure 9.3
Multilevel transmission and lower baud rate.

datastream represented by the received frequencies. Multi-level transmission is invariably used in the design of very high bit-rate modems.

9.4 MODEM 'CONSTELLATIONS'

At this point we introduce the concept of modem constellation diagrams, since these assist in the explanation of more complex amplitude and phase-shift-keyed (PSK) modems. Figure 9.4 illustrates a modem constellation diagram composed of four dots. Each dot on the diagram represents the relative phase and amplitude of one of the possible line signal states generated by the modem. The distance of the dot from the origin of the diagram axes represents the amplitude of the signal, and the angle subtended between the X-axis and a line from the point of origin represents the signal phase relative to the signal state in the preceding instance of time.

Figure 9.4(b) and 9.4(c) together illustrate what we mean by 'signal phase'. Figure 9.4(b) shows a signal of $0°$ phase, in which the time period starts with the signal at zero amplitude and increases to maximum amplitude. Figure 9.4(c), by contrast, shows a signal of $90°$ phase, which commences further on in the cycle (in fact, at the $90°$ phase angle of the cycle). The signal starts at maximum amplitude but otherwise follows a similar pattern. Signal phases, for any phase angle between $0°$ and $360°$ could similarly be drawn. Returning to the signals represented by the constellation of Figure 9.4(a) we can now draw each of them, as shown in Figure 9.4(d) (assuming that each of them was preceded in the previous time instant by a signal of $0°$ phase). The phase angle in this case are $45°$, $135°$, $225°$, $315°$.

We are now ready to discuss a complicated but common modem modulation technique known as *quadrature amplitude modulation*, or *QAM*. QAM is a technique using a complex hybrid of phase (or 'quadrature') as well as amplitude modulation, hence the name. Figure 9.5 shows a simple eight-state form of QAM in which each line signal state represents a 3-bit signal (values nought to seven in binary can be repre-

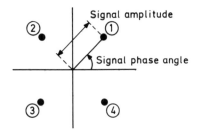

(a) 4-signal state constellation

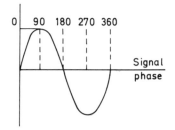

(b) Signal phase, commencing at 0°

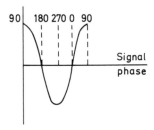

(c) Signal phase, commencing at 90°

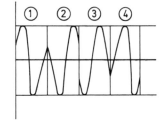

(d) The four line signals

Figure 9.4
Modem constellation diagram.

sented with only 3 bits). The eight signal states are a combination of four different relative phases and two different amplitude levels. The table of Figure 9.5(a) relates the individual 3-bit patterns to the particular phases and amplitudes of the signals that represent them. *Note*: Figure 9.5(b) illustrates the actual line signal pattern that would result if we sent the signals in the table consecutively as shown. Each signal is shown in the correct phase state *relative* to the signal in the previous time interval (unlike Figure 9.4 where we assumed each signal individually had been preceded by one of 0° phase). Thus in Figure 9.5(b) the same signal state is not used consistently to convey the same 3-bit pattern, since the phase *difference* with the previous time period is what counts, not the *absolute* signal phase. Hence the eighth and ninth time periods in Figure 9.5(b) both represent the pattern '101', but a different line is used, 180° phase shifted.

Finally, Figure 9.5(c) shows the constellation of this particular modem.

To conclude the subject of modem constellations, Figure 9.6 presents, without discussion, the constellation patterns of a couple of very sophisticated modems, specified by CCITT Recommendations V22 bis and V32. As in Figure 9.5, the constellation pattern would allow the interested reader to work out the respective 16 and 32 line signal states. We return to the subject of modem constellation diagrams in Chapter 18—when we learn of their use in detecting line errors in analogue data lines. Finally, Table 9.1 lists some of the common modem types and their uses. When reading the table, bear in mind that 'synchronous' and 'asynchronous operation' is to be discussed later in the chapter, and that 'half duplex' means that two-way transmission is possible

Bit combination	Line signal	
	Amplitude	Phase shift
000	low	0°
001	high	0°
010	low	90°
011	high	90°
100	low	180°
101	high	180°
110	low	270°
111	high	270°

(a) Bit combinations and
line signal attributes

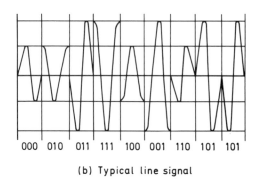

000 010 011 111 100 001 110 101 101

(b) Typical line signal

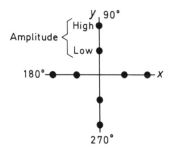

(c) Modem constellation

Figure 9.5
Quadrature amplitude modulation.

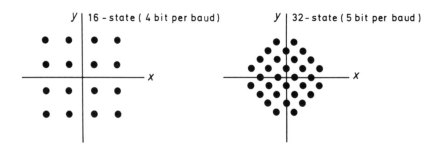

(a) V22 bis (3 amplitudes,
12 phases)

(b) V32 (5 amplitudes,
28 phases)

Figure 9.6
More modem constellations.

Table 9.1
Common modem types.

Modem type (CCITT Recommendation)	Modulation type	Bit speed (bit per second)	Synchronous (S) or asynchronous (A) operation	Full or half duplex	Circuit type required (2w — 2 wire, 4w — 4 wire)
V21	FSK	Up to 300 bps	A	Full	2w telephone network
V22	PSK	1200 bps	S or A	Full	2w telephone network
V22bis	QAM	2400 bps	S or A	Full	2w telephone network
V23	FSK	600/1200 bps	A	Half	4w private circuit
					2w telephone network
V26	PSK	2400 bps	S	Full	4w private circuit
V26bis	PSK	2400/1200 bps	S	Half	2w telephone network
V27	PSK	4800 bps	S	Half	2w private circuit
V27ter	PSK	4800 bps	S	Half	2w telephone network
V29	QAM	9600 bps	S	Full	4w private circuit
V32	QAM	Up to 9600 bps	S or A	Full	2w telephone network
Various 'proprietary' devices	QAM	Up to 19 200 bps	S	Full	4w private circuit
V33	QAM	14 400	S or A	Full	Analogue leaseline
V35	AM	48 000	Wideband	Full	Groupband leaseline
V36	AM	48 000	Wideband	Full	Groupband leaseline
V37	AM	72 000	Wideband	Full	Groupband leaseline
V42	—	—	Convert S or A to synchronous form	—	Error correcting protocol
V42bis	—	Up to 30 kbit/s In association with V32 modem	—	—	Data compression technique

but in only one direction at a time. This differs from simplex operation (as discussed in Chapter 1), where one-way transmission only is possible.

We cannot leave the subject of modems without at least a brief word about the Hayes command set (nowadays also called the AT command set), for many readers will at some time find themselves faced with the question of whether a modem is 'Hayes compatible'. By this, we ask whether it uses the Hayes command set—the procedure by which the link is set up and the data transfer is controlled—if you like, the handshake and etiquette of conversation. The Hayes command set has become the *de facto* standard for personal computer communication over telephone lines.

9.5 COMPUTER-TO-NETWORK INTERFACES

Returning now to Figure 9.1, we can see that we have discussed in some detail the method of data conveyance over the telecommunications line between one DCE and the other (or between one modem and the other) but what about the interface that connects a computer to a DCE or modem? This interface is of a generic type ('between DTE and DCE') and can conform to any one of a number of standards, as specified by various organizations. The interface controls the flow of data between DTE and DCE, making sure that the DCE has sufficient instructions to deliver the data correctly. It also allows the DCE to prepare the distant DTE and confirm receipt of data if necessary. Physically, the interface usually takes the form of a multiple pin connection (plug and socket) typically with 25 or 37 pins arranged in a 'D-formation'. Computer users will be familiar with the type of socket shown in Figure 9.7.

The interface itself is designed either for 'parallel data transmission' or 'serial data transmission', the latter being the more common. The two different methods of transmission differ in the way each 8-bit data pattern is conveyed. Internally, computers operate using the parallel transmission method, employing eight 'parallel' circuits to carry one bit of information each. Thus during one time interval all eight bits of the data pattern are conveyed. The advantage of this method is the increased computer processing speed which is made possible. The disadvantage is that it requires eight circuits instead of one. Parallel data transmission is illustrated in Figure 9.8(a), where the pattern '10101110' is being conveyed over the eight parallel wires of a computer's 'data bus'. (*Note*: nowadays 16-bit, and even 32-bit 'data buses' are used in the most

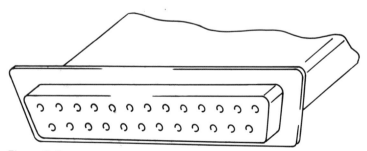

Figure 9.7
Typical 25-pin plug, used for DCE/DTE connection.

advanced computers.) Examples of parallel interfaces are those specified by CCITT Recommendations V19 and V20.

Serial transmission requires only one transmission circuit, and so is far more cost effective for data transmission on long links between computers. The parallel data on the computer's bar is converted into a serial format simply by 'reading' each line of the bus in turn. The same pattern '10101110' is being transmitted in a serial manner in Figure 9.8(b). Note that the baud rate needed for serial transmission is much higher than for the equivalent parallel transmission interface.

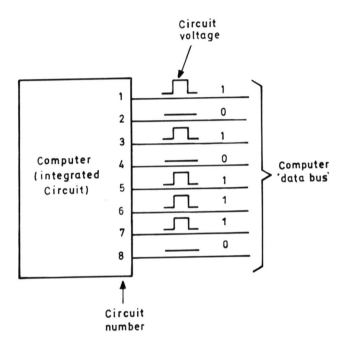

(a) Parallel transmission

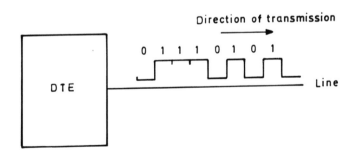

(b) Serial transmission

Figure 9.8
'Parallel' and 'serial' transmission.

Many DTE-to-DCE (or even short direct DTE-to-DTE) connections use a serial transmission interface conforming to CCITT's pair of recommendations, V24 and V28. V24 defines the use to be made of each pin (or channel) of the connector and V28 describes the electrical characteristics. Another couple of interfaces which perform a similar function and which are in common use are the American standards (specified by the Electronics Industries Association (EIA)) called RS-232C and RS-449. RS232C is a 25-pin interface, and the one on which V24/V28 is based. RS449 is an improvement over RS-232C, but requires 37 pins. Both interfaces perform a similar set of basic functions as follows:

- Clocking and synchronizing the data transfer;

- Regulating the bit rate, so that the receiving device does not become swamped with data;

- Providing a common electrical earth (or 'ground') between the devices.

9.6 SYNCHRONIZATION

An important component of any data communications system is the clocking device. The data consists of a square, 'tooth-like' signal, continuously changing between state '0' and state '1', as we recall in Figure 9.9.

The successful transmission of data depends not only on the accurate coding of the transmitted signal (Chapter 4), but also on the ability of the receiving device to decode the signal correctly. This calls for accurate knowledge (or 'synchronization') of where each bit and each message begins and ends.

The receiver usually samples the communication line at a rate much faster than that of the incoming data, thus ensuring a rapid response to any change in line signal state, as Figure 9.10 shows. Theoretically it is only necessary to sample the incoming data at a rate equal to the nominal bit rate, but this runs a risk of data corruption. If we chose to sample near the beginning or end of each bit we might lose or duplicate data

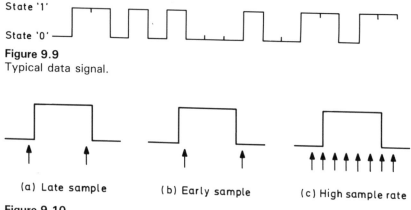

Figure 9.9
Typical data signal.

(a) Late sample   (b) Early sample   (c) High sample rate

Figure 9.10
Effect of sample rate.

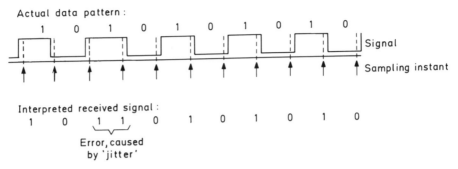

Figure 9.11
The effect of jitter.

as Figures 9.10(a) and 9.10(b) show—simply as the result of a slight fluctuation in the time duration of individual bits. Much faster sampling ensures rapid detection of the start of each '0' to '1' or '1' to '0' transition, as Figure 9.10(c) shows.

Variations in the time duration of individual bits come about because signals are liable to encounter time shifts during transmission, which may or may not be the same for all bits within the total message. These variations are usually random and they combine to create an effect known as 'jitter', which can lead to incorrect decoding of incoming signals when the receiver sampling rate is too low, as Figure 9.11 shows.

Jitter, together with a slight difference in the timing of the incoming signal and the receiver's low sampling rate have created an error in the received signal, inserting an extra '1'. Thus the signal at the top is intended to represent only a ten-bit pattern, but because the bit durations are not exactly correct (jitter), the pattern has been misinterpreted. An extra bit has been inserted by the receiving device.

To prevent an accumulation of errors over a period of time, we use a sampling rate higher than the nominal data transfer rate as already discussed; we also have to carry out a periodic 'synchronization' of the transmitting and receiving end equipments.

The purpose of synchronization is to remove all short-, medium- and long-term time effects. In the very short term, synchronization between transmitter and receiver takes place at a bit level, by 'bit-synchronization', which keeps the transmitting and receiving clocks in step, so that bits start and stop at the expected moments. In the medium term it is also necessary to ensure 'character' (or 'word') synchronization, which prevents any confusion between the last few bits of one character and the first few bits of the next. If we interpret the bits incorrectly, we end up with the wrong characters. Finally there is 'frame' synchronization, which ensures data reliability (or 'integrity') over even longer time periods.

Bit synchronization

Figure 9.12 shows two different pulse transmission schemes used for bit synchronization. The first, Figure 9.12(a), is a 'non-return to zero (NRZ)' code which looks like a string of '1' and '0' pulses, sent in the manner with which we are already familiar. In contrast Figure 9.12(b) is a 'return to zero' (RZ) code, in which '1's are represented

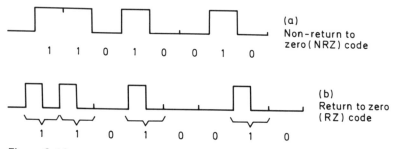

Figure 9.12
Methods of bit synchronization.

by a short pulse which returns to zero at the midpoint. RZ code therefore provides extra 0 to 1 and 1 to 0 transitions, most noticeably within a string of consecutive '1's. By so doing, it maintains better synchronization of clock speeds (or bit synchronization) between the transmitter and the receiver. A number of alternative techniques also exist. The technique used in a given instance will depend upon the design of the equipment and the accuracy of bit synchronization required. More often than not it is the interface standard that dictates which type will be used. The synchronization code used widely on IBM SDLC networks is called NRZI (non-return to zero inverted).

Character synchronization—synchronous and asynchronous data transfer

Data conveyance over a transmission link can be either 'synchronous' (in which individual data characters (or at least a predetermined number of bits) are transmitted at a regular periodic rate) or 'asynchronous' mode (in which the spacing between the characters or parts of a message need not be regular). In asynchronous data transfer each data character (represented say by an 8-bit 'byte') is preceded by a few additional bits, which are sent to mark (or 'delineate') the start of the 8-bit string to the receiving end. This assures that 'character' synchronization of the transmitting and receiving devices is maintained.

Between characters on an asynchronous transmission system the line is left in a 'quiescent state', and the system is programmed to send a series of '1's during this period in order to 'exercise' the line and in order not to generate spurious start bits (value '0'). When a character (consisting of eight bits) is ready to be sent, the transmitter precedes the 8-bit pattern with an extra 'start bit' (value '0'), then it sends the eight bits, and finally it suffixes the pattern with two 'stop bits', both set to '1'. The total pattern appears as in Figure 9.13, where the user's 8-bit pattern 11010010 is being sent. (*Note:* Usually nowadays, only one stop bit is used. This reduces the overall number of bits which need to be sent to line to convey the same information by 9 per cent.)

In asynchronous transmission, the line is not usually in constant use. Typically asynchronous transmission is used between a keyboard and the computer. The idle period between character patterns (the quiescent period) is filled by a string of 1's. The receiver can recognize the start of each character when sent by the presence of the start

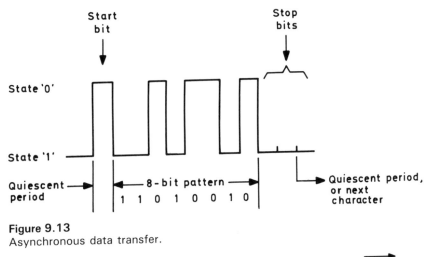

Figure 9.13
Asynchronous data transfer.

Figure 9.14
Synchronous data transfer.

bit transition (from state '1' to state '0'). The following eight bits then represent the character pattern.

The advantage of asynchronous transmission lies in its simplicity. The start and stop bits sent between characters help to maintain synchronization without requiring very accurate clocking, so that devices can be quite simple and cheap. Asynchronous transmission is quite widely used between computer terminals and the computers themselves because of the simplicity of terminal design and its consequent cheapness. Given that human operators type at indeterminate speeds and sometimes leave long pauses between characters, it is ideally suited to this use. The disadvantage of asynchronous transmission lies in its relatively inefficient use of the available bit speed. As we can see from Figure 9.13, out of eleven bits sent along the line, only eight represent useful information. In synchronous data transfer, the data is 'clocked' at a steady rate. A highly accurate clock is used at both ends, and a separate circuit may be used to transmit the timing between the two. Provided all the data bit patterns are of an equal length, the start of each is known to follow immediately the previous character. The advantage of synchronous transmission is that much greater line efficiency is achieved, but its more complex arrangements do increase the cost as compared with asynchronous transmission equipment. Byte synchronization is established at the very beginning of the transmission using a special synchronization (SYN) pattern, and only minor adjustments are needed thereafter. Usually an entire user's data field is sent between the synchronization (SYN) patterns, as Figure 9.14 shows. The SYN byte shown in Figure 9.14 is a particular bit pattern, used to distinguish it from other user data.

9.7 PROTOCOLS FOR TRANSFER OF DATA

Unfortunately the RS232C (V24/V28) and RS449 standards and their inter-DCE equivalents (i.e. the line transmission standards used by modems or digital links) are still not enough in themselves to ensure the controlled flow of data across a network. A number of additional layered mechanisms are necessary, to indicate the coding of the data, to control the message flow between the end devices and to provide for 'error correction' of incomprehensible information. Unlike human beings engaged in conversation, machines can make no sense at all of corrupted messages, and they need hard and fast rules of procedure to be able to cope with any eventuality. When they are given a clear procedure they are able—unlike most human beings—to carry on several 'conversations' at once. These rules of procedure are laid out in 'protocols'.

Originally protocols were relatively unsophisticated like the simple computer-to-terminal networks which they supported and they were contained within other computer 'application' programs. Thus the computer, in addition to its main processing function, would be controlling the line transmission between itself and its associated terminals and other peripheral equipment. However, as organizations grew in size, and data networks became more sophisticated and far-flung, the supporting communications software and hardware developed to such an extent as to be unwieldy and unmaintainable. Many computer items (particularly different manufacturers' equipments) were incompatible. Against this background the concept of layered protocols developed with the objective of separating out the overall telecommunications functions into a layered set of sub-functions, each 'layer' performing a distinct and self-contained task but being dependent on sub-layers. Thus complex tasks would comprise several layers, while simple ones would only need a few. Each layer's simple function would comprise simple hardware and software realization and be independent of other layer functions. In this way we could change either the functions or the realization of one functional layer with only minimal impact on the software and hardware implementations of other layers. For example, a change in the routing of a message (i.e. the topology of a network) could be carried out without affecting the functions used for correcting any corrupted data (or errors) introduced on the line between the end terminals.

Most data transfer protocols in common use today use a 'stack' of layered protocols. By studying such a protocol 'stack' we get a good idea of the whole range of functions that are needed for successful data transfer. We have to consider the functions of each protocol layer as laid out in the International Organization for Standardization's (ISO's) 'open systems interconnection (OSI) model'. The OSI model is not a set of protocols in itself, but it does carefully define the division of functional layers to which all modern protocols are expected to conform.

9.8 THE OPEN SYSTEMS INTERCONNECTION MODEL

The open systems interconnection model, first standardized by ISO in 1983, classifies data transfer protocols in a series of layers. It is meant to set worldwide standards of design for all data telecommunication protocols, so that all equipments produced will be compatible.

To understand the need for the model, let us start with an analogy, drawn from a simple exchange of ideas in the form of a dialogue between two people. The speaker has to convert his ideas into words; a translation may then be necessary into a foreign language which can be understood by the listener; the words are then converted into sound by nerve signals and appropriate muscular responses in the mouth and throat. The listener meanwhile is busy converting the sound back into the original idea. While this is going on, the speaker needs to make sure in one way or another that the listener has received the information, and has understood it. If there is a breakdown in any of these activities, there can be no certainty that the original idea has been correctly conveyed between the two parties.

Note that each function in our example is independent of every other function. It is not necessary to repeat the language translation if the receiver did not hear the message—a request (prompt) to replay a tape of the correctly translated message would be sufficient. The specialist translator could be getting on with the next job as long as the less-skilled tape operator was on hand. We thus have a layered series of functions. The idea starts at the top of the talker's 'stack' of functions, and is converted by each function in the stack, until at the bottom it turns up in a soundwave form. A reverse conversion 'stack', used by the listener, re-converts the soundwaves back into the idea. Figure 9.15 shows our example.

Each function in the protocol 'stack' of the speaker has an exactly corresponding, or so-called 'peer' function in the protocol stack of the listener. The functions at the same layer in the two stacks correspond to such an extent, that if we could conduct a direct 'peer-to-peer' interaction then we would actually be unaware of how the functions of the lower layers' protocols had been undertaken. Let us, for example, replace layers 1 and 2, by using a telex machine instead. The speaker still needs to think up

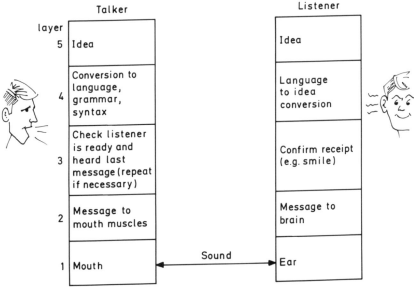

Figure 9.15
A layered protocol model for simple conversation.

the idea, correct the grammar and see to the language translation, but now instead of being aimed at mouth muscles and soundwaves, finger muscles and telex equipment do the rest, provided the listener also has a telex machine. We cannot, however, simply replace only the speaker's layer-1 function (the mouth), if we do not carry out simultaneous 'peer protocol' changes on the listener's side because an ear cannot pick up a telex message. The principle of layered protocols is that as long as the layers interact in a 'peer-to-peer' manner, and as long as the interface between the function of one layer and its immediate higher and lower layers is unaffected, then it is unimportant how the function of that individual layer is carried out. This is the principle of the open systems interconnection (OSI) model. The OSI model sub-divides the function of data communication into a number of layered and peer-to-peer sub-functions, as shown in Figure 9.16. In all, seven layers are defined.

Each layer of the OSI model relies upon the 'service' of the layer beneath it. Thus the transport layer (layer 4) relies upon the 'network service' which is provided by the

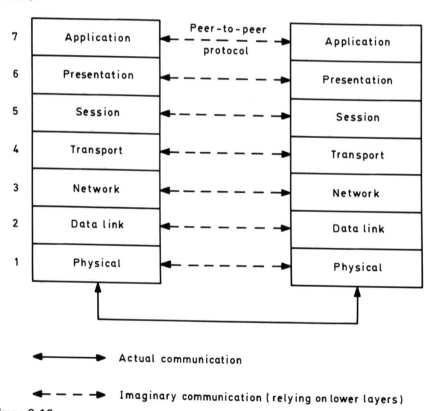

Figure 9.16
The Open Systems Interconnection (OSI) Model. (*Courtesy of CCITT—derived from Figures 12 and 13/X200.*)

'stack' of layers 1–3 beneath it. Similarly the transport layer provides a 'transport service' to the session layer, and so on.

The functions of the individual layers of the OSI model are defined more fully in ISO standards (ISO 7498), and in CCITT's X-200 series of recommendations; in a nutshell they are as follows:

Application layer (Layer 7).   This is the layer that provides communication services to suit all types of data transfer between 'co-operating' computers. It comprises a number of service elements (SEs), each suited for a particular purpose. They may be combined together in various permutations to meet the needs of more complex applications. This does not amount to a full definition of the application layer, but we shall discuss the message handling system (MHS), an example of an application layer protocol, in Chapter 25.

Presentation layer (Layer 6).   As we found in Chapter 4, data may be coded in various forms such as Binary, ASCII, CCITT IA5, EBCDIC, fax encoding etc. The task of the presentation layer is to negotiate a mutually agreeable technique for data encoding and punctuation (called the 'data syntax') and to provide for any conversion that may be necessary between different code or data layout formats, so that the application layer receives the type it recognizes.

Session layer (Layer 5).   When established for a 'session' of communication, the two devices at each end of the communication medium must conduct their 'conversation' in an orderly manner. They must listen when spoken to, repeat as necessary, and answer questions properly. The session protocol thus includes commands such as 'start', 'suspend', 'resume' and 'finish'. The session protocol is rather like a tennis umpire. He cannot always tell how hard the ball has been hit, or whether there is any spin on it, but he knows who has to hit the ball next and whose turn it is to serve, and he can advise on the rules when there is an error, in order that the game can continue. The session protocol negotiates for an appropriate type of session (e.g. full duplex or half duplex data block lengths) to meet the communication need, and then it manages the session.

Transport layer (Layer 4).   The transport service provides for the *end-to-end* data relaying service needed for a communication session. The transport layer itself establishes a transport connection between the two end-user devices by making the network connection that best matches the session requirements in terms of quality of service, data unit size, flow control, and error correction needs. If more than one network is available (e.g. packet network, circuit-switched data, telephone network etc.), it chooses between them. The transport layer must also provide the network addresses needed by the network layer for correct delivery of the message. The 'mapping' function provided by the transport layer, in converting 'transport addresses' (provided by the session layer to identify the destination) into network recognizable addresses (e.g. telephone numbers) shows how independent the separate layers can be: the conveyance medium could be changed and the session, presentation and application protocols could be quite unaware of it. The transport protocol is also capable of some important multiplexing and splitting functions. In its multiplexing mode the transport protocol is capable of supporting a number of different sessions over the same connection, rather like playing two games of tennis on the same court. Humans would get confused about which ball to play, but the 'transport protocol' makes sure that computers do nothing of the kind. Conversely, the 'splitting' capability of the transport protocol allows (in theory) one 'session' to be conducted over a number of parallel network communication paths, like a grand master playing many simultaneous chess games. The transport protocol also caters for the end-to-end conveyance, segmenting or concatenating (stringing together) of the data as the network requires.

Network layer (Layer 3). This layer sets up an end-to-end connection across a real network, determining which permutation of individual links to be used. In other words, the network layer performs overall routing functions. A layer 3 protocol commonly used for data communication is the X25 packet interface standard, discussed later in this chapter. The 1984 and later versions of X25 conform with the OSI model, providing a packet-mode version of the 'OSI network service'.

Datalink layer (Layer 2). The datalink layer operates only within the individual links of a connection, managing the transmission of the data so that the individual bits are conveyed over those links without error. ISO's standard datalink protocol, specified in ISO 3309, is called 'high level data link control' (HDLC). Its functions are to:

● synchronize the transmitter and receiver,

● control the flow of data bits,

● detect and correct errors of data caused in transmission.

Typical commands used in datalink control protocols are thus ACK (acknowledge), EOT (end of transmission), etc.

Physical layer (Layer 1). The physical layer protocol conveys the data over the medium; it is a combination of the hardware and software which converts the data

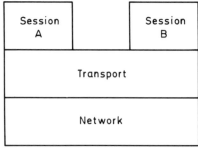

(a) Multiplexing   (b) Splitting

Figure 9.17
Protocol multiplexing and splitting.

bits needed by the datalink layer into the electrical pulses, modem tones, optical signals or other means which are going to transmit the data. The physical layer ensures that the data is conveyed over the link and presented at both ends to the 'datalink layer' in a standard form. Earlier in this chapter we looked at V24/V28 and RS 232C interfaces which form part of CCITT's X21-bis recommendation, the layer 1 standard interface. An alternative layer 1 interface, using only a 15-pin connector is defined by CCITT Recommendation X21.

9.9 DATA MESSAGE FORMAT

The key to any layered protocol lies in its use of headers. Each protocol layer adds a header containing information for its own use. The overall message is thus longer than that received from the higher layer protocol. The headers carry the information the protocol needs to do its work. They are stripped from the received message by the corresponding protocol handler in the distant end device, and the remainder of the message is passed up to the next higher protocol layer. Messages are passed normally in a synchronous manner, and Figure 9.18 illustrates a typical data message format showing some of the fields corresponding to the network service functions of layers 1–3. The exact format depends upon the particular protocols in use.

In order of transmission:

The flag indicates the start of each message. This part of the message includes the synchronization byte described earlier in the chapter, and is used by the layer 2 protocol.

The control bits control the flow of data, passing 'receipt acknowledgement' and other information for data link control (layer 2).

The header contains the network address of the destination. It may include the identity of the origin. This information is provided by the transport layer and is used by the 'network layer' to ensure correct delivery of the message through the network.

The sequence number is a serial number uniquely identifying the message. It enables the individual messages to be collated in the correct order at the receiving station. Further it provides a means of identifying and acknowledging messages which have been received, and can be used for requesting the re-transmission of severely errored or lost messages. The sequence number is used by the datalink layer.

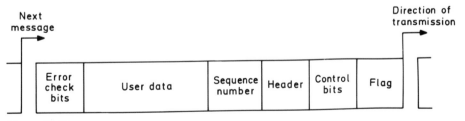

Figure 9.18
Typical data message format.

The user data is the information itself. This field can be further divided to provide higher layer OSI functions (layers 4, 5, 6, 7) if required.

The error check bits are used to verify a successful and error-free transmission of the whole message. As an example, one of the error check bits is usually a 'parity' bit. The parity bit is set so that the total number of bits of binary value '1' within the message as a whole (including the parity bit) is either 'even' or 'odd' (according to whether it is an 'even parity bit' or an 'odd parity bit' respectively). If on receipt, the total number of bits set at binary value '1' does not correspond with the indication of the parity bit, then there is an 'error' in the message. Other error check bits may enable us to correct the error, otherwise we can always request the re-transmission of the message, identifying it by the appropriate 'sequence number'. Other more powerful means of error detection and correction are provided by 'Hamming codes' and 'cyclic redundancy checks (CRCs)' but we do not cover these methods in detail here.

9.10 IMPLEMENTATION OF LAYERED PROTOCOL NETWORKS

In reality, most of the protocol 'layers' of OSI exist only in software (as part of a computer program) and cannot be identified as physical items. None the less the different protocol layers need not all be implemented within the same computer program or indeed be carried out by the same piece of hardware.

We have already seen in the first part of the chapter how it is sometimes convenient that data terminating equipment (DTE) does not have to conduct line communication functions. Instead these activities are usually delegated to a piece of specialized equipment called a data circuit terminating equipment (DCE). In essence this equipment converts between two different physical layer protocols; on the DTE side RS232C or an equivalent interface is used, while on the line side, digital line codes, clocking

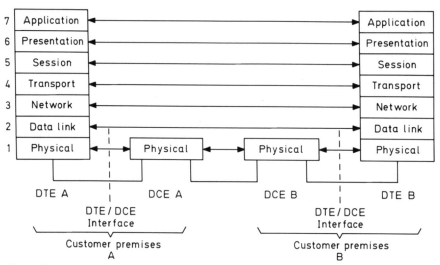

Figure 9.19
OSI model applied to DTE/DCE.

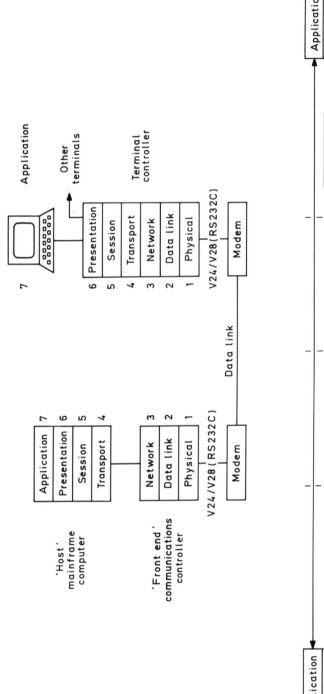

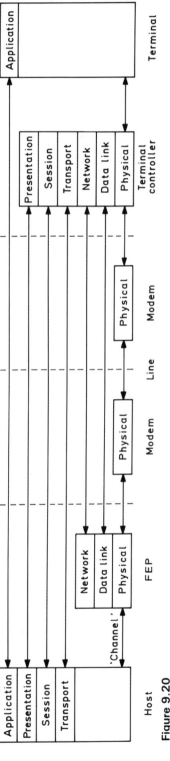

Figure 9.20
OSI model applied to host/terminal connection.

Figure 9.21
Mainframe or host computer. A picture of a big machine known as the AT&T 3B 20A, a high performance and flexible computer. (*Courtesy of AT&T*).

functions etc. have to be carried out. Figure 9.19 now illustrates the application of the OSI model to this example. Notice in the figure how the various protocols are all communicating peer-to-peer. At layer 1, the DCE is converting between different protocols on DTE/DCE and DCE/DCE interfaces, but at all the other layers the peer-to-peer interaction is end-to-end.

A common approach used by many computer manufacturers, for example, is to split the layer 1–3 functions away from the 'host' or 'application processor' and to implement them instead on a specialized 'front end' or 'communications controller'. In this way the host machine is relieved of the burden of communications link supervision, link synchronization, error correction and conversion of data from the internal computer bus format into the standard serial transmission line format. A typical configuration is shown in Figure 9.20. The 'host' or 'mainframe' computer of Figure 9.20 is that on which the main computer software program runs. This program communicates as necessary with other computer devices (e.g. terminals, printers, databases, other computers) for the input or output of information. The communication relies upon an application layer protocol, resident in the host, which is invoked every time a command such as 'PRINT', 'READ', 'INPUT' or 'OUTPUT' appears in the computer program. The application protocol co-operates in a peer-to-peer fashion with a corresponding function in a distant terminal. Layer 4–6 protocols also exist in both the host computer and the distant terminal controller, but layer 1–3 functions (the real spadework of running the telecommunications network) are carried out at the host's end by a collocated 'front end' communications controller. The two inter-communicate through the normal internal computer wiring (called the 'channel'), essentially providing the functions of OSI layers 1–3 for this part of the connection.

Since not all the protocols used between a host and its associated terminals are OSI standard, the various equipments are not fully interchangeable between one manufacturer's equipment and another. In our diagram, however, the protocols are shown fully OSI conformant at all protocol layers used on the line, between FEP (Front End Processor) and modem, and between modem and terminal controller. Thus the host and FEP together could be used with another manufacturer's OSI-conformant terminal controller/terminal combination. Likewise, the terminal controller/terminal combination shown could be used with other host/FEPs.

The actual protocols used in any given case are not defined by the OSI model, though a number of standardized sets of protocols are now appearing which conform with it. Examples include MAP/TOP (Manufacturing Automation Protocol/Technical and Office Protocol) which defines a set of protocols at all seven layers, which together are suitable for manufacturing and office interworking applications; other examples include protocols for the individual layers. CCITT X25 (1984 version or later) supports the OSI network service, ISO 8073 is a transport protocol, ISO 8327 is a session protocol, ISO 8823 is a presentation protocol and ISO 8571 FTAM (File Transfer, Access and Management) is an example of an application layer protocol—intended for the purpose of direct data transfer between mainframe computers.

Because a large number of different application protocols are likely to be conceived, addressing the needs of the many and varied applications possible, ISO has standardized a formal language for the application layer. It is called ASN.1 or 'Abstract Syntax Notation 1'. It defines protocol data structures and encoding rules for application layer protocols and is set out in standards ISO 8824 and ISO 8825.

9.11 THE USE OF NULL LAYERS

An important point to note about the OSI-layered model is that it gives immense scope, allowing very sophisticated networks to be developed. Not all of the very sophisticated functions are required in every case, and their implementation may increase the cost and burden of administration. The model is designed to permit null protocols to be used at some of the layers. For example on a network using similar end devices the syntax conversion capabilities of the presentation layer are unnecessary and can be dispensed with. Likewise on a point-to-point data connection layers 1 and 2 protocols may be sufficient without needing much effort at layers 3 and 4 (the 'network' and 'transport' layers).

9.12 OTHER LAYERED PROTOCOLS

The OSI model defines standard peer protocol functions that ultimately will stimulate the development of wide ranging protocols allowing any two computer devices running any types of applications to intercommunicate without difficulty. However, because some of today's layered protocol sets predate the OSI model, not all conform exactly to it. For example, the 'Systems Network Architecture (SNA)', announced by the IBM computer company as the future direction for its product range in 1973, defines a stack of protocols allowing computers to communicate. The principles and layers of SNA are similar to those of the OSI model, although some significant differences exist that make the two incompatible. The equivalent of HDLC (OSI layer 2) for example, is called SDLC (Synchronous Data Link Control); we shall discuss SNA later in the chapter. Another proprietary layered protocol is the DECNET of the Digital Equipment Corporation.

9.13 DATA NETWORK TYPES

Next we review the different types of real networks that are used for conveyance of data, starting with simple point-to-point technology, and moving on to more complex circuit-switched data networks, packet-switched networks, and local area networks (LANs).

Point-to-point data networks

Point-to-point networks involving the simple interconnection of two pieces of equipment are relatively simple to establish. As we have seen, they may use either digital lines and DCEs, or analogue lines and modems. The DCE itself (either digital version or modem) may take the form of a distinct 'box' of kit, or it may be provided as part of a 'line controller' built into the computer itself. Figure 9.22 shows a typical set-up.

Provided the protocols at both ends of the link match, the data terminal equipments (DTEs) converse easily. In the point-to-point case most of the protocol layers (e.g.

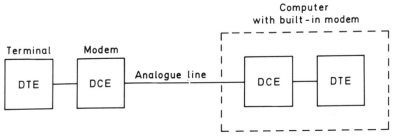

Figure 9.22
Point-to-point data network. DTE, data terminal equipment; DCE, data circuit terminating equipment (e.g. a modem for an analogue line).

layers 3–7) described in the OSI model are redundant, and relatively simple communications software is necessary in the two DTEs. At its simplest a point-to-point network can be run in an asynchronous, character-by-character mode. This is a common method of connecting remote terminals to a computer. The use of such a simple data transmission technique considerably reduces the complexity and cost of hardware and software required in the (remote) computer terminals. This type of connection does not conform to the OSI ideal, as is plain from the fact that few manufacturers' computer terminals of this type can be used with other manufacturers' computers. One of the drawbacks of the OSI dream is the burden of software and hardware required on every transmitting and receiving device.

Circuit-switched data networks

In our discussion of circuit-switched networks, in Chapter 6 we were chiefly interested in making a connection on demand between any two customers of a telephone network. Each connection would be established for the length of their call, and provide a clear transmission path for their conversation. At the end of the call the connection would be cleared. Subsequent connections could then be made to other destinations as required.

The benefit of circuit-switched networks lies in the great range of destinations available on different calls, and this means that circuit-switched networks are just as useful for data conveyance, an example being the telex network. Once a telex connection is established, characters are transmitted between the two telex terminals at a 50 baud line rate, using the IA2 alphabet (Chapter 4 refers).

Other examples of circuit-switched data networks include AT&T's 'Accunet Switched-56 kbit/s' network and the German Bundespost's 'Datex-L' network. Both of these are public circuit-switched digital data networks. ISDN (Integrated Services Digital Network), discussed in Chapter 24, is another example of a network offering circuit-switched data capability.

Figure 9.23 shows the usual configuration of a circuit-switched network. Notice how the network 'starts' at the DCE, which consequently is usually provided by the public telecommunications operator (PTO), though it may be located on customers' premises.

Connections in a circuit-switched data network are established in the same way as telephone connections. Once the connection is established, the data protocols in the two end terminals control the flow of data across what is then a point-to-point connection.

An alternative means of providing circuit-switched data communication to a range of destinations is by the use of an ordinary telephone network, as Figure 9.24 shows.

On the left-hand side of Figure 9.24 a modem sits between the DTE and the telephone line. On the right-hand side the DCE is a combination of an ordinary telephone and a device called an 'acoustic coupler'. The combination performs the same function as the modem. The acoustic coupler converts the data from the DTE into a series of soundwave tones, and the telephone converts these tones into their electrical analogue signal equivalents. The modem simply converts the original data directly into the electrical equivalent of the acoustic tones, and is slightly more efficient and reliable. The acoustic coupler may be more suited to some (mobile) customers who may use a portable terminal with dial-up computer access from a number of different locations.

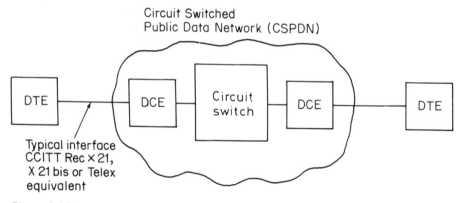

Figure 9.23
Circuit switched data network. DCE, Data circuit terminating equipment; DTE, Data terminal equipment.

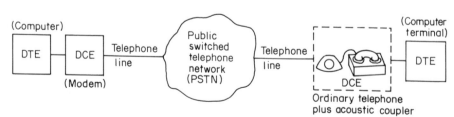

Figure 9.24
Conveying data over the telephone network.

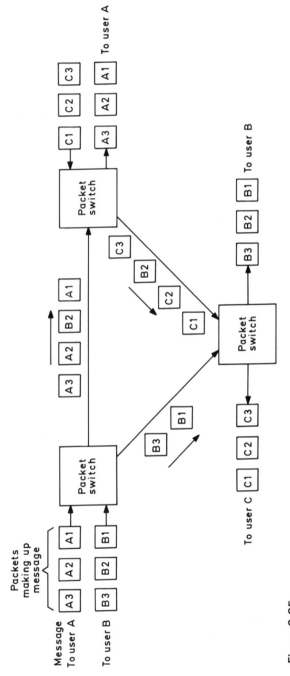

Figure 9.25
Packet switching.

Packet-switched data networks

A limitation of circuit-switched networks when used for data is their inability to provide variable bandwidth connections. This leads to inefficient use of resources when only a narrow bandwidth (or low bit-rate) is required. Conversely, when short bursts of high bandwidth are required, there may be data transmission delays. A more efficient means of data conveyance, packet switching, emerged in the 1970s.

Packet switching is so called because the user's overall message is broken up into a number of smaller 'packets', each of which is sent separately. We illustrated the concept in Figure 1.10 of Chapter 1. Each packet of data is labelled with the address of its intended destination, and a number of control fields are added (see Figure 9.18) before it is sent. The receiving end re-assembles the packets in their proper order, with the aid of the sequence numbers.

With packet switching, variable bandwidth becomes possible because the exchanges are allowed to operate in a 'store-and-forward' fashion, each packet being routed across the network in the most efficient way available at the time. Packet-switched exchanges (PSEs) are computers with a large data storage capacity and a wealth of communications software.

The principle of packet switching is illustrated in Figure 9.25. Notice how each of the individual packets, even those with the same overall message, may take different paths through the network. Thanks to this flexible routing, variable bandwidth can be accommodated, and transmission link utilization is optimized.

None of the links are dedicated to the carriage of any one message. On the contrary, the individual packets of a message may be jumbled up with packets from other messages. The effect is as if a permanent channel existed between the two ends, although in reality this is not the case. The result is known as a 'logical' or 'virtual channel'.

Packets are routed across the individual paths within the network according to the prevailing traffic conditions, the link error reliability, and the shortest path to the destination. The routes chosen are controlled by the (layer 3) software of the packet switch, together with 'routing information' preset by the network operator.

Packet switching gives good end-to-end reliability—with well designed switches and networks it is possible to bypass network failures (even during the progress of a 'call'). Packet switching is also efficient in its use of network links and resources—sharing them between a number of calls thereby increasing their utilization.

CCITT Recommendations X25 and X75

Many packet-switched networks nowadays have adopted the protocol standards set by CCITT's Recommendation X25, which sets out how a data terminal equipment (DTE) should interact with a data circuit terminating equipment (DCE), forming the interface to a packet-switched network. The relationship is shown in Figure 9.26.

The X25 recommendation defines the protocols between DTE and DCE corresponding to OSI layers 1, 2, and 3. At layer 1, X21 may be used in conjunction with a digital access line, or X21-bis with an analogue line. At layer 2, a CCITT variant of ISO's HDLC, called the Link Access Protocol (LAP) is defined. The layer 3 (network

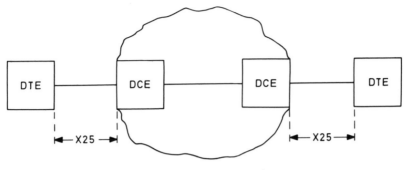

Packet switched network

Figure 9.26
CCITT Recommendation X25.

layer) protocol is also defined. The recommendation does not stipulate how the packets of data should be conveyed through the network between DCEs, giving some scope for network and equipment designers to design and use their own proprietary practices. In practice either CCITT's recommendation X75 (a super set of the X25 protocol originally developed for international interconnection of packet networks) or a very similar proprietary protocol is usually employed as shown in Figure 9.27. X75 protocol is the standard protocol used for interconnecting different packet-switched networks—for example, those in different countries. But where all the packet-switched exchanges in a network are of a single manufacturer's type, then a proprietary equivalent of X75 protocol is often used. Examples include those used by 'Telenet', 'BBS', 'Tymnet' and France's 'Transpac'.

Like X25, recommendation X75 defines all the protocols for layers 1–3. At layer 1, a 64 kbit/s bearer conforming to Recommendation G703 is stipulated. At layer 2, the 'Single Link procedure (SLP)' and the 'Multilink Procedure (MLP)' are defined. SLP is based on LAP (the 'Link Access Protocol' used in X25) and should be used where packet-switched exchanges (PSEs) are interconnected by a single datalink. Where more than one parallel link is available between exchanges the more advanced MLP must be used. The protocol of X75 is similar to X25 at layer 3, with a few extra messages added to cater for inter-exchange signalling requirements. Communication using X75 occurs between logical functions called Signalling TErminals (STEs) which are software programs within the packet-switched exchanges (PSEs).

Packet assembler/disassembler (PAD)

One more practical consideration before we leave the subject of packet-switched networks: how to interface a slow and relatively unintelligent asynchronous computer terminal to an X25 packet network? Given the very large numbers of asynchronous equipments in use, it seems a pity for them to miss out on the economies of packet switching!

Necessary interworking is achieved by means of an 'intelligent' (computerized) piece of equipment provided in place of the DCE at the packet-switched exchange. This

186

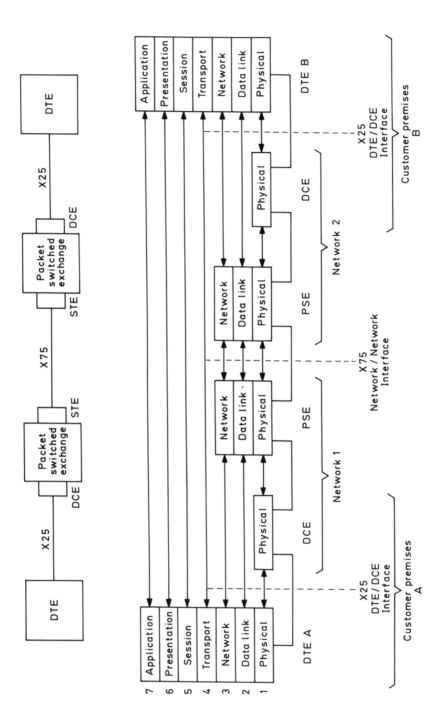

Figure 9.27
CCITT Recommendation X75 and its interrelationship with X25. DTE, Data terminal equipment; DCE, Data circuit terminating equipment; PSE, Packet-switched exchange; STE, Signalling terminal.

equipment is called a Packet Assembler/Disassembler, or PAD for short. As its name suggests it assembles packets from the asynchronous terminal data for onward transmission, and it disassembles packets received from the far end DTE, conveying them on to the asynchronous terminal as individual characters. Other types of PAD cater for the conversion of other types of data (e.g. IBM's proprietary SNA format data) into X25 packet format. IBM SNA format is discussed later in this chapter.

Three CCITT recommendations define the operation of PADS:

(i)   X3 defines the functions of the PAD;

(ii) X28 defines the interface between PAD and asynchronous terminal;

(iii) X29 defines the interface between PAD and remote X25 DTE (usually some form of 'host' computer).

We shall not discuss these recommendations in detail, but Figure 9.28 illustrates their interrelationship.

Typically, the X25 DTE of Figure 9.28 would be the front-end processor (FEP) of a mainframe computer (e.g. an IBM 3725 running NPSI (Network Packet Switching Interface)). In this case, the X25 DTE is connected directly to the packet network. If the packet network is a private one, then a packet network switch may be sited alongside the FEP in the computer centre. Otherwise, a permanent connection to the public packet network may be provided using a DCE on site, connected using a direct line, typically operating at 4.8 kbit/s, 9.6 kbit/s or 48 kbit/s.

The asynchronous terminal might be a normal computer keyboard and screen, but equally could be some other type of terminal—for example, a credit card validating telephone for electronic funds transfer in retail stores (see Chapter 30). The PAD may be equipment located on the asynchronous terminal user's site, as might be the case for a number of mainframe users in a remote office. Alternatively, for single remote users it is often more economic to make use of PADs made available on public network operators' premises and accessed via a local telephone connection. In this case the X28 connection of Figure 9.28 actually comprises a local telephone line and modems. The

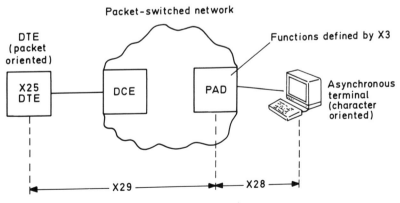

Figure 9.28
The Packet Assembler/Disassembler, or PAD.

latter arrangement avoids the cost of purchase of a PAD but instead incurs local telephone charges. It is economic where only occasional use is necessary and where connections to the computer centre are always established from the remote site.

A major limitation of the dial-up PAD configuration is that few public packet networks are capable of outdialling to the terminal. This means that the mainframe computer is unable to initiate the establishment of a connection to the remote site. Thus, for example, if the remote terminal was a personal computer in a retail store recording 'Electronic Point of Sale (EPOS)' information about sales, then this information could only be collated at a central mainframe site using a public packet network if the PC initiated the connection. The central mainframe could not poll the stores via a packet network at its own initiation. To make such polling possible, thereby smoothing the mainframe's receipt of data, the mainframe-to-PC connection might have to employ a dial-up telephone connection instead.

Local area networks

We have seen how packet switching has contributed greatly to the efficiency and flexibility of 'wide area' data networks, involving a large number of devices spread at geographically diverse locations. Packet switching according to the X25 and related recommendations, however, is not so efficient for smaller scale networks, those limited say to linking almost identical personal computers within a single office; that is the realm of an alternative type of packet-switched-like network called a 'local area network' or 'LAN' for short.

LANs have emerged in the last few years as the most important means of conveying data between different machines within a single office, office building, or small campus. They are usually constrained by their mode of operation to a geographically limited area, but they are ideally suited for short-distance data transport. Some of the more sophisticated LANs are also able to cope with the stringent demands of voice and video signalling. (We discuss the difficulties of carrying voice and video with data in Chapter 29 on 'Broadband Networks'.)

A high bit speed LAN can carry high volumes of data with rapid response times. Such performance is crucial for most office applications, and has made them the ideal foundation for the new generation of 'electronic offices' comprising electronic workstations, word processors, shared printers, electronic filing cabinets, electronic mail systems and so on.

Most LANs conform to one of the different types specified in the Institution of Electrical and Electronic Engineers' IEEE 802 series of standards. All the types have been developed from proprietary LANs, developed earlier by individual companies or organizations, but have now achieved American and worldwide recognition, as 'ISO 8802' standards.

LAN topologies and standards

The different types of LAN are characterized by their distinctive topologies. They all comprise a single transmission path interconnecting all the data terminal devices, with

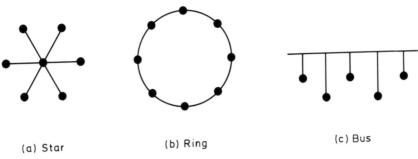

(a) Star   (b) Ring   (c) Bus

Figure 9.29
Alternative LAN topologies.

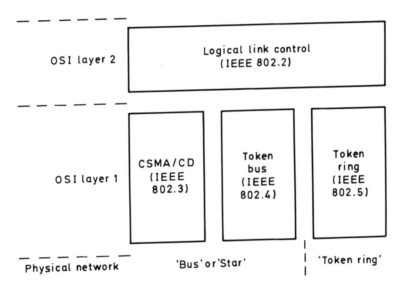

Figure 9.30
The IEEE 802 LAN standards.

a bit speed typically between 1 and 30 Mbit/s, together with appropriate 'protocols' to enable data transfer. The three most common topologies are illustrated in Figure 9.29, and are called the 'star', 'ring' and 'bus' topologies.

Slightly different protocol standards apply to the different topologies. For example, IEEE 802.3 defines a physical layer protocol called CSMA/CD (Carrier Sense Multiple Access with Collision Detection) which may be used with a bus or star topology. In the 'bus' form it is often called 'Ethernet'. IEEE 802.4 (ISO 8802.4) defines an alternative layer 1 protocol for a 'token bus', again suitable for either a bus or star topology. IEEE 802.5 defines a layer 1 protocol suitable for use on a 'token ring' topology. Finally, IEEE 802.2 (ISO 8802.2) defines a 'logical link control' protocol (equivalent to the OSI layer 2) that can be used with any of the above. Figure 9.30 shows the relationship of the various standards.

Which physical layer protocol and what topology of LAN to use depends largely upon individual preference and compatibility of existing computer kit needing to be connected to the LAN. To a lesser degree, the geographic circumstances and the network's performance requirements are also factors. All the possible protocols transfer data between the nodes, using a packet mode of transmission; they differ in how they prevent more than one terminal using the bus or ring at the same time. The various protocols and their relative merits are now considered in turn.

CSMA/CD (IEEE 802.3 ISO 8802.3) — 'Ethernet'

CSMA/CD stands for 'carrier sense multiple access with collision detection'. It is a 'contention' protocol. On a CSMA/CD LAN the terminals do not request permission from a central controller before transmitting data on the transmission channel; they 'contend' for its use. Before transmitting a packet of data, a sending terminal 'listens' to check whether the path is already in use, and if so it waits before transmitting its data. Even when it starts to send data it has to continue checking the path to make sure that no other stations have started sending data at the same time. If the sending terminal's output does not match that which it is simultaneously monitoring on the transmission path, it knows there has been a 'collision'. To receive data each terminal monitors the transmission path, decoding the destination address of each packet passing through to find out whether it is the intended destination. If it is, the data is read and decoded, if not the data is ignored.

A popular type of network very similar to the 802.3 standard, and also employing CSMA/CD is called 'Ethernet'. Ethernet is a proprietary LAN standard (predating the IEEE 802.3 standard) developed by the Xerox corporation of USA. Historically, Ethernet LANs have nearly all been installed using a length of coaxial cable, with 'tee-offs' to individual work stations, typically up to a maximum of around 500 stations. More recently it has become possible to use simple 'twisted pair' telephone cabling, in a star configuration.

LANs using CSMA/CD protocols are usually resilient to transmission line failures. The fact that any station may use the transmission path, as long as it was previously idle, means that fairly good use can be made of the LAN even when some destinations are unavailable because of a transmission path break, a capability which is not enjoyed by LANs employing more sophisticated data transmission as we shall see later. LANs using CSMA/CD do have the significant disadvantage that they perform poorly under overload; the data throughput of the network as a whole drops off rapidly as a result of widespread and frequent 'collisions'.

Token bus (IEEE 802.4, ISO 8802.4)

A 'token bus' LAN controls the transmission of data on to the transmission path by the use of a single 'token'. Only the terminal with the 'token' may transmit packets onto the bus. The token can be made available to any terminal wishing to transmit data. When a terminal has the token it sends any data frames it has ready, and then passes the token on to the next terminal. To check that its successor has received the token correctly the terminal makes sure that the successor is transmitting data. If not, the successor is assumed to be on a failed part of the network, and in order to prevent 'lock-up' of the LAN, the original terminal creates a new successor by generating a new token. Transmission faults in the LAN bus can therefore be circumvented to some extent. However, those parts of the LAN that are isolated from the token remain cut off.

Token bus networks are not commonly used in office environments—where ethernet and token ring networks predominate. Token bus networks are most common in manufacturing premises—often operating as 'broadband' (high bandwidth) networks for the tooling and control of complex robotic machines.

Token ring (IEEE 802.5)

The 'token ring' standard is similar in operation to the token bus, using the token to pass the 'right to transmit data' around each terminal on the ring in turn. The sequence of token passing is different: the token itself is used to carry the packet of data. The transmitting terminal sets the token's 'flag', putting the destination address in the 'header' to indicate that the token is full. The token is then passed around the ring from one terminal to the next. Each terminal checks whether the data is intended for it, and passes it on; sooner or later it reaches the destination terminal where the data is read. Receipt of the data is confirmed to the transmitter by changing a bit value in the token's flag. When the token gets back to the transmitting terminal, the terminal is obliged to empty the token and pass it to the next terminal in the ring.

The feature of IEEE 802.5 protocol is its capability for establishing priorities amongst the ring terminals. This it does through a set of priority indicators in the token. As the token is passed around the ring, any terminal may request its use on the next pass by putting a 'request' of a given priority in the reservation field. Provided no other station makes a higher priority request, then access to the token is given next time around. The reservation field therefore gives a means of determining demand on the LAN at any moment by counting the number of requests in the flag, and in addition the system of prioritization ensures that terminals with the highest preassigned authority have the first turn. High-speed operation of certain predetermined, time-critical devices is likely to be crucial to the operation of the network as a whole, but they are unlikely to need the token on every pass, so that lower priority terminals get a chance to use the ring when the higher priority stations are not active.

Token rings, like Ethernets, are common in office environments, linking personal computers for the purpose of data file transfer, electronic messaging, mainframe computer interaction or file sharing. Users tend to be emotive about whether 'Ethernet' or 'Token ring' offers the best solution, but in reality, for most office users, there is little to choose between them. Token ring LANs perform better than Ethernets at near full capacity or during overload but can be more difficult and costly to install—especially when only a small number of users are involved. In most cases, the choice between Ethernet and token ring comes down to the recommendation of a user's computer supplier, since hardware and software of a particular computer type may have been developed with one or other type of LAN in mind. Token ring was developed by IBM, and is common amongst IBM personal computer users. It is available in either a

4 Mbit/s form (which may use either coaxial cable or 'twisted pair' telephone cable) or alternatively in a 16 Mbit/s version (which needs higher quality cable).

'Slotted ring' and other types of LAN also exist, but are not covered in detail in this book, since they are rare. ISO 8802.7, for example, describes a slotted ring LAN used primarily by the UK academic community.

Interconnection of LANs

The interconnection of numerous LANs, perhaps of different types, or the connection of a LAN to a mainframe computer requires the use of 'bridges' or 'gateways'. A bridge is used to link two separate LANs together as if they were a single LAN—typically enabling the maximum capacity of a single LAN to be surpassed. A gateway, meanwhile, provides access to an external service, such as a mainframe computer. Typically a 'bridge' or 'gateway' consists of a personal computer (PC) on the LAN used entirely to run 'bridge' or 'gateway software'. This PC is not usually available for normal interactive use. The 'gateway' or 'bridge' connection is made directly to and from the PC. A commonly used layer 4 data transport protocol used in LAN gateways and across LAN bridges is called TCP/IP (Transmission Control Protocol/Internet Protocol). TCP/IP allows file transfer, electronic messaging and file sharing to occur across the bridge. It is an interim transport protocol (OSI layer 4) defined by the US government as a stepping-stone to ISO standards.

Another function performed by some LAN gateways is that of 'terminal emulation'. To explain—terminals connected directly to mainframe computers communicate in a special manner, usually proprietary to the computer manufacturer. If the terminal's place is to be taken either by a PC or by a LAN network, then that PC or LAN (via its gateway) must 'emulate' the actions and responses of the terminal. This is called terminal emulation. 'IBM 3270', for example, is the communication protocol used between the host computer and the terminal of an IBM mainframe. The protocol allows the computer to interpret keyboard interaction at the terminal and control the image appearing on the terminal screen, without there having to be 'intelligence' in the terminal. Thus 3270 type terminals are sometimes described as 'dumb terminals'. The conversion necessary to make an intelligent terminal (such as a personal computer) appear to talk to a host computer like a dumb terminal is carried out by 'terminal emulation' or '3270 emulation' software. Where this software resides in a LAN gateway, then it is often loosely called a '3270 gateway'.

9.14 PRACTICAL COMPUTER NETWORKS

We complete the chapter with a short discussion of the components that go to make up a real computer network. A number of other ideas for practical application can be found in Chapter 30.

A typical data network configuration, linking a mainframe computer to a number of terminals is illustrated in Figure 9.31, mainly with the idea of familiarizing readers with some of the terminology used to describe the various equipments, but also to describe their functions in relation to what we have discussed in this chapter.

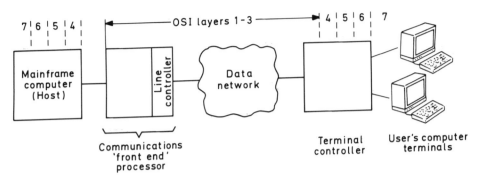

Figure 9.31
Typical computer data network.

Five components are illustrated in Figure 9.31. The mainframe computer (sometimes also called the 'host') is on the left-hand side. This is interconnected via three other components ('front-end processor', 'data network', and 'terminal controller') to the user's computer terminals, which are shown on the right-hand side of the diagram.

The 'front end' processor contains all the necessary communications software for setting up the connections and conversing with the terminals (i.e. protocols equivalent to OSI layers 1 to 3). A particular piece of equipment which is integrated with the 'front end' processor is called the line controller. This provides the physical (layer 1) interface to the data network. It is possible to integrate the front end functions into the host itself, but the advantage of using a dedicated front end processor is that it removes a considerable burden from the 'host', allowing it to get on with the main processing job in hand.

The data network might be one of a variety of types. It might be a pair of modems, a digital circuit, a LAN, a packet-switched network, or a circuit-switched network.

Finally, the 'terminal controller' interfaces a number of remote users' terminals to the data network. Like the communications 'front end' processor, it incorporates a line controller and some communications software to perform the functions of the data protocols. Most terminal controllers connect with terminals using the simple asynchronous, and character-oriented technique with which we are already familiar. The application (layer 7 function) thus exists to convey information to the computer about which terminal keys have been pressed, and relay back to the terminal the information to be displayed on the screen.

IBM's 'Systems Network Architecture'

Like the OSI model, IBM's 'system network architecture' defines the functions of a layered protocol set, but unlike OSI (which SNA pre-dates), specified protocols exist for each of the layers. One of the best known of these protocols, SDLC (Synchronous Data Link Control) is the datalink control protocol of SNA.

SNA was first introduced by IBM in 1974 and has evolved through several generations to its current sophisticated form. It was developed as an all-purpose data networks architecture, a replacement for the growing number of incompatible communication products which IBM had at the time. The latest SNA networks allow a large number of 'host' machines, 'communications controllers', and 'cluster controllers' (IBM's name for terminal controllers) to share a sophisticated communications network. By so doing SNA creates:

(i)   an ability for terminal users to switch dynamically between different application programs or even different host computers;

(ii) an ability to attach dissimilar devices to the same network;

(iii) an ability to concentrate (or 'multiplex') a number of data connections to share a common circuit or network;

(iv) an ability for direct interconnection between 'host' computers, PCs and/or minicomputers—so enabling 'distributed' processing and storage of data.

SNA caters for communication between five different classes of physical units (PUs) such as terminals and computers, classified as shown in Table 9.2.

The communication itself takes place between logical units (LUs) which are software programs resident within the PUs. A 'session' of communication ultimately takes place to serve the communications needs of these programs. Thus an LU is the equivalent of the OSI application layer.

Serving a logical unit (LU) at either end of an SNA network there are up to seven layered protocols, as shown in Table 9.3. The protocols are similar in purpose to those of the OSI model but there are some significant differences in the functions of individual layers.

The layered nature of SNA is immediately apparent in the message structure, as we show in Figure 9.32. Just as in the OSI case, each progressively lower layer adds an information field, called a 'header' or 'tailer', to the original data which is held in the 'request/response unit'. At the receiving end each progressively higher layer removes its appropriate header and passes to the next higher layer all that is left.

The layered nature of SNA is intended to provide for modular design of computer network software, giving greater ease of maintenance and more easily modifiable

Table 9.2
SNA physical unit (PU) classes.

Class type	Class name
5	'Host' computer or 'System Services Control Point'
4	'Communications controller' or 'Network control program (NCP)'
3	Never existed
2.1	'Independent' cluster controller (i.e. can operate in a peer-to-peer network of PU2.1s)
2	'Dependent' cluster controller—i.e. one that requires a type 4 or type 5 'controlling' PU
1	Remote devices, including 'cluster controllers' and 'terminal nodes'. Now virtually obsolete

Table 9.3
Systems network architecture (SNA).

SNA layer name	Equivalent	Function
NAU services (NAU = Network Addressable Unit)	7	Data exchange between logical units (LUs)
FMD services (FMD = File Management Data)	6	Syntax. Data compression and compaction type. ASCII, EBCDIC code type
Data flow control	5	Dialogue control. (Session control)
Transmission control	4	Activating and deactivating data flow within a session. Multiplexing
Path control	2/3	Routing and flow control. Message Packetization
Datalink control	2	Managing bit oriented data flow. SDLC
Physical	1	X21. RS232-C

implementations. Between them the layers are intended to meet all the data communications requirements of an entire distributed data-processing network. The layers themselves may thus be realized in a number of different forms. Usually the layers are either all realized in the same physical unit (PU), or as is done with many host computers, the protocols are implemented in two parts; the lowest three layers (physical, datalink and path control), making up the 'path control network', are conducted by the communications controller (front end processor) while the higher layers together form a 'network addressable unit' (NAU) which resides in the host processor itself. The NAU is held in software in the host, and it relies on the 'network control program' (NCP) software in the 'front-end processor' or 'communications controller'. A number of alternative IBM software products are available for installation as NAUs on IBM or compatible equipment. The most commonly used communications product on IBM mainframe computers is ACF/VTAM (Virtual Telecommunications Access Method). An older, less frequently used product is ACF/TCAM (Advanced Communication Function/Telecommunications Access Method). Both provide the necessary host software for telecommunications in support of a logical unit (LU) in the application program of a 370-series computer.

Both TCAM and VTAM include a function called the SSCP (System Services Control Point). This is a software function, and each SNA network must have at least one of them to control the overall network configuration—activating and deactivating the network and establishing communication sessions. This it does by interacting with appropriate software in each of the other physical units (PUs) of the network.

The communications controller, if provided as a separate physical entity from the host, can be any of a number of IBM 37XX series machines. These are specialized telecommunications hardware devices which conform to the SNA model when loaded with IBM 'network control program' (NCP) software. A cluster controller (e.g. IBM 3174 or 3274) allows the connection of dumb terminals to the network. A typical configuration is shown in Figure 9.33.

As already discussed, one of the most common SNA communications regimes is

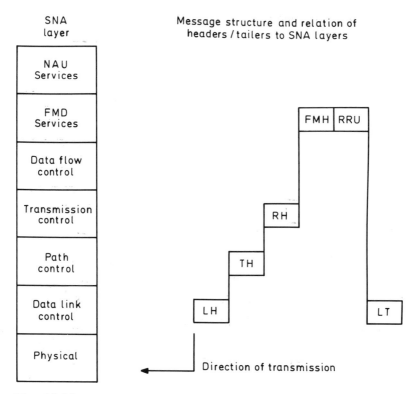

Figure 9.32
SNA message structure. LH, Link Header; TH, Transmission Header; RH, Response Header; FMH, Function Management Header; RRU, Request Response Unit; LT, Link Trailer.

often referred to as 3270 communication. This is the method used for communication between a mainframe computer and a dumb, 327X-type terminal. But the emergence of the personal computer (PC) in recent years has strained the capabilities of this method. Nowadays it is usual to use PCs as part-time mainframe computer terminals, but to do so requires that they 'emulate' the dumb terminals they substitute. Step forward '3270 emulation'. Either by hardware means (a communications board added to the back of the PC) or under software control, the PC pretends to be a dumb terminal. Interaction may progress—but not without its constraints—the session can only be established from the PC end of the connection. New OSI protocols should alleviate this constraint.

Where instead the PC (or LAN gateway) emulates the cluster controller rather than just the terminal, and by so doing can be connected directly to the front end processor, then we refer to it as a 'SDLC adapter' or 'SDLC gateway'.

An alternative to SDLC, predating SNA but still used today, is BSC—Binary Synchronous Communication or 'BISYNC'. Invented in 1964, this is a protocol in which the bit strings representing individual characters are delineated by control character sequences, rather than being synchronous as SDLC. And for completeness why not also mention 2780 and 3780 protocols? These were the pre-SNA protocols originally

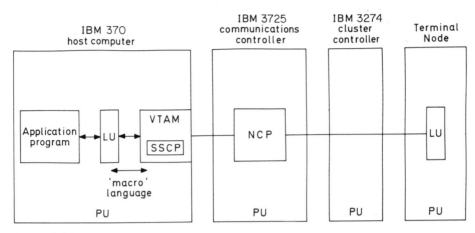

Figure 9.33
A typical SNA network.

Table 9.4
SNA logical user (LU) session types.

LU session class	End users
0	Session conducted between end users of unknown types
1	Session conducted between an application program and an input/output device in an interactive, 'batch' or 'distributed data processing environment'
2	Session conducted between an application program and a single interactive operator display terminal (3270 type)
3	Session conducted between an application program and a single printer terminal in an interactive environment (3270 type printer)
4	Session conducted between an application program and multiple interactive operator display terminals, or between two operator display terminals
6	Session conducted for file or data transfer between application programs in a 'cooperative' distributed data-processing environment

used in the days of batch computing—when 'jobs' were submitted by users to remote data centres. At these centres, computer staff would schedule and run the jobs in a priority order. Protocol 2780 (or the enhanced version, 3780) was used for 'remote job entry' by these staff. Protocol 2780/3780 remains in common use between IBM mainframes and operation/maintenance devices.

Finally, back to SNA and a mention of the classification for the various different types of logical user (LU) sessions which is made possible by SNA. There are six classes of LU session, characterized in terms of the 'session partners' (Table 9.4). Each session type has slightly different needs and therefore puts slightly different demands on the protocol layers. These demands are met by using different 'options' within the protocols. The overall set of options used is called the session 'profile'. The six different session types are listed in Table 9.4.

Of the LU session types shown in Table 9.4 we mention only LU6 specifically, and this only to record the widespread interest among computer systems designers for the second generic version LU6.2 which has set the *de-facto* standard for program-to-program communications.

The future of IBM's architecture

The year 1988 saw the announcement by IBM of a new concept, an even more embracing computer and telecommunications architecture. Called SAA, or 'Systems Applications Architecture', ultimately it will allow maximum flexibility of hardware, software and telecommunications within the IBM environment by bringing together four standard components:

- A common user access (CUA) comprising common screen layouts, keyboard layouts etc. (i.e. the same layout for PCs, mainframe terminals and other devices);

- A common programming interface, thus allowing software to be moved easily from PC to mainframe or vice versa;

- A common communications interface (either the SNA or OSI);

- Common, modular hardware and interfaces.

BIBLIOGRAPHY

Bacon, M. D. and Bull, G. M., *Data Transmission*. MacDonald/London and American Elsevier, 1973.

Curne, W. S., *The LAN Jungle Book—A Survey of Local Area Networks*. Edinburgh University/Kinesis Computing, 1988.

Fitzgerald, J. and Eason, T. S., *Fundamentals of Data Communications*. John Wiley & Sons, 1978.

Fritz, J. S., Kaldenbach, C. F. and Progar, L. M., *Local Area Networks—Selection Guidelines*. Prentice-Hall, 1985.

Gandy, M., *Choosing a Local Area Network*. NCC Publications, 1986.

Gee, K. C. E., *Proprietary Network Architectures*. NCC Publications, 1981.

Government Open Systems Interconnection Profile (UK). Central Computer and Telecommunications Agency, 1988.

Guruga, A., *SNA Theory and Practice*. Pergamon Infotech, 1984.

Held, G., *Data Communications Networking Devices—Characteristics Operation, Applications*, 2nd edn. John Wiley & Sons, 1989.

Henshall, J. S. and Shaw, S., *OSI Explained—End to End Computer Communications Standards*. Ellis Horwood, 1988.

Inglis, A. F., *Electronic Communications Handbook*. McGraw-Hill, 1988.

Jennings, F., *Practical Data Communications; Modems, Network and Protocols*. Blackwell Scientific Publications, 1986.

Jones, V. C., *MAP/TOP Networking*. McGraw-Hill Manufacturing & Systems Engineering Series, 1988.

Judge, P., *Open Systems—The Basic Guide to OSI and its Implementation*. Computer Weekly/ Systems International (Reed Business Publishing), 1988.

Knowles, T., Larmouth, J. and Knightson, K. G., *Standards for Open Systems Interconnection*. BSP Professional Books, 1987.

Lent, J. D., *Handbook of Data Communications*. Prentice-Hall, 1984.

Local Area Networks (LANs) Standards IEEE 802 Series. (ISO 8802 series.)

Maynard, J., *Computer and Telecommunications Handbook*. Granada, 1984.

Meijer, A., *Systems Network Architecture: A Tutorial*. Pitman, 1989.

NCC Handbook of Data Communications. NCC Publications, 1982.

Poulton, S., *Packet Switching and X25 Networks*. Pitman, 1989.

Pye, C., *What is OSI?* NCC Publications, 1988.

Sarch, R., (ed.), *Basic Guide to Data Communications*. McGraw-Hill, 1985.

Systems Network Architecture Format and Protocol Reference Manual: Architectural Logic—SRL no. SC30-3112. IBM Corporation.

The Open Systems Interconnection Model. ISO 7498.

X3, 'Packet assembly/disassembly facility (PAD) in a public data network'.

X21 bis, 'Use on public data networks of data terminal equipment (DTE) which is designed for interfacing to synchronous V-series modems'.

X21, 'Interface between data terminal equipment (DTE) and data circuit-terminating equipment (DCE) for synchronous operation on public data networks'.

X25, 'Interface between DTE and DCE for terminals operating in the packet mode and connected to public data networks by dedicated circuits'.

X28, 'DTE/DCE interface for a start-stop mode DTE accessing the PAD facility in a public data network situated in the same country'.

X29, 'Procedures for the exchange of control information and user data between a PAD facility and a packet node DTE or another PAD'.

X75, 'Terminal and transit call control procedures and data transfer system on international circuits between packet-switched data networks'.

X200, 'Reference model of open systems interconnection for CCITT applications'.

PART 2

RUNNING
A NETWORK

10

TELETRAFFIC THEORY

Telecommunications networks, like roads, are said to carry 'traffic', consisting not of vehicles but of telephone calls or data messages. The more traffic there is, the more circuits and exchanges must be provided. On a road network—the more cars and lorries, the more roads and roundabouts are needed. In either kind of network, if traffic exceeds the design capacity then there will be pockets of congestion. On the road this means traffic jams; on the telephone the frustrated caller gets frequent 'busy tones'.

Short of providing an infinite number of lines, it is impossible to know in advance precisely how much equipment to build into a telecommunications network so as to meet demand without congestion. But there is a tool for 'dimensioning' network links and exchanges. It is the rather complex statistical science of 'teletraffic theory' (sometimes also called 'teletraffic engineering') and is the subject of this chapter.

We begin with the teletraffic dimensioning method used for circuit-switched networks, first published in 1917 by a Danish scientist, A. K. Erlang. Erlang defined a number of parameters and developed a set of formulae, which together give a framework of rules for planners to design and monitor the performance of telephone, telex and circuit-switched data networks. The latter part of the chapter deals with the dimensioning of data networks, reviewing the techniques in such a way as to offer the reader a practical method of network design.

10.1 TELECOMMUNICATIONS TRAFFIC

We have accepted 'traffic' as the term to describe the amount of telephone calls or data messages conveyed over a network, but this could cover any number of different scientific definitions. Possible definitions of 'traffic' include:

(i)   the total number of calls or messages;

(ii) the total 'conversation time' (i.e. the number of calls multiplied by the 'conversation time' on each);

(iii) the total circuit 'holding time'. This is number of calls multiplied by the 'holding time' on each. The holding time includes the time period during which the parties are in conversation and also the time prior to conversation when the call is being 'set-up'. Holding time is the total time for which the network is in use;

(iv) the total number of data characters conveyed.

Only one of these definitions is correct as far as the science of teletraffic is concerned. In particular the 'traffic volume' is normally defined by definition (iii) above. In other words, the total traffic *volume* is equal to the total network holding time. None of these definitions, however, helps in the determination of exactly how many circuits or what size of exchanges will have to be provided. What we need is some idea of the maximum usage of the network at any one time. For this reason it is normal to measure the 'traffic intensity' of circuit-switched networks.

10.2 TRAFFIC INTENSITY

Traffic intensity is, by definition, the average number of calls simultaneously in progress during a particular period of time. It is measured in units of 'erlangs'. Thus an average of one call in progress during a particular period would represent a traffic intensity of one erlang. The traffic intensity on any route between two exchanges can also be quoted in erlangs. It is measured by first summing the total holding time of all the circuits within the route and then dividing this by the time period, *T*, over which the measurement was made.

In some countries, including the United States, traffic intensity is measured not in erlangs but in units called CCS (100 call seconds). CCS is a measure of the total call holding time during the network or route busy hour. The two units, CCS and erlang are very simply related, since:

$$1 \text{ erlang} = 3600 \text{ call seconds} = 36 \text{ CCS}.$$

The definition of traffic intensity is not restricted to traffic between exchanges. Even cross-exchange traffic (that passing across an exchange from incoming ports to outgoing ports) can be measured and quoted in erlangs. As an example, the route between exchanges A and B in Figure 10.1 consists of five circuits. The chart in Figure 10.1 also shows the periods of usage of each of these circuits during a 10-min period.

It will be seen from the chart in Figure 10.1 that the total number of minutes of circuit usage during the 10-min period was 35 min. This is the sum of the individual circuit holding times, respectively: 6 min, 7.5 min, 7.5 min, 8 min, 6 min. Dividing the total of 35 min by the length of the monitoring period (10 min) we may deduce that the traffic intensity on the route was 3.5 erlangs. In other words, an average of 3.5 circuits were in use throughout the whole of the period.

It would be easy to succumb to the mistaken belief that a traffic intensity of 3.5 erlangs (3.5 simultaneous calls) could be carried on four circuits. Figure 10.1 clearly illustrates an example in which this is not the case. All five circuits were simultaneously in use twice—between time 2 and 3, and between time 5.5 and 6. The general principle is that more circuits will be needed on a route than the numerical value of the route traffic intensity.

Here we must digress briefly to examine the concepts of 'offered' and 'carried traffic', though their names may give them away. 'Offered traffic' is a theoretical concept. It is a measure of the unsuppressed traffic intensity that would be on a particular route if all the customers' calls were connected without congestion. 'Carried traffic' is that resultant from the carried calls—and is the value of traffic intensity actually measured.

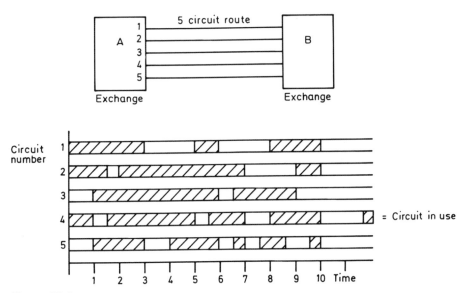

Figure 10.1
Circuit usage on a route between exchanges.

For a network without congestion the 'carried traffic' is equal to the 'offered traffic'. But if there is congestion in the network, then the offered traffic will be higher than that carried.

Teletraffic theory leads to a set of tables and graphs which enable the required network circuit numbers to be related to the 'offered traffic' demand (in erlangs).

10.3 PRACTICAL TRAFFIC INTENSITY (ERLANG) MEASUREMENT

In practice, the 'traffic intensity' on a given route may be measured using one of two main methods, either by sampling or by an absolute measurement. The latter method was used in the example of Figure 10.1. Historically, however, electro-mechanical exchanges did not lend themselves to easy calculation of absolute value of the total holding time. Instead, it was common practice to use a sampling technique. We can get an estimate of the total circuit holding time by looking at the instantaneous state of the circuits at number of sample points in time. If we take ten samples, after $1/2$ min, $1 + 1/2$ min, $2 + 1/2$ min, ..., $9 + 1/2$ min, we could assume that this was representative of the respective time periods, 0–1 min, 1–2 min, ..., etc. Thus, because after $1/2$ min has elapsed, we find that three circuits are in use; we assume that three circuits will be in use for the whole period of the first minute. By this method we get ten 'snapshots' of the number of circuits in use, corresponding to the ten individual one-minute periods which we sampled. The results are as shown in Table 10.1.

Averaging the sample values will give us an *estimate* of the traffic intensity over the whole period. This value comes out at an *estimated* 3.9 erlangs. This is calculated as the sum of the ten sample values (39), divided by the overall duration (10). Compare this with the actual value of 3.5 erlangs. The error arises from the fact that we are only sampling the circuit usage rather than using exact measurement. The error can be reduced by increasing the frequency of samples. Table 10.2 gives the values calculated at a 1/2 min (as opposed to a 1-min) 'sampling rate' (sometimes also called 'scan rate').

The increased sampling rate of Table 10.2 estimates the traffic correctly as 3.5 erlangs ($70 \times 0.5/10$). The exactness of the estimate on this occasion is fortuitous. None the less, the principle is well illustrated that too slow a sampling rate will produce unreliable traffic intensity estimates and that the estimate is improved in accuracy by a higher sampling rate. A good guide is that the sample period length should be about one third of the average call holding time. In the example of Figure 10.1, the average call holding time is 35 min/17 calls = 2.06 min, so a sample should be taken about every 0.7 min in order to derive a reliable estimate of the traffic intensity. The minimum monitoring period should be about three times the average call holding time. This helps us to avoid unrepresentative peaks or troughs in call intensity.

Nowadays, stored program controlled (SPC) exchanges have made the absolute measurement of call holding times relatively easy. So today many exchanges produce measurements of traffic intensity which are exact.

Table 10.1
Samples of number of circuits in use.

Elapsed time	Circuits in use
$\frac{1}{2}$	3
$1\frac{1}{2}$	5
$2\frac{1}{2}$	5
$3\frac{1}{2}$	3
$4\frac{1}{2}$	4
$5\frac{1}{2}$	5
$6\frac{1}{2}$	4
$7\frac{1}{2}$	2
$8\frac{1}{2}$	4
$9\frac{1}{2}$	4

Table 10.2
$\frac{1}{2}$ min scan rate.

Elapsed time	Circuits in use	Elapsed time	Circuits in use
$\frac{1}{4}$	3	$5\frac{1}{4}$	4
$\frac{3}{4}$	3	$5\frac{3}{4}$	5
$1\frac{1}{4}$	4	$6\frac{1}{4}$	2
$1\frac{3}{4}$	4	$6\frac{3}{4}$	4
$2\frac{1}{4}$	5	$7\frac{1}{4}$	1
$2\frac{3}{4}$	5	$7\frac{3}{4}$	2
$3\frac{1}{4}$	3	$8\frac{1}{4}$	4
$3\frac{3}{4}$	3	$8\frac{3}{4}$	3
$4\frac{1}{4}$	4	$9\frac{1}{4}$	3
$4\frac{1}{4}$	4	$9\frac{3}{4}$	4

10.4 THE BUSY HOUR

In practice, telecommunications networks are found to have a discernible 'busy hour'. This is the given period during the day when the traffic intensity is at its greatest. Traditionally measurements of traffic intensity have been made over a full 60-minute period at the 'busy hour' of day, in order to calculate the 'busy hour traffic' (a shortened term for the 'busy-hour traffic intensity'). And by taking samples over a few representative days, future 'busy-hour traffic' can usually be predicted well enough to work out what the equipment quantities and route sizes of the network should be. The dimensioning is done by using the predicted busy hour traffic as an input to the 'Erlang formula' (presented later in this chapter). When that formula has pronounced the scale of circuits and equipment that are going to be needed at the busy hour, we can feel secure at less hectic times of day.

In more recent times, it has been found that exchanges are developing more than one busy hour; maybe as many as three, including morning, afternoon and evening 'busy-hours'. The morning and afternoon busy hours are usually the result of business traffic. The evening one results from residential and international traffic. Some examples of daily traffic distribution are given in Figure 10.2.

The first distribution in Figure 10.2 is a typical business-serving exchange, with morning and afternoon busy hours. The second example shows the traffic in a residential exchange, where morning and afternoon 'busy hours' are less than the evening

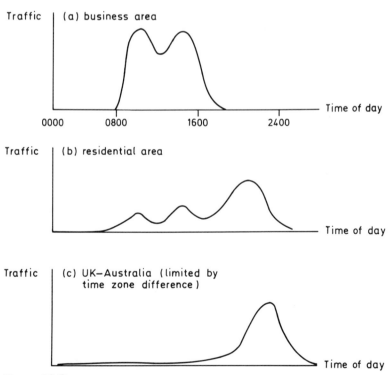

Figure 10.2
Some typical daily traffic distributions.

busy hour. The third example is of an international route, where there is often a single peak, constrained in duration by the time zone difference between the countries.

Multiple busy hours necessitate the calculation of separate busy-hour traffic values for each individual busy hour, and design of the network to meet each one. It is not enough to use only one of the sets of busy-hour traffic values. In Figure 10.2 for example, the overall effect of distributions (a) and (b), when combined, is a two-humped pattern with overall busy hours in morning and afternoon of approximately equal magnitude. But if we were to design the whole network on the basis of its morning and afternoon busy hour traffic values alone, we would not provide enough circuits at the residential exchange to meet its evening busy hour. Normal practice is therefore to dimension each exchange matrix according to its own 'exchange busy hour', and each route accordingly to its individual 'route busy hour'.

Another point to consider when dimensioning networks is that a route or exchange is unlikely to be equally busy throughout the entire 60-min busy-hour period. In order to meet the peak demand, which might be significantly higher than the hour's average traffic, it is sometimes convenient to redefine the busy hour as being of less than 60 min duration. It may seem paradoxical to have a busy hour of less than 60 min, but Figure 10.3 illustrates a case of traffic which has a short duration peak of between 15 and 30 min.

In the example of Figure 10.3, if we were to use a 60-min busy-hour measurement period, we would estimate a busy-hour traffic of value Q as shown on the diagram. This would be a gross underestimate of the actual traffic peak, and would guarantee that the busiest period would be heavily congested. By redefining the 'busy-hour' to be of only 30 min duration we get an estimate P which is much nearer the actual traffic peak. Surprisingly, however, the use of much shorter busy-hour periods and attempts to pinpoint peaks of traffic which last only a few minutes are not really important; customers who suffer congestion at this time are more than likely to get through on a repeat call attempt within a few minutes anyway!

To sum up, it is usual to monitor 'exchange' and 'route busy hour' traffic values as a gauge of current customer usage and network capacity needs. Future network design and dimensioning can then be based on our forecasts of what the values of these parameters will be in the future. (Forecasting methods are covered in Chapter 11.)

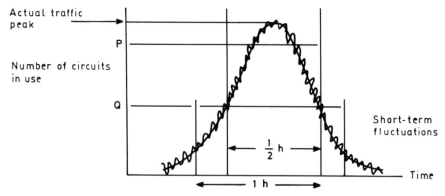

Figure 10.3
A short duration busy hour.

10.5 THE FORMULA FOR TRAFFIC INTENSITY

Our next step is to develop the formula for traffic intensity, as a basis for subsequent discussion of the Erlang method of network dimensioning. Recapping in mathematical terms, the traffic intensity is given by the expression:

$$\text{Traffic intensity (carried traffic)} = \frac{\text{the sum of circuit holding times}}{\text{the duration of the monitoring period}}$$

Now, let A = the traffic intensity in erlangs
T = the duration of the monitoring period
h_i = the holding time of the ith individual call
c = the total number of calls in the period
Σ = mathematical summation

Then, from above

$$A = \frac{\Sigma_1^c h_i}{T}$$

Now, since the sum of the holding times is equal to the number of calls multiplied by the *average* holding time, then

$$\sum_1^c h_i = c\bar{h}$$

where $\bar{h}$ = average call holding time and therefore

$$A = \frac{c\bar{h}}{T}$$

It is interesting to calculate the call arrival rate, in particular the number of calls expected to arrive during the average holding time. Let N be this number of calls, then

N = number of call arrivals during a period equal to the average holding time

$= h \times$ call arrival rate per unit of time

$= h \times c/T$

$= c\bar{h}/T = A$

In other words, the number of calls expected to be generated during the average holding time of a call, is equal to the traffic intensity, A. This is perhaps a surprising result, but one which sometimes proves extremely valuable.

10.6 THE TRAFFIC-CARRYING CAPACITY OF A SINGLE CIRCUIT

In this section we discuss the traffic-carrying capacity of a single circuit. This leads on to a mathematical derivation of the Erlang formula, which is the formal method for calculating the traffic carrying capacity of a circuit group of any size.

For our explanation let us assume we have provided an infinite number of circuits, laid out in a line or 'grading', as shown in Figure 10.4. The infinite number provides enough circuits to carry any value of traffic intensity. Now let us further assume that each new call scans across the circuits from the left-hand end until it finds a free circuit. Then let us try to determine how much traffic each of the individual circuits carries.

First, let us consider circuit number 1. Figure 10.5 shows a timeplot of typical activity we might expect on this circuit, either 'busy' carrying a call, or 'idle' awaiting for another call to 'arrive'. The timeplot of Figure 10.5 starts with the arrival of the first call. This causes the circuit to become busy for the duration of the call. While the circuit is busy a number of other calls will arrive, which circuit number 1 will be incapable of carrying. These other calls will scan across towards a higher-numbered circuit (circuit number 2, then 3, and so on) until the first free circuit is found. Finally at the end of the call on circuit number 1, the circuit will be returned to the idle state. This state will prevail until the next new call arrives.

Let us try to determine the proportion of time for which circuit number 1 is busy. For this purpose, let us assur e each call is of a duration equal to the average call

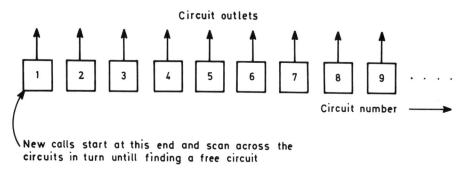

Figure 10.4
Scanning for a free circuit.

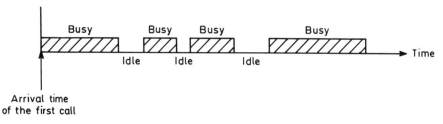

Figure 10.5
Activity pattern of circuit number 1.

holding time, $\bar{h}$. This is not mathematically rigorous but it makes for simpler explanation. Let us also invent an imaginary cycle of activity on the circuit.

Our imaginary cycle is as follows:

After the arrival of the first call, we expect circuit number 1 to be busy for a period of time equal to $\bar{h}$. As we learned in the last section we can expect a total of A calls to arrive during the average holding time, where A is the offered traffic. $(A - 1)$ of these calls (i.e. all but the first) will scan over circuit number 1 to find a free circuit among the higher numbered circuits. At the end of the first call circuit number 1 will be released, and an 'idle' period will follow until the next call arrives (i.e. the $(A + 1)$th). We can imagine this cycle repeating itself over and over again.

As we can see from our imaginary 'average' cycle, shown in Figure 10.6, the total number of calls arriving during the cycle is $A + 1$. The total duration of the cycle is therefore:

$$\bar{h} \times \frac{A + 1}{A}$$

But we also know that circuit number 1 is busy during each cycle for a period of duration, $\bar{h}$. Therefore the average proportion of the time for which circuit number 1 is busy is given by:

$$\text{Occupancy of circuit no 1} = \frac{\bar{h}}{\bar{h}(1 + A)} = \frac{A}{1 + A}$$

This value is the so-called 'circuit occupancy' of circuit number 1. It is numerically equal to the average number of calls in progress on circuit number 1, and is therefore the intensity of the traffic 'carried' on circuit number 1, and is measured in erlangs accordingly. Thus if one erlang were offered to the grading $(A = 1)$, then the first circuit would carry half an erlang. The remaining half erlang is carried by other circuits.

Taking one last step, if we assume that new calls 'arrive' at random instants of time, then the proportion of calls rejected by circuit number 1 is equal to the proportion of time during which the circuit is busy, i.e. $A/(1 + A)$. In our simple case, if only one

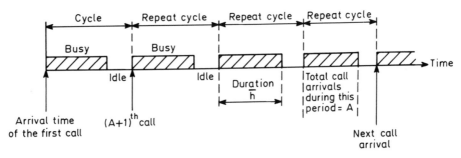

Figure 10.6
Average activity cycle on circuit number 1.

circuit were available, then $A/(1+A)$ proportion of calls cannot be carried. This is called the blocking ratio B, and is usually written:

$$B \text{ (blocking ratio—for one circuit)} = \frac{A}{1+A}$$

where A = offered traffic

Though not proven above in a mathematically rigorous fashion, the above result is the foundation of the Erlang method of circuit group dimensioning. Before going on, however, it is worth studying some of the implications of the formula a little more deeply for two cases:

First, take the case of one erlang of offered traffic to a single circuit. Substituting in our formula $A = 1$, we conclude that the blocking value is $1/(1+1) = 1/2$. In other words, half of the calls fail (meeting congestion), and only half are carried. This confirms our earlier conclusion that the circuit numbers needed to carry a given intensity of traffic are greater than the numerical value of that traffic.

With traffic intensity of $A = 0.01$, then the proportion of blocked calls would have been only $0.01/1.01$ ($= 0.01$), or about one call in 100 blocked. This is the proportion of lost calls targeted by many network operating companies. So in practical terms the carrying capacity of a single circuit in isolation is around 0.01 erlangs.

Next let us consider a very large traffic intensity offered to our single circuit. In this case most of the traffic is blocked (if $A = 99$), then the formula states that 99 per cent blocking is incurred. But the corollary is that the traffic carried by the circuit (equal to the proportion of the time for which the circuit is busy) is 0.99 erlangs. In other words the circuit is in use almost without let-up. This is what we expect, since as soon as the circuit is released by one caller, a new call is offered almost immediately.

In the appendix to this chapter the full Erlang 'lost-call' formula is derived using a more rigorous mathematical derivation in order to gain an insight into the traffic-carrying characteristics of all the other circuits in Figure 10.4. For the time being, however, Figure 10.7 simply states the formula.

To confirm the result from our previous analysis, let us substitute $N = 1$ into the formula of Figure 10.7. As before, we get a proportion of lost calls for a single circuit

$$E(N, A) \text{ or } B(N, A) = \frac{A^N/N!}{\left(1 + A + \frac{A^2}{2!} + \frac{A^3}{3!} + \cdots + \frac{A^N}{N!}\right)}$$

where $E(N,A)$ = proportion of lost calls and probability of blocking
A = offered traffic intensity
N = available number of circuits
$N!$ = factorial N

Figure 10.7
The Erlang lost-call formula.

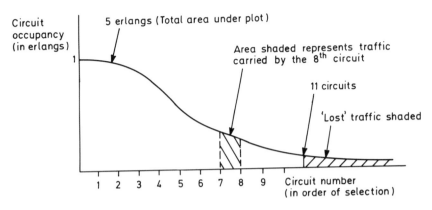

Figure 10.8
Circuit occupancies.

(offered traffic A) of:

$$B(1, A) = \frac{A/1}{(1 + A)} = \frac{A}{1 + A}$$

This of course is also equal to the circuit 'occupancy' (the traffic carried by it). The advantage of our new formula is that we may now calculate the occupancy of all the other circuits of Figure 10.4. By calculating the lost traffic from two circuits we can derive the carried traffic. Subtracting the traffic carried on circuit number 1 we end up with that carried by circuit number 2. In a similar manner, the traffic carrying contributions of circuit numbers 3, 4, 5 etc. can be calculated. Eventually we are able to plot the graph of Figure 10.8, which shows the individual circuit occupancies when 5 erlangs of traffic is offered to an infinite circuit 'grading'.

As expected, the low numbered circuits carry nearly one erlang and are in near-constant use, while higher numbered circuits carry progressively less traffic. The eleven circuit line is also shown. From the formula of Figure 10.7, this is the number of circuits needed to guarantee less than 1 per cent proportion of calls lost. Thus the right-hand shaded area in Figure 10.8 represents the small proportion of lost calls if 11 circuits are provided. The ability to calculate individual circuit occupancies is crucial to grading design (Chapter 6 refers).

10.7 DIMENSIONING CIRCUIT-SWITCHED NETWORKS

The future circuit requirements for each route of a circuit-switched network (i.e. telephone, telex, circuit-switched data) may be determined from the Erlang lost-call formula. We do so by substituting the predicted traffic intensity, A, and using trial-and-error values of N to determine the value which gives a slightly better performance than the target blocking or 'grade of service', B. A commonly used grade of service for interchange traffic routes is 0.01 or 1 per cent blocking.

It is not an easy task to determine the value of N (circuits required), and for this reason it is usual to use either a suitably programmed computer or a set of 'traffic tables'.

In recent years, numerous authors and organizations have produced modified versions of the erlang method—more advanced and complicated techniques intended to predict accurately the traffic-carrying capacity of various sized circuit groups for different 'grades of service'. All have their place but in practice it comes down to finding the most appropriate method by trying several for the best fit for given circumstances. In my own experience, the extra effort required by the more refined and complicated methods of dimensioning is unwarranted. In practice the traffic demand may vary greatly from one day or month to the next and the practicality is such that circuits have to be provided in whole numbers—often indeed in multiples of say 12 or 30. The decision then is whether 1 or 2; 12 or 24; 30 or 60. It is rather academic to decide whether 23 or 24 circuits are actually necessary when at least 30 will be provided.

Table 10.3 illustrates a typical traffic table. The one shown has been calculated from the Erlang lost-call formula. Down the left-hand column of the table the number of circuits on a particular route are listed. Across the top of the table various different grades of service are shown. In the middle of the table, the values represent the maximum offered erlang capacity corresponding to the route size and grade of service chosen. Thus a route of four circuits, working to a design grade of service of 0.01, has a maximum offered traffic capacity of 0.9 erlangs.

We can also use Table 10.3 to determine how many circuits were required to provide a 0.01 grade of service, given an offered traffic of 1 erlang. In this case the answer is five circuits. The maximum carrying capacity of five circuits at 1 per cent grade of service is 1.4 erlangs—slightly greater than needed, but the capacity of four circuits is only 0.9 erlangs.

The problem with traffic routes of only a few circuits is that only a small increase

Table 10.3
A simple Erlang traffic table.

Number of circuits	Grade of service $(B(N,A))$ 1 lost call in			
	50 (0.02)	100 (0.01)	200 (0.005)	1000 (0.001)
	erlangs	erlangs	erlangs	erlangs
1	0.020	0.010	0.005	0.001
2	0.22	0.15	0.105	0.046
3	0.60	0.45	0.35	0.19
4	1.1	0.9	0.7	0.44
5	1.7	1.4	1.1	0.8
6	2.3	1.9	1.6	1.1
7	2.9	2.5	2.2	1.6
8	3.6	3.2	2.7	2.1
9	4.3	3.8	3.3	2.6
10	5.1	4.5	4.0	3.1
11	5.8	5.2	4.6	3.6
12	6.6	5.9	5.3	4.2

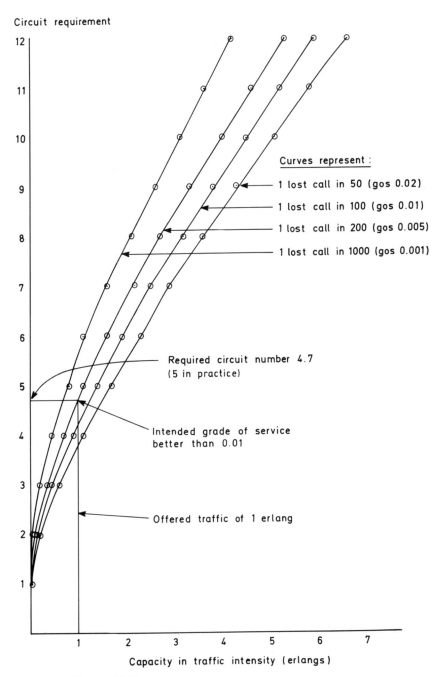

Figure 10.9
Graphical representation of the Erlang formula.

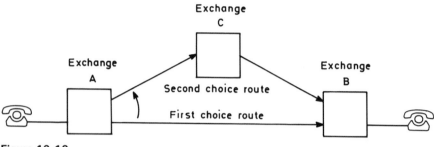

Figure 10.10
Overflow routing.

in traffic is needed to cause congestion. It is good practice therefore to ensure that a minimum of circuits are provided on every route (say 5).

The information held in Table 10.3 is sometimes presented graphically, and Figure 10.9 illustrates this. The traffic offered in erlangs is usually plotted along the horizontal axis, and the circuit numbers up the vertical axis. A number of different curves then correspond to different grades of service. To determine how many circuits are required for a given offered traffic, the offered erlang value is read along the horizontal axis, then a vertical line is drawn upwards to the curve corresponding to the required grade of service, and a horizontal line is drawn from this point to the vertical axis, where the circuit requirement can be read off. Our earlier example is repeated on the figure, confirming that five circuits are needed to carry 1 erlang at 1 per cent grade of service.

The traffic-to-circuit relationship shown in Table 10.3 and Figure 10.9, and the Erlang lost-call formula on which they are based are only suitable for dimensioning 'full-availability' networks (as defined in Chapter 6). For 'limited availability' networks, slightly different formulae and traffic tables must be used, but they are available. Further, another slightly different set of traffic tables is called for when dimensioning networks which use overflow routing. Figure 10.10 shows an example of overflow routing in which traffic from exchange A to exchange B first tries to route directly, but is allowed to 'overflow' via exchange C if all direct circuits are busy. Only link A–B in this network can be dimensioned by the simple Erlang lost-call formula (provided we specify an overflow grade of service—e.g. 10 per cent of calls), but as we shall find out in Chapter 12, only a slight modification to the method is necessary in order to permit us to dimension all the other links.

10.8 EXAMPLE ROUTE DIMENSIONING

As an example of the use of the Erlang lost-call formula in route dimensioning, let us calculate the number of circuits required to carry offered traffic $A = 55$ erlangs at a grade of service, $B = 0.01$.

We shall show in Chapter 12 that an estimate of the number of circuits, B, required for 1 per cent grade of service is given by the formulae:

$$N = 6 + (A/0.85) \quad \text{where } A < 75$$
$$N = 14 + (A/0.97) \quad \text{where } 75 < A < 400$$
$$N = 29 + A \quad \text{where } 400 < A$$

so that, for our case, $A = 55$;

$$N \text{ (approx)} = 70 \text{ circuits}$$

So far so good, but from here on we rely on trial and error and the formula in Figure 10.5

substituting $A = 55$, $N = 70$ we get $B = 0.007$

This is slightly better than we need, so let us try

$A = 55$, $N = 69$. This time $B = 0.009$

Again, $A = 55$, $N = 68$, then $B = 0.012$

68 circuits give a slightly worse grade of service than we need, so 69 circuits must be provided!

10.9 CALL WAITING SYSTEMS

Whereas in telephone and other circuit-switched networks the dimensioning is carried out to a given grade-of-service (or call loss), this technique is not suitable for dimensioning all types of network. Operator switchrooms and data networks provide two examples where a queueing (or 'buffer') system allows callers (or data packets) to wait for a free operator (or for free transmission capacity). The queueing minimizes, and may even eliminate, call or data packet loss. In such networks our dimensioning techniques must be oriented towards ensuring that the number of operators (or the capacity of a datalink) is sufficient to keep the queue waiting time within acceptable bounds. If there are insufficient operators the queue and the consequent waiting time will get longer and longer. We must also ensure that the queue capacity is long enough.

Consider the operator switchroom first, since what happens there follows on from the Erlang lost-call formula already discussed. Figure 10.11 illustrates a network in which an operator switchroom is serving customers of the telephone network.

In all, N operators (called 'servers' in teletraffic jargon) are working and K places (confusingly also called 'servers') are available in the queue. The switchroom manager wants to make sure that the number of operator positions staffed (N) at any point of time during the day is great enough to cope with the traffic and so keep down the queue waiting time. Further, the switchroom designer wants to ensure that the queue capacity (K) is great enough to minimize the number of lost calls.

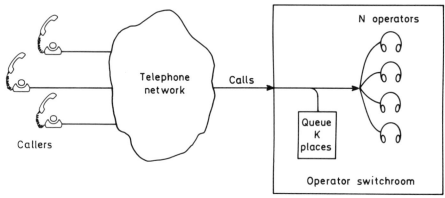

Figure 10.11
Call queueing in an operator switchroom.

Both values N and K can be determined from an adapted form of the Erlang 'lost-call' method, using various forms of the 'Erlang waiting-call' formula, as shown in Table 10.4.

The formulae of Table 10.4 are unfortunately complex and this is not the place to describe their derivation of any of the more complex versions of them that have also been developed. We can, however, discuss their practical use, as follows. A tolerable delay, t, is set for the time which telephone callers in Figure 10.11 will have to wait for an operator (e.g. 15 s), and we then decide what percentage of calls will be answered within this delay time (a typical target would be 95 per cent). It is then a trial-and-error exercise, using formulae (i) and (iv) of Table 10.4, to give the number of servers (N) which will meet the target constraints.

The value N will always be greater than the offered traffic A. If it were not so, the operators would be reducing the queue at a rate slower than that at which the queue was forming. The queue and the delay could only get longer (Substitute $N = A$ in formula (ii) and you will see that the average delay is infinite.)

Let us use an example to illustrate the method. We shall assume that the switchroom handles an average of 3000 calls per hour, and that they take an average of $d = 60$ seconds of operator's time (i.e. 'service time') to deal with. Then the offered traffic,

$$A \text{ (average number of calls in progress)} = 3000 \times 60/3600 = 50.$$

Guessing a value of $N = 55$ to meet our target that 95 per cent calls will be answered within 15 s, we calculate in order:

$A = 50$ erlangs	offered traffic
$N = 55$	active servers (operators)
$B = 0.054$	from the lost-call formula
$C = 0.388$	from formula (i)
$t = 15$ s	target answer time
$d = 60$ s	average service time

Probability that delay exceeds 15 s from formula (iv) $= 0.11$

Since this proportion is higher than the target of 0.05, we will need more servers—but probably not too many more since we are not too far adrift.

Repeating for $N = 56$ we find the probability of answer in 15 s is 0.07. Almost, but not quite! Repeat again for $N = 57$ and bingo!—the probability of answer taking longer than 15 s is only 0.04. Thus 57 operators will be needed in our switchroom.

Table 10.4
The Erlang waiting call formula.

(i)   Probability of delay (i.e. that a customer will have to wait to be served)

$$= C = \frac{NB}{N - A\,(1 - B)}$$

This is called the Erlang call-waiting formula.

(ii)   Average delay, (Held in queue)

$$= D = \frac{C}{N - A} \times d$$

(iii)   Average number of waiting calls

$$= \frac{AC}{N - A}$$

(iv)   Probability of delay exceeding t seconds

$$= C_e^{-(N - A)t/d}$$

(v)   Probability of j or more waiting calls

$$= C\left(\frac{A}{N}\right)^{j}$$

(vi)   Probability of x servers being busy

    (a)   No queue, $X =$ number of operators busy

$$p(x) = \frac{A^x}{X!}\, p(0) \qquad 0 \leqslant x \leqslant N$$

    (b)   Queue, all operators busy, j queue places taken

$$x = N + j$$

$$p(N + j) = C\left(1 - \frac{A}{N}\right)\left(\frac{A}{N}\right)^{j}$$

Notation

$N =$ number of servers (those processing the calls, e.g. operators)
$j =$ number of calls in the queue
$B =$ lost call probability if there were no queue—derived from the Erlang lost call
    formula $= E(N,A)$
$A =$ offered traffic in erlangs
$d =$ average service time required by an active server to process a call

---

Having determined the number of operators (N), the number of queue places needed is found by using formula (v) of Table 10.4. The planner first decides what small proportion of calls it is acceptable to lose because there are no free places in the queue. Then he finds the value j from the formula which satisfies the condition that the probability of lost calls (more waiting calls than target) is met.

Going back to our example, let us aim to lose only 1 per cent of calls because of inadequate queue capacity. Then the probability of j or more waiting calls must be less than 0.01, where j is the designed capacity of the queue. The calculation is as follows:

$A = 50$ erlangs	offered traffic
$N = 57$	active operators
$B = 0.039$	Erlang lost-call formula
$C = 0.246$	Erlang waiting-call formula

So, from formula (v) $0.01 = C(A/N)^j$

rearranging $j = (\log 0.01 - \log C)/(\log A - \log N)$
$$j = 24.4$$

rounding up, 25 queue places will be needed.

Formulae (ii), (iii) and (vi) of Table 10.4 are not normally used in dimensioning, but out of curiosity, why not calculate the average waiting time and the average number of callers in the queue? The average waiting time is 2.1 s, and the average number of waiting calls is 1.76.

Calculating the number of operators needed to answer calls on a company's office switchboard is carried out in an identical way to that used above. Theoretically, different formulae must be used on occasions where the offered traffic distribution does not follow Erlang's detailed mathematical assumptions (given in the appendix). Inappropriate use of the formulae may lead to errors in the available network capacity, but on the other hand, most practitioners prefer to use simple dimensioning methods and rely upon their practical experience rather than worry too much about complex statistical theory. If interested, a number of more complex methods and their detailed derivations are given in numerous specialist texts on queueing theory, some of which are listed at the end of the chapter.

10.10 DIMENSIONING DATA NETWORKS

Terminals and other devices sending information on data networks generally do so in one of two ways:

(i) In a bit-oriented fashion, along a direct connection between transmitting and receiving devices. The dedicated path can either be a leased circuit, or a circuit-switched connection.

(ii) In a packet-oriented format over a packet-switched network.

In the former case two factors affect the 'success' of the call. First, if the connection is established over a circuit-switched network then the grade of service of lost connections is important. But having set up the dedicated path, the bit speed of the link is also important. In both case (i) and case (ii) the network designer will wish to ensure that the link between the terminal devices can cope with a greater overall speed of data transfer than the average demand generated by the terminals. If the link were not fast enough, a build-up of data would occur at the transmitting end, slowing down the whole process.

In an extreme case there may be very 'peaky' demand in which high volumes of data are required to be transferred—but only in short bursts, as shown in Figure 10.12(a). In this case, the link may not be able to cope with the peak demand (Figure 10.12(b)) and the profile of the carried data may have to be flattened by storing it in a buffer pending available transmission capacity. The example is typical of data networks and demands the use of a 'call-waiting' dimensioning technique to decide the required data-link bit rate and buffer capacity.

Some of the formulae used to dimension data networks are derived from the Erlang waiting-call method already given in Table 10.4, but recent research has led to the use of other methods.

The formulae of Table 10.4 may be adapted for data network dimensioning by substituting $N = 1$. This corresponds to a single connection between points. It is equivalent to a single data-link between terminals or packet-switched data exchanges. The link will have a given bit speed capacity, and the data flowing over the link could be thought of as being measured in erlangs. For example an average bit rate of 1250 bit/s being carried on a 2400 bit/s link is equivalent to 1250/2400 or 0.52 erlangs.

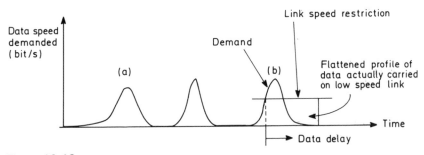

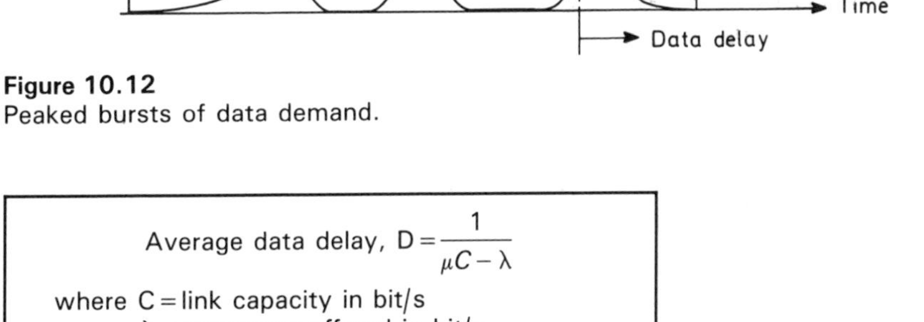

Figure 10.12
Peaked bursts of data demand.

$$\text{Average data delay, } D = \frac{1}{\mu C - \lambda}$$

where C = link capacity in bit/s
λ = messages offered in bit/s
$1/\mu$ = average number of bits per message

Figure 10.13
Average data delay.

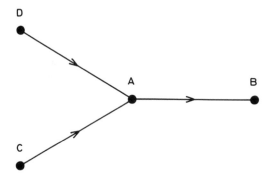

Figure 10.14
Dimensioning meshed data networks.

Setting $N = 1$ in formula (ii) (Table 10.4), by rearranging it, and by renaming some of the parameters to some more appropriate for our understanding, we obtain the formula shown in Figure 10.13. Knowing the message characteristics λ and μ of the offered traffic, and having predetermined a tolerable delay (D), the minimum line capacity of the point-to-point data-link (C) may be determined. If the bit speed of a single link is exceeded, then two or more will be needed.

In more complex networks, as exemplified by Figure 10.14, the same delay dimensioning method may be used for the links of packet and store-and-forward type data networks. Thus link D–A of Figure 10.14 can be dimensioned according to the needs resulting from the sum of D–A and D–B traffic. Likewise link A–B is dimensioned according to the aggregate needs of D–B, C–B and A–B traffic.

10.11 POLLACZEK–KHINCHINE DELAY FORMULA

The formulae presented so far in this chapter all derive from Erlang's work of 1917 and are therefore subject to the assumption that the offered traffic follows a particular 'Poisson distribution' and has a 'negative exponential' distribution of holding times. While in practice these assumptions are found to give reasonable modelling of the performance of many networks, they are not always applicable, especially in some types of data network. We therefore conclude the chapter by presenting without further

$$\text{Average delay, } D = \frac{A(1 + V^2)\,\bar{h}}{2(1 - A)}$$

A = probability of delay (= traffic in erlangs)
$V = s/\bar{h}$ = coefficient of variation of the holding time
s = standard deviation of the holding time.
$\bar{h}$ = average holding time

Figure 10.15
Pollaczek–Khinchine delay formula.

discussion the more general 'Pollaczek–Khinchine' formula for the average data delay. The formula, given in Figure 10.15, is always applicable (and can be rearranged to the 'special form' of Figure 10.15 by the substitutions which embody the earlier assumptions: $V = 1$, $A = \lambda t \simeq 1$ and $t = 1/\mu C$).

APPENDIX: THE DERIVATION OF ERLANG'S FORMULA

The dimensioning of telephone and other circuit-switched networks is based upon a technique developed by a Danish mathematician, A. K. Erlang, in 1917. The 'Erlang formula' or 'Erlang lost-call formula' and its derivatives enable network designers to determine how many circuits are required on a given route in order to control the proportion of blocked calls (a typical target is for losses of one-call-in-one-hundred or less).

In arriving at his statistical formula, Erlang made a number of assumptions, based upon his observations of telephone callers' behaviour, as follows:

(i) Calls are generated at random and only one-at-a-time. (This kind of behaviour has been described as the 'Poisson' process, after the mathematician who invented it.)

(ii) Statistical equilibrium exists. By this we mean that the probability of a given number of circuits being in use does not change over the relevant period.

(iii) Calls 'arriving' in the system when all the circuits are busy are lost, their holding time is zero, and they do not generate 'repeat attempts'. (This is the 'lost-calls cleared' assumption that lends its name to the 'lost-call' formula. Other more advanced formulae model the behaviour when calls which cannot be imme- diately completed are either queued up, or are considered to stimulate customer repeat attempts.)

(iv) Call holding times follow a negative exponential distribution.

Let us imagine that at a particular time, x circuits are in use. Now, let us consider a time only a tiny fraction of a second later (which we denote as dt). Given assumption (i) above, only one of three things can have happened:

(a) One more circuit has become busy (raising the number to $x + 1$).

(b) One circuit has become idle (reducing the number busy to $x - 1$).

(c) The number of busy circuits has remained constant.

Now, the probability that event (a) has happened is equal to the probability of a single call arriving during the period dt. From the earlier part of the chapter we know this to be AdT/h. Conversely, the probability of any *particular one* of the established calls terminating during the period dt is independent of the new traffic being offered, and is equal to dt/h. Multiplying by the number in progress we obtain the probability that any one of the calls clears as xdt/h.

Next we apply assumption (ii). Since the system is in statistical equilibrium, then the probability that there were x busy circuits to start with and that one more circuit became busy must equal the probability that there were $x + 1$ busy circuits to start with, and that one circuit was released. If this were not the case we would find that the circuits either became generally busier over a period of time, or generally less busy, depending on the state of the statistical unequilibrium. Assuming $p(x)$ to be the probability of x circuits being busy, then the probability of $(x + 1)$ busy circuits will be $p(x + 1)$.

From our results above, the probability that there were x busy circuits to start with, and that there has been a transition from x to $x + 1$ busy circuits is:

$$p(x)A \ dt/h$$

And the probability that there were $(x + 1)$ busy circuits to start with followed by a transition from $x + 1$ to x busy circuits is:

$$p(x + 1)(x + 1) \ dt/h$$

As we said, we may equate the two values. Hence:

$$p(x)\frac{A}{h} \ dt = p(x + 1) \ \frac{(x + 1)dt}{h}$$

In other words,

$$p(x + 1) = \frac{A}{x + 1} \ p(x)$$

By substituting different values for x, we determine that:

$$p(1) = \frac{A}{1} \ p(0)$$

$$p(2) = \frac{A}{2}\frac{A}{1} \ p(0) = \frac{A^2}{2!} \ p(0)$$

$$p(3) = \frac{A}{3}\frac{A}{2}\frac{A}{1} \ p(0) = \frac{A^3}{3!} \ p(0)$$

$$p(x) = \frac{A^x}{x!} \ p(0)$$

The value $p(x)$ is the probability that x circuits are in use. From it we derive the 'probability density curve' shown in Figure 10.16, which gives the relative probabilities of different circuit numbers being in use. The curve is also called the 'Erlang distribution'. As can be seen from the curve, and as we would expect, the most likely number of circuits in use is nearly equal to the offered traffic intensity A.

Finally, we use assumption (iii); that calls arriving when all circuits are busy are lost. They may lead to delayed customers re-attempts but we assume that these do not distort the statistical distribution of call arrivals. By doing so, we can thus determine the value of $p(0)$. We add all the probabilities and adjust them so that their cumulative probabilities are equal to 1. Thus:

$$p(0) + p(1) + p(2) + p(3) + \cdots + p(N) = 1$$

From this, and our earlier result we conclude that:

$$p(0) = \frac{1}{1 + A + (A^2/2!) + \cdots + (A^N/N!)}$$

and

$$p(x) = \frac{A^x/X!}{1 + A + (A^2/2!) + \cdots + (A^N/N!)}$$

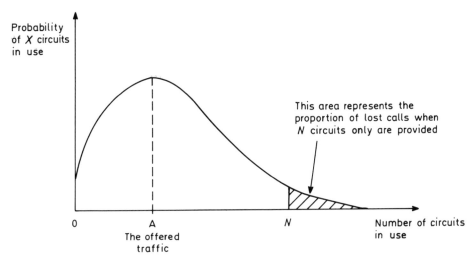

Figure 10.16
The Erlang distribution.

$$E(N, A) \text{ or } B(N, A) = \frac{A^N}{N!} \bigg/ \left\{ 1 + A + \frac{A^2}{2!} + \frac{A^3}{3!} + \cdots + \frac{A^N}{N!} \right\}$$

where $E(N,A)$ = proportion of lost calls, and probability of blocking
A = offered traffic intensity
N = available number of circuits

Figure 10.17
The Erlang lost-call formula.

This is the formula of the Erlang distribution, already illustrated in Figure 10.7. Its most useful application is in derivation of a further formula for calculating the proportion of lost calls offered to an N circuit group. This is the Erlang lost-call formula, usually denoted $E(N, A)$, or sometimes $B(N, A)$. It calculates the probability of call blocking, given N available circuits and an offered traffic intensity of A erlangs. Numerically, the value is equal to the probability of finding all N circuits busy, and it is derived from the Erlang distribution by substituting $x = N$. Figure 10.17 presents the formula.

BIBLIOGRAPHY

Bear, D. *Principles of Telecommunication Traffic Engineering*, 3rd edn. Peter Peregrinus, 1988.

Cooper., R. B., *Introduction to Queueing Theory*, 2nd edn. North-Holland, 1981.

Erlang, A. K., *Solution of some Problems in the Theory of Probabilities of Significance in Automatic Telephone Exchanges*. 1918.

O'Dell, G. F., An outline of the trunking aspect of automatic telephony, *J. IEE.*, 1927, 65, 185–222.

Schwartz, M. *Computer-Communication Network Design and Analysis*. Prentice-Hall, 1977.

Teletraffic Engineering Manual. Standard Electrik Lorenz AG, Stuttgart, 1966.

11

TRAFFIC MONITORING AND FORECASTING

Having explained the underlying principles of electrical communication and the statistical 'laws' of telecommunications traffic, we can now consider the practical design and operation of networks. A prime concern is to ensure that there are adequate resources to meet the traffic demand, or to prioritize the use of resources when short-falls are unavoidable. Two things have to be done in order to keep abreast of demand. These activities are the monitoring, and future forecasting of network use. In this chapter we shall review the parameters to be applied in measuring traffic activity and go on to provide an overview of some forecasting models for the prediction of future demand. Used as an input to the teletraffic engineering formulae of Chapter 10, the forecast values of future traffic can be used to predict the future network equipment requirements on which the overall planning process will be based.

11.1 MEASURING NETWORK USAGE

We discussed different methods of defining the volume of network usage in Chapter 10. These methods in various ways served to measure, over a predetermined period of time, any of the following parameters:

  (i)   The total number of calls or messages.

  (ii) The total conversation time.

  (iii) The total holding time (holding time includes both conversation time and call set-up time).

  (iv) The total number of data characters or 'packets' conveyed.

We have covered the concept of 'traffic intensity', which is a measure of the average call demand. We have shown how traffic intensity measured in erlangs determines the number of circuits needed on a route in a circuit-switched network to maintain a given 'grade of service'. For this reason it is usual practice to monitor the magnitude of the traffic intensity, and to predict its future values by forward forecasting. This provides the basis for planning a future circuit provision programme.

Forecasting, however, need not be confined to predicting future traffic intensity. The measuring and forecasting of other parameters can also be valuable predictors of network traffic demand. For example, in telephone networks where customers pay for calls based on the total time used in conversation, the measurement of conversation time allows not only the preparation of customer bills but also enables the network operator to calculate the revenue due and the resulting profits, immediately after the end of each financial period. The actual amount of use and predicted future revenue may well be vital to the network operator on the business and financial side.

Forecasting also allows investments to be planned to meet future demand, commensurate with predicted levels of profits and financial resources. Forecasting also helps to determine when established networks or services will become unprofitable, and therefore when it may be appropriate to consider run-down or discontinuation. No commercially-minded company can tolerate loss-making products or services for long.

11.2 USAGE MONITORING IN CIRCUIT-SWITCHED NETWORKS

The parameters commonly used to monitor the usage of circuit-switched or point-to-point (leased) networks are listed below:

- The traffic intensity.

- The total number of minutes of usage.

- The total number of calls completed.

- The total number of calls attempted.

The reason for monitoring each of these parameters, and the methods of measurement, are described in the following sections.

Traffic intensity

'Traffic intensity', as we learned in Chapter 10, is the measure of the average number of simultaneous calls in progress on a route between two exchanges, or across an exchange, during a given period of time. It is normally measured during the hour of greatest traffic activity (or so-called 'busy hour') and is quoted in 'erlangs'. There are two principal methods of measuring traffic intensity: either by frequent sampling of the number of circuits actually in use and calculation of the statistical average, or by using a call-logging system which records the start and finish time of each individual call, so allowing the total circuit usage time to be measured accurately. Total usage during the period is divided by the duration of the period to give the average number of circuits in use.

A check at least once per month of the busy-hour traffic intensity on each route in a network is essential to make sure there are enough circuits on each individual route. Circuit requirements for each route are calculated from these values according to the Erlang formula, using the forecast traffic and the desired grade of service as inputs to the formula.

On their sampling days, many network operators are not content to monitor the traffic intensity during the anticipated busy hour (though this is an accepted practice); they also monitor the traffic activity profile on each route throughout the whole day. The profile is invaluable as an indicator of overall calling patterns, revealing short or long periods during the day when the traffic on a particular route is either very heavily congested or is under-utilizing the available capacity. In the latter case, the capacity may be useful as a means of relieving a congested route (by transit routing in the manner discussed in the next chapter). Alternatively, an advertising campaign could be used to stimulate extra network usage.

Traffic profiles on different routes vary greatly according to the nature of the route (e.g. local, trunk or international) and of its users, and they may change slowly over a period of months. Some profiles slowly distort towards a highly peaked, busy-hour oriented pattern, while others reflect a steadier traffic load throughout the day.

Highly peaked and busy-hour oriented profiles, of which Figure 11.1 is an example, can be problematic for network operators for two reasons. First congestion is highly likely during the period of peak traffic demand (even if circuits are relatively well provided), and second, over the day as a whole there will be relatively little network use, with correspondingly low revenue.

Highly peaked traffic profiles can sometimes be flattened by charging a heavy price premium for calls made at the peak time. The idea is to persuade customers to make their calls either a little earlier or later during the day. If it can be achieved, a very flat profile such as that shown in Figure 11.2, makes highly efficient use of the network. Charges which depend upon the time of day tend to have an effect on domestic call demand, but the effect is less marked on business calls, because few of the callers worry about the cost.

Figure 11.2 illustrates an extreme redistribution of the traffic previously shown in Figure 11.1. In the course of the day the same number of calls have been completed in Figure 11.2 as in Figure 11.1 and thus a similar revenue has been earned, but the flatter traffic profile has meant a lower busy-hour traffic value and a reduced circuit

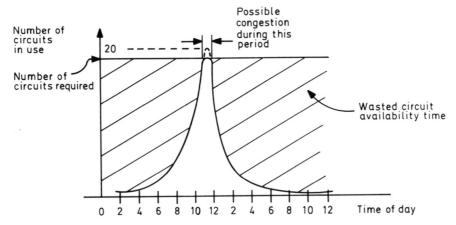

Figure 11.1
Busy hour oriented traffic profile.

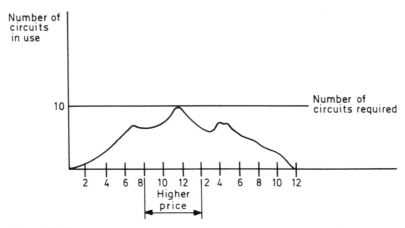

Figure 11.2
Traffic redistribution.

requirement. The profile has been flattened by stimulating 'time-shift' of some of the peak traffic, by charging a higher price for the peak period 8 am–2 pm.

There are times when significantly higher busy-hour prices fail to dissuade customers from calling at the busiest time, and none of the traffic is timeshifted. When this happens the only way to improve overall network utilization and revenue is by stimulating new off-peak traffic.

Total usage monitoring

Besides coping with its route busy-hour demand, a network must be designed to carry the total volume of demand. Exchanges may be limited in their capability to record the details of individual calls, so that above a certain limit, billing or other call detail information may be lost. Also, the amount of exchange equipment (switching points, registers etc.) which is required will depend on the total volume of demand. There are two related measures of the total usage of a circuit-switched network or a leased point-to-point connection. The two measures are the 'paid' time and the 'holding' time.

The 'paid' time equals the total number of usage minutes for which customers have been charged. In telephone network terms this equates to the number of 'conversation minutes'. In a telex, or a circuit-switched data network, the paid time equates to the 'connected' time, the duration of which has a direct relationship with the total amount of textual information that could be carried.

By contrast, the total 'holding' time is the time for which the network itself has been in use and is always slightly greater than the 'paid' time, as Figure 11.3 shows. It includes not only the 'connected' or 'conversation' time, but also the time needed for call set-up and cleardown and the ringing time of the destination telephone. The holding time is the better indicator of network resource requirements, but network operators may choose to monitor either 'paid' or 'holding' time, or both.

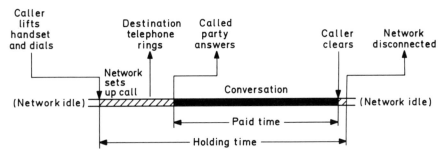

Figure 11.3
'Holding' and 'paid' time.

Paid time

The paid time is a direct measure of customers' actual usage. This is the time during which the customer actually communicates, and could be considered to be the best measure of 'real' demand. The more paid-time use of a network is made, the greater is the volume of communication. This contrasts with the traffic intensity, which is only a measure of how many people choose to communicate at the same time. It is paid time that attracts revenue and is therefore the most important financial parameter. The paid time accumulated by particular customers or on particular routes between individual exchanges gives a strong indication to network operators as to where they should concentrate their resources and efforts, and where they are most at risk from competitors. Operators can ill afford to lose valuable customers or the traffic on profitable routes.

There are three methods of either measuring or estimating paid time. The first and most obvious method is to sum the usage made by individual customers. However, while summation gives valuable route-by-route paid minutes statistics when derived from *itemized* customer bills, in those networks which only record their customers' usage as a bulk total number of units on simple cyclic meters (as Chapter 17 discusses), summation only reveals total network use, without any route-specific paid minute information. But all is not lost, since the paid time on each route may be measured directly by sampling the traffic intensity on the route itself at a number of regular time intervals, and applying one of the following conversion formulae for estimating the paid minutes:

Busy hour paid minutes = Busy hour traffic in erlangs × 60 × efficiency factor

where efficiency factor = $\dfrac{\text{Paid time}}{\text{Holding time}}$

(the value is estimated from historical information).

Daily paid minutes = 5 or 6 times busy-hour paid minutes

(again the value is obtained from historical records).

232

Total monthly paid minutes = approx. $22 \times$ daily paid minutes

(It is a good enough assumption to estimate an entire month's traffic as equivalent to that of 22 working days.)

Holding time

In contrast to the paid time, the longer period of 'holding' time represents the total time when network resources are in use, and the difference between holding and paid time volumes is the time when the network 'common equipment' (used for call set-up and cleardown) is in use. This quantity is called the common equipment holding time. The total holding time, the common equipment holding time, and the traffic intensity values all help to determine how many of each of the various equipments must be provided in the network.

The amount of a particular type of common equipment (e.g. signalling receivers) needed in an exchange is usually calculated by the normal Erlang method, with the common equipment traffic load calculated by multiplying the expected number of calls by the average common equipment holding time per call. Typical grades of service demanded of common equipment might be 0.5 per cent lost calls, 0.1 per cent or even 0.05 per cent.

Number of calls attempted

The total number of calls attempted (also called the number of bids) is the best measure of unconstrained customer demand, since unlike the paid minute volume or the traffic intensity actually carried by a network, it is not limited by network congestion. At a time of network congestion, the busy hour call attempt (BHCA) count may continue to increase (as unsatisfied customer demand continues to grow) while paid minute volumes and traffic intensities may saturate at the maximum capacity of the network. A very large number of unsatisfied call attempts is an almost certain sign of congestion—either through underprovision of equipment or resulting from short-term network failure, and is a good means of spotting suppressed traffic within a network.

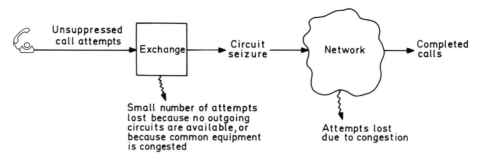

Figure 11.4
Call attempts and congestion.

Unfortunately the exact amount of suppressed traffic is difficult to estimate, since the 'persistence' of customers in repeating call attempts many times over affects the overall bid count. Figure 11.4 illustrates an example of the failure of some call attempts.

The number of call attempts can only be measured accurately by monitoring each individual customer's line. A value measured at any point deeper in the network will not be an accurate measure, since it will have been reduced by the effects of any congestion at the network fringe.

In cases where the call attempt count suggests that traffic is being suppressed, then the unexpurgated busy-hour traffic intensity and paid minute demand can be estimated from one of the following conversion formulae:

Busy hour traffic (in erlangs) = No. of call attempts in the busy hour × (average call holding time)/60

Busy hour paid time = No. of call attempts in the busy hour × average paid time per call.

Sometimes exchange monitoring limitations make it impossible for network operators to measure accurately the number of call attempts from a given source to a given destination. If so, the number of call attempts may be estimated as the number of outgoing circuit 'seizures'. As shown in Figure 11.4, however, this estimate will be lower than the actual since a small number of 'call attempts' ('bids') fail in the first exchange due to congestion, and so do not mature into an outgoing circuit seizure.

Number of calls completed

The number of calls completed in a network-sense (i.e. reaching ringing tone or 'answer'), when compared with the number of calls attempted, gives another measure of the state of network congestion. The proportion of busy hour calls completed, when expressed as a percentage of the number of calls attempted, should equal the design grade of service.

Hence:

Grade of service

$$= \frac{\{(\text{number of busy hour call attempts}) - (\text{number of busy hour call completions})\} \times 100\%}{(\text{number of busy hour call attempts})}$$

The grade of service is a measure of the frustration that a customer will experience when trying to complete a call during the busiest hour of the day. We could calculate the average daily percentage of lost call attempts, but this is not so usually done, because psychological analysis of customer behaviour suggests that it is of no relevance.

The number of calls completed by the network is a difficult quantity to measure, since not all signalling systems indicate the 'network-completed' state. In the American network a 'peg-count overflow' monitors this value but its absence in other networks means that it is also usual to measure only the number and proportion of 'answered' calls. This number is clearly lower than the number completed by the network since some calls are bound to encounter either a 'subscriber busy' state or a 'ring tone—no reply' condition. The proportion of answered to attempted calls, and of answered calls to 'seizures', are known as the 'answer bid ratio' (ABR) and the 'answer seizure ratio' (ASR) respectively—they are defined mathematically below. Measured over relatively short periods of time (5–15 min), both are good indicators of instantaneous network congestion. The higher the network congestion, the lower the ABR or ASR. Conversely, the higher the ABR or ASR the lower the congestion, is not necessarily true since calls may remain unanswered for a range of other reasons (e.g. people are simply not answering their phones). These same uncertainties mean that no conclusion can be drawn from the actual value of the ABR or ASR. A conclusion can only be drawn from the value relative to its 'normal'.

$$\text{Answer bid ratio (ABR)} \quad = \frac{\text{no. of answers}}{\text{no. of call attempts}}$$

$$\text{Answer seizure ratio (ASR)} = \frac{\text{no. of answers}}{\text{no. of seizures}}$$

In Chapter 19 the use of ABR and ASR statistics as tools for short-term 'network management' surveillance of the network will be described further.

11.3 MONITORING USAGE ON PACKET-SWITCHED NETWORKS AND LANS

There are two prime factors of importance to data network users. These are the overall throughput capacity and the 'response' (or 'network transaction time') of the network.

The throughput capacity is the amount of data that can be sent over the network during a given period. Usually a network is designed to cope with the maximum demand of the peak hour, although network cost savings can be realized by queueing up less important data, for transmission outside the peak hour. Such queueing leads to more effective 24-hour use of the network.

Throughput capacity can be measured in bits per second (bit/s), messages per second, packets per second, or segments per hour (1 segment = 512 bits). In designing and upgrading network capacity, it is important to consider the practical network throughput rather than just the transmission line speed capacity. The practical network throughput will always be less than the line speed capacity, first because not all messages are received without errors and some have to be retransmitted, and second because some of the available line capacity is effectively 'wasted' in gaps between messages. The two effects are summarized in the formulae below:

$$\text{Successful message throughput (TP)} = \frac{\text{correct messages received}}{\text{time duration taken}}$$

Writing it another way:

$$TP = \frac{\text{total messages} \times \text{proportion with no errors}}{\text{time taken for message transmission} + \text{wasted time between messages}}$$

$$TP = \frac{M\,(1-P)}{(M/R) + T}$$

where M is the average message length in bits, P is the probability of errors in the received message, R is the line speed in bits per second, T is the average line time 'wasted' between messages and M/R is the time required to transmit average message.

Rearranging the formulae, so that we can relate the required network line speed to the required user data throughput:

$$R = TP \left(\frac{1 + TR/M}{1 - P} \right)$$

This formula ensures an adequate average throughput capacity of the network, but another equally important characteristic of data networks is their response time. The computer systems linked by data networks will have been designed to carry out a given set of functions within a given period of time, to maintain for example a database of share prices updated at least once every hour, or—more onerously in terms of response time—to control the trains on a metropolitan underground railway.

The overall response time of a computer and data network includes the time taken to transmit an error-free message along the line and back, plus the computer processing time at the far end of the communication link, and the line turn-around time (if the link is half duplex). The line transmission time includes not only the period during which the line is conveying data, but also any time spent waiting for any previous messages to be transmitted. If the response time of the network is too long, then the network designer must increase the network throughput capacity, which he can do either by upgrading line speeds (as might be appropriate on a LAN or on a point-to-point leased circuit network, using say 9600 bit/s modems rather than 4800 bit/s ones) or by providing a larger number of circuit connections (as might be appropriate in a circuit or packet-switched data network).

The dimensioning method of calculating what overall bit throughput is required to meet a given response time constraint is based upon the Erlang waiting-time formula described in Chapter 10. As might be expected, the faster the required response time, the greater is the transmission linespeed required.

In conclusion, no matter which dimensioning method is used, if the network is to have adequate capacity to meet the user's throughput demand, the transmission line speed must be chosen with a relatively higher capacity, sufficient to counteract the effect of errors and the 'wasted' time between messages. The 'wasted' time, incidentally, includes not only periods of instantaneous non-use, but also includes some of the data overheads which accompany each message (e.g. the header and destination codes of the 'network protocol' discussed in Chapter 9).

Because they have a marked impact on the performance of data networks, it is worth digressing for a moment to discuss the effect of data 'packet', 'block' or 'frame lengths'. As we learned in Chapter 9, the protocol used to convey data messages breaks these messages down into a number of 'frames', 'blocks' or 'packets', and transmits each with some other 'overhead' information which is needed to ensure delivery to the correct destination and to control the frequency of errors. During periods of line noise disturbance, longer packets of data are more likely to be affected by the noise than shorter ones. The result is a lower effective data throughput and a slower response time because of the extra workload imposed by the need for large-scale data re-transmission. On the other hand, shorter blocks have the disadvantage of decreasing the throughput at all times (whether the network is busy or not). This is because of the overhead of protocol 'headers', one each for the larger number of packets or frames.

11.4 FORECASTING MODELS FOR PREDICTING FUTURE NETWORK USE

The operation of networks, their efficient utilization, carriage of traffic, and their successful evolution all depend critically upon accurate estimation of future needs. It is vital for network operators to have an efficient method of forecasting the traffic which the network will need to carry in order that a network of sufficient size can be maintained to meet users' needs.

Forecasts have to cover both short- and long-term periods so that the whole business of providing for the future, the planning and the capital investment, can be put in hand in good time. For straight circuit provision a twelve-month forecast may be adequate, but for normal extensions of switching and transmission equipment it is necessary to look from two to five years into the future, in line with the ordering lead time. Even longer forecasts may be needed when planning new cables, laying new ducts, and setting up major new sites. The longer-term forecasts are inevitably less accurate than the shorter, but since there is more time for adjustment, this is acceptable.

Forecasting uses historic measurements of network usage, and particularly of growth in usage, to predict likely future customer demand. A number of different forecasting models have been developed, but none of the methods are 100 per cent reliable. This is hardly surprising since they all involve attempts at predicting future human behaviour. The reliability of any forecast relies upon the accuracy of the historical information and the period over which it has been collected. The shorter the period of historical knowledge and the lower the number of observations on which it is based, the less reliable the forecast. In general, the selection of an appropriate forecasting method ensures that the forecast is accurate for a period into the future approximately equal to one third of the period of the historical information. Thus a one year forward forecast should be based on at least three years historical information. The best way to select a particular forecasting method for use in any given circumstance is by experience. If the model appears to give reliable results in practice and therefore is a 'good predictor' then use it, if not, try another method.

In this chapter we shall not attempt to cover all the available forecasting methods, but instead review a few simple methods, particularly those described in CCITT's Recommendation E506 on 'forecasting of traffic'.

As we have seen in the earlier part of this chapter the most important parameters measuring the overall network usage are the maximum instantaneous traffic demand (i.e. the traffic intensity, or its equivalent) and the maximum volume demand (i.e. the paid minutes or their equivalent). These parameters, more than any others, govern how much capacity is needed in the network and the revenues that result. As we have also seen, the two values are related by the daily traffic profile, so that paid minutes may be estimated from the busy-hour traffic intensity or vice versa. Because of this fact, either or both of the parameters can be forecast, and a forecast for the other parameter can be calculated using the conversion factors presented earlier. Forecasting the values directly is called 'direct forecasting', while deriving them from conversion formulae is called 'composite forecasting'. By using both methods, the forecast can be double checked.

In preparation for a traffic forecast, the forecaster should identify any regular or irregular disturbances which have affected past traffic and may affect the future forecast. These may include discontinuities in traffic growth or decline, resulting from disturbances such as changes in tariffs, modernization of the network, or removal of congestion. The use of composite forecasting as a double check comes into its own when a large number of discontinuities or irregular jumps in volume are present. Figure 11.5 shows an example of a leap in traffic growth caused by the stimulation of traffic resulting from the introduction of automatic service. Such leaps in telephone traffic growth—even growth of many times over—are not uncommon.

The process of developing a forecast comprises five steps. First, one of a number of different types of forecasting models must be chosen. Examples include simple models which might assume linear, quadratic or exponential growth in traffic as shown in Figure 11.6. Alternatively, more complex forecasting models such as econometric models may be used. Econometric models attempt to predict future demand by assuming that the traffic depends upon a number of socio-economic variables such as the retail price index, the population count etc. The equations of an econometric model reflect a relationship between these variables and the traffic parameters.

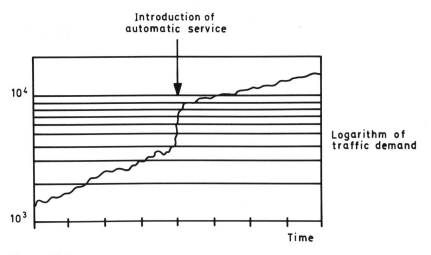

Figure 11.5
Discontinuity in traffic growth.

The second step is to guess exactly which model within the class fits the historical values most accurately, and the third step is to evaluate the unknown coefficients in the appropriate formula (i.e. values A, B and C in Figure 11.6).

The accuracy of the model should be confirmed in the fourth step by a method known as 'diagnostic checking'. This ensures that there are no major discrepancies between the model and the historical information. If there are discrepancies, it attempts to remedy them by adding correction factors or by adjusting the formula coefficients. Finally, the future forecast can be made. The work process is shown in the flow diagram of Figure 11.7.

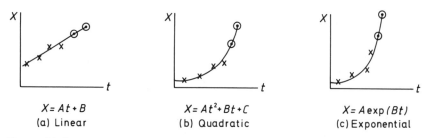

$X = At + B$
(a) Linear

$X = At^2 + Bt + C$
(b) Quadratic

$X = A \exp(Bt)$
(c) Exponential

Figure 11.6
Simple forecasting models. X, historic values; ⊙, forecast future values.

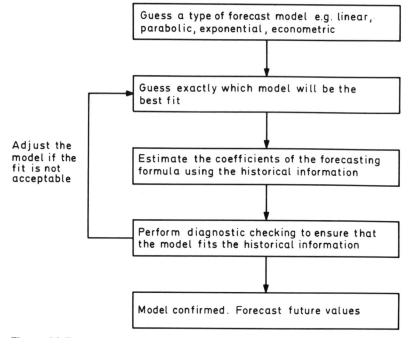

Figure 11.7
Forecasting procedure.

11.5 FITTING THE FORECASTING MODEL

Identification of the appropriate model and of the explanatory variables is the most difficult aspect of forecasting. Even when a suitable form of equation for estimating the behaviour has been determined, the evaluation of the coefficients is complex and a number of methods are available. Perhaps the simplest method of estimating the coefficient values is called 'regression'. This is the only method we discuss here, and the example that follows shows how it can be applied to a linear forecasting model. In such an instance the method is known as 'linear regression'.

In linear regression we assume that the growth in traffic demand (or the change in value of some other parameter) will change over time in a linear, or straight line manner. Linear regression therefore attempts to 'fit' the best straight line possible on to the historically recorded values of the parameter. When this has been done, the straight line is extended into the future to give a forecast of subsequent values, as Figure 11.8 illustrates.

In Figure 11.8 the value of a parameter has been measured every year from 1982 to 1988; based on this information, a forecast of future values corresponding to the years 1989 to 1992 is required. (The forecast is likely to be valid for around two years (one third of the period of historical information).) The forecast values are those predicted by the straight line shown in Figure 11.8. This is the 'best-fit' straight line through the historical values.

The best fit straight line can be determined by minimizing the total of the vertical distances of each of the historic parameter values from the 'best-fit line' as illustrated by Figure 11.9. In this example the best-fit line is taken to be that line for which the sum of the errors $(a + b + c + d)$ is minimized. This is the so-called 'least-difference' regression line.

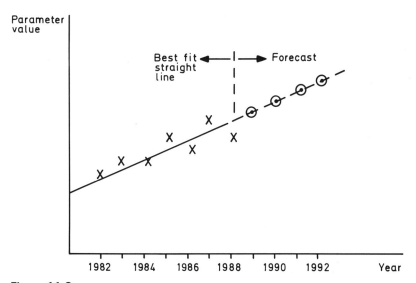

Figure 11.8
Forecasting by linear regression. X, measured historic values; ⊙, forecast future values.

It is more normal, however, when using the regression method of line fitting, to work on 'least squares regression' rather than 'least-difference'. In least squares regression the 'best-fit' line is that corresponding to the minimum value of the squared errors, in other words the minimum value of $(a^2 + b^2 + c^2 + d^2)$.

The method produces an equation for the 'best-fit' or 'regression' line as follows:

$$Y = a + bX$$

In this book we shall not discuss in detail how to derive the values of a and b, and if interested the reader should refer to an advanced statistics or mathematics book. However, for the benefit of those of a mathematical mind the formulae are as shown in Figure 11.10.

Future values of the parameter Y are predicted one at a time by substituting relevant values of x (the future dates), a, and b into the equation shown in Figure 11.10.

Forecasting and regression methods need not be confined to straight line predictions of future events, and Figure 11.11 demonstrates the application of either the 'least

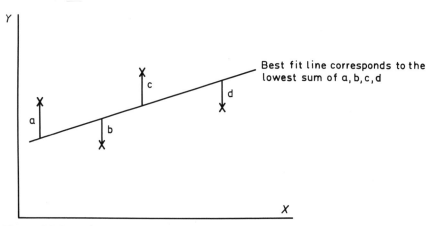

Figure 11.9
Least-difference regression.

Best fit line equation $Y = a + bX$

$a = M_y - bM_x$

$b = \dfrac{x_i y_i - M_x M_y}{S_x^2}$

where M_x = mean of the x-values
M_y = mean of the y-values
S_x = standard deviation of the x-values

Figure 11.10
Least square regression equation.

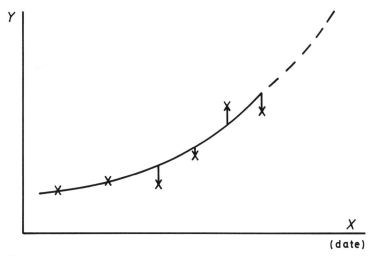

Figure 11.11
Using a curve for forecasting.

difference' or 'least squares' regression technique for determining the coefficients of a curved forecasting model.

11.6 OTHER FORECASTING MODELS

Many parameters exhibit a curved growth pattern similar to that already shown in Figure 11.11, and this is typical of a graph of total traffic demand on many different types of network. Almost any shape of curved forecasting model may be used, and some examples are given in Figure 11.12. The method remains the same—the forecaster needs to make a 'hypothesis' as to the shape of curve that best predict future events, and then to conduct regression analysis to fix the coefficients of the corresponding formula which represents the forecast curve.

As we said earlier, the type of curve that is used is entirely up to the forecaster, and previous success with a given curve and forecasting method is usually the best guide. There is no 'correct' method, and only time will tell whether the prediction is correct. The only recourse is to try something different next time. The forecaster must check that any forecast he makes is sensible. Simple questions give a good guide. For example a 10 000 line local exchange is unlikely to generate 3000 erlangs of traffic—that would require 0.3 erlang per exchange line (18 min of use each day at the peak time).

Among the large public telecommunications companies, econometric forecasting models are becoming the most popular. By using such models, the forecasters assume that the total volume of public network traffic is related to the overall economics, social conditions and trade of the company or nation. There is no doubt about their value since the companies continue to use them. However, to end on a gloomy note, it is important to recognize that forecasts are seldom accurate, and it is no disgrace to be under or over forecast. What is inexcusable is for plans to be drawn up without

242

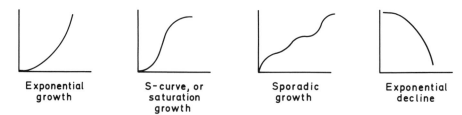

| Exponential growth | S-curve, or saturation growth | Sporadic growth | Exponential decline |

Figure 11.12
Possible prediction curves.

sufficient contingency and flexibility to be adapted as time goes on and the real level of demand makes itself known.

BIBLIOGRAPHY

CCITT Recommendation E401, 4th edn. 'Statistics for the International Telephone Service (Number of Circuits in Operation and Volume of Traffic)'.

CCITT Recommendation E506, 'Forecasting International Traffic'.

CCITT Recommendation E507, 'Models for Forecasting International Traffic'.

CCITT Recommendation E508, 'Forecasting New International Services'.

Hunter, J. M., Lawrie, N. and Peterson, M., Tariffs, *Traffic and Performance—The Management of Cost Effective Telecommunications Services*. Comm Ed Publishing, 1988.

Mayer, R. R., *Production and Operations Management* (chapter on Methods of Forecasting Demand). McGraw-Hill, 1982.

Morgan, T. J., *Telecommunications Economics*, 2nd edn. Technicopy, 1976.

Radford, J. D. and Richardson, D. B., *The Management of Manufacturing Systems* (chapter on Forecasting and Resource Planning). Macmillan, 1977.

12

NETWORK TRAFFIC CONTROL

Traffic control, we shall find, is as much art as it is science, and the profits are gratifying. The right control mechanisms and routing algorithms in the right places will determine the overall performance and efficiency of our network. What is more they will give us simpler administration, better management, and lower costs. This chapter describes a number of these admirable devices: first the simple methods commonly used for optimizing network routing under typical 'normal-day loading' conditions; then we look at recent more powerful and complex techniques, together with some of the practical complications facing telecommunications today. Finally we note how several network operators have found ways of interconnecting with the multiple carriers that have emerged lately as a result of market deregulation.

12.1 NETWORKS

Any metropolitan, international, trunk or transit network, by virtue of its nature and sheer size has to include a considerable number of large exchanges. If there are many interconnections between these exchanges, any individual call crossing the network will have plenty of alternative paths open to it. It is in fact the number of path permutations which sets the level or grade of service provided (i.e. the probability of successful call connection), and offers the customer what he most wants, an acceptable chance of making a successful call every time. Particular attention must be paid to choosing the call-control mechanisms which are going to determine the routing of our customers' individual calls, bearing in mind that in an ideal network we aim to achieve a controlled flow of traffic throughout the network at reasonable cost and without the penalty of complicated administration. Various methods and preliminary calculations can now be considered in turn.

12.2 SIZING TELECOMMUNICATIONS NETWORKS

First of all, to find out how many circuits a telecommunications network will need in order to meet traffic demand, a mathematical model is required for predicting network performance and it will comprise at least two parts:

- a statistical distribution to represent the number of calls in progress at any given time, and,

- a forecast of the overall volume of traffic.

Erlang's model provides a statistical method for approximating telephone traffic, which is based upon his measurements of practical telephone networks. A short reminder follows.

Erlang devised a method of calculating the probability of any given number of calls being in progress at any instant in time. The 'probability density function' used for the calculation is termed the 'Erlang distribution'. The 'Erlang formula', a related and complex iterative mathematical formula, can be manipulated into a number of different forms. The most common form is used to calculate the required number of circuits (or circuit group size) to carry a given traffic volume between two exchanges, under the constraint of having to meet a given grade of service. The traffic value input into this formula is the measurement of 'traffic intensity'. The 'intensity' of traffic on a route between any two exchanges in a network is equal to the average number of calls in progress, and is measured in *erlangs*.

In Chapter 10 we discussed how, for planning purposes, it was normal to measure the route traffic intensity during the busiest hour of activity, or so-called 'route busy hour'. Using this value and the 'Erlang formula', the number of circuits required for the route is calculated according to a given target grade of service.

The grade of service (GOS) of a telephone route between any two exchanges in a circuit-switched network is the fractional quantity of calls which cannot be completed owing to network congestion. The lower the numerical value of GOS, the better the performance. A typical target value used in many trunk and international networks is 1 per cent, or 0.01 (in other words 1 per cent of calls cannot be completed owing to network congestion—the other 99 per cent can). (*Note*: the calling customer, however, probably only perceives an end-to-end grade of service, on a call-by-call basis, of around 5 per cent (i.e. 5 per cent lost calls). This is because most connections comprise a number of links and exchanges, each of which is likely to be designed to inflict a 1 per cent loss.)

The common method is to dimension circuit groups within a network using a target grade of service and the Erlang formula; so that

Circuits required $= E(A, GOS)$

where E represents the appropriate form of the Erlang formula or function with inputs as follows:

A = route busy hour traffic

GOS = target grade of service

Alternatively, for packet or message-switched networks, we discussed an alternative formula, which sought to model the waiting time distribution of packets, rather than the proportion of lost calls; instead of dimensioning in accordance with grade of service, we can dimension according to waiting time. The method was again covered by Chapter 10.

12.3 HIERARCHICAL NETWORK

The Erlang model is a good predictor of the traffic behaviour of most telecommunications networks. A feature of telephone routes dimensioned using the Erlang method is that routes comprising a large number of circuits are proportionately more efficient (i.e. require proportionately fewer circuits per carried call) than smaller ones, when designed to the same grade of service, and are therefore more economic. This is

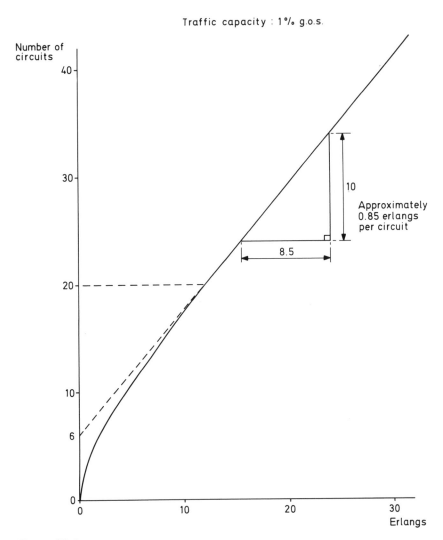

Figure 12.1
The inefficiency of small routes.

apparent from the general shape of the graph of circuits against traffic carrying capacity. Figure 12.1 illustrates the graph of 1 per cent loss, or 1 per cent grade of service. Note that the graph is almost linear for circuit groups with circuit numbers above 20 circuits, each extra circuit adding approximately 0.85 erlangs of traffic capacity. Thus by adding ten circuits to a group of 20 circuits we make it suitable for carrying 20.5 erlangs rather than 12 (an increase in capacity of 8.5 erlangs). Notice that an isolated group of 10 circuits is only suitable for 4.5 erlangs of offered traffic. Smaller circuit groups are thus far less efficient, as Figure 12.1 shows.

Looking at the graph it is useful to imagine that the first six circuits carry no traffic at all, and that all subsequent circuits have a capacity of 0.85 erlangs each. It is almost as if there is a 'penalty' price of six circuits for having a route at all! The line representing this assumption is shown on Figure 12.1. The equation of this line provides a useful method of estimating the number of circuits required to carry a given offered traffic load, but it is only valued up to about $A = 75$ erlangs. Above this value, other estimating formulae can be used as follows:

$$N = 6 + (A/0.85) \qquad \text{for } A < 75$$
$$N = 14 + (A/0.97) \qquad \text{for } 75 < A < 400$$
$$N = 29 + A \qquad \text{for } 400 < A$$

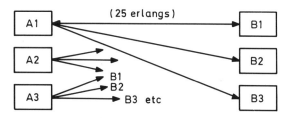

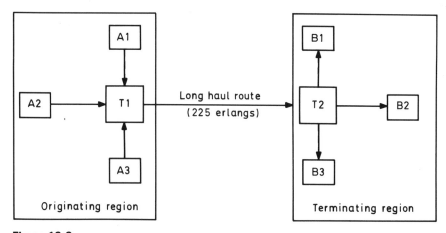

Figure 12.2
The benefit of combining small routes via transit switches.

The easiest way to minimize the circuit penalty which arises on small routes is to adopt a network which has a hierarchical structure, similar to the networks shown in Figures 12.2(b) and 12.3. In such a structure, a small number of main exchanges are fully interconnected with one another, while various tiers of less important exchanges have progressively less direct connections to other exchanges. If no direct route is available from one given exchange to another, then the call is referred to an exchange in the next higher tier. By using such a structure we can reduce the number of circuits on long-haul routes. Under such a hierarchical scheme, routes are combined so that groups may be dimensioned (or sized) to carry multiple-traffic streams, and so reap the economic benefits of the larger scale. The hierarchical network technique works as shown in Figure 12.2.

Let us assume that the callers on each originating exchange (A1, A2, A3) in Figure 12.2 are generating calls equivalent to a traffic intensity of 25 erlangs (i.e. an average of 25 simultaneous calls in progress) to each terminating exchange (B1, B2, B3). Let us also assume that exchanges A1, A2, A3 are quite close together, as are exchanges B1, B2, B3, but that region A and region B are a long way apart.

Without a hierarchical network structure, each A exchange would need a set of long-haul circuits to connect with each B exchange, as shown in Figure 12.2(a). This would give a total of 9 routes (A1–B1, A1–B2 etc.). Using Erlang's formula for 1 per cent GOS, each of these routes or 'circuit groups' would require 36 circuits, amassing a total requirement of $9 \times 36 = 324$ long-haul circuits.

Alternatively, if the traffic is concentrated through collecting exchanges T1 and T2 (called 'transit' or 'tandem' exchanges) within the originating and terminating regions, only a single route is required. That route has to carry all 225 erlangs, but only 247 long-haul circuits (again using Erlang's formula) are needed. Figure 12.2(b) illustrates this alternative network structure. Of course, when comparing the cost of the two structures we must not overlook the additional switches T1 and T2, as Comparison 1 below indicates.

Comparison 1: Direct versus hierarchical structure

(Traffic between each A–B pair: 25 erlangs)

All direct routes (as Figure 12.2(a))	324 long-haul circuits (9×36)
Hierarchical structure (as Figure 12.2(b))	247 long-haul circuits $2 \times$ transit switches (T1, T2); 225 erlangs each 324 short-haul circuits (A–T1 and T2–B)

So then, if 77 long-haul circuits (324-247) costs more than two 225 erlang switches (T1, T2) plus the short-haul access circuits (A–T1 and T2–B), hierarchical structure (Figure 12.2(b)) is the most cost effective.

In individual cases the particular circumstances will decide whether direct or hierarchical network structure is cheaper, though the general rule applies that in very-long-haul situations (international networks for example) hierarchical structure usually wins. Hierarchical structure can also be cost-effective when point-to-point route traffic is very small (as in rural networks). To illustrate this point, in Comparison

2 the example of Figure 12.2 has been repeated with much lower traffic values (and with circuit numbers re-calculated, again using the Erlang method).

Comparison 2: Direct versus hierarchical structures: low traffic only

(Traffic between each A–B pair: 3 erlangs only)

All direct routes	72 long-haul circuits (9×8)
	(8 circuits required for each 3 erlang route)
Hierarchical structure	38 long-haul circuits (for 27 erlangs)
	2 transit switches (T1, T2); 27 erlangs each
	72 short-haul circuits

Notice how in Comparison 2 the proportion of long-haul circuits saved as the result of a network hierarchy, is much greater than it was in Comparison 1 (47 per cent saving as opposed to 24 per cent). This is because the inefficiency of very small routes is very marked, as we saw in Figure 12.1. There are therefore greater proportional savings to be made from combining very small routes.

Hierarchical structure is common in many of the world's networks, in which a large number of local exchanges (or end offices) route their trunk traffic via a smaller number of trunk (or toll) exchanges. At the highest tier in the hierarchy there are probably only one or two international exchanges, while each lower tier has a greater number of exchanges, but each with only a restricted degree of long-haul interconnection. Figure 12.3 illustrates a typical national hierarchy.

An added advantage enjoyed by hierarchical networks lies in their capacity for getting the most out of their circuits at times when the busy hours of various transmission routes do not coincide. For an example look back again at Figure 12.2(b), and we can see that if A1–B1 is busy in the morning and A3–B3 in the afternoon, then thanks to hierarchical structure the same circuits can be used for both traffic streams. On the other hand, separate direct circuit groups (as in Figure 12.2(a)) would be inefficient, since one or other of the groups would always be idle.

A disadvantage of hierarchical networks when compared with direct-circuited networks, is their greater susceptibility to congestion under network overload. There are two causes:

(i) fewer overall circuits are available in hierarchical than in equivalent direct-circuited networks, and

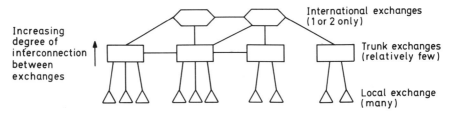

Figure 12.3
Hierarchical network structure.

(ii) congestion between only one pair of exchanges (e.g. A1–B1 of Figure 12.2(b)) will result in congestion on all other routes (e.g. A2–B3), since all calls have to compete for the same circuits (T1–T2). This can rapidly lead to further congestion, as customers dial merrily on regardless.

Overflow or 'automatic alternative routing' (AAR)

A simple means of improving purely hierarchical networks is to apply the technique of 'overflow', and for this we require the automatic alternative routing (AAR) mechanism.

AAR is also known as 'overflow' or 'alternative routing', and most exchanges are capable of it. It can be understood as a priority listing of the route choices leading to any given destination. Turning back to our example of Figure 12.2(b), let us assume that we introduce an additional link between exchanges A1 and B1. This might be a first choice, or 'high-usage' (HU) route (as described below). In this case, an AAR table for traffic from A1 to B1 might be as shown in Figure 12.4.

Cost benefits can be gained by judicious application of AAR (as Figure 12.4) rather than using the simple hierarchical structure (as Figure 12.2(b)), particularly if the

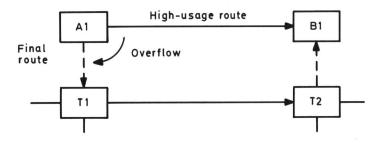

For traffic from A1 to B1 AAR table is :

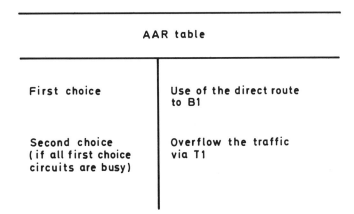

AAR table

First choice	Use of the direct route to B1
Second choice (if all first choice circuits are busy)	Overflow the traffic via T1

Figure 12.4
Automatic alternative routing (AAR).

route traffic between points A1 and B1 is large. The saving is achieved by making quite sure that the first choice route between A1 and B1 does not have enough circuits to carry the whole traffic. A remainder is then left which is 'forced' to overflow via T1. Because of the way they are used, the two routes A1–B1 and A1–T1–T2–B1 are called 'high-usage' (HU) and 'final' routes respectively. In fact, A1–B1 is a 'primary high-usage route', since it is first choice; however, secondary, tertiary, and so on, high-usage routes (i.e. extra route choices between 'primary HU' and 'final') may also be used, and they have their place in the AAR table.

To evaluate the savings of high-usage working, let us assume we provide high-usage routes of 25 circuits between each pair of exchanges in Figure 12.2(b) (i.e. 9 HU routes; A1–B1, A1–B2 etc.). Then the network adopts the pattern shown in Figure 12.5.

In order to compare the overall circuit requirement with earlier examples, we must dimension the network to the same overall 1 per cent grade of service performance. However, the dimensioning of the overflow links (A–T1 and T2–B), as well as of the overflow or 'final' route T1–T2, presents a special problem for which the Erlang formula has to be modified. This is because overflow traffic does not have a random nature as the Erlang dimensioning method requires. Consequently, in Comparison 3, the more complex Wilkinson–Rapp 'equivalent random' method (explained later in the chapter) has been employed, and the network has been dimensioned so that all customers get a grade of service equivalent to or better than 1 per cent. By this method, we determine that 43 'final route' circuits are required on T1–T2. We thus need $(9 \times 25) + 43 = 268$ circuits in total. Comparison 3 below compares this value with the direct and the hierarchical network structures.

Comparison 3: Direct versus hierarchical versus HU structures

(Traffic between each A–B pair: 25 erlangs)

All direct routes: 324 long-haul circuits

Hierarchical structure: 247 long-haul circuits; 2 × transit switches (225 erlangs each); 324 short-haul circuits

Overflow structure: 268 long-haul circuits; 2 × transit switches (32 erlangs each); 99 short-haul circuits

Comparison 3 shows that it is possible (by choosing the optimum size of high-usage (HU) circuit groups) to reduce significantly the size of the transit switches required (between the 'hierarchical' and 'overflow' structures), without significantly increasing the long-haul circuit requirement. This clearly reduces the cost. However, our example is not very realistic. In practice, if any of the individual traffic streams are too small, the benefit of providing high-usage circuits for the corresponding direct route disappears. In practice then, it is often useful to provide high-usage routes only for those traffic streams exceeding a given erlang threshold value. In other words 'if more than X erlangs of traffic exist between any two nodes, then a direct HU route is justified'. The value of X will depend upon the relative costs of lineplant and exchanges equipment and upon whether the exchanges are local, trunk, or international ones.

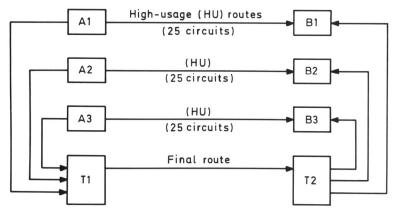

Figure 12.5
The benefit of high-usage working.

A problem facing planners using overflow in practical networks is that the number of circuits required cannot always be cut to a minimum without adding another control mechanism such as trunk reservation, as explained later in this chapter.

Furthermore, the high-usage structure is not as advantageous as the hierarchical when route busy hours of the various traffic streams do not coincide, since the circuits are 'less available' for shared use, e.g. by one stream in the morning and by another in the afternoon.

HU structures do, however, have a better overload performance, in that individual traffic streams are to some extent protected from congestion on other routes (caused for example by very high seasonal or short-term demand).

12.4 WILKINSON–RAPP EQUIVALENT RANDOM METHOD

In the previous section, when we were attempting to dimension routes carrying overflow traffic we ran into the difficulty that since the distribution of traffic which is 'overflowed' from a high usage circuit group is not random, the Erlang formula may not be used directly. So, use the Wilkinson–Rapp 'equivalent random' method instead. This is in fact the Erlang method slightly adapted.

Imagine two traffic streams a and b which have high-usage routes of A and B circuits available respectively. These two traffic streams overflow amounts of traffic a' and b' on to a common final route of N circuits, which is also the first choice for the 'first-offered' traffic stream c. Figure 12.6 illustrates this schematically. The problem is to find the value of N which will guarantee the appropriate grade of service on all traffic streams.

Circuit group N is subjected to overflow traffic $a' + b'$ and first-offered traffic c, and we dimension it by the Wilkinson–Rapp 'equivalent random' method as follows. The method assumes that the N circuits will behave in the same way as a set of N circuits part way down a larger group of $(N_{EQ} + N)$ circuits when subjected to an imaginary single stream of 'equivalent random' traffic, A_{EQ}. Figure 12.7 illustrates the set of N

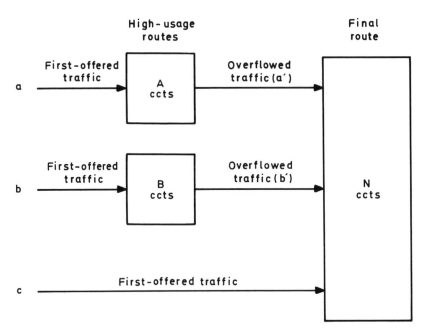

Figure 12.6
Dimensioning final routes.

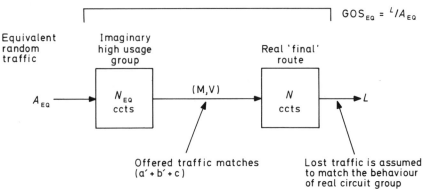

Figure 12.7
The Wilkinson–Rapp equivalent random method.

circuits and the imaginary set N_{EQ}. The problem is finding the values of N_{EQ} and A_{EQ} so that the traffic overflowed from the N_{EQ} circuits exactly matches the characteristics of the real traffic $a' + b' + c$. The mean (M) and variance (V) characterize the traffic $a' + b' + c$ which we imagine to overflow from the N_{EQ} circuits, and by choosing the values of A_{EQ} and N_{EQ} carefully we can equate the values exactly with the corresponding mean and variance values of the real traffic.

By knowing the values A_{EQ} and N_{EQ}, the traffic lost by the N circuit group can easily be determined. It is done by using the normal Erlang formula, assuming that an 'equivalent random' traffic value A_{EQ} is offered to a total number of circuits $N + N_{EQ}$. The 'grade of service' determined by the formula is the overall lost traffic (L in Figure 12.7) quoted as a proportion of the imaginary original traffic value A_{EQ}.

It is normal to dimension the group of N circuits by first calculating the permissible traffic loss in erlangs (i.e. the amount of the real traffic which need not be carried—L in Figure 12.7), and then calculating this as a proportion of A_{EQ}. This is the imaginary or 'equivalent' grade of service required when traffic A_{EQ} is offered to $N + N_{EQ}$ circuits, and allows us to calculate the value of $(N + N_{EQ})$ using the Erlang method as follows:

maximum permissible lost traffic $= L$

$L = GOS$ required $\times$ traffic on $(a' + b' + c)$

imaginary GOS required on $N + N_{EQ}$ circuits $= GOS_{EQ} = L$

$N + N_{EQ} = E(A_{EQ}, GOS_{EQ})$, where E represents the Erlang formula

Having determined the value of $N + N_{EQ}$, the real number of circuits required for the real traffic (N) is found by subtracting value N_{EQ}.

The values N_{EQ} and A_{EQ} are relatively straightforward to calculate, but the process requires a number of mathematical steps. An appendix at the end of the chapter shows how to calculate these values and gives an example of the whole Wilkinson–Rapp overflow route dimensioning method for those readers who are interested.

The Wilkinson–Rapp method is now quite old (1956) and in consequence more complex methods have evolved which seek to improve it. As with the alternatives to the basic Erlang method it rests with the user to decide the method which suits his circumstance the best.

12.5 DIMENSIONING FINAL ROUTES

It is normal practice to dimension final routes for only a 1 per cent loss of the mean traffic offered to them. This ensures a 1 per cent grade of service for any traffic which is 'first offered' to the final route. This practice was adopted in the previous section when mixed overflow and first-offered traffic shared the same circuit group.

In instances where there is no 'first offered' traffic on the final route, then the circuit numbers may be reduced, commensurate with an overall 1 per cent loss on each of the individual traffic streams. Thus if only 10 per cent of traffic overflows from the high usage route (a typical value), only 10 per cent of the original traffic is offered to the final route. The caller experiences a net 1 per cent grade-of-service even if as much as 10 per cent of the traffic offered to the final route is lost. Thus the grade of service of the final route need only be 10 per cent in this case!

12.6 TRUNK RESERVATION

To some trunk reservation is a relatively new technique, for although it has been around for more than 25 years, only recently has the advent of stored program control (SPC) exchanges made it more widely available. It provides a method of 'priority ranking' traffic streams, which can be particularly useful in conjunction with overflow network schemes, when the network planner may wish to give higher priority to 'first-offered' traffic than to 'overflow' traffic.

As we saw in the example shown in Figure 12.6, final routes must be dimensioned large enough to ensure the chosen grade of service (usually 1 per cent) for any 'first-offered' traffic streams. The consequence of this is an unavoidably and unnecessarily good grade-of-service for the overflow traffic (far better than 1 per cent), and the penalty regrettably is the need to provide extra circuits over and above the minimum. Trunk reservation, however, gives a way out.

Developing our example of Figure 12.2(b) a little further to illustrate the principle of trunk reservation, let us now recognize that in practice the route traffic between individual A and B nodes is not identical in value and varies widely. Let us assume that exchanges A1, A2, B1, B2 and B3 are in large towns, while A3 is in a much smaller town and consequently generates and receives much less telephone traffic. It might well be that while A1, A2, B1, B2 and B3 justify direct interconnection by high-usage (HU) routes, overflowing via T1 and T2, the traffic originated at A3 might be insufficient to justify such direct HU groups. In this case the traffic passing over the circuit group from T1 and T2 is a mixture of:

- overflow (i.e. not all) traffic generated by A1 and A2, but
- *all* traffic generated by A3.

Under such circumstance, it seems reasonable to take some precaution to ensure that the A3-originated traffic has some type of priority for use of the T1–T2 circuits in order that the grade of service to A3 customers is on a par with the performance available to customers at A1 and A2. Such a mechanism is provided by one form or another of 'trunk reservation'.

Numerous slight variants of trunk reservation exist. In the simplest, each traffic stream competing for a given route is allocated a trunk reservation value (typical values range from 0 to 15). The value corresponds to a number of idle circuits. If at any given moment only this number of idle circuits are available within the group (the others being busy), then new calls from the particular traffic stream to which the trunk reservation value applies, are rejected (i.e. are given network-busy tone and not completed). In other words, the trunk reservation value 'disadvantages' the traffic stream, to the benefit of other streams; the larger the value set, the larger the 'disadvantage'.

Each traffic stream competing for the circuit group may be allocated a different trunk reservation value (or disadvantage) but at least one stream should be given value 0 (top priority), otherwise some circuits will always be idle.

By using trunk reservation we can ensure that the various traffic streams are completed according to a priority order during periods when the circuit group is heavily loaded (i.e. few circuits are idle). In this manner we can nearly equalize the grades of service of the traffic streams, by disadvantaging overflow traffic streams which have had previous route options.

When further circuits become free, fewer traffic streams are disadvantaged, so that more of the traffic streams are allowed to complete calls. It is important to note that the algorithm does not take into account *which particular* circuits within the group are free, only the *total number* which are idle.

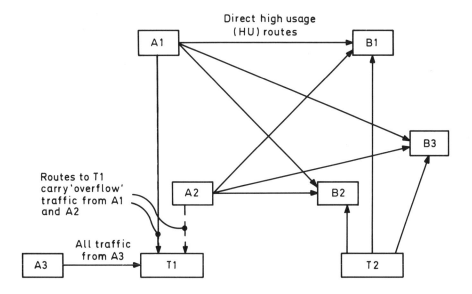

Figure 12.8
Typical trunk reservation values.

Traffic source	Trunk reservation value (applied at exchange T1, according to the source of traffic)
A3	0 (top priority)
A1 or A2	3 or 4 (typical value)

For our example illustrated in Figure 12.8 it would probably be appropriate to set values as shown in the table in Figure 12.8.

The mathematics needed to calculate trunk-reservation values are exceptionally complex. However, it is found in practice that values of 3 or 4 establish a significant priority and can be used in trial-and-error fashion. More accurate calculations are possible by the use of mathematical tables or computer programs, but since any one value of trunk reservation has much the same effect on all sizes of traffic streams, whether large or small, this only strengthens the case for trial-and-error use of values 3 and 4.

Trunk reservation is highly effective in providing a priority list based on 'disadvantage', and a value as low as 15 may virtually cut off a given stream.

As we shall see later in the chapter, routing techniques are moving from hierarchical (i.e. structured 'overflow') schemes towards non-hierarchical (i.e. dynamic) schemes. In consequence, new teletraffic modelling methods are necessary, capable of modelling end-to-end congestion rather than simple link-by-link analysis. Simple capacity flow models are used for this analysis, but the technique of trunk reservation is no less relevant in protecting priority streams on particular links. In a dynamic routing environment, trunk reservation values need to be higher than in simple overflow schemes (i.e. AAR)—typically 10.

12.7 'CRANKBACK' OR 'AUTOMATIC RE-ROUTING' (ARR)

A more powerful form of overflow routing is that of automatic re-routing (ARR), also called 'crankback'. This mechanism, which relies upon sophisticated switch and signalling interaction, enables an alternative transit route to be chosen even if it is the second link of a previous transit route choice that is busy. Figure 12.9 shows how it works.

This mechanism is particularly useful in highly interconnected networks; giving much better grade-of-service where most of the available paths between two points are via transit exchanges, and made up of a multiple number of links.

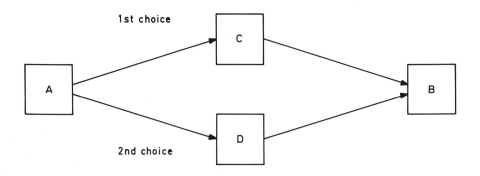

Alternative routes A to B	
ARR first choice	Via : C (i) if A_C congested then ARR via A_D (ii) if A_C free and C_B free then forward call to B otherwise if C_B busy go on to ARR second choice
ARR second choice	Via : D (if D_B busy then fail the call)

Figure 12.9
Automatic re-routing (ARR).

12.8 PROPORTIONATE BIDDING FACILITY (PBF)

When there is more than one route to a given destination, it is sometimes an advantage to be able to control the exact proportion of traffic offered to each route. Proportionate bidding provides a mechanism for this purpose. Call attempts are 'dealt' in turn around all the available routes according to a set percentage regime (one for you, three for you, two for you, etc.). The 'dealing' is done by an exchange software program, which works in almost the same way that a croupier deals a pack of cards. Both calls and cards get subdivided into random fractional samples. Figure 12.10 provides an example.

Useful applications of this algorithm might be as follows:

(i) *Route balancing*: Suppose exchange A is a trunk exchange, while B and C are international exchanges with say 60 per cent and 40 per cent respectively of circuits to a given overseas country. Traffic can be split by A and routed to exchanges B and C in appropriate fractions (60/40), using PBF as a routing mechanism at exchange A.

(ii) *Multiple-carrier environment*: Suppose this time that exchanges A, B, C, D are international exchanges, A being in one country while B, C and D are owned by separate carriers (competing telecommunications companies) in another. Outgoing traffic from exchange A can be split, using PBF, in the same proportion as incoming traffic is received from carriers B, C and D. PBF in this case provides a fair 'proportionate return' solution to the problems of connecting with multiple carriers. The method has already been adopted by a number of telephone network operators as a means of interconnecting with multiple telecommunications carriers operating in a destination country. UK and USA are examples of telecommunications markets where deregulation has allowed such competition between carriers, and has in consequence induced telephone network operators in other countries to connect to the many new British and American carriers precisely in this way.

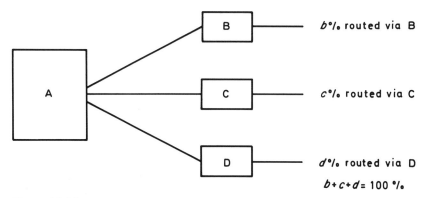

Figure 12.10
Proportionate bidding facility (PBF).

12.9 DYNAMIC ROUTING

Another new technique is that of dynamic routing. A number of slightly different 'dynamic routing' techniques are currently being developed, but all of them work on the same general premise: that the traffic pattern in a large network is continually changing according to the time of day, the day of the week, and the season of the year. At any given time some routes will be busy while others are slack. If, then, one could employ some of the slack capacity to serve the busy routes, there would be an overall saving in network resources, even though some of the individual calls might have to be routed circuitously via exchanges in other time zones. Figure 12.11 shows an example of traffic being routed from New York to Washington via Los Angeles.

In the Figure 12.11 example, as customers in Los Angeles are waking up and are starting to ring New York, the overall routing pattern has to be adjusted dynamically, in order to prevent the New York/Washington traffic from causing congestion on the trans-USA links. At this time of day European and African exchanges will be downloading as their business days end, and these switches could be used as alternative transit points for the New York/Washington traffic (instead of Los Angeles). Clever control mechanisms are needed to achieve optimum routing patterns, and to keep adapting the network so as to maintain the optimum. A number of such mechanisms have been patented. They include American Telephone and Telegraph's (AT&T's) 'dynamic non-hierarchical routing' (DNHR) and British Telecom's (BT's) 'dynamic alternative routing' (DAR).

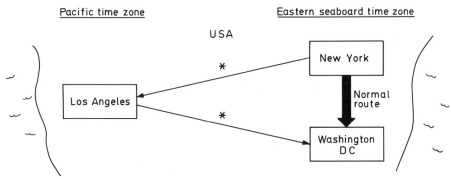

*New York to Washington route supplemented in early morning by slack capacity via Los Angeles (still asleep)

Figure 12.11
Dynamic routing.

12.10 NETWORK DESIGN

A number of traffic-controlling mechanisms have been presented in this chapter. In practice, any one or any combination of mechanisms may be used by network planners in striving towards the perfect network; and always the optimum topology makes some form of compromise between least cost, overall performance, and robustness under overload. The constraints will arise from the control mechanisms actually

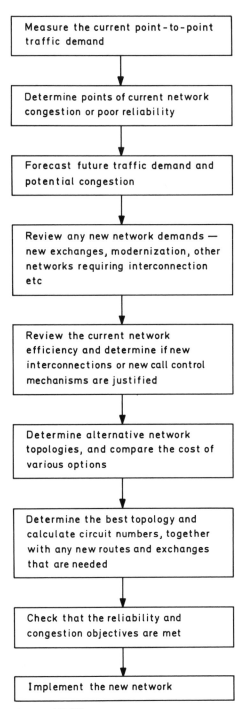

Figure 12.12
Network design.

available in any given case. None of the constraints and impositions are ever the same. Overall network design is indeed a complex and imprecise art!

Figure 12.12 provides a flow chart for network design, outlining the steps leading to the determination of a suitable topology. It starts by assuming that a network already exists and that the current point-to-point traffic demand between exchanges can be obtained by measurement. If this is not the case, current demand will have to be estimated. Alternatively, the scope for using an entirely new network configuration might be considered.

Having measured the current traffic, the next step is to identify any problems of congestion or unreliable equipment. We must then estimate all the future requirements of the network, first by forecasting the future traffic demand (Chapter 11 refers), and then by reviewing any other reasons for undertaking network changes.

The network may need to be changed in order to accommodate routes to new exchanges, or to connect to networks in previously unserved areas. Alternatively, its topology may need to change for economic or service reasons. Instead of being one made up of a large number of small outdated exchanges it may become one comprising a small number of large, modern technology ones. Often these factors have the most impact on the evolving network structure.

When the future demands on the network are known, we review the efficiency and suitability of the current network. For its intended future purpose we determine as many different options as possible for its evolution. At simplest, an option may only require adjustment to the capacity of the exchanges or of the transmission links interconnecting them. More complex options might demand the use of new call control techniques (such as ARR, PBF or dynamic routing). However, the deployment of such techniques is likely to incur considerable cost and needs to be weighed against the benefits in *overall* network performance and in *overall* network cost.

Once the choice has been made, the final circuit numbers can be determined and a plan worked out for the new circuits, routes and exchanges that will be needed. Then it is up to the implementors.

Depending on the nature of the network and its customers, the plan will need periodic, if not frequent, review and readjustment. None the less, the more comprehensive and forward looking is the plan, the greater is the likelihood of smooth network administration and performance. Conversely, insufficient forward forecasting or lack of anticipation of future needs can lead to short-term 'firefighting', with a need to adjust continually the structure of the network. Under such circumstances few networks reach their optimum of performance and economic efficiency.

APPENDIX: THE WILKINSON–RAPP ROUTE DIMENSIONING METHOD

In the main part of the text we briefly described the Wilkinson–Rapp method for dimensioning overflow circuit groups, but while we described the method we did not demonstrate how to calculate the values A_{EQ} and N_{EQ}. For those interested readers we now cover the necessary calculations and give an example of the use of the method. The explanation refers back to the diagrams of Figures 12.6 and 12.7 in the main text and describes the method in mathematical notation.

In Figure 12.6, the traffic offered to the N circuit group is:

$a' + b'$: overflow traffic
c: first-offered traffic

The question is 'how many circuits (N) are required in order that a 1 per cent GOS is achieved?' The value can be calculated as below.

First, the separate streams within the offered traffic are characterized in mathematical terms by the statistical means and variances of their traffic intensities, which are calculated using the normal Erlang formula. Provided the streams are statistically independent of one another, then the mean and variance of the combined stream $a' + b' + c$ can be calculated by simple summation. These values in turn allow A_{EQ} and N_{EQ} to be worked out.

The method is then as explained in the main text of the chapter. Refer to Figure 12.7. The actual traffic which may be lost, L, is calculated as $1\% \times (a' + b' + c)$ and from this the equivalent grade of service GOS_{EQ} can be worked out.

Finally, the total size of the equivalent circuit group $N + N_{EQ}$ is obtained from the normal Erlang formula and the value of N deduced by subtraction of value N_{EQ}.

There now follows a short summary in mathematical notation:

(i)   First, the mean of the traffic overflowing from the group of A circuits is worked out by multiplying the traffic offered, a, by the grade of service experienced. The grade of service is obtained using the Erlang method (although the Erlang formula must be used in the manipulated form that allows the grade of service from the offered traffic and the number of circuits).

grade of service, $GOS = E(A, a)$
(the Erlang formula, Chapter 10 refers)

thus mean overflow traffic is $M(a') = a \times E(A, a)$

(ii)   The statistical 'variance' of the same traffic comes from the following formula

$$V(a') = M(a') \times \{1 - M(a') + a/(A + 1 + M(a') - a)\}$$

(iii) Similarly $M(b')$ and $V(b')$, the mean and variance of the b' stream are calculated as:

$$M(b') = b \times E(B, b)$$

and

$$V(b') = M(b') \times \{1 - M(b') + b/(B + 1 + M(b') - b)\}$$

(iv)   The mean and variance of the first offered traffic are both equal to c. This fact results from the assumption that the arrival rate of first-offered calls is entirely random. This contrasts with the two cases above, where the variance is greater than the mean because the overflowed calls tend to arrive only in short bursts of 'peaky' activity, the HU having removed the more predictable and steady 'base-load'. Thus:

$$M(c) = V(c) = C$$

(v) The method requires us to calculate the mean and variance of the total traffic offered to the N circuit group. This is done by addition. Let us use M and V to represent the total. These values are shown on Figure 12.7.

$$M = M(a') + M(b') + M(c)$$
$$V = M(a') + V(b') + V(c)$$

(vi) This now allows us to calculate an important characteristic of the offered traffic stream known as the 'peakedness'. The peakedness measures the tendency of calls to arrive together, or in 'peaks', rather than in a more uniform rate of arrival. The peakedness is the value obtained by dividing the offered traffic variance, by the offered traffic mean. Thus:

$$\text{Peakedness} = Z = V/M$$

As we saw in (iv) above, normal or 'first-offered' traffic has a 'peakedness' of value 1 (i.e. $M = V$), while overflowed traffic has a peakedness value higher than 1 (i.e. $V > M$).

(vii) Finally, the equivalent random traffic A_{EQ} and the equivalent (imaginary) high usage group size N_{EQ} (as shown in Figure 12.7) can now be determined as:

$$A_{EQ} = V + 3Z(Z - 1)$$
$$N_{EQ} = \frac{A_{EQ}(M + Z)}{M + Z - 1} - M - 1$$

where M = mean traffic load offered
       V = offered traffic variance
       Z = peakedness = V/M

(viii) Next we calculate the imaginary grade of service required of the imaginary circuit group $N + N_{EQ}$. This is equal to the permissible lost traffic ($L = 0.01 \times M$, for 1% GOS) divided by the equivalent random traffic A_{EQ}.

$$GOS_{EQ} = L/A_{EQ}$$

(ix) Finally we determine the total number of circuits required in $N + N_{EQ}$ to meet the grade of service GOS_{EQ} when offered traffic A_{EQ} using the normal Erlang method. We then deduce the number of real circuits required (N) to carry the real traffic M by simple subtraction:

$$GOS_{EQ} = E(A_{EQ}, \{N + N_{EQ}\})$$

(normal trial-and-error Erlang method used to determine value $\{N + N_{EQ}\}$; Chapter 10 refers)

$$N = (N + N_{EQ}) - N_{EQ}$$

BIBLIOGRAPHY

Berry, L. T. M., An explicit formula for dimensioning links offered overflow traffic, *Australia Telecommunications Res.*, 1974, 8, 13–17.

Gibbens, R. J., Kelly, F. P. and Key, P. B., Dynamic alternative routing—modelling and behaviour. *12th International Teletraffic Congress, Turin, 1988.* Paper 3.4A.3.

Rapp, Y., Planning of junction network in a multi-exchange area. *Ericsson Technics No. 1*, 1964, 77–129.

Songhurst, D. J., Protection against traffic overload in hierarchical networks employing alternatives routing. *Telecommunications Network Planning Symposium, Paris,* 1980, pp. 214–220.

Stacey, R. R. and Songhurst, D. J., Dynamic alternative routing in the British Telecom trunk network. *International Switching Symposium, Phoenix, 1987.*

Wilkinson, R. I., Theories for toll traffic engineering in the USA. *Bell System Technical Journal*, 1956, 35, 421–514.

13

PRACTICAL NETWORK TRANSMISSION PLANNING

Transmission media exist to convey information across networks. No matter what form the information takes (be it voice, video, data, or some other form), the prime requirement is that the information received at the destination should as closely as possible match that originally transmitted. The signal should be free from noise, echo, interference and distortion, and should be of sufficient strength (or volume) to be clearly distinguishable by the receiver. The reliability of the service is also important. Meeting these objectives requires careful network transmission planning.

This chapter covers two aspects of planning, first describing the electrical engineering design guidelines usually laid out in a formal network 'transmission plan', and then going on to discuss the general administrative and operational practicalities of 'lining-up' and running a transmission network. Resource management is crucial for ensuring that lineplant and circuits are available when needed, that radio bandwidth is available without interference, that cables have been laid and satellites put in orbit.

13.1 NETWORK TRANSMISSION PLAN

A set of guidelines, laying down simple 'rules' for planning and commissioning new line systems or new circuits, are usually set out in a formal 'transmission plan'. These guidelines are intended to ensure that the electrical principles of telecommunications theory are adhered to. The transmission plan should include, for example, stringent rules on the use, positioning, and strength of amplifiers to correct the effects of attenuation. The plan needs to give guidelines on minimization of noise and interference, on the use of echo control devices to eliminate echo, and on the use of equalizers to rectify frequency attenuation distortion. A good transmission plan is a guideline for network design, ensuring that circuits are electrically stable and perform to a standard acceptable to the majority of their users. A typical target is 'to ensure that 90 per cent of people consider the performance to be 'fair', or 'better''. Acceptable values of attenuation and distortion can be established by subjective tests, and then the network can be designed to accord with them.

CCITT's recommendations on transmission planning include design guidance on:

(i) The overall signal volume at all points through the network.

(ii) The control of signal loss and the circuit's electrical stability.

(iii) The limits on acceptable signal propagation times. (Excessive propagation times manifest as a delay in transmission, a potential cause of slow data throughput and response times, or of unacceptable gaps and pauses in conversation.)

(iv) The limits on acceptable noise disturbance.

(v) The control of sidetone and signal echo (both these effects manifest themselves to speakers, who hear their own words echoed back to them).

(vi) The minimization of signal distortion, crosstalk and interference.

The recommendations lay down a standard reference system, against which national and international transmission networks may be designed or calibrated. These facilitate the correct placing of circuit conditioning equipment, and the establishment of correct signal volumes and controlled levels of signal distortion at all points along a connection.

13.2 SEND AND RECEIVE REFERENCE EQUIVALENTS

When designing telephone or other voice transmission networks, the strength of the electrical signal induced by the microphone at the sending end and the final sound volume produced by the receiver is all important. It varies according to the proximity of the speaker to the microphone and the listener to the earpiece, the total loss being composed of three basic parts; the loss incurred when inducing the electrical current in the microphone, the electrical signal loss across the transmission network itself, and finally the sound induction loss in the receiver. The relationship is straightforward to understand, as Figure 13.1 illustrates, but the problem for the network designer lies in deciding just how much signal loss is acceptable in the main part of the network itself.

With no overall loss, the listener will hear too loud a signal, equivalent to a talker at normal volume speaking directly into his ear. On the other hand, too much loss will result in an inaudible signal.

To ease the design dilemma, the system of 'reference equivalents' was developed by CCITT. The 'send reference equivalent' or 'SRE' of a telephone microphone is the signal loss expressed in dB when the electrical signal volume is compared with the original speech volume. Similarly, the 'receive reference equivalent' (RRE) of the earpiece is the signal loss when the final (heard) sound volume is compared with the electrical signal volume input to the earpiece. Both SRE and RRE are shown on Figure 13.1. To measure the sending reference equivalent (SRE) of a handset, a special 'NOSFER' equipment such as that shown in Figure 13.2 is used. This compares the equipment with a standard device, named after this system of calibration, 'Nouveau Système Fondamental pour la détermination des Equivalents de Reference (New Fundamental System for determining reference equivalents)'.

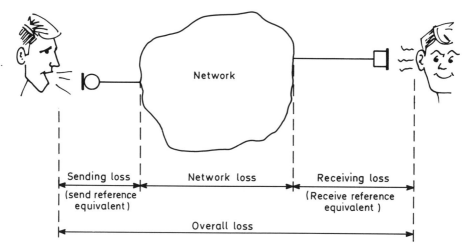

Figure 13.1
Send and receive reference equivalents.

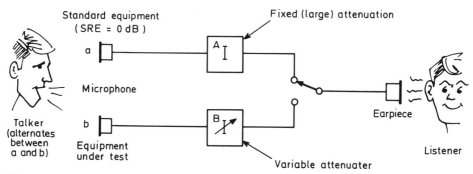

Figure 13.2
Measuring send reference equivalent (SRE).

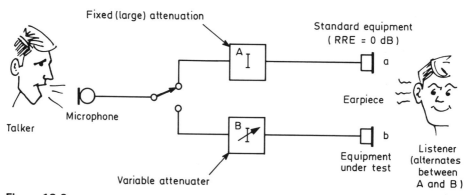

Figure 13.3
Measuring receive reference equivalent (RRE).

The NOSFER equipment in Figure 13.2 consists of a high-quality microphone with a device which maintains it at a fixed distance from the speaker's mouth. This equipment is connected to the listener's earpiece via a large attenuator, of strength A dB. To perform the calibration, a person talks alternately into microphone a and microphone b, while the listener adjusts the variable attenuation B to a value at which the volumes appear to be the same. The difference between values B and A (i.e. A − B) will then give a measure of the relative microphone efficiencies. The more that attenuation B must be reduced (i.e. the lower the value of B) the less efficient the microphone and the larger the numerical value of A − B, which is termed the 'send reference equivalent (SRE)'. In other words, the higher the value of SRE, then the lower the efficiency of the microphone and the greater the loss of signal strength in it. Thus SRE is a measure of the signal loss in the sending device.

Receive reference equivalents (RREs) similarly are a measure of the signal loss in the receiving device, and they may be measured by an equivalent NOSFER equipment; but in this case the comparison must be made against a standard high-quality receiver, as Figure 13.3 shows.

13.3 CONNECTION REFERENCE POINTS AND OVERALL REFERENCE EQUIVALENT

For the design of the network itself, all telephone handsets are assumed to have typical nominal SRE and RRE values. The values chosen are usually quoted for the handset loss relative to an imaginary 'reference' point on the customer's line side of the telephone exchange, and they take account not only of typical handset SREs and RREs, but also of the loss encountered on the local line. Typical values are 13 dB SRE and 2−3 dB RRE. In practice, customer lines have SRE and RRE of varying values, depending upon the exact handset in use and upon the length of the line. However, choosing a nominal design value is important as a guide for handset design and also as a bench-mark for the required performance of the network itself.

In Figure 13.4 two reference points (a) and (b), have been marked, one at each end

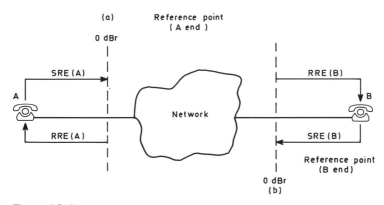

Figure 13.4
Connection 'reference point'.

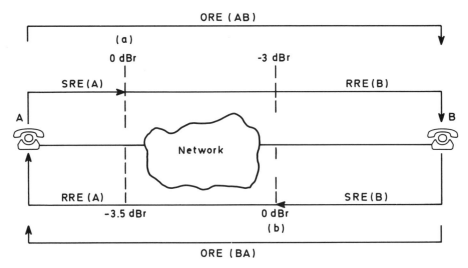

Figure 13.5
Overall reference equivalent.

of the connection, corresponding to the points relative to which SRE and RRE values apply. These points are marked on the diagram as '0 dBr'. (The nomenclature dBr stands for 'decibels-relative', and is used to indicate the received signal loudness at any point, measured *relative* to the signal strength at the reference point. Hence, at the reference point itself, where the relative strength is equal to the signal strength at the reference point, the value will be zero dBr.)

If the network introduces a loss of 3 dB in the transmit direction from A to B, and 3.5 dB in the transmit direction from B to A, then the received signal strengths at points (b) and (a) respectively will be -3 dBr and -3.5 dBr (i.e. signals 3 dB and 3.5 dB weaker than those sent from the opposite reference points). These values have been marked on to the diagram in Figure 13.5.

Now we can meaningfully discuss the end-to-end loss, called the 'overall reference equivalent' (or ORE). Surprisingly, the ORE may not be precisely equal to the sum of the SRE, the RRE and the cross-network loss. This is because both the SRE and RRE are only subjective measurements made in isolation on the network's general speech carrying performance, while the ORE is a more stringent measure of the actual end-to-end loss, usually measured using a single frequency tone of 800 Hz. Thus the formula,

$$ORE = SRE + Network\ loss + RRE$$

does not apply.

Initially, network planners thought the discrepancy was small enough to be ignored, thereby greatly easing the task of network design, and CCITT used to recommend that the network loss be adjusted to confirm with an overall reference equivalent not exceeding 40 dB. However, a number of network operators found difficulty with the system, reporting discrepancies of up to 5 dB, so in 1976 CCITT developed a new measure of performance called the 'loudness rating'. Still using the NOSFER equipment as a fundamental reference, and based upon similar principles and reference points, the 'loudness rating' method of network design was developed in such a way that loudness ratings (LRs) added to give a reliable and accurate algebraic sum. Send loudness ratings (SLRs), receive loudness ratings (RLRs) and overall loudness ratings were defined in a similar manner to the equivalent REs, but now the formula holds true:

$$OLR = SLR + Network\ loss + RLR$$

The method achieves greater reliability because it uses standard equipment, which although similar in principle to a NOSFER equipment, more closely responds to the frequencies of speech. The equipment is called an 'intermediate reference system (IRS)'. First the IRS is calibrated against NOSFER, then the system under test is calibrated. The difference in the two attenuation values (needed for each system individually to give performance equal to NOSFER) is the 'loudness rating' (LR). Values of loudness rating differ from the reference equivalent by varying amounts, up to 3 dB. Loudness ratings were adopted by CCITT in 1976 and are covered in recommendation P76. The 'intermediate reference system' (IRS) used in their measurement is covered in Recommendation P48.

Following on from the successful introduction of its 'loudness rating' method, CCITT in 1980 upgraded its original method of 'reference equivalents', applying a 'correction factor' to them to enable the values to be added algebraically. A simple mathematical formula converts 'reference equivalents' (REs) into 'corrected reference equivalents' (CREs). Corrected reference equivalents are the current standard method for designing networks against transmission loss. The following relationship applies:

$$OCRE = SCRE + Network\ loss + RCRE$$

Corrected reference equivalent (CRE) values typically differ by a variable amount (up to 3 dB) from corresponding reference equivalents (REs), but are usually a fixed amount (5 dB) greater than corresponding loudness ratings (LRs).

CCITT recommends that the traffic weighted mean of CREs throughout a telephone network should not exceed 25.5 dB and has set a longer-term objective that this mean should not exceed 16 dB nor be less than 13 dB. Such targets should be easily achievable with the excellent performance of modern digital networks. In practice the OCRE may differ slightly even from the planned value, but this will not matter provided sufficient safety margin is allowed during planning, and provided that the differential losses in the two directions of transmission are not too dissimilar. In Figure 13.5, for example, the OREs in the two directions are not quite the same. CCITT recommends that the difference be limited to 8 dB.

(*Note*: A similar reference connection transmission plan is also specified for other types of network. For example CCITT Recommendation X92 sets out the 'hypothetical reference connections for packet-switched data transmission'.)

13.4 MEASURING NETWORK LOSS

That the loss in decibels incurred by a signal traversing a network will depend upon the frequency of that signal is a fact we learned in Chapter 3. With that in mind, how can we meaningfully quote a value for network loss? The answer is that it is defined as the signal loss incurred by a standard tone of either 800 Hz or alternatively 1000 Hz frequency. It is measured using a standard 800 Hz tone source which is set to generate a known absolute signal power at a reference point known as the 0 dBr point. Reference (dBr) values can then be measured at any other required points in the same transmit path, and compared with the nominal values laid down by the formal network transmission plan.

13.5 CORRECTING SIGNAL STRENGTH

The transmission plan will lay out a rigid framework for the expected signal strength at all points along a connection. Any necessary adjustments in strength are achieved by the use of amplifiers and variable attenuators (also called pads). Weak signals are most common and are boosted by inserting amplifiers into the connection. The position of the amplifiers is critical. If insufficient amplification is used the signal strength gets too weak, becomes inaudible, and is affected by noise interference. Subsequent amplification (if attempted) does not correct the situation, since it amplifies the noise as much as the original signal. Conversely, if over-amplification is used it is likely to cause the electrical circuit to saturate (i.e. overload), resulting in signal distortion and electrical instability (manifested as a whistling and loud 'feedback' signal). Furthermore, over-amplification can lead to interference in adjacent circuits. Unduly strong signals are easily corrected either by reducing the amplification or by the introduction of attenuators.

There are standard positions in the circuit, where amplifiers or attenuators may be used. The strength necessary is determined by comparing the actual signal strength with the predetermined value appropriate for that particular reference point. As well as there being a reference point in the customer's local exchange, reference points can also be defined at other points in the network, at least at each exchange. This allows each individual section of an overall transmission link to be designed and 'lined-up' accordingly. The sub-sections may then be joined to form a high quality, end-to-end connection. If instead, the circuit is lined up once (as a single long section) there could be a counteracting effect between the performance of the various sub-sections which would mask some of the impairment that exists. We have, for example, already noted that it is not acceptable for us to allow the signal to fade to a strength equal to that of the interfering noise before we start amplifying it. Hence the need for a segmented approach to transmission planning. Figure 13.6 illustrates the transmission plan for the North American telephone network while it was analogue. (Unfortunately the plan has been superseded by the digital transmission plan, but is none the less no less useful in illustrating the use of such a plan.) Each link is marked with its maximum permitted loss.

Any new circuit between any two exchanges in the network must be lined up to have an overall loss within the ranges shown on the plan. Thus the maximum end-to-end

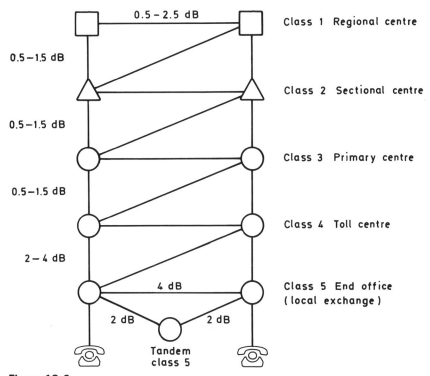

Figure 13.6
The North American telephone network transmission plan.

network loss is assured to be 19.5 dB. The Americans call this end-to-end loss between end offices the 'via net loss' (VNL).

It is best for longer links to be 'lined-up' with loss values towards the upper end of the permitted range. This ensures greater electrical stability. Another important feature of the plan is that greater losses are permitted on the links at the bottom of the network hierarchy (i.e. between Class 4 and Class 5 exchanges). This is because the vast majority of lines are in this category, and considerable savings are possible if amplification can be avoided.

Analogue and digital networks both need transmission plans, but they take different forms. For example, in digital networks, the signal strength in decibels is only an important consideration in mixed networks at the points of analogue-to-digital signal conversion, since the regeneration of digital signals by amplification is quite unnecessary in pure digital networks. The digital transmission plan concerns itself instead with such factors as maximum permissible bit error rates (BER) and the extent of quantization limit, which we shall go into more fully later in the chapter. In the meantime, Figure 13.7 shows the signal loss transmission plans for the UK analogue and new UK digital networks.

Figure 13.7(a) shows the connection of two normal telephones via 2-wire analogue lines to their respective local exchanges, and then over 4-wire digital transmission across both exchanges and the interconnecting link. At the originating exchange the speech signal is converted from a 2-wire analogue form into a 4-wire digital form by an analogue/digital conversion device, adjusted to produce a net loss of 1 dB. The signal is then switched through both exchanges in a digital form without further loss until it reaches the digital-to-analogue converter at the other end. Unlike the first converter, this one adds a further 6 dB loss, giving an overall network performance of 7 dB loss between reference points. The ORE can then be calculated by adding the relevant SRE and RRE values. This part of the transmission plan is very similar to the analogue plan (Figure 13.7(b)) which it replaces, except that although the total 4-wire

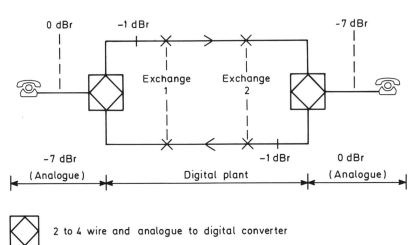

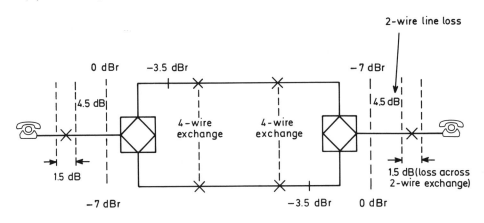

Figure 13.7
(a) The UK digital network transmission plan. (b) The UK analogue network transmission plan.

section loss is 7 dB the recommended signal strengths at the various intermediate reference points in this part of the connection are different. Another difference is that the analogue plan allows a small number of 2-wire links to be used in the connection.

13.6 THE CONTROL OF SIDETONE

'Sidetone' is the name given to an effect on 2-wire systems where the speaker hears his own voice in his own earphone while he is speaking. (We first introduced it in Chapter 2.) Too little sidetone can make speakers think their telephone is dead, while too much leads them to lower their voices. CCITT recommends 'sidetone reference equivalents' of at least 17 dB.

13.7 THE CONTROL OF ECHO

By ensuring that the CCITT recommended loss of 6 to 7 dB at least is encountered between the two reference points at either end of the 4-wire part of a connection, we not only safeguard the circuit against the effects of instability; we also make it unlikely that any signal echo is generated by the 2-to-4-wire converter (the so-called 'hybrid' converter). These echoes are caused by reflection of the speaker's voice back from the distant receiving end, and Figure 13.8 shows a 'hybrid' converter causing a signal echo of this kind. The problem of echo arises whenever a hybrid converter is used, and it results from the non-ideal electrical performance (i.e. the 'mismatch') of the device. No matter whether the 4-wire line is digital or analogue, the result is an echo.

To understand the cause of the echo we have first to consider composition of the hybrid itself. It consists of a 'bridge' of four pairs of wires; one pair correspond to the 2-wire circuit, two more pairs correspond to the 'receiver' and 'transmit' pairs of the 4-wire circuit, and the final pair is a balance circuit, the function of which is explained below. The wires are connected to the 'bridge' in such a way as to create a separation of the receive and transmit signals on the 2-wire circuit from or on to the corresponding receive and transmit pairs of the 4-wire circuit. It is easiest to understand the type of hybrid which uses a pair of 'cross-coupled' transformers as a bridge. This is shown in Figure 13.9.

Any signal generated at the telephone in Figure 13.9 appears on the transmit pair

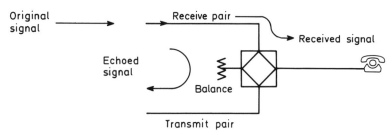

Figure 13.8
Echo caused by a hybrid converter.

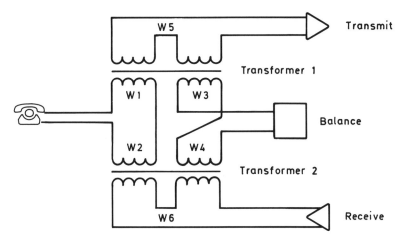

Figure 13.9
A hybrid transformer.

(but not on the receive pair), while any incoming signal on the receive pair appears on the telephone (but not on the transmit pair). It works as follows.

Electrical signal output from the telephone produces equal magnetic fields round windings W_1 and W_2. Now the resistance in the balance circuit is set up to be equal to that of the telephone so as to induce equal fields around windings W_3 and W_4, but the cross-coupling gives them opposite polarity. The fields of windings W_1 and W_3 tend to act together and to induce an output in winding W_5. Conversely, the fields of windings W_2 and W_4 cancel one another (due to the cross-coupling of W_4), with the result that no output is induced in winding W_6. This gives an output on the transmit pair as desired, but not on the receive pair.

In the receive direction, the field around winding W_6 induces fields in W_2 and W_4. This produces cancelling fields in windings W_1 and W_3 because of the cross-coupling of windings W_3 and W_4. An output signal is induced in the 2-wire telephone circuit but not in the 'transmit' pair, winding W_5.

Unfortunately, the balance resistance of practical networks is rarely matched to the resistance of the telephone. For one thing, this is because the tolerance on workmanship in real networks is much greater. Also, the use of exchanges in the 2-wire part of the circuit (if relevant) means that it is impossible to match the balance resistance to the resistance of all the individual telephones to which the hybrid may be connected. The fields in the windings therefore do not always cancel out entirely as intended. So for example, when receiving a signal via the receive pair and winding W_6, the fields produced in windings W_1 and W_3 may not quite cancel, and a small electric current may be induced in winding W_5. This manifests itself to the speaker as an echo. The strength of the echo is usually denoted in terms of its decibel rating relative to the incoming signal. This is a value called the 'balance return loss', or sometimes, the 'echo return loss'. The more efficient the hybrid, the greater the balance return loss (the isolation between receive and transmit circuits). A variety of other problems can be caused by echo, the two most important of which are:

- electrical circuit instability (and possible 'feedback')

- talker distraction.

If the returned echo is nearly equal in volume to that of the original signal, and if a rebounding 'echo' effect is taking place at both ends of the connection, then the volume of the signal can increase with each successive echo, leading to distortion and circuit overload. This is circuit instability, and as we already know, the chance of it occurring is considerably reduced by adjusting the 4-wire circuit to include more signal attenuation. The total round-loop loss in the UK digital network shown in Figure 13.7 is at least 14 dB, probably inflicting at least 30 dB attenuation on echoes even if the hybrid has only a modest isolating performance. Talker distraction (or data corruption) is another effect of echo, but if the time delay of the echo is not too long, then distraction is unlikely, because all talkers hear their own voices anyway while they are talking.

The echo delay time is equal to the time taken for propagation over the transmission link and back again, and is thus related to the length of the line itself. The longer the line, the greater the delay. Should the one-way signal propagation time exceed around 8 ms, giving an echo delay of 15 ms or more, then corrective action is necessary to eliminate the echo which most telephone users find obtrusive. A one-way propagation time of 8 ms is inevitable on all long lines over 2500 km, so that undersea cables of this length and all satellite circuits usually require echo suppression. Further propagation delay can also be caused by certain types of switching and transmission equipment. Indeed, a significant problem encountered with digital transmission media is that the time required for intermediate signal regeneration (detection and waveform reshaping) means that the overall speed of propagation is actually reduced to only 0.6 times the speed of light. This means that even quite short digital lines require echo suppression. Two methods of controlling echoes on long-distance transmission links are common. These are termed 'echo suppression' and 'echo cancellation'.

An echo suppressor is a device inserted into the transmit path of a circuit. It acts to suppress retransmission of incoming 'receive path' signals by inserting a very large attenuation into the transmit path whenever a signal is detected in the receive path. Figure 13.10 illustrates the principle.

The device in Figure 13.10 is called a 'half-echo suppressor' since it acts to suppress only the transmit path. A full echo suppressor would suppress echoes in both transmit and receive paths.

It is normal for a long connection to be equipped with two half-echo suppressors, one at each end, the actual position being specified by the formal transmission plan. Ideally half-echo suppressors should be located as near to the source of the echo as possible (i.e. as near to the 2-to-4-wire conversion point as possible), and work best when near the ends of the 4-wire part of the connection. In practice it may not be economic to provide echo suppressors at all exchanges in the lower levels of the hierarchy, and so they are most commonly provided on the long lines which terminate at regional (class 1 of Figure 13.6) and international exchanges.

Sophisticated inter-exchange signalling is used to control the use of half-echo suppressors. Such signalling ensures that on tandem connections of long-haul links intermediate echo suppressors are 'turned off' in the manner illustrated by Figure

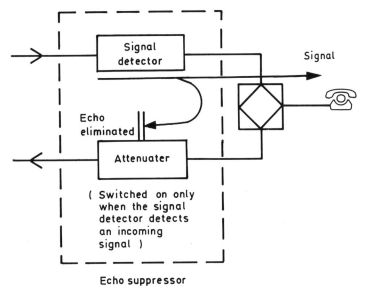

Figure 13.10
The action of an echo suppressor.

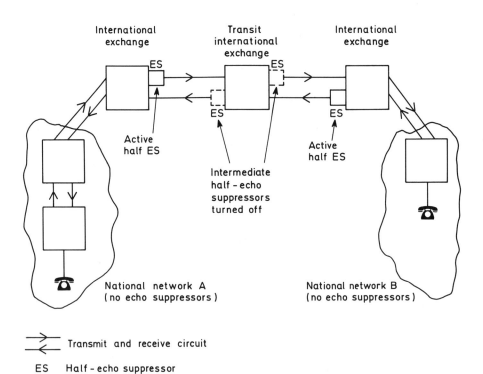

Figure 13.11
Controlling intermediate half-echo suppressors.

13.11. This ensures that a maximum of two half-echo suppressors (one at each end of the 4-wire part of the connection) are active at any one time.

The amount of suppression (i.e. attenuation) required to reduce the subjective disturbance of echo depends upon the number of echo paths available, the echo path propagation time, and upon the tolerance of the telephone users (or data terminal devices). CCITT recommends that echo suppression should exceed $(15 + n)$ dB, where n is the number of links on the connection.

Unfortunately echo suppressors cannot be used in circuits carrying data, because the switching time between attenuation-on and attenuation-off states is too slow and can itself cause loss or corruption of data. Most data modems designed for use on telephone circuits are therefore programmed to send an initiating 2100 Hz tone over the circuit, in order to disable the echo suppressors.

Another form of echo control device, called an 'echo canceller', can be used on either voice or data circuits. Like an echo suppressor, an echo canceller has a signal detector unit in the receive path. However, instead of using it to switch on a large attenuator, it predicts the likely echo signal and literally subtracts this prediction from the transmit signal, thereby largely 'cancelling' out the real echo signal. Other signals in the transmit path should be unaffected. Listeners rate the performance of echo cancellers to be better than that of echo suppressors. This, coupled with the fact that they do not corrupt data signals, is making them a popular alternative to suppressors, common on digital line systems and exchanges, and standard equipment in some countries (e.g. USA).

13.8 SIGNAL (OR 'PROPAGATION') DELAY

An important consideration of the network transmission plan is the overall signal delay or 'propagation time'. Excessive delay brings with it not only the risk of echo, but also a number of other impairments. In conversation, for example, long propagation times between talker and listener can lead to confusion. In the course of a conversation, when we have said what we want to say, we expect a fairly prompt response. If we are met with a silent pause, caused by a propagation delay, then we may well be tempted to speak again, to check that we have been heard. Inevitably, as soon as we do that, the other party starts speaking, and everyone is talking at once.

On video the effect of signal delay is even more revealing. For example, on live satellite television broadcasts, whoever is at the far end always gives the impression of pausing unduly before answering any question.

Nothing can be done to reduce the delay incurred on a physical cable or satellite transmission link. Thus intercontinental telephone conversations via satellite are bound to experience a one-way propagation delay of about 1/4 second, giving a pause between talking and response of 1/2 second. Furthermore, the extremely rapid bit speeds and response times that computer and data circuitry is capable of can be affected by line lengths of only a few centimetres or metres. Line lengths should therefore be minimized and circuitous routings avoided as far as possible. It is common for maximum physical line lengths to be quoted for data networks. Similarly, in telephone networks, rigid guidelines demand that double or treble satellite hops or other excessive delay paths (i.e. those of 400 ms one-way propagation time or longer) are avoided whenever possible. Excessive delays can be kept in check by appropriate network routing algorithms, as we shall see in Chapter 14.

13.9 NOISE AND CROSSTALK

Noise and crosstalk are unwanted signals induced on to the transmission system by adjacent power lines, electrostatic interference, or other telecommunication lines. The only reliable way of controlling them is by careful initial planning and design of the transmission system and the route. One source of noise results from the induction of signals on to telecommunications cables which pass too close to high-power lines. Another source of noise is poorly soldered connections or component failures. Both of these are easily avoided. But by far the most serious source of noise in telecommunications networks and the type most difficult to contain is that caused by electromechanical exchanges themselves. This type of noise results from the electrical noise associated with the electrical pulses needed to activate the exchange, and also from the mechanical chatter.

Noise is minimized by ensuring that the signal strength is never allowed to fade to a volume level comparable with that of the surrounding noise. Thus a minimum signal-to-noise (S/N) ratio of signal strengths is maintained throughout the connection. This ensures that the real signal is still perceptible amongst all the background. If the signal gets too weak in comparison with the noise, unfortunately amplification is then of little value since it boosts signal and noise strength equally. It is difficult to remove noise without affecting the signal itself, but there is some scope for removing noise which lies outside the frequency spectrum of the signal itself by simple filtering. This can slightly improve the signal-to-noise ratio (S/N).

The noise caused by atmospheric interference and magnetic or electrostatic induction has a random nature and is heard as low level hum, hiss, or crackle. When this type of noise is measured using a special type of filter, weighted to reflect the human audible range, then the noise strength can be quoted in 'psophometric' units. CCITT recommends that the strength of 'psophometric noise' should not exceed an 'electromotive force (emf)' at the receiving end of 1 millivolt (1 mV). Depending upon the length of the connection, the transmission network designer may decide the maximum allowable noise disturbance per kilometre of line. Typically this value is around 2–3 picowatts (a very small unit of power) per kilometre. This might be written 2 pW0p/km, where pW stands for picowatts, 0 indicates the value is measured relative to the 0 dBr reference point, and p denotes psophometric noise.

Crosstalk is the interference caused by induction of telecommunications signals from adjacent transmission lines (Chapter 3 refers). The presence of crosstalk is usually an indication that the circuit has a fault which is causing it to perform outside its design range. Perhaps an amplifier is turned up too much or some other fault has caused an abnormally high volume signal in the adjacent transmission line, or perhaps there is a short circuit or an insulation breakdown caused by damp. The transmission plan normally seeks to minimize crosstalk by ensuring that the 'transmit' and 'receive' signal levels at any point along the connection never differ by more than 20 dB.

13.10 SIGNAL DISTORTION

In Chapter 3 we discussed the use of 'equalizers' to counteract the 'attenuation distortion' of signals which is cased by differential attenuation of the component frequencies. Equalizers are generally located in transmission centres—alongside amplifiers. The equalizers which correct frequency (or 'attenuation') distortion make sure that the attenuation of component frequencies is less than a specified maximum. CCITT's recommendations for speech channels suggest a maximum differential attenuation of 9 dB, using appropriate equalization to ensure that the frequency response of the circuit is within the 'mask' illustrated in Figure 13.12.

But frequency distortion equalizers are not the only type of equalizer. Another type is called a 'group delay equalizer'. This acts by removing the distorting effects of group delay, which is an effect of signal phase distortion of the received signal caused by slightly different propagation times of the component frequencies within the signal. Group delay is particularly harmful to data signals, and it can be corrected by a device which adds delay to those frequency components which are received first, in order to even up the delay incurred by all the frequency components.

Both normal frequency attenuation equalizers and group delay equalizers are calibrated when the circuit is initially lined up. A range of different frequencies of known phase and signal strength is sent along the transmission line, and the characteristics of the received signals are carefully measured. The equalizers are then adjusted accordingly and placed in the circuit. A quick re-test should reveal perfect circuit equalization.

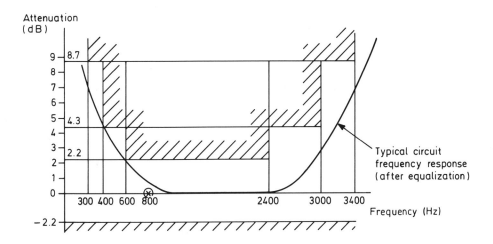

Figure 13.12
Frequency distortion: CCITT's G132 'Mask'.

13.11 TRANSMISSION PLAN FOR THE DIGITAL NETWORK

Transmission planning of a digital network, as compared with its analogue equivalent, is relatively straightforward primarily because the technique of regeneration practically eliminates the problems of crosstalk, noise, attenuation, and interference. However, three new factors have to be taken into account in a digital transmission plan; they are:

(i)   the digital line 'error rate';

(ii) the synchronization of the network;

(iii) the quantization distortion;

(iv) (during the period of network transition) the number of analogue-to-digital and digital-to-analogue conversions.

Taking them in order:

'Regeneration' of the digital bit pattern (as described in Chapter 5) must be undertaken frequently along the length of the line to ensure that 'marks' ('1's') and 'spaces' ('0's') are not confused by the receiver. Also, although less prone than analogue lines, it is still prudent to protect a digital line as far as possible from noise interference and crosstalk in order to prevent spurious bit errors. The frequency of such error is measured in terms of the error rate (the 'bit error rate' BER). Typical acceptance values of BER range from around 1 bit error in 10^5 bits ($BER = 10^{-5}$) to one error in 10^9 bits, depending upon the application. The error rate is usually checked when the digital line system is first established. All channels in the same line system (e.g. each 64 kbit/s channel of a 2 Mbit/s line system) will experience the same 'bit error rate' (BER).

The synchronization of digital networks ensures that there is no build-up or loss of information in the line system. Without synchronization, if a transmitter sent data faster than the receiver was ready to receive it, the result would be a build-up, and ultimately loss of information. Conversely, if the receiver was expecting data to arrive at a rate faster than the transmitter could send it, then imaginary 'fill-in' data would have to be created for the missing bits. Synchronization of digital networks is usually carried out in a hierarchical manner, with one exchange designated to house the 'master clock', providing 'synchronization' for other exchanges. Figure 13.13 illustrates a typical three-tier synchronization plan. At the top of the hierarchy, a single exchange has a master clock. In the second tier a number of main exchanges receive synchronization (i.e. a data transmitting and receiving rate) from the master clock exchange; additionally they are locked together by two way synchronization links, keeping them all rigidly in step. Finally, at the bottom of the hierarchy the smaller exchanges merely receive synchronization sources from the higher level exchanges, but do not have two-way synchronization links. Thus all exchange clocks are kept in lock-step with the synchronization clock. And just in case the main clock fails it is usual for one of the second tier exchanges to act as a standby master clock, ready to take over should the exchange with the master clock go off-air.

The third important element of a digital transmission plan is the control of quantization distortion (also called 'quantizing distortion'). As we learned in Chapter 5, quan-

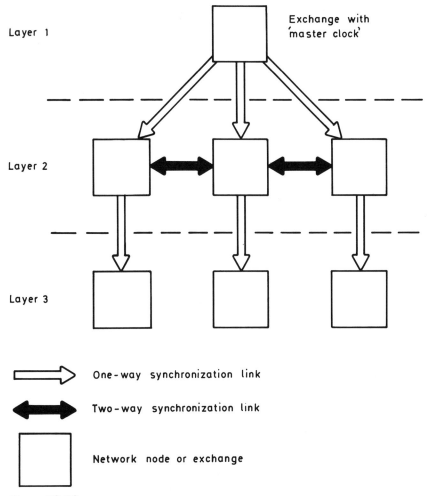

Layer 1 — Exchange with 'master clock'

Layer 2

Layer 3

⟹ One-way synchronization link

⟷ Two-way synchronization link

▢ Network node or exchange

Figure 13.13
Hierarchical synchronization plan.

tization distortion arises because the quantum amplitude values which are available to represent the signal always differ slightly from the actual value. The relationship between the representative digital signal and the original is therefore less than perfect. In turn, there will be differences between the final signal and the original, manifested as slight (maybe even imperceptible) distortion. This type of quantization distortion occurs during the initial conversion of any analogue (e.g. speech) signal into its digital equivalent and cannot be recovered on re-conversion. Further quantization distortion can occur should certain other transmission equipments be installed on the digital transmission path. The use of such equipments may be unavoidable, but the transmission plan should clearly set out how much extra distortion is permissible. Either the equipments have to be designed to conform with these limits, or the quality of transmission will suffer. Examples are echo cancellers (discussed earlier in this chapter) and circuit multiplication devices (discussed in Chapter 20).

Quantization distortion is measured in 'quantization distortion units or (qdus)'. One qdu is equivalent to a difference in amplitude between the digital signal and the original analogue signal equal to one digital quantum level (for explanation of quantum levels, return to Chapter 5). Thus if the final signal is reproduced with amplitudes varying from that of the original signal by a whole quantum amplitude step, then 1 qdu of distortion has been encountered. A digital transmission plan needs to lay out strict limits on the maximum quantization distortion that can be allowed in any link or part of the network. CCITT's 1984 recommendations suggested a maximum end-to-end quantization distortion of 14 qdus, allocating 5 qdus for each national network and a 4 qdu limit for the international connection in between. Typical values of quantization distortion are given in Table 13.1. They reflect the 'planning values' given in CCITT Recommendation G113.

From Table 13.1 it is clear that quantization distortion occurs as a result of any form of signal processing, and is particularly sensitive to analogue/digital conversion processes. For this reason, connections of interleaved analogue and digital sections should be avoided as far as possible. CCITT states this requirement by recommending the limitation of the number of unintegrated PCM digital sections to three or four and no more than seven. Of course, as the network evolves and digital transmission becomes more widespread this problem will disappear.

Digital data signals should not be recoded into an analogue form using a normal analogue/digital speech conversion device. On the other hand it is permissible to pass analogue encoded data over PCM lineplant, provided that the number of analogue/digital conversions is limited as stated above.

13.12 INTERNATIONAL TRANSMISSION PLAN

As always, the CCITT has some advice for international network transmission planning. This advice is to be found in its G-series recommendations. The principles described therein are exactly as discussed in this chapter, although it is briefly worth explaining CCITT's concept of a 'virtual switching point' or 'VSP'. This is a hypothetical reference point, like our other reference points. In particular it is the point at which a national network is assumed to be connected to the international network. The VSP concept allows CCITT to lay down recommendations ensuring an acceptable transmission performance of the individual parts (national and international) of a con-

Table 13.1
Typical quantization distortion values.

Digital process	Quantization distortion units
A/D conversion by 8-bit PCM	1
A/D conversion by 7-bit PCM	3
A-law to Mu-law conversion	0.5
Digital attenuator	0.7
ADPCM (Chapter 20 refers)	3.5

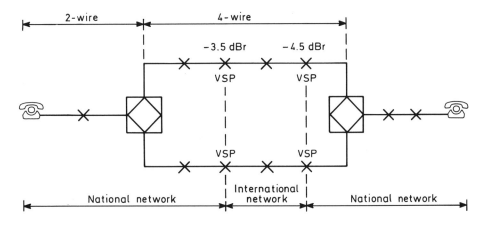

X – exchange

VSP – virtual switching points

Figure 13.14
National and international networks and VSPs.

nection. By so doing, the overall acceptability of the end-to-end network is assured without unnecessary strain on the internal plans adopted in any of the individual sub-sections. This is clearly important for international interconnection of networks, making it possible for dissimilar networks to work together. Figure 13.14 illustrates the concept, and the 'national' and 'international' network sub-components. Individual recommendations in CCITT's G-series may apply to one or more of the network sub-sections. By convention, the transmit VSP resides at the -3.5 dBr point in analogue networks, and a nominal 0.5 dB loss is allocated for each international transmission link. The signal reference value at a receiving international gateway in a direct link connection is thus 4.0 dBr. In Figure 13.14, however, the international connection comprises two links, so the receive level should therefore be set up to -4.5 dBr as shown.

13.13 PRIVATE NETWORK TRANSMISSION PLANNING

In the foregoing analysis we have been preoccupied with transmission planning for the international public telephone network, but the principles apply in exactly the same way to national and international private telephone networks, and to other forms of data or telecommunications networks as well. The ideal parameter values, however, may vary from network to network.

13.14 CIRCUIT AND TRANSMISSION SYSTEM LINE-UP

When setting up new circuits, a procedure called 'lining-up' ensures that they conform with the requirements of the transmission plans. They are normally lined up first on a section-by-section basis, and then end-to-end. This ensures that the performance is as near to the design as possible. For leased circuits, the end-to-end check can literally be from the terminal in one customer's premises to the other. However, in the case of switched networks, the network operator usually has to be content with 'lining-up' and checking only the individual links between adjacent exchanges. Practical constraints, the huge number of possible permutations of interexchange connections, prevent verification of each of the end-to-end permutations of the various links and exchanges. Figure 13.15 shows diagrammatically the difference in emphasis of the line-up procedure, comparing point-to-point links with switched networks.

The usual line-up procedure is first to install and line-up any higher order transmission systems (e.g. an analogue FDM system, or a high bit rate digital system). This ensures that the overall bandwidth broadly meets requirements. Then each individual circuit is carefully lined up. This ensures correct circuit alignment even when a circuit is actually routed over a concatenation of different higher order transmission systems. Figure 13.16 shows such a complicated circuit routed from exchange A to exchange B over different higher order systems A–C and C–B, and directly connected with wire ('hard-wired') within the building of exchange C.

The equipment needed and the operational techniques and procedures for line-up are covered in more detail in Chapter 18 (Maintenance).

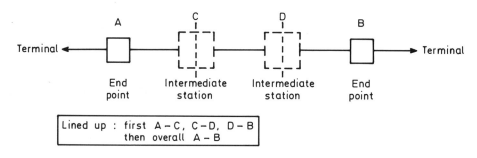

(a)   Line up of point-to-point (e.g. leased) circuits

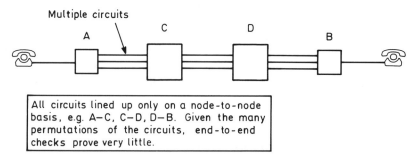

(b)   Line up of switched network links

Figure 13.15
Circuit line-up procedures.

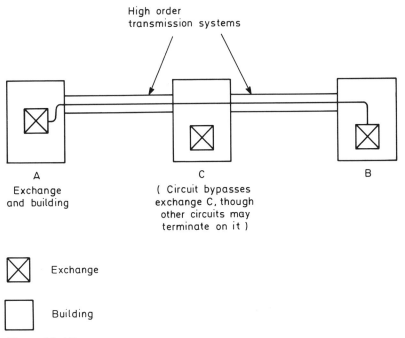

High order
transmission systems

A
Exchange
and building

C
(Circuit bypasses
exchange C, though
other circuits may
terminate on it)

B

⊠ Exchange

☐ Building

Figure 13.16
The need for end-to-end circuit line-up.

13.15 NETWORK RESOURCE MANAGEMENT

One of the key elements of successful administration of a large-scale network is the proper planning and management of resources. The remainder of this chapter discusses some of the practical factors pertinent to the operation of cable, radio and satellite systems, and it starts with a few tips on planning circuit provision.

13.16 CIRCUIT PROVISION PLANNING

'Circuit provision' is the terminology applied to the continuous process of preparing new and reconfiguring old circuits in a network. In any expanding network, there is a continuous programme of circuit provision and rearrangement. Unplanned circuit provision can lead rapidly to such a mass of entangled circuits that any future evolution or reconfiguration of the network becomes exceedingly complex and laborious. Pre-allocation of lineplant and of exchange capacity helps to avoid this unnecessary tangle, and allows shortfalls of equipment to be identified earlier, so that extra equipment can be ordered in good time. A computerized database of circuit information is an extremely useful aid to this process.

In cases where a single transmission network is used to carry a range of different circuit types, a detailed circuit routing policy should be determined. For instance it should be decided whether telephone and telex circuits should be treated separately as far as possible, or mixed together haphazardly, on common plant. Other circuit types which might need to be covered by the routing policy include circuits between packet-switched exchanges and point-to-point circuits between data terminal equipment.

The disadvantage of mixing all the circuits up with carefree abandon is that it is difficult to rearrange all the circuits being used for one particular use. For example, it makes it more difficult to shut down a telephone exchange at a particular site and re-direct all the telephone circuits to the replacement exchange at a different site, while leaving a number of telex circuits to a telex exchange behind. The re-direction is much easier if all the telephone circuits are generally 'groomed' together. Figure 13.17 illustrates two networks in which different circuit provision policies have been adopted. In each network there is a single site which has a telex, a telephone, a data exchange, and three transmission cables to interconnect the site with another. In the first, the 'groomed' network (Figure 13.17(a)), each of the transmission cables has been reserved for a particular type of circuit; one is dedicated to telex, one to telephone and one to data. If the telephone exchange is now closed down, then the cable can be extended very simply to the replacement exchange at a new site. These groomed networks are liable to breakdown; if one of the cables goes out-of-service for any reason, then one or other of the telephone, telex or data exchanges will be isolated. Another disadvantage of a groomed network is the proportionately large amount of spare capacity lying idle within it.

By contrast, Figure 13.17(b) shows a similar configuration of exchanges and cables, but in this case each cable is used to carry all types of circuits. This makes the job of reconfiguring particular circuit types more difficult, but reduces the probability of a complete cut-off of one or more of the telephone, telex, and data services. Mixing circuit types also reduces the overhead of spare capacity, since a fourth cable need be purchased only when all three of the cables are full (an improvement on Figure 13.17(a) where a fourth cable is needed as soon as *any one* of the cables is full).

In practice, it is usual to use an amalgam of both methods of circuit routing, the groomed and the ungroomed. Each of the services passing over more than one cable

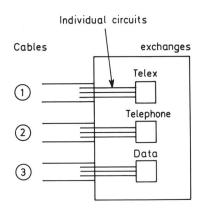
(a) Groomed network

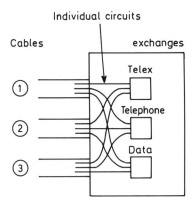
(b) Ungroomed network

Figure 13.17
Circuit provision policies.

is diversely routed as far as is practical and economic (in order to reduce service susceptibility to failure), but chunks of bandwidth within each cable are dedicated to the use of a particular circuit type (to ease reconfiguration). So for example, one FDM group within an analogue cable could be dedicated to telephone use, while another group on the same cable might be dedicated for data circuits. Likewise a whole 2 Mbit/s worth of capacity on a higher bit-rate digital cable could be dedicated to a particular circuit type. Where, however, only a few circuits are required, or where there is not much spare lineplant, it is inevitable that different circuit types have to be more closely mixed together in the 'ungroomed' way.

13.17 NEW CABLE PLANNING

New cable planning has to take into account not only the forecast of circuit numbers required, but also the technology to be used (e.g. digital, analogue, optical fibre etc.),

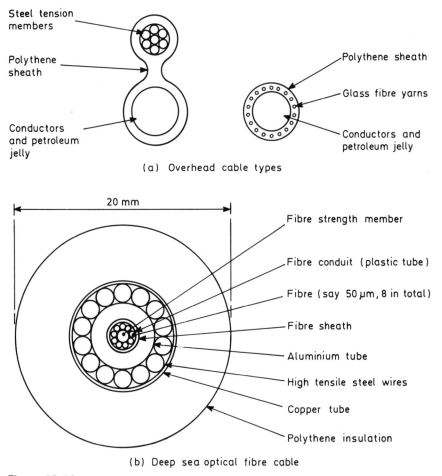

Figure 13.18
Robust cable constructions.

the route to be taken, and the physical strength the cable is going to need. Thus customers' individual lines and office networks require different treatment from street backbone wiring of local telephone exchanges. Inter-exchange and international cables need to be planned on an even grander scale.

Routing any type of telecommunications cables too close to high-power transmission lines can lead to noise interference; routing of data network cables around an office needs similar care, and accurate measurement is frequently needed to ensure that the maximum permissible line length is not exceeded.

The physical strength and robustness of cables must be attuned to the working conditions. All cables are covered with a sheath to protect them from physical damage and

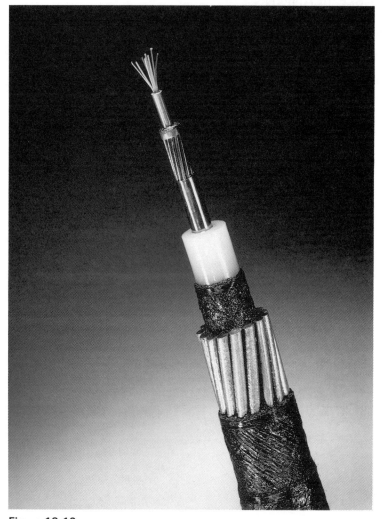

Figure 13.19
A deep sea optical fibre cable. Only four pairs of fibres in this one, but a huge amount of protection and tensile support to stand up to the rigors of deep-sea cable laying, and a life on the very uneven sea bed. (*Courtesy of British Telecom.*)

provide electrical insulation, but some are additionally 'screened' with metal foil to suppress electrical noise. A light plastic sheath may be adequate where the cable is supported along its entire length in an unexposed cable run (such as an office conduit), but where conditions are more harsh, as in a street conduit, extra protection may be required against mud and damp. For this purpose many cables are often packed with grease, or pressurized to keep water out. (Water can result in undesirable 'fried egg' circuit noise.)

The tension applied to the cable, either during installation (as it is pulled through a conduit) or during use (if it is strung as an overhead wire between 'telegraph poles'), may damage the cable if it does not have sufficient tensile (pulling) strength. Optical fibre cables, for instance, usually comprise not only the fibres and the sheath, but also a carbon fibre rope, which provides tensile strength and prevents the optical fibres from being stretched during installation in a conduit.

Cables with inadequate tensile strength gradually sag, gaining resistance, and so become less reliable in performance and more prone to attenuation. An undersea cable, laid on the ocean floor, will not lie completely flat on the sea-bed, but may rest on rocks or lie across caverns and ravines. It has therefore to be strong enough to carry its own weight, and it needs protection against damage from fishing trawlers and sharks etc. As examples of robust cable designs, Figure 13.18 illustrates two designs of overhead cables designed to carry their own weight, and also a possible design for a deep-sea optical fibre cable. Notice how large a proportion of the structure of each of the cables is there to provide strength and protection.

But even the under-sea cable in Figure 13.18(b) is not the ultimate in robust cable, since where an under-sea cable comes on to the 'continental shelf' near land, a further sheathing of 'rock armour' is added to give protection against fishing trawlers nets etc. This can nearly double again the diameter of the cable.

13.18 LOCAL LINE PLANNING

The local line is that part of a telephone or other public network which provides connection between a customer's premises and the local exchange (or 'end-office'). Traditionally, local lines for telephone networks have been 2-wire copper or aluminium 'audio' circuits, with a maximum attenuation of about 10 dB and a resistance no greater than 1000 ohms. Such limits are practical guides which enable local line planners to design circuits which conform with the transmission plan and the signalling needs, while allowing scope for long lines to be run out from the exchange to relatively remote customers, several kilometres away. Planning the local line network involves careful positioning of distribution cables, and appropriate choice of their capacity (in terms of the number of pairs of conductors). Usually, a number of high capacity 'distribution' (10–600 pair cables) cables lead out in a star fashion from a 'distribution frame' in the exchange (where they are connected to exchange equipment) to a hierarchy of cross-connection points (street-side jointing boxes). At each of these boxes a multiplicity of smaller capacity cables (or single 'dropwire' pairs) can be run on to further cross-connection points (streetside 'cabinets' or 'pillars') or direct into customer premises. Figure 13.20 illustrates the typical layout of a local line network.

The planner must ensure that the conductors in each of the cables are of sufficient

290

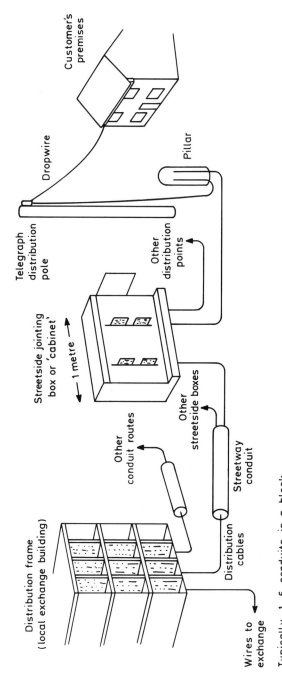

Customer's premises

Dropwire

Telegraph distribution pole

Pillar

Streetside jointing box or 'cabinet'

← 1 metre →

Other distribution points

Other streetside boxes

Other conduit routes

Streetway conduit

Distribution frame (local exchange building)

Distribution cables

Wires to exchange

Typically 1–6 conduits in a block
1–3 cables per conduit
10–600 pairs per cable

Figure 13.20
Local line network.

Figure 13.21
Streetside cabinet and jointing box. An engineer checking the wiring on the cross-connection frame inside a streetside cabinet. This provides for flexible connection of the many cables and individual pairs out to customers' premises and back to the exchange. In the foreground, the cables running away into the street conduits via a jointing box. (*Courtesy of Telecom Technology Showcase, London.*)

Figure 13.22
Streetside joint repair. Checking a joint during the installation of telephone service in 1935. Phoning the exchange. (*Courtesy of Telecom Technology Showcase, London.*)

gauge (i.e. diameter) to enable the resistance and attenuation limits to be met for each of the circuits to all the customer premises. Usually the gauge of distribution cables is the greatest, since these cover the longest part of each access line and therefore need the lowest resistance per kilometre. The shortest section, the dropwire, need only be fairly narrow gauge. The difference in resistance between narrow and wide gauge wire is not important over this short distance. For 4-wire communication paths, including high grade point-to-point data circuits, two pairs of wire are needed, and where a particularly high bandwidth or digital bit speed is required, coaxial or some other special type of cable may be called for.

Like trunk networks, local line networks are being revolutionized by the introduction of optical fibre. For customers with very high digital bit speed requirements, the fibre can be run directly into digital multiplexors on their premises, but even for those with more modest requirements, optical fibres offer a cheap and reliable alternative to metal distribution cables. They provide for high capacity, low attenuation links from the exchange to a digital multiplexor in the streetway box, and in some of the world's major cities, fibre-optic cable used in this way will alleviate the physical conduit congestion consequent on the size and number of individual metal pair conductors.

Figure 13.23
Metropolitan transmission system. In a sunlit manhole in New Jersey, USA, an installer checks out the wiring in a maintenance case—part of the Bell System's metropolitan transmission system. (*Courtesy of AT&T.*)

The advent of optical fibres usage in local exchange networks has brought with it new topologies of local line networks. Recently developed by British Telecom in the United Kingdom, the Telephony Passive Optical Network (TPON) and the Broadband Passive Optical Network (BPON) are examples of the new techniques. These networks use either a single fibre ring (Figure 13.24(a)) or a multiplexed star arrangement (Figure 13.24(b)) to backhaul the customer lines into the local exchange.

TPON is a network designed for basic telephony—including normal leased lines,

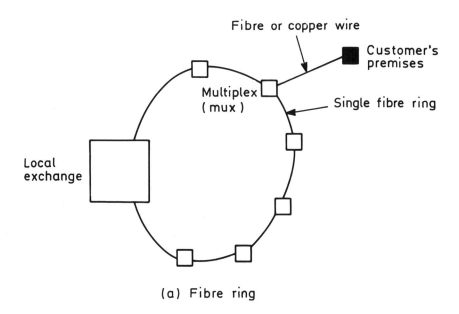

(a) Fibre ring

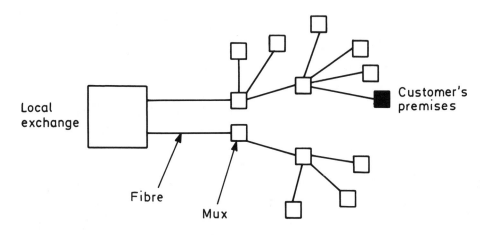

(b) Multiplexed star

Figure 13.24
Optical fibre in the local network.

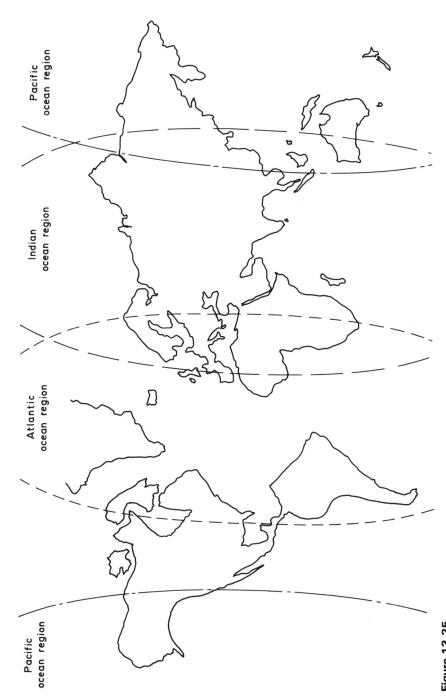

Figure 13.25
Approximate INTELSAT coverage regions.

speech and data networks and ISDN. But BPON goes one stage further—delivering video and cable television to the home. But who knows, maybe radio will evolve to the point where a fixed local line is unnecessary (Chapter 27 refers).

The local exchange is also evolving—and miniaturizing. Already some small exchanges can actually be located within streetside cabinets and manholes.

13.19 TRUNK AND INTERNATIONAL LINE PLANNING

Basically, the planning of trunk and international lines is no different from planning local ones. It is simply that the capacity needed of the individual cables and the transmission plan constraints on cable performance mean that much higher quality equipment has to be used.

13.20 RADIO TRANSMISSION SYSTEMS

The most important factors in management and operation of radio systems are the allocation of the radio bandwidth and the location of radio antenna sites.

The International Telecommunications Union (ITU) administers the registration of radio users and allocates radio bandwidth to ensure that interference of signals is minimized. The job is carried out by the World Administrative Radio Council (WARC), a subgroup of the ITU's 'International and Frequency Registration Board' (IFRB). Within each country, the allocation is further delegated to government bodies, to whom individual users must apply for registration. In the UK, for example, this is carried out by the DTI (Department of Trade and Industry), while in the United States it is the responsibility of FCC (the Federal Communications Commission).

However, not even the proper allocation of radio bandwidth is enough to prevent interference, so the sites for radio stations need to be away from strong electrical fields and radio path obstruction.

Radio transmission systems have been used in all types of networks. Historically, of course they have been used for public broadcast service, for land, sea and air mobile services. Today they also provide the basis for local, trunk and international telecommunications links, and additionally, as cellular and cordless telephones become more popular, are taking over from the domestic fixed telephone line.

13.21 SATELLITE TRANSMISSION MANAGEMENT

Satellite transmission is only a specialized form of microwave radio, and so like other types of radio transmission it requires proper frequency management and careful siting of both ground-based antennas (called 'earth stations') and orbit vehicles (called 'satellites'). International meetings and conferences administer the position of satellites in orbit, their coverage areas, radio frequencies and permitted signal power outputs. Without such management, there could be severe interference. Separation in orbit gives physical separation. Radio frequency separation prevents radio interference. Regulating power outputs governs the coverage (or footprint) area, also preventing interference. A high signal power increases the coverage area on the ground and has the benefit of reducing the size of the receiving antennas needed on the ground, but increases the risk of interference between different systems. The radio frequencies used in satellite systems are in the so-called C-band and Ku-band ranges of microwave frequencies. The shorter wavelength, Ku-band systems need only relatively small antennas (1–3 m diameter) and have thus become the favoured systems for corporate, office rooftop use.

The satellites used for telecommunications are generally 'geostationary'. By this we mean that they orbit around the earth's equator at a speed equal to the spin speed of the earth itself. They therefore appear to be *geographically stationary* above a point on the earth's equator, which allows them to be tracked easily by earth station antennas. Any given earth station will either be able to 'see' a satellite for 24 hours a day, every day (dependent on the satellite's longitudinal position), or will never be able to see it. The world's leading organization for the development and exploitation of satellite communications is called INTELSAT (the International Telecommunications Satellite consortium). INTELSAT is a jointly owned consortium, set up in 1964 by the world's leading public telecommunications operators to provide a means for worldwide transmission. It is not the only satellite body. Others include INMARSAT (the International Maritime Satellite organization), EUTELSAT (the European equivalent of INTELSAT) and a number of privately owned and run companies which operate satellites for a range of national telecommunications purposes (e.g. direct broadcast by satellite, or telephone circuit access to remote areas).

The geographical coverage area, or 'footprint' of a satellite in orbit (i.e. the area on the ground in which earth stations are able to work to it) is affected by its geographical position above the equator and by the design and power of its antennas. The position above the equator is usually stated as the geographic longitude. The antennas may either be directed at a small area of the earth's surface in a 'spot beam', or broadcast to all the hemispherical zone on the earth's surface that the satellite can see. The latter type of antenna needs to be much larger and it demands radio signals of considerably greater power. The INTELSAT system comprises a number of satellites, positioned in clusters over the middle of the world's three great oceans, the Atlantic, the Pacific and the Indian Ocean, and give the coverage areas shown in Figure 13.25. Between them the satellites allow most countries to have telecommunications with almost any other country in the world.

European countries have access to the Atlantic and the Indian Ocean INTELSAT satellites, while countries in North America can view the Pacific and Atlantic satellites (from west and east coasts respectively).

A major benefit offered by satellite transmission over cable is the fact that a single earth station in one country and a single satellite in space may be used to provide transmission links to not just one, but a number of other countries. The signal is sent up by the transmitting earth station to the satellite (the 'up-link'). The satellite responds by amplifying and re-transmitting a downpath radio signal over the whole coverage area. Any earth station within the coverage area can pick up the signal, and select from it the information intended for that particular destination. This type of operation is called 'multiple access' working.

The responding and transmitting radio equipment on board the satellite is usually referred to as a 'transponder'. A number of transponders usually share the same dish antennas on board the satellite. Transponders are either analogue or digital (i.e. frequency division multiplexed or time division multiplexed), and may be used either for point-to-point use or for multiple access as explained above. Every distant earth station receives all the radio carrier signals but selects only the appropriate carrier frequencies for demodulation.

The major disadvantage of satellite transmission is the one-way propagation delay of around 270 ms. This is the minimum time it takes for the radio signal to travel the distance up to the satellite and back down to earth, and in a two-way telephone conversation it causes a half-second silence between asking a question and hearing the first part of the answer. Allocation of radio bandwidth and planning for new satellites in the INTELSAT system are carried out annually at the 'Global Traffic Meeting', held in Washington, DC, United States. At the meeting, all constituent operators and users of the INTELSAT system declare and specify their forward five-year bandwidth requirements. The main types of radio carriers used by INTELSAT are called FDMA, TDMA, SCPC, IDR and IBS. Explanation of these terms is as follows.

FDMA (Frequency Division Multiple Access) uses analogue 'frequency modulated' carriers on up and downlinks. 'Broadcast' of the downlink signal is the key to multiple access operation, allowing the same signal to be picked up at a number of distant end earth stations, but only the relevant receive channels are picked out by each destination earth station. Thus a contiguous uplink bandwidth must be allocated to each earth station to meet only its outgoing transmission requirements.

TDMA (Time Division Multiple Access) uses digitally modulated, time division multiplexed signals. As with FDMA, when used in the 'multiple access' mode, one transmit carrier is used by each earth station, and a number of received carriers have to be scanned for the appropriate return circuits. The difference is that the transmit 'carrier' is in reality a pre-allocated 'burst' of a high bit speed signal. The bursts from the different earth stations are interleaved in a time-shared fashion like ordinary TDM.

SCPC (Single Channel per Carrier). As the name suggests, single channel per carrier systems only support one channel on each carrier. They are generally used for point-to-point circuits when only a small number of overall circuits are required between end points.

IDR (Intermediate Data Rate). These are the latest type of carriers, of varying digital bandwidths from 1.5 Mbit/s and 2 Mbit/s upwards.

IBS (International Business System). IBS transponders support point-to-point business applications between small dishes and are ideal for bandwidths up to 2 Mbit/s.

BIBLIOGRAPHY

CCITT Recommendation G101, 'General Characteristics for International Telephone Connections and International Telephone Circuits—The Transmission Plan'.
Cook, N. P., *Microwave Principles and Systems*. Prentice-Hall, 1986.

Connection of Terminal Equipment to the Telephone Network. US Code of Federal Regulations, title 47, part 68 (47 CFR 68).

Digital Data Special Access Service—Transmission Parameters and Interface Combinations. Bellcore TR-NPL-000341. March 1989.

Freeman, R. L., *Telecommunication System Engineering—Analogue and Digital Network Design*. John Wiley & Sons, 1980.

Freeman, R. L., *Telecommunications Transmission Handbook*, 2nd edn. John Wiley & Sons, 1981.

Gunston, M. A. R., *Microwave Transmission-Line Impedance Data*. Van Nostrand Reinhold, MARCONI Series, 1972.

Inglis, A. F., *Electronic Communications Handbook*. McGraw-Hill, 1988.

Jordan, E. C., *Reference Data for Engineers: Radio, Electronics, Computer and Communications*, 7th edn. Harold W. Sams & Co., 1985.

Maslin, N., *HF Communications—A Systems Approach*. Pitman, 1987.

Maynard, J., *Computer and Telecommunications Handbook*. Granada, 1984. (Gives radio frequencies.)

Morgan, W. L. and Gordon, G. D., *Communications Satellite Handbook*. John Wiley & Sons, 1988.

Sachdev, D. K., *Intelsat Network Evolution in the 1990s and Beyond*. INTELSAT Forum 87. 5th World Telecommunications Forum Conference Proceedings Part 2, pp. 179–189. ITU, Geneva, 1987.

Voicegrade Special Access Service—Transmission Parameter Limits and Interface Combinations. Bellcore TR-NPL-000335. June 1986.

Voicegrade Switched Access Service—Transmission Parameter Limits and Interface Combinations. Bellcore TR-NPL-000334. June 1986.

Wickersham, A. F., *Microwave and Fibre Optic Communications*. Prentice-Hall, 1988.

14

NETWORK ROUTING AND NUMBERING PLANS

We might choose, laudably as it may seem, to attempt to run our telecommunications network by completing as many calls as possible or delivering the greatest proportion of data messages. The problem is that if we attempt to do so, we are bound to affect the intelligibility of messages and how long it takes to deliver them. It can in fact be tried, provided you are prepared to programme the exchanges to route all messages 'any way possible', rather than ever fail anything. Unfortunately in the attempt, your network will eventually perish of congestion and suffer appalling signal quality. Studying the scenario, however, is highly instructive and we shall look at an example or two shortly. The rational and rewarding alternative to this regime is to have a network routing plan, together with a supporting numbering plan. The appropriate routing algorithms laid out by the routing and numbering plans are selected to control

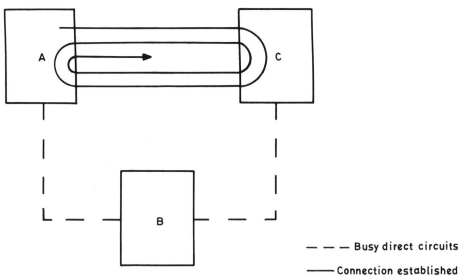

Figure 14.1
Uncontrolled circular routing.

network traffic and to comply with the overall constraints which the transmission plan imposes upon all end-to-end connections across the network. To work within these various constraints, and still to achieve a network that is reasonably cheap as well as highly efficient is an arduous test of planning and administration; but it is well worth while when we consider the alternative of uncontrolled network routing and its disastrous effect on network congestion and the quality of connections, which the examples in Figures 14.1 and 14.2 will illustrate.

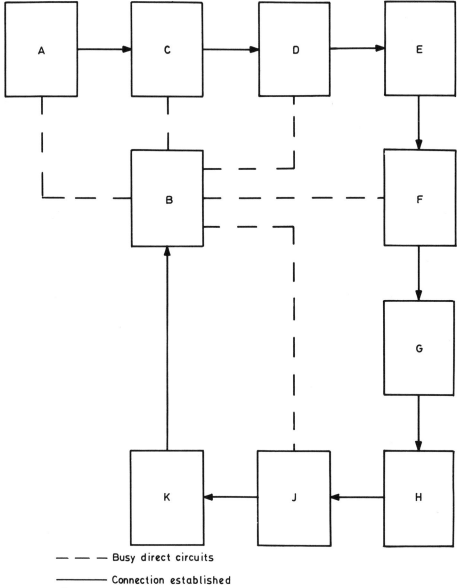

— — — Busy direct circuits

———— Connection established

Figure 14.2
Uncontrolled transit routing.

In the example shown in Figure 14.1, a circuit-switched connection is desired from exchange A to exchange B. Exchange A has direct circuits to B, but these are currently busy, so exchange A has made a connection to exchange C and passed the call on, intending to transit this exchange and connect via the direct circuits from C to B. Unfortunately these circuits are also busy, and taking no cognizance of the call's previous history, exchange C extends the connection back to exchange A using a similar logic. The process continues in a circular fashion until either all the circuits between A and C also become busy, so that the call eventually fails, or finally a circuit becomes available on either of the routes A–B or C–B, in which case the call eventually completes. In either eventuality the circular routing between A and C ties up network resources, restricting communication between customers on exchanges A and C. Furthermore, even if the call does eventually complete, the transmission quality is likely to be appalling or the delay in packet data delivery may be intolerably long.

A second effect of uncontrolled routing is shown in Figure 14.2, where a communication path has finally been completed over nine individual links, transiting eight intermediate exchanges. As in the circular routing example of Figure 14.1, even though the call has completed, an undue quantity of network resources have been tied up, causing possible congestion for other traffic. Also transmission quality or intolerable delay is repeated. The quality will be particularly poor if one or more of the nine links passes over a satellite connection. In this case the end-to-end propagation time may be several seconds. On a packet mode data connection, the data throughput rate is likely to be severely limited by such long propagation delays, particularly if the protocol requires acknowledgement of individual packets.

14.1 NETWORK ROUTING OBJECTIVES AND CONSTRAINTS

To maintain high transmission quality and to ensure the minimization of time delays both on call set up and on message or speech propagation, it is desirable to minimize the overall number of links and exchanges making up a connection. In addition it is desirable to limit the number of concatenated links of certain transmission types (e.g. satellite links, or links using low rate speech)—since the tandem connection of such devices can lead to unacceptable transmission impairment, as we saw in Chapter 13.

Historically networks were designed in a hierarchical fashion. This enabled all endpoints to be interconnected. Also the number of links required on a 'maximally adverse connection' could be limited during design by ensuring that exchanges in the top layer of the hierarchy were fully interconnected and that each lower level exchange was connected with at least one exchange in the next higher layer. Thus a hierarchical network structure consisting of n layers needed, at most, only $(2n-1)$ links in order to interconnect any two exchanges. For example, in a network comprising a three-layer hierarchy, any exchange may be connected to any other without the need to use more than $(3 \times 2 - 1) = 5$ links, as Figure 14.3 shows.

In more modern networks, the strict hierarchical method of network design is becoming less common, in favour of simpler and more flexible routing and numbering guidelines. Less rigidly structured networks prevail in which each exchange recognizes:

● The need to select an economical routing conforming with transmission plan requirements.

- The need to contain the likelihood of rapidly escalated network congestion.

- The need to charge for calls in line with the incurred costs.

- The need for flexibility of routing to be available to network operators in order to take advantage of the non-coincidence of route busy hours via transit exchanges.

- The need to minimize the overall number of links in a connection, and in particular to limit the number of satellite links or bandwidth compression equipments which may be used in tandem.

- The policy of preferred transmission media, say when alternative satellite and cable links are available to the same destination.

- The need to avoid circular routings.

Careful network design and a shrewd call routing programme at each exchange will ensure conformance to the routing plan; recent developments in signalling systems and exchange call control functions have fortunately made this job a little easier. Now signals which are passed between exchanges and accompany the dialled number of the

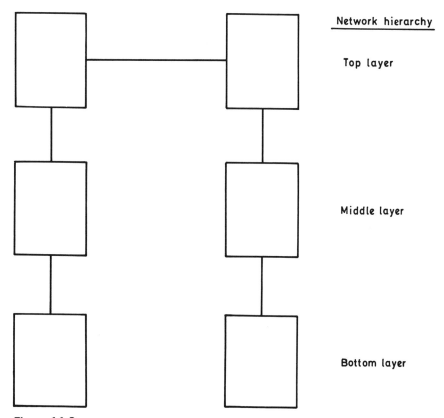

Figure 14.3
Five-link connection in a three-tier network hierarchy.

desired destination help to convey the previous history of the call or packet (for example, the existence of a previous satellite link).

In the example of Figure 14.4, caller 1, connected to exchange A and wishing to call B, has reached exchange C by means of a satellite link. Although both cable and satellite links are available from exchange C to exchange B, the call is only allowed to mature if a circuit is available on the cable link. If instead the call were to be permitted to overflow to the satellite link, then the connection would not meet the required transmission quality standard. (If, however, the connection was only possible by the use of a double satellite link, then the call could have been permitted to mature.)

By contrast, caller 2 (on exchange C) may be connected either over the satellite or the cable link. Two alternative routing policies are available to the owner and operator of exchange C to ensure optimum routing of both callers' calls. In the one shown, the operator has chosen to make the satellite first choice for caller 2's calls. This inflicts the propagation delays associated with satellite links on a large proportion of caller 2's calls to exchange B, but has the advantageous effect of maximizing the availability of cable circuits for connection of caller 1's calls to exchange B, so preventing the failure of calls in the instance when otherwise only an unacceptable 'double satellite path' were available.

In the alternative scheme the operator of exchange C could have chosen to make the cable link to exchange B first choice, even for caller 2's calls. This would have the effect of minimizing the propagation time of caller 2's calls, and may be desirable when caller 1 is a customer of a different network operator.

A common feature of all good routing schemes is their simplicity. Complicated routing schemes can lead to administrative difficulties and oversights. Apart from network congestion and poor transmission quality, slow call set-up and a burden of exchange data maintenance can also result.

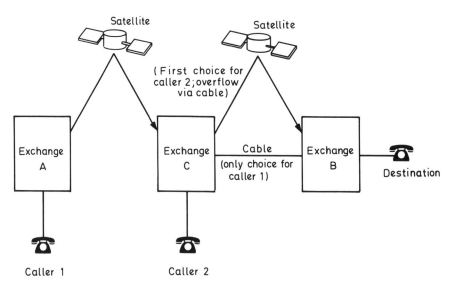

Figure 14.4
Routing based on call history.

All routing schemes rely upon the exchanges to analyze the 'dialled number' or 'packet address' to determine the destination of the call. Additionally, signalling information about the call's previous history (e.g. 'previous satellite link') help to determine the selection of an appropriate route to the destination and an appropriate charge. The exchange therefore needs to analyze a minimum amount of information. For example, at an outgoing international exchange at least the 'country code' indicator digits of the dialled number need to be inspected in order to select the appropriate route to the country concerned. Similarly a trunk exchange must inspect a sufficient number of digits of the area code to determine the onward route selection, and a destination local exchange needs to examine all the digits of the destination customer's local number in order to select the exact line required.

Most signalling systems allow the number analysis and route selection to be carried out in one of two ways, either in an *en bloc* manner, or in the alternative 'overlap' manner. In the *en bloc* manner, the first local exchange waits for the customer to dial all the digits of the destination number before the number analysis is completed and the outgoing route is selected. All necessary digits of the dialled number and other information is then sent together (or *en bloc*) to the subsequent exchange. The subsequent exchanges are thus not bothered with setting up calls until all the information about the call and its destination is available. The exchange processor load on subsequent exchanges is thereby minimized. Figure 14.5(a) illustrates *en bloc* call set-up.

In the alternative 'overlap' manner of call set-up, each exchange in the connection selects the outgoing route as soon as it has sufficient information to do so (even if not

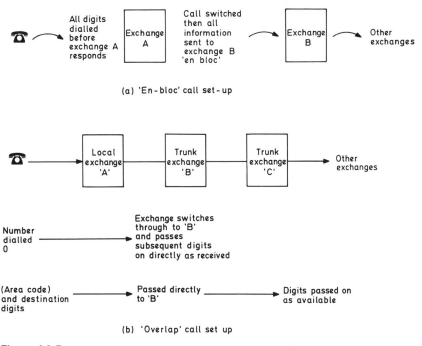

(a) 'En-bloc' call set-up

(b) 'Overlap' call set up

Figure 14.5
'*En bloc*' and 'overlap' signalling at call set-up.

all the information about the destination has been received) and passes on subsequent information as it receives it.

Thus in the diagram of Figure 14.5(b), the connection may already have been made right through to exchange C even before the customer has finished all the digits of the destination number. The same would not be true in the *en bloc* case. It is this feature that gives the overlap signalling method its edge over *en bloc* signalling, in being faster at setting up calls. The disadvantage of the method is the greater processing time wasted at all exchanges waiting to receive and relay digits of the dialled number.

A common failing of network operators, and one which may seem attractive to those using the *en bloc* method, is to carry out undue 'plausibility checks' on the dialled number. Thus, for example, exchanges could be made to look up and check whether a valid number of digits have been dialled, or whether a particular area code within a destination country is valid etc. Such 'plausibility checks' can have the benefit of removing the burden of spurious traffic from the network. Unfortunately, however, the updating of these 'plausibility checks' is often overlooked when new area codes are made available in the destination country, or when number length changes are made. The result is that the exchanges may fail calls to newly valid numbers, with understandable customer annoyance. Had the 'plausibility check' never been used, the problem would not have arisen. Furthermore where plausibility checks are instigated in a network using overlap signalling, the call set-up may be unnecessarily delayed. Special administrative care is therefore required in the use of such checks.

14.2 NUMBERING PLANS

Administration of the numbering plan is an important part of the network routing plan. In isolated networks there is considerable scope for inventing very flexible and well-adapted number plans, but the principles are always much the same. CCITT recommendations list four different numbering schemes for the different types of networks:

● Recommendation E163 is the numbering plan for the international telephone service.

● Recommendation E164 is an extension of Recommendation E163 to cover the needs of the ISDN era. It allows longer numbers and slightly modified procedures.

● Recommendation X121 is the numbering plan for the international packet-switched network.

● Recommendation F69 is the numbering plan for the international telex service.

The principle of each of the above numbering schemes is similar, and only that for the telephone network is described in the remainder of this chapter, to provide an illustration. The geographic distributions and exact country code allocations can be found in the relevant CCITT recommendation.

The international telephone numbering plan

The international telephone numbering plan is defined by CCITT's Recommendation E163, which lays down the principles of numbering pertinent to public telephone networks. The numbering plan is intended to ensure the allocation of a unique number (string of digits) to identify each individual telephone line connected to the worldwide telephone network. E163 numbers are analysed by telephone exchanges to determine the appropriate call routing and the appropriate call charging rate. They are designed to allow exchanges to select an economical and satisfactory onward connection by analyzing only a minimum number of digits. The recommendation does not control each individual country's numbering plan, but allocates instead a large series of numbers for use in each individual national network. This flexibility allows each national network operator to prepare a 'national numbering plan', optimized for their own particular purposes.

Most network operators choose to adopt a three-tier numbering plan. The three tiers correspond to overseas (i.e. international) calls, long-distance (i.e. 'toll' or 'trunk') calls and local calls, and the procedures for each of these types will be discussed in order.

To place an international call over the automatic telephone network, a customer must dial first the 'international prefix code' (to indicate that the digits immediately following indicate a destination overseas). The next digits will be a 1, 2 or 3-digit 'country code (CC)' to indicate the particular country required, then follows the area code and destination customer number. Figure 14.6 illustrates the component parts of a full international number.

The example of Figure 14.6 shows a London, UK telephone number, when dialled from Switzerland. The CCITT recommended international prefix '00' is followed by country code digits '44' to signify the United Kingdom, digits '71' to identify Central London, digits '234' to specify the destination exchange and digits '5678' to earmark the customer. Successive exchanges within the connection gradually discard the early

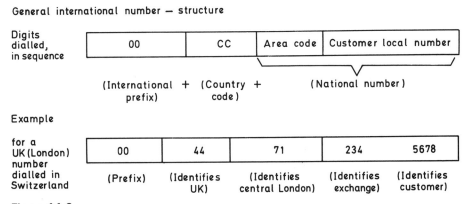

Figure 14.6
International telephone number.

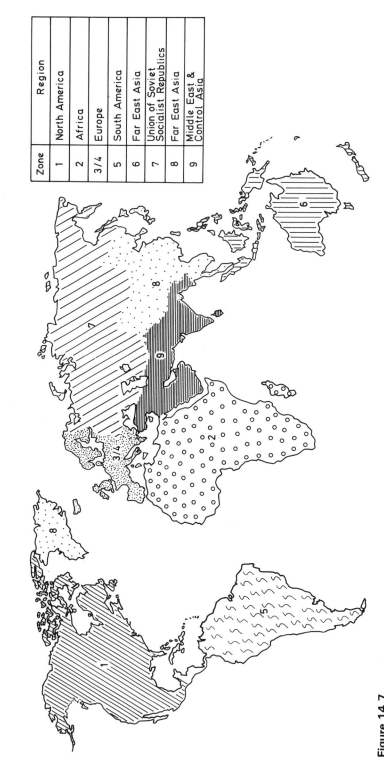

Zone	Region
1	North America
2	Africa
3/4	Europe
5	South America
6	Far East Asia
7	Union of Soviet Socialist Republics
8	Far East Asia
9	Middle East & Control Asia

Figure 14.7
World telephone numbering zones (CCITT Recommendations E163/E164).

digits and analyze progressively more of the later ones to determine the required routing.

The standard '00' prefix is used in many countries, including Switzerland, but some countries, for historical reasons currently use other prefixes. For example, the international prefix for calls made from UK is currently '010', from USA it is '011', and from France '19'. The CCITT recommends that all countries eventually migrate to the common '00' prefix—to ease travelling customers' dialling difficulties.

The country codes of E163 numbers, identifying each individual country, are allocated by CCITT, and the same country code is used to identify a particular country no matter where in the world the telephone caller is situated. Figure 14.7 illustrates the demographic location of world numbering zones.

The entire list of country code allocations is too lengthy to reproduce here, but as an example the country codes of the European region are shown in Table 14.1.

The North American country code is only one digit. This is because there is an integrated area code scheme, called the 'North American dial plan' or 'number plan of America'. This plan covers the whole of the United States, Canada and the Caribbean islands. Telephone calls placed within this area need only be dialled as 'toll' numbers. Table 14.2 shows the allocation of the 3-digit area, or correctly 'number plan of America' (NPA), codes which identify the various regions. A typical New York, USA number is thus + 1 212 345 6789.

Recommendation E163 sets the maximum number of digits allowed in an international number at 12 digits plus the international prefix. This limits each national

Table 14.1
European country codes (CCITT Recommendations E163/E164).

World numbering Zones 3 and 4			
Country	Country code	Country	Country code
Greece	30	Denmark	45
Netherlands	31	Sweden	46
Belgium	32	Norway	47
France	33*	Poland	48
Monaco	33*	Germany	49
Spain	34	Gibraltar	350
Hungarian People's Republic	36	Portugal	351
German Democratic Republic	37	Luxembourg	352
Yugoslavia	38	Ireland	352
Italy	39	Iceland	354
Romania	40	Albania	355
Switzerland	41*	Malta	356
Liechtenstein	41*	Cyprus	357
Czechoslovak Socialist Republic	42	Finland	358
Austria	43	Bulgaria	359
United Kingdom of Great Britain and Northern Ireland	44		

* Integrated numbering plan.

Table 14.2
The number plan of America.

201 New Jersey	202 Washington DC	203 Connecticut
204 Manitoba (Canada)	205 Alabama	206 Washington
207 Maine	208 Idaho	209 California
212 New York City	213 Los Angeles	214 Dallas
215 Pennsylvania	216 Ohio	217 Illinois
218 Minnosota	219 Indiana	301 Maryland
302 Delaware	303 Colorado	304 West Virginia
305 Florida	306 Saskatchewan (Canada)	307 Wyoming
308 Nebraska	309 Illinois (311 Calling Card Service)	312 Chicago
313 Detroit	314 Missouri	315 New York
316 Kansas	317 Indianapolis	318 Louisiana
319 Iowa	401 Rhode Island	402 Nebraska
403 Alberta (Canada)	404 Georgia	405 Oklahoma City
406 Montana	408 San Jose	412 Pittsburgh
413 Springfield, Mass.	414 Milwaukee	415 San Francisco
416 Toronto (Canada)	417 Missouri	418 Quebec (Canada)
419 Ohio	501 Arkansas	502 Kentucky
503 Oregon	504 Louisiana	505 New Mexico
506 New Brunswick	507 Minnesota	509 Washington
510 4 Row TWX (USA)*	512 Texas	513 Ohio
514 Montreal (Canada)	515 Iowa	516 New York
517 Michigan	518 New York	519 Ontario (Canada)
601 Mississippi	602 Arizona	603 New Hampshire
604 British Columbia (Canada)	605 South Dakota	606 Kentucky
607 New York	608 Wisconsin	609 New Jersey
610 4 Row TWX (Canada)*	612 Minneapolis	613 Ottawa (Canada)
614 Columbus, Ohio	615 Tennessee	616 Michigan
617 Massachusetts	618 Illinois (700 Value Added Services)	701 North Dakota
702 Nevada	703 Virginia	704 North Carolina
705 Ontario (Canada)	707 California	709 Newfoundland
710 4 Row TWX (USA)*	712 Iowa	713 Houston, Texas
714 San Diego	715 Wisconsin	716 New York
717 Pennsylvania	718 New York City	800 800 Service‡
801 Utah	802 Vermont	803 South Carolina
804 Virginia	805 California	806 Texas
807 Ontario (Canada)	808 Hawaii	809 Caribbean
810 4 Row TWX (USA)*	812 Indiana	813 Florida
814 Pennsylvania	815 Illinois	816 Kansas City
817 Fort Worth, Texas	819 Quebec (Canada)	900 900-Service ‡
901 Memphis	902 Nova Scotia (Canada)	903 Mexico
904 Florida	905 Mexico City	906 Michigan
907 Alaska	910 4 Row TWX (USA)*	912 Georgia
913 Kansas	914 New York	915 Texas
916 Sacramento	918 Oklahoma	919 North Carolina

* TWX is the American equivalent of the Telex service, introduced by the Bell company in 1931. Telex was not available until Western Union introduced it in the 1950s.
‡ 800 and 900 service are described in Chapter 26.

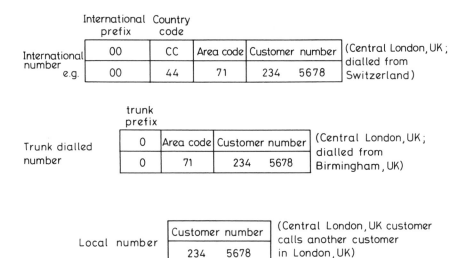

Figure 14.8
International, trunk and local numbers.

numbering plan (area code plus customer number) to a maximum length of $12 - n$ digits, where n is the length in digits of the corresponding country code. Limiting the number length in this way ensures that each network in the world can be designed to cope with the maximum number length. The recommendation also states that it should not be necessary for an outgoing international exchange to analyze more than four digits after the international prefix in order to determine the routing and charging information for any call. (Four digits corresponds to the country code and part or all of the 'area code'.) The example of Figure 14.6 showed how the UK numbering scheme conformed with the E163 recommendation.

Returning to the example shown in Figure 14.6, Figure 14.8 illustrates the different digit strings for dialling the same London customer from either Switzerland (i.e. overseas calling), Birmingham, UK (i.e. trunk calling), or from a 'local' customer in London. Note how the trunk number comprises only the latter part of the full international number, the international prefix and country code having been replaced with a simpler 'trunk prefix'. The standard CCITT trunk prefix '0' is used in the UK, but as is the case with the international prefix, some countries have not yet moved on to the use of the standard trunk prefix, for various historical reasons. (Example: the trunk prefix used in North America is '1'.) Finally, for local calling the customer's number is normally dialled without prefix, as Figure 14.8 shows.

Numbering for the ISDN era (Recommendation E164)

With the growth in demand for telephone lines throughout the world and with the advent of new techniques such as ISDN (see Chapter 24), CCITT has had to revise its numbering recommendations (E163), and (E164) now lays out a new 'numbering plan for the ISDN era'. Four key differences which will affect the design of new

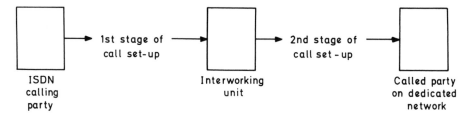

Figure 14.9
Two-stage call set-up.

exchanges are the introduction in E164 of:

(i) Extended international numbers (up to 15 digits following the international prefix). This will expand 1000-fold the quantity of numbers available in each country and so meet the needs of the foreseeable future.

(ii) The concept of 'direct dialling-in (DDI)'. In DDI the last few digits at the end of the ISDN subscriber number are transferred to the customer's PBX or other equipment, so enabling the call to be completed direct to the recipient's desk telephone without the assistance of a human PBX operator.

(iii) The concept of 'sub-addressing' (also called 'network address extension'). A sub-address comprises up to 40 additional decimal digits on top of the ISDN number, allowing routing with established local area networks on customer premises at the distant end of a public ISDN.

(iv) The concept of two-stage call set-up for the support of interworking. The very nature of Integrated Services Digital Networks (ISDNs) demands that different services can interwork over a common network. Sometimes, however, this is not possible without the use of a specialized interworking unit between the ISDN and the dedicated network. As Figure 14.9 shows, two-stage dialling can be invaluable in this case.

Two-stage dialling may require the return of a second dial tone and the sending of extra digits after the completion of the first stage, and it can often be necessary on calls passing between two normal telephone networks, either from or into some other specialized network.

BIBLIOGRAPHY

CCITT E-Series recommendations.
CCITT Recommendation E160, 'Numbering Plan of the International Telephone Service'.
CCITT Recommendation E163, 'Numbering Plan for the International Telephone Service'.
CCITT Recommendation E164, 'Numbering Plan for the ISDN Era'.
CCITT Recommendation E170, 'International Routing Plan'.
CCITT Recommendation F69, 'Plan for Telex Destination Codes'.
CCITT Recommendation X110, 'International Routing Principles and Routing Plan for Public Data Networks'.

CCITT Recommendation X121, 'International Numbering Plan for Public Data Networks'.

Flood, J. E., *Telecommunication Networks*. Peter Peregrinus (for IEE), 1975.

Frank, H. and Frisch, I. T., *Communication, Transmission and Transportation Networks*. Addison-Wesley, 1971.

Vestmar, B., *Design Considerations for the Grid Type of Communication Network*. Bronder Offset BV, 1975.

15

OPERATOR ASSISTANCE AND MANUAL SERVICES

Early telephone networks were all manually operated. In the 1950s automatic networks began to take over, but even today they have failed to supplant all manual 'assistance services'. In the public network human operators provide a 'safety net' of assistance and advice for customers, while in some private networks human PBX operators are still employed to answer incoming calls from the public network and to connect them to the required extension.

Today, users take automatic switched communications networks for granted, and they expect to be able to dial directly to almost any other telephone in the world.

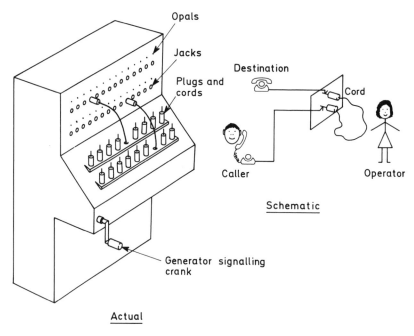

Figure 15.1
Early manual, or 'sleeve control' switchboard.

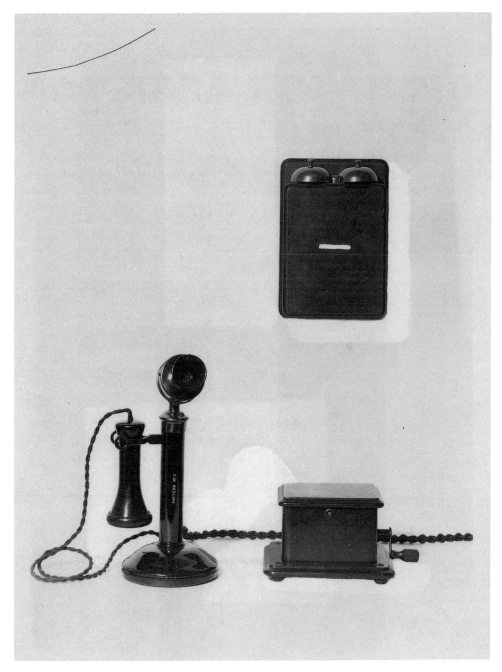

Figure 15.2
A manual telephone system, the customer premises equipment. To the left is the telephone. On the wall is the bellset and to the right is the hand generator signalling set. (*Courtesy of Telecom Technology Showcase, London.*)

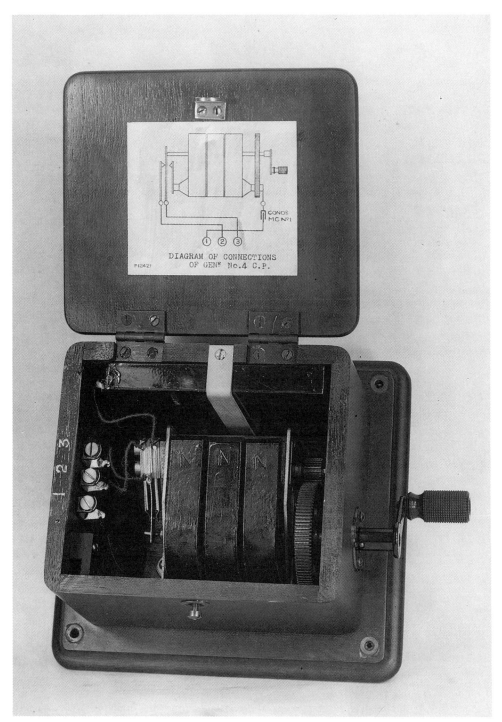

Figure 15.3
Inside of a hand generator signalling set, showing the magneto coils. (*Courtesy of Telecom Technology Showcase, London.*)

Figure 15.4
Early operator switchboard. (*Courtesy of British Telecom.*)

THE EXCHANGE AT WORK.

*An explanation which will prevent
many common misunderstandings.*

The telephone lines in a large city are divided
into groups, called Exchanges.

A Museum A-side operator answers and gets into
touch with a Hop B-side operator, who finds out if
Hop 3000 is free and helps the Museum operator to
connect the lines. If conversely Hop 3000 wants
Museum 605, it is a Hop A-side operator who
answers and gets into touch with a Museum B-side
Operator.

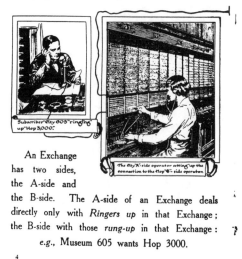

An Exchange
has two sides,
the A-side and
the B-side. The A-side of an Exchange deals
directly only with *Ringers up* in that Exchange ;
the B-side with those *rung-up* in that Exchange :
e.g., Museum 605 wants Hop 3000.

4

If Central
6000 wants
Central 3000,
a Central A-side operator gets into touch with a
Central B-side operator.

In every call, therefore, two Post Office operators*

* In reality, of course, there are four operators : two Post Office servants
and two amateurs—that is, the two subscribers connected. We, in fact, are
members of the unpaid Post Office telephone staff. Hence these notes.

5

Figure 15.5
The exchange at work. An extract from an early British Post Office publication providing
an explanation which will prevent many common misunderstandings. (*Courtesy of
Telecom Technology Showcase, London.*)

15.1 MANUAL NETWORK OPERATION

In a manual network, the connection of caller to destination is carried out by human
operator. It is done by 'plugging' cords into individual line sockets or 'jacks', one 'jack'
corresponding to each possible destination user. Figure 15.1 illustrates an early
manual switchboard, and Figure 15.2 a typical telephone used on such a manual network. Instead of a numbered 'dial' there is just a cradle for the handset and a magneto
generator to call the operator.

The routine for making a call on a manual network is as follows. The caller lifts
the handset, and rings the magneto generator by turning the handle. This has the effect
of alerting the operator and lighting an 'opal' (a light) on the operator's switchboard
(Figure 15.1). In some cases, the operator was alerted merely by rattling the cradle.
This had the effect of flashing the opal. There is an opal above each incoming line jack,
indicating precisely which caller wishes to make a call. To answer the request, the

operator uses one of the cords mounted on the console part of the switchboard, which is pulled out and plugged into the relevant 'jack' socket immediately below the 'opal'. The operator is now able to speak to the caller and ask for the name of the person he wishes to call. The operator records the caller's name, the destination number and the time of day, on a 'ticket' for later 'billing' of the caller. The destination party is then alerted by the operator, who rings their telephone with another hand-cranked generator. The connection is completed by plugging the other end of the cord into the 'jack' of the destination party. In this way the pair of plugs and the cord connect the two corresponding line jacks to caller and destination. At the end of the call the caller replaces the handset, extinguishing the 'opal'. On noticing this, the operator removes the plugs and cord from the jacks, ready for use on another call.

For making calls to customers on other exchanges the operator has a number of 'trunk line jacks'. To use them, the operator must relay the call details to the operator on the second exchange, and forward the connection. The second operator either completes the call, or forwards it to another operator, as necessary. In manual networks, setting up telephone calls is highly labour intensive, and in the early days the majority of the workforce in public telephone companies were 'telephone operators'.

15.2 SEMI-AUTOMATIC TELEPHONY

'Semi-automatic telephony' is the term used to describe connections which are set up by an operator across an automatic network. Semi-automatic telephony was common when telephone networks were first being automated, especially when some exchanges had been automated while others remained manual. Callers on the manual exchange, who wished to call others already connected to an automatic exchange, would get their calls connected 'semi-automatically' by the operator. The caller would first contact the operator, and the operator would then connect the call onward using special equipment to control the automatic network rather than routing through further operators.

Ironically, in the reverse direction any caller who was connected to an automatic exchange would have to dial a code to get hold of an operator so as to get a manual connection to any destination customer still connected to a manual exchange.

Manual exchanges have progressively given way to today's predominantly automatic networks, but even today callers resort to dialling for 'assistance' from the operator in a number of instances:

- to call a user on a residual manual exchange;
- to get assistance following difficulty on an automatic connection;
- to get the answer to a general enquiry;
- to enquire for the directory number of another user;
- to make a special service call, such as a 'reverse-charge' (also known as a 'collect') call, or a 'personal' call, etc.

15.3 CALLING THE OPERATOR

If all we want is to get through to the operator and tell him we need assistance, the effect of cranking the magneto is very much the same as dialling the right number on an automatic network. On a 'sleeve controlled' switchboard (one using plugs, cords and jacks, as illustrated in Figure 15.1), incoming calls are indicated to the operator by lighting the 'opals'. In modern 'cordless' operator switchrooms the call is administered by 'call queueing equipment'. This equipment stacks-up calls in the order in which they are received, and allocates them to telephone operators in the switchroom as they become free from dealing with previous calls. Operators simply press a button to indicate that they are ready to handle another call.

While waiting in the queue for an operator to become free, the caller may hear 'ringing tone', or may instead be given a recorded message, something like 'this is the

Figure 15.6
Switchroom operators at work. A picture giving an idea of the tangle of hands and leads—the frenetic operation of manual exchanges. (*Courtesy of British Telecom.*)

assistance service—an operator will deal with your enquiry shortly'. Recorded messages have the benefit of confirming that callers have 'got through', giving reassurance that they are not waiting in vain. The recorded message also allows callers who have accidentally dialled the number for the operator service (when meaning to dial some other number), to hang up their calls and try again.

Because connections made via the operator are multi-link rather than single link connections, special measures are required, to ensure that the end-to-end quality of the connection is acceptable, and to charge the customer correctly. Callers are normally connected to the nearest switchroom. This ensures that the end section connection is of the best available quality. This part of the connection (i.e. from caller to operator) is not automatically metered for charging purposes; the call charges are derived either from paper 'tickets' written by the operator, or from 'electronic tickets' produced on the operator's computer consoles.

Operator switchrooms are designed to be efficient workplaces, and staff numbers and rosters are planned to meet customers' call demand. Just as an automatic telephone network must be provided with sufficient circuits to meet the traffic, so must the number of 'positions' manned by operators at any given time of day match the traffic demand at that time. A useful quality target for staff providing an operator service is to aim to answer all calls within a given time (say 25 seconds), or perhaps to aim to answer 90 per cent of calls within say 15 seconds. The latter statistic is often written in shorthand as PCA15 = 90%, i.e. the Percentage of Calls Answered in 15 s = 90 per cent. PCA25 can also be used as a performance statistic (measuring the percentage of calls answered in 25 s), but the average caller on a public network who wants operator assistance, is not satisfied with such a long wait. There is a relationship between the measured value of PCA and the staffing level of the switchroom, the use of more staff generally increasing the PCA value. Going to one extreme, to employ a very large number of operators queueing up to answer calls would ensure almost instantaneous answer. At the other extreme, with too few operators it is the caller who does the queueing and waiting. A modification of the Erlang formula presented in Chapter 10 is called the Erlang waiting-time formula. It can be used to calculate the number of operators required in a switchroom in order to keep the waiting time down to a target figure.

15.4 OPERATOR PRIVILEGES

In the onward connection of calls, operators are given a number of special networking privileges. They may have exclusive use of particular routes or of certain circuits within a route, in order to give their callers a better chance of getting through than normal customers (especially when the network is busy). This enables the operator to be of real assistance to the caller in cases of difficulty, and furthermore, reduces the likelihood of wasted operator time spent in futile repeat attempts. Other privileges explained below include 'manual hold', circuit monitoring and interruption, and forward transfer. However, with the increasing development and automation of networks, and the small number of human operators available for network 'policing', these features are becoming obsolete.

Manual hold allows the operator to 'hold' the connection even after the calling subscriber has replaced the handset. This makes it possible to trace the origin of a malicious call in a case when a caller has given a false identity. It also prevents the caller making any further calls. This use of the facility is now largely superseded by 'calling line identity' information in automatic networks and many networks today no longer have the facility. An alternative use of the 'manual hold' facility permits tracing the cause of faulty connections. The ability to trace emergency service calls (e.g. fire, police, ambulance) is of especial importance.

Circuit monitoring and interruption   Sometimes the operator is given the facility to monitor or interrupt customers' calls while in progress. This can be useful in investigating customer complaints, including account discrepancies. It can also be used to break in on conversation already in progress, when an important incoming call is received, and this would have been done historically if a trunk call was received while only a local was in progress. (Hence the term for this facility—'trunk offer'.)

Forward transfer   Another facility becoming largely obsolete, the forward transfer, allows the operator who has *previously* established a semi-automatic connection for the call, to request assistance from the operator at the destination exchange; international operators used it to provide 'language assistance' (i.e. translation). The operators might speak an intermediate language (typically French or English) between themselves, and their mother tongue to their own customers in order to resolve any difficulties during call set-up. (Example for person-to-person calls.) The desired language of assistance is indicated by a special digit called the 'language digit', which is inserted by the originating operator's exchange into the called customer's dialled digit string. This is discussed in more detail later in the chapter.

15.5   TYPICAL ASSISTANCE SERVICES

Since most calls are made automatically nowadays, operators have to provide only a range of 'assistance' services to complement the automatic service; here are some of the more common ones:

'Station call' service   A 'station' call is the name given to an 'ordinary' call between two telephone stations, when it is made via the operator. A station call may be made (via the operator as opposed to automatically) either because the call cannot be dialled directly, or because customers prefer it, or perhaps because customers have had difficulty in getting through. Another reason may be that the customer wishes to be rung back immediately after the call has finished to be Advised of the Duration and Charge (ADC) (also called 'time and charges').

Reverse charge', or 'collect call' service   For any of a variety of reasons, callers when travelling may not wish to pay for calls themselves, preferring to transfer the charges to the call recipients. The service which does this for them is the 'reverse charge' or 'collect call' service. The reason for transferring the charge may be shortage of change when using a payphone, or to avoid leaving one's host with a large telephone bill when staying away from home. Whatever the cause, collect call service must always be made via the operator. Before receiving a 'collect call', recipients are asked by the operator whether they are willing to accept the call charges. If so, the call is connected and an operator 'ticket' records the call details in the normal way, except that the bill is sent to the recipient rather than to the caller. A similar service, self-explanatory, is also sometimes available—'Bill call to third party' (this might allow a payphone caller to charge the call to his home account).

'Personal call' service (also called 'person-to-person')   Sometimes callers may wish to contact particular people who share their telephone with a number of others. A caller may not want to make an automatic call and pay for a connection, only to find out that the desired individual is not available. In these circumstances it is appropriate to make a 'personal call', via the operator. The caller gives the operator the name and telephone number of the individual required, and the operator then makes the call and checks that the right recipient is available to come to the telephone. If so, the caller is charged from the moment when the operator allows conversation to commence. Usually either a surcharge or a higher charge per minute of conversation is levied on personal calls. If the person wanted is not available, the connection is cleared without conversation and the caller is not charged.

'Directory enquiry' service (also called 'directory assistance')   To make an automatic call a caller must have the number of the destination telephone station. Without that number the network has no indication what connection the caller wants. If the caller does not know it, perhaps because he has not called that particular person or company before, then one way of 'looking-up' numbers is to use the paper directory, issued to telephone customers. This gives an alphabetic list of the names of all customers with their numbers. However, the sheer number of customers nowadays has tempted many telephone companies to issue only a telephone directory covering the immediately surrounding geographical area. The operator 'directory enquiry service', or 'DQ' service provides a more comprehensive nationwide service. Access to the service in the normal way is by dialling an access code and waiting in a queue at the nearest directory enquiry switchroom. The operator looks up the number in the appropriate paper directory or by 'querying' a computerized directory system. When the number is found it is given verbally to the caller, who then presumably follows up the enquiry by placing a call over the automatic network. Alternatively the operator can immediately put a call through. Some network companies charge customers for each directory enquiry; others give a free service, which they believe stimulates more automatic calls anyway.

General enquiry service   Whether publicized directly or not, the operator is also often called upon by customers to answer general enquiries about the telephone or other service. A typical enquirer might ask for the area code to be used for a particular town, or an international operator may be asked for the time-of-day difference in hours between originating and destination countries. The answers to these enquiries, and further reference information to help operators in the undertaking of all their services is usually provided in the form of a 'visual index file', or 'VIF'. This is a handy reference book, kept on each operator's console. Operator 'tickets' may or may not be made out to record general enquiries, and charges may or may not be levied.

15.6 CO-OPERATION BETWEEN INTERNATIONAL OPERATORS—CODE 11 AND CODE 12 SERVICE

When networks belonging to more than one company are interconnected, there are times when the operator in one network requires assistance from the operator in the other. This usually comes about because neither of the operators has quite the same privileged control over the other network, as they have over their own. Alternatively, in international networks operators may require no more than language assistance (i.e. translation of the distant end mother tongue into a comprehensible intermediate language).

In order to give operators of different networks a means of calling one another, CCITT's telephone signalling systems are provided with three signals specifically for the use of operators, these are:

- Code 11 signal.
- Code 12 signal.
- 'Language digit' signal.

The *code 11 signal* is used by the operator of the originating network to get assistance from the operator of the terminating network, in cases where the call cannot be connected semi-automatically because the terminating network is a manual one. The signal gives access to appropriately equipped operators in the terminating country. In countries where the network is already fully automatic, code 11 service will have been withdrawn. The code 11 signal itself is single digit signal, just as the values 1, 2, 3, 4, 5, 6, 7, 8, 9, 0 are. The difference is that code 11 cannot be dialled by ordinary telephone customers. This precaution prevents ordinary telephone subscribers from masquerading as telephone operators, and from using this privileged role to defraud overseas network companies.

The *code 12 signal*, like the code 11 signal, is also used by the operator of an originating network to get assistance from an operator in the terminating network. Also like the code 11 signal, the code 12 signal is a single digit signal which cannot be dialled by ordinary telephone customers. Code 12 is to be used when difficulty has previously been experienced in establishing an automatic or semi-automatic connection, and it gives access to an appropriately equipped operator in the terminating country. Unlike the code 11 signal, code 12 is often used with extra digits.

e.g. (Code 12) + ABCD

The extra digits ABCD may be used to identify an individual or particular group of operators within a switchroom—so helping overseas operators to return to the same assisting operator.

Code 11 and code 12 signals are always sent immediately following the country code and language digit of the international number.

The *language digit* is an extra digit, inserted into the digit train of all semi-automatic operator controlled telephone calls in order to indicate the language that the calling operator would prefer the distant operator to speak in, should language assistance be required. Not all countries' telecommunications companies offer language assistance appropriate to all the language digits, but the following values are allocated as follows:

1 = French
2 = English
3 = German
4 = Russian
5 = Spanish
6 to 9 (Spare)
(0 = dimensioning digit—for automatic working)

The language digit is always inserted immediately after the country code of the number dialled, or in the alternative position demanded by the international signalling system. Thus on a call to the number

44 71 234 5678 (a number in Central London, UK)

the train

44 (1) 71 234 5678

might be used by an operator overseas requiring language assistance from the UK operator in French. The digit actually is not inserted by the operator manually, but rather is systematically added by the exchange on a route by route basis. Thus LD = 2 for English might set on the route to Japan while LD = 1 for French might be appropriate on a route to a French African colony. When the language digit is set at value 0, it is called the 'discriminating' digit, and is used to identify automatic (i.e. non-operator-controlled) calls. The discriminating digit (value 0 of the language digit) is inserted systematically by automatic telephone exchanges. Thus the receiving network can distinguish between automatic origin calls and operator controlled (i.e. semi-automatic) calls. If necessary, and in response to either a code 11, code 12, or 'forward transfer' signal, language assistance can be given. Having examined the language digit and in so doing, performed appropriate routing and call accounting, the international exchange in the destination country (but not an international transit exchange) will remove the language digit before sending out the destination national number into the network of the terminating country. The language digit therefore only exists on international telephone links.

15.7 A MODERN OPERATOR SWITCHROOM

Since the early days of operator-controlled manual telephone networks, operator switchroom equipment has moved through several phases of technology, first to cordless boards and latterly to fully computerized systems. It may seem ironic that a computer system should be designed to work in a mode requiring human intervention rather than setting out to eliminate the human element; but the fact is, telephone customers still find comfort in spoken assistance.

Today, computerized operator exchanges give optimized ergonomic conditions for the human operators without compromising the network efficiency. Some of these exchanges have a central 'switching' equipment, where the switching and control of circuits takes place; but in addition, more luxurious remote switchrooms, where the operators sit, are connected via computers and computer data links. From their remote switchroom the operators can issue commands, using the keyboard of a computer terminal, in order to instruct the central switch which call routing, or other control action should be performed. Meanwhile the computer can automatically generate its own electronic tickets, storing the A number (caller's number), B number (destination number), call duration, time of day etc. Some of the information can be typed in by the operator when talking to the caller (A-party) (e.g. the B number), and some of the information can be derived automatically (e.g. time of day and call duration). Figure 15.7 illustrates schematically the network arrangement of this type of computer-controlled operator exchange.

As well as the ergonomics and comfort benefits of this type of switchroom (more

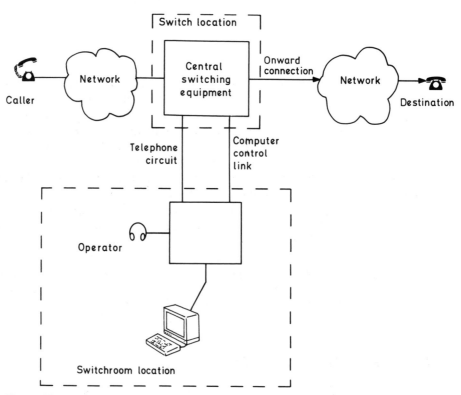

Figure 15.7
Network schematic of modern computer controlled operator exchange.

like an office, and less like the factory-like switchrooms of the past), there are a number of other benefits:

- the operator time required to handle individual calls is reduced;

- call re-attempts can be made more easily (by pressing a 'last number re-dial' button);

- the accuracy of operator tickets is improved, and the information is directly available in a computer format for subsequent computerized bill calculation;

- the central switching equipment can be located with other automatic exchanges, and maintained by a common technical staff, while the switchrooms may be located near an available workforce;

- if more than one remote switchroom is connected to the same central switching equipment better staff rostering can be achieved, say by closing some switchrooms at night, and handling all the overnight traffic at one switchroom. This was not generally possible in the past.

15.8 OPERATOR ASSISTANCE IN TELEX NETWORKS

The principles of telex operator assistance networks are similar to those of telephone operator assistance networks. The difference is that the operator has to 'speak' to the caller through a telex machine and not by voice. Other slight differences exist in the actual menu of services offered; for example, the telex operator might offer a multiple destination service to telex callers.

15.9 OPERATOR ASSISTANCE ON DATA NETWORKS

Operator assistance is not normally available on packet-switched and other data networks. However, for the purpose of point-to-point data connection over the telephone network, some companies (declining in number) allow their telephone operators, when asked by callers, to select particularly high grade circuits designed specifically for the transmission of voice band data. The customer uses such circuits in conjunction with data modems located on the customer premises at either end of the connection.

BIBLIOGRAPHY

Chapuis, R., *One Hundred Years of Telephone Switching (1878–1978), Part 1: Manual and Electromechanical Switching (1878–1960's), Volume 1.* North-Holland Studies in Telecommunication, 1982.

Flood, J. E., *Telecommunication Networks.* Peter Peregrinus (for IEE), 1975.

Rhodes, F. L., *Beginnings of Telephony.* Harper & Brothers Publishers, 1929.

16

QUALITY AND THE MANAGEMENT OF SERVICE

The maintenance of good quality for any product or service (i.e. its 'fitness for purpose' and its price) is of supreme importance to the consumer and therefore requires utmost management attention. But while it is easy enough to test a tangible product to destruction, measurement of the quality of a service is more difficult. In telecommunication the customer is left with nothing more tangible than his or her own perception of how well the communication went: on a datalink the errors that have had to be corrected automatically are barely appreciated, while in conversation, unobtrusive bursts of line noise may go unnoticed. This is not to say that loss of data throughput on a datalink caused by continual error correction is of no concern to the customer, nor does it mean that noises which disturb conversation are acceptable. What we really mean is that in measuring service quality, due regard must be paid to anything of importance that the customer perceives and remembers. Concentration on setting and meeting quality targets should be paramount in planning and administration. Insufficient attention to them is the road to customer dissatisfaction and loss of business.

This chapter reviews some aspects of communications quality and the practices of management, with examples of the commoner quality parameters and control measures used by the world's major operators.

16.1 FRAMEWORK FOR PERFORMANCE MANAGEMENT

Good management, of whatever industry, demands the use of simple, structured and effective monitoring tools and control procedures to maintain the efficiency of the internal business processes and the quality of the output. When all is running smoothly a minimum of effort should be required. But in order to be able quickly to correct defects or cope with abnormal circumstances, measurable means of reporting 'faults' or 'exceptions' and rapid procedures for identifying actionable tasks are required. The framework for doing so needs to be structured and comprehensive. In Table 16.1,

Table 16.1
Framework for performance management.

Dimension	Measures/controls
Formal plan and delivery cycle (organizational framework)	Formal *presentation* and *agreement* of company strategy, policy, guidance and plans. Regular review. Framework for business 'value' assessment and assurance.
User support (fit for purpose)	User comprehension of services (by questionnaire). Swift elimination of problems (full complaint log).
Supplier performance (quality of supply)	Maintaining professional attitude with suppliers. Monitoring and demanding swift lead times. Meeting targets for repair performance.
Quality performance (effectiveness)	Meeting user needs (number of complaints). Number of lost messages (e.g. percentage of data lost). Percentage of 'down-time' during normal business hours. Measure of delay (e.g. propagation delay, or backlog of orders/messages). Measure of congestion (e.g. percentage calls lost-callers frustrated). Repair performance (e.g. percentage not repaired in target time). Connection quality (e.g. percentage customers satisfied). 'Cost of poor quality' (i.e. that correcting avoidable faults).
Financial performance (efficiency)	Return on capital investment. Shareholders financial return. Cost actually spent can be compared with competitors or alternative suppliers. Internal chargeout rate for telecom services (e.g. 'pounds per bit-mile').
Technical performance (efficiency)	Compatibility of networks can be adjudged by measuring the money spent on interworking or upgrading and comparing it with total system value.
Resource management (efficiency)	Asset inventory and management (needs to be accurate and up to date) Cost per task. Adequate resourcing to meet target service lead times. Head count and man-hours should be maintained on target.
Evolution of networks (responsiveness)	Network upgrades should be properly planned and meet their budget. There should be steps to give 'benefits' assurance. The response time of new software and computer networks should be according to plan. The networks should be adaptable.

a number of simple management 'dimensions' together with possible management performance tools/parameters within each dimension are presented. As you will note the dimensions cover a range of areas, some requiring more tangible monitoring measures and control procedures than others.

16.2 QUALITY—A MARKETING VIEW

To look at networks from the customer's viewpoint we have to understand his reasons for using telecommunications services. It is a mistake to generalize about all customers as 'wanting to convey information over long distances', since that takes no account of the service in use, or of each customer's individual purpose or 'application'. It is, after all, a fundamental of marketing that products should be attuned to the markets and customers they are intended to serve. Thus the motivations and interests of someone who is out for the evening in calling home from a payphone are quite different from the needs of a large multinational company, wishing to convey vast volumes of computer data around the world on sophisticated and dedicated networks at any hour of the day or night.

It may be that the same network infrastructure can accommodate the service needs of a number of different markets or customer 'clubs' but it must do so without compromise. One example of a network capable of supporting a number of services is the public telephone network, and it must run at optimum quality not only for human conversation, but also for facsimile machine interconnection. A more advanced example of integrated services networks is the Integrated Services Digital Network (ISDN) discussed in Chapter 24.

Customers' communication needs can often be met in a number of alternative fashions, ranging from travelling in person, using the postal service, the telephone, telex, packet switching, and high-speed data telecommunications services. Pitching a given service to meet a given need is the job of telecommunications marketing specialists who determine its relevant qualities, expressed as a number of measurable parameters. An example of such a 'quality parameter', used commonly by telephone network operators is the 'percentage of calls completed'. Not only is this a fair measure of network congestion, it is also a good predictor of the level of customer frustration.

16.3 QUALITY OF SERVICE (QOS) AND NETWORK PERFORMANCE (NP)

You may have realized in our example above that the percentage of answered calls is not dependent only on network congestion, but also upon the availability of somebody at the destination end to answer the telephone. Thus the 'quality of service' enjoyed by the customer depends upon a number of factors in addition to the performance of the network. CCITT has also drawn this distinction, and its general recommendations on the quality of telecommunications services recognize two separate categories of performance measurement:

● Quality of service (QOS)

● Network performance (NP)

Quality of service measurements help a telecommunications service or network provider to gauge customers' perceptions of the service. Network performance parameters, on the other hand, are direct measurements of the performance of the network, in isolation from customer and terminating equipment effects. Thus 'quality

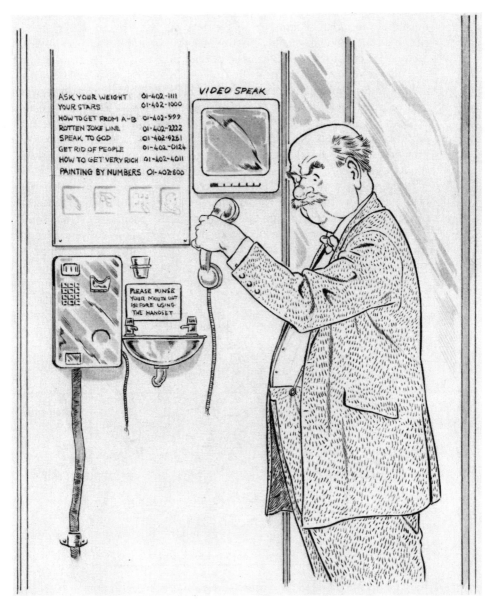

Figure 16.1
A customer's perception of quality. Get the basics right before worrying with the frills. Drawing by Patrick Wright. (*Courtesy of M. P. Clark.*)

of service' encompasses a wider domain than network performance, so that it is possible to have a case of poor overall quality of service even though the network performance may be excellent. The relationship of 'quality of service' to 'network performance' is shown in Figure 16.2.

The distinction between quality of service (QOS) and network performance (NP) is somewhat artificial, but it has become necessary as the result of recent regulatory change in some countries, where the public telephone operator (PTO) is allowed to provide service only up to a socket in the customer's premises, while end-user equipment (i.e. 'terminal') manufacturers slug it out in the customer premises equipment (CPE) market. Government regulation should keep PTO network performance up to the mark, but overall service may be of poor quality if it is let down by faulty or badly designed customer apparatus.

Parameters should be chosen to reflect high quality end-to-end service and should be expressed quantitatively as far as possible. These QOS parameters should then be correlated with one or a number of directly related network performance (NP) parameters, each NP parameter reflecting the performance of a component part of the network, and therefore contributing to the end-to-end quality. In this way quality of service problems can be traced quickly to their root cause.

In the example of Figure 16.3, QOS criteria have laid it down that the signal loudness received by the listener should not be less than 14 dB below the original signal transmitted. The end-to-end QOS parameter is therefore 'signal loudness' and is measured in dB. The network between the two people in conversation comprises two telephone exchanges and three linking circuits, which are designed to individual 'network performance' loudness losses of 1 dB. The overall network performance is therefore $3 + 1 + 3 + 1 + 3 = 11$ dB, and when the losses in the telephone handsets are

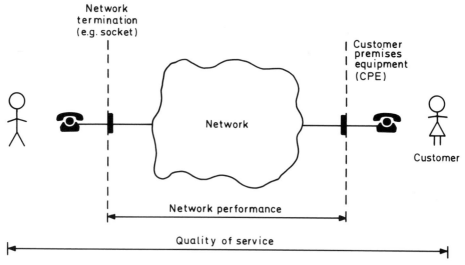

Figure 16.2
Quality of service and network performance.

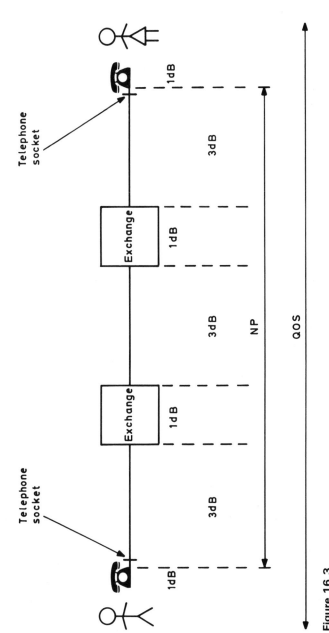

Figure 16.3
Relating QOS and NP parameters.

added, the end-to-end (QOS) loudness is 13 dB, i.e. giving a 1 dB margin within the maximum allowed loss of 14 dB.

Considering only the network performance (NP) between the two telephone sockets of Figure 16.3, an acceptable end-to-end QOS is delivered even if one of the links or exchanges in the network degrades in performance by up to 1 db (provided all the other parts continue to work at their designed losses). In other words, a network performance of 12 dB is sufficient to deliver an acceptable end-to-end QOS of 14 dB loss. Likewise, if either of the customer telephone sets degrades in performance to the extent of adding up to 1 dB, then the overall QOS of 14 dB is still maintained, provided other components remain stable.

In normal operation it is difficult to monitor the end-to-end QOS performance of all connections, particularly when a large number of permutated 'switched connections' are possible, and the concept of network performance parameters is invaluable. Each element of the network may be monitored in isolation and maintained within a predetermined network performance range. Thus in the example of Figure 16.3 we might aim to keep the individual exchange losses below 1.5 dB, and the individual link losses below 3.5 dB. Although theoretically this might mean that the overall network performance could include a loss of $3.5 + 1.5 + 3.5 + 1.5 + 3.5 = 13.5$ dB, which is 1.5 dB outside the maximum allowed network performance of 12 dB, it is highly unlikely that all the links and exchanges will simultaneously be performing in the most adverse manner. As a safeguard against such an adverse occurrence, small occasional samples of the end-to-end quality of some representative connections will have to be taken.

Sometimes the relationship between the end-to-end quality of service parameter measurements and the network performance parameters may not be as direct as in the example just discussed. In that example we were able to use similar parameters, all measured using the same scientific units (in our case dB, or decibels). But in the QOS parameter 'percentage of calls reaching answer', there is a much more complex relationship with measurable network performance parameters, depending not only on the state of network congestion, but also upon the availability of alternative routes, the incidence of faults, and other abnormalities such as 'misrouting'. Some of the contributions may have negligible effect under normal conditions, so that reliable approximations may be used. For example, if the percentage of calls being answered at a given destination starts to deteriorate then the network operator's first action is to look for any network links that are congested. If any are found, then these are more than likely the cause of the congestion, and extra resources should be provided accordingly. If no individual link or exchange congestion is immediately apparent then the operator will have to look for less common causes and faults.

There are no hard-and-fast rules as to which parameters should be used to measure quality of service and network performance, although CCITT has standardized some parameters and is currently studying the subject. The remainder of the chapter proposes a pragmatic approach for establishing QOS and NP parameters for telecommunications services. The approach is designed to generate a balanced mix of parameters that will prove valuable in monitoring all aspects of service quality, including not only the quality of the connections themselves, but also the important support activities such as maintenance and provision of service.

Categories of QOS and NP

By defining a standard set of categories relating to telecommunications service quality, we can ensure that selected parameters within each category give a balanced picture of current network performance and quality of service as against target. Without such discipline in the past, many network operators have made extensive measurements of the technical aspects of performance with too little respect for the broader 'service attributes' giving them a distorted view of their quality, and they have suffered accordingly. For while they may have been returning exemplary technical performance statistics, their customers may have been far from happy, because they could never get a reply from the fault reporting bureau, or could never get assistance about how to use a service. A possible set of categories from which to choose a balanced set of performance parameters is

- Customer service

- Service availability

- Provision and alteration of service

- Service reliability

- Making calls (and 'clearing' them)

- Quality of conversation or information transfer

- Fairness, reliability and accuracy of service charges

Each category is explained, and some examples of actual parameters are given in the paragraphs that follow.

Customer service   Parameters in this category measure general customer perceptions of the 'helpfulness' of network operator staff, their courteousness, their 'understanding' of problems and their 'willingness to help'. Some parameters might additionally reflect customer feeling about the availability and usefulness of literature, operating instructions and advertisements. Parameters in this category have historically been the most overlooked, perhaps because it is difficult to measure them accurately by market research, and perhaps because they are sensitive to swings in public opinion. Typical parameters might well include 'percentage of satisfied customers'. Great care is required when selecting the precise wording of market research questions.

Service availability   A customer's perception of the quality of service may depend upon its availability. For example the geographic coverage may be too restricted, either prohibiting the customer from service subscription in the first place, or being too constraining in the number of egress points available. One's view of a payphone service, for example, is much affected by the time it takes to find a kiosk at the time when wanting to make a call. It is also influenced by the number of payphones available at busy sites such as stations and airports where customers face intolerable delays. A third factor influencing the quality of payphone service might be the destinations which are available from it. For example, payphones which only accept small denomination coins have little appeal for foreign tourists making international calls to their home country. Typical parameter measures in this category might thus be:

- percentage of customers desiring service who are not in the coverage area (waiting list length);

- average queueing-time (for a payphone);

- percentage of desired destinations not available (from market research).

Provision and alteration of service   Having ordered a service, the customer expects it to be provided promptly and conveniently. Vitally worth measuring is the time taken to complete customer orders, and perhaps the percentage of appointments kept. Other parameters could also measure the standard of installation work. Typical measures might be:

- percentage of contracts for the provision of service completed within a target number of days;

- percentage of appointments for provision of service kept;

- percentage of installations meeting customers' quality expectation;

- number of installation revisits required.

Service reliability   Having had a line established on the premises, a customer will wish to make and receive messages or calls according to whim. Loss of service occurs every time the user is unable to make or receive messages or calls for any reason other than customer mis-operation. Furthermore, when he has suffered a loss of service, the customer is anxious to be restored to full service within an acceptable period of time, and with minimum inconvenience. Thus 'time to repair' or 'time for repair' are important. Other measures in this category might be:

- total period of lost service within one year;

- number of occasions of service loss within one year;

- average time lapse between reporting a fault and satisfactory restoration of service;

- percentage of faults cleared within target time period;

- average number of faults per customer line per year;

- percentage of payphone time out of order;

- percentage of payphone time out of order due to full coinbox.

Making calls (and clearing them)   Perhaps the most over-indulged category from the historical point of view. Having had the line and terminal equipment installed, the customer will be interested in establishing calls, or sending messages over the network.

The customer expects prompt service, and correct delivery of messages, or connection of calls to the right number. Network congestion is a frustration. A plethora of easily quantifiable and measurable parameters are available:

- percentage of calls receiving dial tone within target time (fractions of a second);

- percentage of calls connected to the correct number;

- percentage of calls failed owing to network congestion (a network performance measure);

- average time taken between commencement of dialling and receipt of ringing tone;

- average delivery time required (for a packet data message);

- percentage of calls not clearing correctly at the end of a call;

- answer bid ratio—the ratio of calls answered to the total number of call 'bids';

- connection accessibility (availability) objective (CCITT Recommendation G180 refers).

Conversation or information transfer The prime reason for the service to convey conversation or other information. The customer demands clarity of speech, lack of noise, sufficient loudness, absence of interruptions or interference, and security of the connection (from cut-off, or intrusion by a third party). For data, accurate conveyance of information is required (e.g. low jitter (variance in speed), low bit error ratio (BER) (introduced errors) etc.). Measurements of parameters in this category have historically been made by taking a small statistical sample of human 'service observations'. A small percentage of calls are listened into by human operators, to confirm intelligibility of speech, to detect any apparent dissatisfaction of customers, and to trace the cause of call failures during set up. With the growth in telephone traffic and the increased sophistication of services there is now a demand for exchange manufacturers to develop more accurate computer controlled sampling methods. Typical parameters in this category include:

- percentage of calls with too faint conversation;

- percentage of calls with excessive distortion;

- percentage of data calls with excessive jitter;

- percentage of calls cut off prematurely;

- average time to deliver a data packet;

- proportion of misrouted calls, or undelivered data packets.

Service charges The charge invoiced to a customer reflects the accuracy and adequacy of the invoicing procedure as well as that of the exchange call metering device. Errors can be introduced by the mechanism of collecting call data, by the transfer of that data to the billing system, by the aggregation of individual call charges, and by the transfer of that information from electronic to printed form. All of these functions must work at an acceptable standard. Typical check measures might be:

- percentage of (correctly) disputed bills;
- percentage of individual calls incorrectly charged;
- percentage bill inaccuracy on a per customer basis;
- probability of an individual customer's bill being wrong.

16.4 SUMMARY

This chapter has stressed the importance of good quality in telecommunications services and has outlined the two concepts 'quality of service' and 'network performance', categorizing both so as to allow easy establishment of a balanced set of parameters for continuous monitoring and correction of service performance.

BIBLIOGRAPHY

Adam, E. E. Jr, Hershauer, J. C. and Ruch, W. A., *Productivity and Quality—Measurement as a Basis for Improvement*. Prentice-Hall, 1981.

Caplen, R. H., *A Practical Approach to Quality Control*, 2nd edn. Business Books, 1972.

CCITT Recommendations on Quality of Service (QOS) and Network Performance (NP). Q800 Series.

CCITT Recommendation E420, 'Checking the Quality of the International Telephone Service—General Considerations'.

Forum 87—5th World Telecommunication Forum. Conference Proceedings. Part 2—Technical Symposium. IV.5 User in Focus—Quality of Service. pp. 223–258. ITU, Geneva, 1987.

17

CHARGING AND ACCOUNTING FOR NETWORK USE

So far we have concerned ourselves entirely with the technical and operational side of running networks and providing telecommunications services. Network operators, if they want to stay in business, also have to think about capital and operational costs, and how to recover them from their customers. For a network to be financially sound and independent its tariff structure needs to cover total expenditure made up of:

- provision of lineplant, exchanges and buildings (i.e. capital costs);

- service maintenance and administration costs;

- interest on capital;

- depreciation of equipment;

- research and development;

- future capital investment;

- profit and shareholders' earnings.

Government regulation in some countries set tight financial limits on profits and targets, while in other countries the tariff structure is tailored to meet political ends, and financing is direct from government coffers. None the less, the financial practices of charging and accounting for network use are common throughout the world, and they reflect the principles explained in this chapter.

17.1 RECOMPENSE FOR NETWORK USE

Public telecommunications customers will be familiar with the arrival of their monthly or quarterly bill, comprising both subscription charges (paying for the rental of their line), and a charge for usage (the number of calls made, or packets of data sent etc.).

The practice of charging for network use is not confined to public networks; private network operators, especially the large corporations, are increasingly levying internal 'transfer charges' on individual departments for their use of telecommunications services. This is in order to keep close control of areas of company activity representing either 'profit centres' or 'cost centres'. Additionally, accounting between telecommunications operators is becoming more common, as a result of the recent large increase in the interconnection of networks.

CCITT defines two different types of financial transaction, to each of which different practices and a very particular vocabulary apply. The two types of financial transaction are described below, and a number of key items of vocabulary are highlighted.

Charging   (also called *billing*) This is the name given to the practice of recovering financial costs from 'end-users' for their use of the network. *Charges*, including *subscription* charges and *usage* charges, are accumulated over a standard period, typically three months, and are *billed* to the customer. Monies are then *collected* from the customer, typically a telephone line subscriber.

Accounting   The term *'accounting'* is normally applied only to the practice of financial *'settlement'* between different network operators, where they mutually co-operate to provide services. An accounting settlement is usually made by the owner of the originating network (i.e. the one collecting the customer charges) to all the other operators participating in the connection. This reimburses the other operators for the use of their networks.

17.2 CUSTOMER SUBSCRIPTION CHARGES

Customer charges, also known as 'tariffs', usually comprise two parts, a subscription charge and a usage charge. The subscription charge is made to offset the 'standing' or 'fixed' costs, for example the rental of a telephone or other end terminal, the provision and upkeep of the local exchange and of the access line, and the operator's general administrative overheads. The actual rate of the subscription charge depends upon the type of access line, the end equipment provided, and the services to which the customer subscribes. They are easily calculable from a fixed formula and are usually charged by quarterly bill. Examples of services which attract a higher subscription charge are: circuits specially designed for carriage of data, or lines with supplementary services such as diversion of incoming calls (on 'customer busy' or 'customer absent'), three-party-call facility, etc.

17.3 CUSTOMER USAGE CHARGES

Customer usage charges are made in addition to subscription charges, to offset the use-dependent (the 'variable') costs of network provision. As we have seen in Chapter 10 on teletraffic theory, the greater the amount of traffic between any two exchanges, the greater the number of trunks that are required between them. The cost of these extra trunks is recovered in the usage charge, which is usually calculated on a 'cost-plus-mark-up' basis to reflect the network resources required to provide the service. The charge for any given call (or for the carriage of a data message) thus usually depends upon the following parameters:

- the type of network used;
- the duration in time of the call, or;
- the volume of data carried, and;
- the distance of the communication.

In addition, more sophisticated factors may also be taken into account:

- whether operator assistance, or some other 'value-added service' was provided;
- whether the call (or message) was conveyed during the network's peak usage period (the charge may be time-of-day dependent);
- how many call attempts (whether abortive or not) were made. Making a charge for each call attempt dissuades users from congesting the network by incessantly re-calling busy numbers.

To calculate the charge due from each individual customer, the network operator must have a means of monitoring and recording each incidence usage. In most networks charges are levied only on the call (or message) originator (so that the receiver does not pay), and usage is normally monitored by the originating exchange, which must determine the origin and destination of each call (or message), its duration, and the time-of-day when conveyed. In a circuit-switched network the origin of a call can be determined simply by detecting which incoming line is in use. The destination is then determined by analysis of the number dialled during call set-up. The chargeable duration of the call (the 'call duration') is normally taken to be the 'conversation time', which is determined by timing the period between the 'answer' signal and the 'clear-down' signal at the end of the call. Call duration on manual networks, or of calls established 'semi-automatically', can be measured directly by the human operator, but for automatic networks different procedures are used. There are two principal means for automatic customer charge monitoring: 'pulse metering' and 'electronic ticketing', as explained next.

Pulse metering

The older and simpler charging method (used in early telephone networks) is to levy call charges on the basis of the number of predetermined intervals of time used in the course of the call. This method is called 'pulse metering'. In it a fixed price is charged for a small time period or 'unit' of the conversation phase. When the call is first answered, the first unit (of time) commences. After a predetermined period of time (the length of which will depend upon the location of origin and destination and the time of day), the first unit will expire and a second 'unit' of time will commence. The charge payable is calculated by multiplying the number of units used, by the fixed price per unit. The customer is usually unaware when each unit ends and the next begins, or indeed how much of a partially used unit remains.

On different calls, e.g. to destinations at different distances away, or on calls to the same destination made at different times of day, the time duration of a unit may be varied. In this way, the average per-minute-charge made for different calls can be varied. One caution however. It is not generally understood that the upper limit of metering pulse rate may not actually be determined by the capability of the pulsing unit, but upon the ability of the transmission line to carry the pulse correctly and the meter to clock properly.

Over a given period of time (say three months) the number of 'units' used by the customer is accumulated on a counting device (called a 'cyclic meter'), located at the exchange. Location at the exchange eases the job of reading the meters, and reduces the scope for customer fraud.

During the course of individual calls, the cyclic meter is triggered to step onward one count at the beginning of each new 'unit' of time. This is done by sending a periodic electrical pulse to the meter. The customer's invoice shows the total number of units used, and a financial charge which is calculated by multiplying this figure by the fixed price per unit.

While calls are being set up by the exchange, the exchange control and routing system ('common equipment') obtains the destination of the call from the number dialled, in order to determine the appropriate outgoing route and charge rate. For routing, as described in Chapter 14, a fair number of digits may need to be examined to decide which precise exchange the call should be routed to next. However, although the same piece of common equipment may determine both the route and the call charge rate, it is likely that fewer digits will have to be analysed for charging purposes. This is because most network operators choose to run 'chargeband' schemes, in which destinations within a fairly wide geographical zone (or 'chargeband') are charged at the same rate. This eases the administration of the scheme and the customer's comprehension of it. As an example of a simple chargeband scheme, a public telephone company might opt to levy charges for international calls based only on the first digit of the 'country code'. This would divide the world into eight distinct zones, as we saw in Figure 14.7 of Chapter 14 (North America, Africa, Europe, South America, Far East Asia, Union of Soviet Socialist Republics, Middle East, and Central Asia). A similar national chargeband scheme could equally well be developed corresponding to the area code digits of the national telephone number.

Each of the chargebands needs a 'charging programme'. The 'programme' divides the day up into a number of time periods, for each of which a different per-minute-charge will apply. At the busiest times, we expect higher rates, while cheap rates at off-peak times may help to stimulate extra traffic and revenue without needing more equipment. As an example, within the United Kingdom, British Telecom sub-divides its chargebands into times of day called 'peak rate' (corresponding to the busiest times of day, when the highest per-minute-charge is levied), 'cheap rate' (corresponding to periods of low activity, when the lowest per-minute-charge is made) and 'standard rate' (for which a medium charge is made). Over the course of a weekday, peak rate applies from 9 am to 1 pm; standard rate from 1 pm to 6 pm, and from 8 am to 9 am;

and cheap rate applies each evening from 6 pm to 8 am. The per-minute-charge during the peak rate period is made more expensive by reducing the time duration of the unit (remember that the 'units' have a fixed price). Figure 17.1 illustrates the relationship between price-per-minute and unit duration on the British Telecom scheme, and shows the pulsing signal required to step the customer's meter.

Exchanges which employ pulse metering to charge their customers usually have a number of machines or electronic devices to generate the meter pulse signals, one for each of the pulse supplies. A selector device, controlled by the dialled number analysis and by the time-of-day, connects the customer meter to the appropriate pulse supply for any particular call. More than one pulse meter supply may exist. For example local calls will be pulsed from the local exchange (if metered at all). But trunk and international call charge pulses will most likely be generated at the trunk or international exchange which analyses the destination 'country code' digits. A simplified diagram of a pulse metering device is shown in Figure 17.2.

A very simple form of pulse metering, in which only one meter pulse is generated for each call, independent of call duration, is used in some networks—particularly for

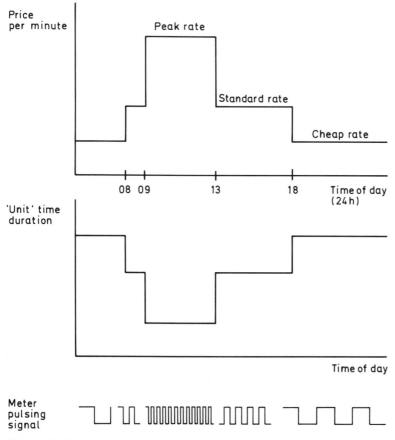

Figure 17.1
Pulse metering.

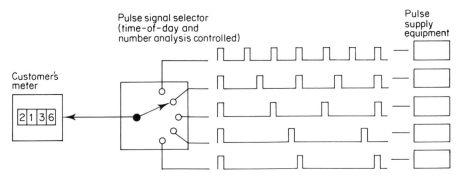

Figure 17.2
Pulse metering selector or 'tariff equipment'.

metering local calls. The advantage of such a charging philosophy is the simplicity of the equipment required and the ease of administration. However, as the need to recover usage dependent costs becomes more acute with the boom in demand, so the method is falling into obsolescence. Another drawback is the tendency to promote network congestion and phone box queues!

Electronic ticketing

Electronic ticketing (also called automated message accounting) is far more accurate and reliable than pulse metering and is becoming common in consequence of the spread of stored program control exchanges. In this method the exchange control computer monitors and records information about each call, including the number dialled, the duration, the time of call set up, the time of call clear etc. The information is stored in an electronic 'call record', usually on computer storage disk or magnetic tape, until the time when the customer is due to be invoiced. The invoice is calculated by adding the amount due from each individual call, derived from the details on the individual call records. Electronic ticketing provides the scope for more complex chargeband structures, and the capability (much demanded by customers) for an itemized record of the details and cost of each call made.

17.4 ACCOUNTING

'Accounting' is the method of financial settlement between network operators, when more than one network is traversed by a call. Accounting has historically been necessary to allow reimbursement between the networks of different countries in recompense for their handling of international calls, but in addition, accounting is also becoming common even between networks operating in the same country, as deregulation allows or even demands ever greater interconnection.

Most accounting is nowadays undertaken using computer monitoring devices to record the details of individual calls in a manner resembling 'electronic ticketing'. The

345

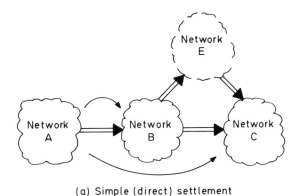

(a) Simple (direct) settlement

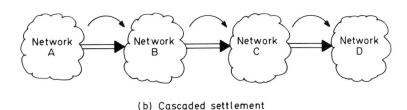

(b) Cascaded settlement

$\Longrightarrow$ Call path, $\longrightarrow$ Accounting settlement

Figure 17.3
Cascaded settlement of account.

financial settlement is made on the basis of total usage over a given period of time, and is paid from the call originating network direct to all other network operators. Alternatively, the accounting settlements may be 'cascaded' between network operators, starting at the originating end, if more than two networks have been used. (Figure 17.3 refers.)

The direct method of accounting settlement (Figure 17.3(a)) is only possible when the owner of network A knows the exact path taken by calls in reaching network C, after traversing network B. For example, if there were also a route from network B to network C via a fourth network E, then the direct settlement method may not be possible, since owner A may not be able to calculate an apportionment in settlement for E. This is the main reason for the use of the 'cascade' method of settlement shown in Figure 17.3(b). In this method the transit party settlements are proportioned and cascaded in sequence.

Accounting settlements depend upon total usage measured in 'paid minutes'. As we learned in Chapter 11, the number of accountable or 'paid minutes' on each call is equal to the total useful time that has passed between answer and call cleardown. The settlement is calculated by multiplying the paid minute total by the 'accounting rate'. The accounting rate itself is fixed by negotiation directly between the interconnected network operators, and is intended to reflect the costs associated with:

- the destination of the call and the distance;
- the services used;
- the call duration;
- the time of day;
- the type of switching;
- the route taken etc.

The accounting rate may be set in any accepted international currency (e.g. US Dollars, Sterling, 'gold Francs', 'special drawing rights' (SDRs)), and may either be negotiated by the bilateral agreement of the parties, or in case of difficulty, paid at standardized rates such as those recommended by the CCITT D series recommendations.

17.5 ROUTE DESTINATION ACCOUNTING

When there is more than one route to a given destination, 'route destination' accounting must be undertaken; an example is shown in Figure 17.4, where calls between network A and network B may either be routed directly or may 'overflow' via network C.

Correct reimbursement of both operators B and C is possible only if operator A records total paid minutes by route and destination (further sub-divided into peak, standard and cheap rates according to time of day if necessary). 'Destination settlement' is due to the operator of network B for all calls, whether routed directly or not, but the payment may be slightly lower for calls transiting network C. This would

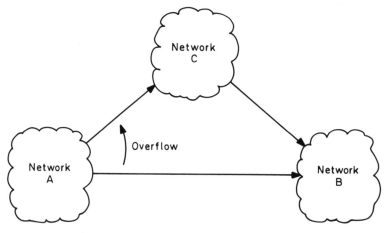

Figure 17.4
'Route destination' accounting.

reflect an apportionment of revenue to operator C and also the effective cost saving resulting from the reduced number of direct (A–B) circuits that are required.

CCITT recommendations demand that the operator of network A periodically 'declares' two 'route-destination' (or RD) accounts, corresponding to the total originated paid minutes sent:

● to Destination B; by route direct;

● to Destination B; by route via C.

Declaration of accounts between network operators consists of exchanging information about each other's total originated paid minutes of traffic, categorized as route/destination and peak/standard/cheap time-of-day (as necessary). The due payment is calculated as the difference between outpayments due for outgoing traffic and revenue due from incoming calls; and one of the operators pays the outstanding balance accordingly. When incoming and outgoing traffic is roughly balanced, very little settlement is required, but usually there is a net financial gain for settling accounts with an incoming balance of traffic, and a net outpayment when most of the traffic is outgoing. For some developing countries the net balance can provide a significant service of hard currency—influencing the speed of development of their networks. For others, the debt is a severe problem and embarrassment.

17.6 CHARGING AND ACCOUNTING ON DATA NETWORKS

When data is carried by a circuit-switched network, customer charges and inter-operator accounts are calculated according to the total time for which the circuit was available to the end users. The usage time, or 'paid time', is the time that elapses between the answer signal and the circuit cleardown. This period is equivalent to the conversation time of circuit-switched telephone network, and the same principles apply as have already been described. Data network types charged and accounted in this way include telex and circuit-switched data networks.

In cases where data is carried by a packet-switched network, the accounting and charging principles must be amended, since the duration of communication is no longer a meaningful and easily measurable parameter. Instead, it is usual to charge and account according to the total volume of data carried by the network. The volume is quoted in *segments*. One segment is equal to 64 octets (or bytes), equivalent to $64 \times 8 = 512$ bits. Methods of charging and invoicing customers, and for accounting and settling inter-network accounts are in general similar to the electronic ticketing technique used in circuit-switched networks.

17.7 CHARGING AND ACCOUNTING FOR MANUAL (OPERATOR) ASSISTANCE

Calls set up on behalf of the end user, and controlled by the network operator (i.e. manual or 'semi-automatic' calls), are normally charged and accounted according to the details of the call, as recorded on the operator's paper 'ticket' or electronic equivalent.

The chargeable (and accountable) time on an operator call is usually less than the total 'conversation time' (the time between the called-subscriber-answer and the call-clear). This is because some time is taken up by the operator before the two end parties can speak. According to the nature of the call, the unchargeable conversation time may have been used by the operator:

● to confirm that a particular person has answered the call, if a *personal call* has been requested;

● to check that the receiver is willing to accept call charges if a *collect* (i.e. *reverse-charge*) call has been requested;

● for some other similar reason.

Operator controlled calls often carry a premium charge. This offsets the cost of the operator's time, and the extra unchargeable network time required at the commencement of each call. Operator assisted calls are also accounted between network operators at a slightly higher rate than automatic calls, for similar reasons. The 'manual' accounting rate is used when a second operator is also used in the terminating country, otherwise the 'semi-automatic' accounting rate applies.

As we learned in Chapter 14, semi-automatic calls which route via automatic exchanges may be recognized by the presence of a language digit in the dialled number train, rather than by the automatic call 'discriminating digit' (i.e. language digit value '0'). This fact prevents the possibility of accounting the same call twice, once on the operator's ticket and once on the automatic exchange accounting record.

17.8 CHARGING AND ACCOUNTING FOR LEASED CIRCUITS

Besides providing switched services, network operators are often willing to provide transmission bandwidth on a point-to-point basis. Circuits conforming to any of the standard bandwidths may be leased from the network either by private individuals as 'private circuits' or by other network operators as 'administration leases'. For both types charges and accounting rates are normally applied according to the crow's flight distance either between the end points of the leased section of the circuit or between the countries' capital cities.

For private circuits, commercial market rates are set by the network operators, but for administration leases, a lower accounting rate (nearer to cost) may be used. Within Europe, TEUREM accounting rates, set by CCITT, govern the charges made between network operators for administration leases.

17.9 CHARGING PAYPHONE CALLS

Payphones present a particular challenge for network designers. Unlike normal subscription telephones, the call charges must be collected prior to or during the call, since the attempt to collect the money afterwards, or by invoice, is an invitation to fraud. Special telephones are required for the purpose; they collect the money 'up-front', calculate when the period paid for is about to run out and then they prompt the caller to insert more money. Forced clearing is applied to the call if the caller fails to insert more money when the paid time has expired. As with operator controlled calls, charges for payphone calls are generally slightly higher than normal calls. The extra charge offsets the greater costs of the telephone, its maintenance, and upkeep of the kiosk, which clearly cannot be defrayed against a subscription charge.

17.10   SETTING CUSTOMER CHARGES AND ACCOUNTING RATES

Setting customer charges (tariffs) and accounting rates is a skilled task in which network operators have not only to ensure coverage of their costs, a reasonable profit, and a return on investment, but give customers a 'fair deal'. The setting of charging and accounting rates therefore demands careful consideration, together with acute awareness of market expectations and willingness-to-pay. Anticipation of competitors' reactions is important, since tariff or accounting rate changes, particularly when increased, tend to have a deterrent effect on demand, and may have an adverse political or public relations consequence, factors which network operators cannot afford to ignore.

Figure 17.5 illustrates revenue breakdowns for the 'originated' and 'terminated' traffic of an imaginary telephone network. The calls have been connected in co-operation with another operator and relevant account settlements are shown.

In Figure 17.5(a), the overall revenue (area of the pie chart) is large, reflecting the call collections from telephone user tariffs. Unfortunately, however, over half of this

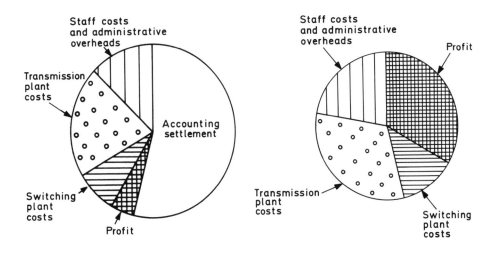

(a) Originated call revenue   (b) Terminated call revenue

Figure 17.5
Apportionment of revenue.

revenue has had to be paid out to the terminating network operator in settlement of accounts. You might think it fairer that the settlement should be almost exactly half of the collection revenue, and historically this was so; but while the cost of equipment has fallen and the tariffs charged for calls have been forced down by market pressures, accounting rates have been slow to adjust, not helped by the fact that an existing high accounting rate is difficult to negotiate down if the balance of traffic (and therefore the net payment) favours the other network operator. Thus where there is a network outflow of traffic, the high accounting rates which exist for historic reasons can be heavily penalizing. This is a particularly acute problem for some developing countries—where the outdated and congested nature of the national network can result in more successful outgoing international calls than incoming ones. The net outpayment settling the account can be an extreme embarrassment and problem for the country's overall balance of trade. Surprisingly perhaps, even between more advanced nations, a PTO (public telecommunications operator) may actually make more money from incoming calls received from the distant PTO than from outgoing ones. This even despite the fact that his own outgoing customer tariff may be nearly twice his compatriot PTO's price for incoming calls! In time this accounting rate injustice is likely to be redressed, but as you can imagine it is a very sensitive issue between world governments.

The next major element of expenditure shown in the pie chart of Figure 17.5 is the transmission plant costs, then the company overheads and, finally the switching costs. All this leaves only a small margin available for profit.

For a contrast, look at the terminated (i.e. incoming) call revenue breakdown shown in Figure 17.5(b). The overall revenue (size of the pie chart) is smaller, but it is greater than half of the outgoing call revenue (provided the accounts settlement is more than half of the outgoing collections, and there is rough balance of traffic). Meanwhile the costs are similar (for similar traffic demand), so a much larger amount of the revenue goes to profit.

The example of Figure 17.5 is typical of the revenues and profits an international telephone network operator may earn from a given overseas route. Note the extreme sensitivity of the profits and revenue to the call charge tariff and the accounting rate. Note also the difference in profitability between 'originated' and 'terminated' accounts.

17.11 FUTURE ACCOUNTING AND CHARGING PRACTICES

More and more new telecommunications services emerge and the world becomes littered with new and competing network operators, new charging and accounting methods and new monitoring equipments. They are bound to be more complicated than those we have already and they will demand skilful administrators and engineers to realize them and set the rates.

BIBLIOGRAPHY

CCITT D-Series recommendations.
Flood, J. E., *Telecommunication Networks*. Peter Peregrinus (for IEE), 1975.

Forum 87—5th World Telecommunication Forum. Conference Proceedings. Part 4—Symposium on Economic and Financial Issues of Telecommunications. ITU, Geneva, 1987.

Hunter, J. M., Lawrie, N. and Peterson, M., *Tariffs, Traffic and Performance—The Management of Cost Effective Telecommunications Services.* Comm. Ed. Publishing, 1988.

18

MAINTAINING THE NETWORK

No matter how much careful planning goes into the design of a network, and no matter how reliable the individual components are, corrective action will always be required in some form or another, in order to prevent or make good network and component failures, and maintain overall service standards. But attitudes towards maintenance and the organization behind it vary widely, ranging from the 'let it fail then fix it' school of thought right through to 'prevent faults at any cost'. This chapter describes a typical maintenance regime in its philosophical, organizational, and procedural aspects.

18.1 THE OBJECTIVES OF GENERAL MAINTENANCE

As succinctly stated by CCITT, the objective of a general maintenance organization is to 'minimize the occurrence of failures, and to ensure that in case of failure:

- the right personnel can be sent to
- the right place with
- the right equipment at
- the right time to perform
- the right corrective actions'.

18.2 MAINTENANCE PHILOSOPHY

In pursuing these objectives, the wise network operator establishes a maintenance philosophy closely linked with overall targets for network quality and for the proportion of time the network is intended to be fault free ('available'). In this task he will take due account of network economics and the most likely causes of failure. Networks fail for all sorts of reasons; common examples are:

- cable damage;

- equipment overheating;

- electronic component failure;

- mechanical equipment jamming, or other failure;

- mechanical wear;

- dirty (high resistance) relay or switch contacts (a diminishing problem as electromechanical exchanges are withdrawn);

- power supply loss;

- vandalism (e.g. to public payphones);

- software errors;

- erroneous exchange data;

- poor connections between cables or other components (e.g. 'dry' soldered joints).

Each cause of failure has its own cure, but broadly speaking, there are three main approaches:

- *corrective* maintenance;

- *preventive* maintenance;

- *controlled* maintenance.

Corrective maintenance is carried out after the failure has been diagnosed and it consists of the repair or replacement of faulty components.

Preventive maintenance is to eliminate the accumulation of faults. Preventive maintenance usually consists of 'routine' testing and 'correction' of working equipment (as opposed to failed equipment) to prevent degradation in performance before any failure actually occurs.

Controlled maintenance is a more systematic approach, combining both the corrective and preventive methods. The underlying philosophy of controlled maintenance is to prevent network failure. This is done by using special analysis techniques to monitor day-to-day network performance and degradation, thereby avoiding maintenance work. The advantage of controlled maintenance is that it concentrates upon areas where the customer is likely to benefit most, and it reduces the extent of preventive maintenance and the complications of corrective maintenance. When new networks are being designed or extended, or when new capabilities are being added to existing networks, consideration needs to be given to the maintenance philosophy and to the organization, the maintenance facilities, and the test equipment that will support it. The best 'controlled' mix of both corrective and preventive philosophies depends upon the number and nature of problems. These, in turn, depend upon the overall network structure and the component equipment types. So, in the days of widespread electro-mechanical switches and relays when a frequent cause of failure was mechanical wear, a good deal of time was spent on preventive type maintenance—oiling the moving parts as it were. Nowadays, when hardware faults in modern electronic equipment are relatively rare and software faults often take some time to present themselves (the exchange operating fault-free for extended periods between occurrences), a corrective philosophy is adopted. Faulty component boards are completely replaced without even attempting a diagnosis, and software faults are debugged as they arise.

18.3 MAINTENANCE ORGANIZATION

Real networks are in a constant state of change throughout their lives. To match traffic demand, new circuits are continually established between exchanges. Established circuits may have to be rearranged as transmission systems are upgraded or taken down and faulty equipment has to be repaired, replaced or avoided by a diversion. For optimum efficiency the organizations set up to establish these maintenance tasks, and the tools with which they are provided, should be planned in such a way that lifetime costs will be minimized. There is a choice, for instance, between paying more at the start for expensive but reliable equipment, or using cheaper equipment and incurring higher ongoing running costs. Lifetime cost analysis must include:

- initial cost of equipment;
- cost of spares and test equipment;
- ongoing running and maintenance costs;
- costs associated with periods of lost service.

High wages and skills shortages in recent years have weighed the scales in favour of using more reliable equipment and a smaller maintenance workforce. Indeed, in some instances the field workforce has been pared to the minimum of two people—one worker plus a stand-in to cover annual leave and periods of sickness. Some observers question the sense of this, pointing to the fact that so complex are the devices, so computerized the routine activities and so rare the faults that the field maintenance staff often do not have the experience to cope.

For this reason a comprehensive headquarters 'maintenance support' or back-up organization is needed in addition to the direct maintenance staff, to perform the following functions:

- to provide detailed equipment and maintenance documentation;
- to provide maintenance training on new equipment and document 'fail-safe' maintenance procedures (explaining the general methods to be adopted, and making sure that unintended disturbance to other customers is not caused by maintenance action);

- to develop and put in place a fault reporting procedure;

- to repair complicated items of equipment or resolve complex software problems;

- to develop and procure necessary test equipment;

- to maintain an appropriate store of spare parts and to call for re-design of poor equipment, taking into account the failure rate of each item, the number of items in operation, and the actual repair turn around time (e.g. repair time, or delivery time for a part not held in stock), and calculating the service and revenue risk if no spare part is available;

- to develop and maintain an equipment identification and inventory scheme for tracking equipment in use and spare equipment either in stock or on order. (In addition, maintenance staff in different exchanges need to be able to indicate faulty lines or circuits to one another);

- to prepare a list of 'contact points' and telephone numbers through which the maintenance staffs in different maintenance centres may communicate.

The direct maintenance workforce is usually collocated with the exchange, some staff being switching experts while others have transport so that they can go out and deal with exterior plant problems. The staff located within the exchange provide a 'control point' for new circuit 'line-ups', and for the initial reporting and diagnosis of faults.

The number of staff located at any exchange depends upon the size, complexity, and reliability of the exchange. Not every exchange can justify its own on-site maintenance staff, and out-stationed staff may be posted to the exchange either on a regular preventive maintenance schedule, or just when there is a fault.

Centralized Operation and Maintenance

A modern practice, aimed at reducing the maintenance workforce, is to leave exchanges unmanned. Computer technology and 'extended alarms' allow staff in a single 'centralized operation and maintenance' (CO&M) room to monitor and control a number of different exchanges in 'real time' (giving instantaneous and 'live' control of each exchange). Figure 18.1 illustrates a typical centralized operation and maintenance scheme.

Centralized operation and maintenance (CO&M) can be introduced only when the remote exchanges are computer controlled, and designed to be capable of self-diagnosis of faults. A 'datalink' back to a computer at the centralized operation and maintenance centre allows the maintenance staff to monitor the exchange performance, noting any problems and applying any necessary controls.

Under a CO&M scheme, the exchanges are designed with duplicated items of equipment, which remain idle until they are 'activated' electronically to take over the function of a failed item. The exchange can thus continue to work at full load, while a member of the maintenance team is sent out to the exchange site, to repair the faulty equipment, or to replace it completely by a 'circuit-board' change.

356

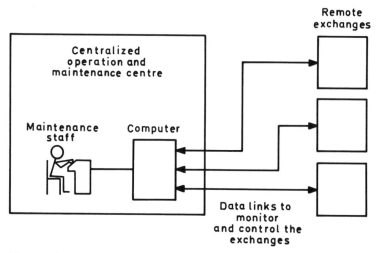

Figure 18.1
Centralized operation and maintenance.

18.4 LINING UP ANALOGUE AND MIXED ANALOGUE/DIGITAL CIRCUITS

The transmission links between exchanges are commissioned (or 'lined up') using a two-stage method as follows. First, the lineplant itself is established in sections which are tested and calibrated in turn, and then connected together. A number of reference measurements are made along the entire length to check and calibrate the overall end-to-end performance. When the line system as a whole has been established, multiplexing and other terminal equipment is applied to its ends to obtain the individual circuits or groups which may be tested individually. The individual circuit testing is necessary because, as Figure 18.3 shows, a real circuit (or group) is likely to traverse a number of line systems, which may interact adversely. So while each section in isolation may be within limits, the combination may not be so. The calibration of each group and circuit is thus carried out on an end-to-end basis.

Figure 18.3 shows four exchanges A, B, C, D, located in transmission centres a, b, c, d. From the inset diagram of Figure 18.3 we see that the exchanges are configured topologically in a fully interconnected manner. Each exchange has a circuit to each other exchange but the total of six circuits has been achieved with the use of three line systems only: a–b, b–c and b–d. Where the transmission centres are not directly interconnected by a line system, a circuit has been provided by the concatenation of two line systems, with a 'jumper wire' completing the connection across the intermediate transmission centre. Thus, for example, the circuit from exchange A to exchange C uses line systems a–b and b–c, and a jumper wire across transmission centre 'b'.

The number and diversity of calibration measurements and adjustments necessary during circuit line-up, and the amount of deviation allowed in the ongoing values depend upon the type of circuit (e.g. analogue or digital), and upon the use to which the circuit is being put (e.g. voice or data). High grade data circuits, for example, have

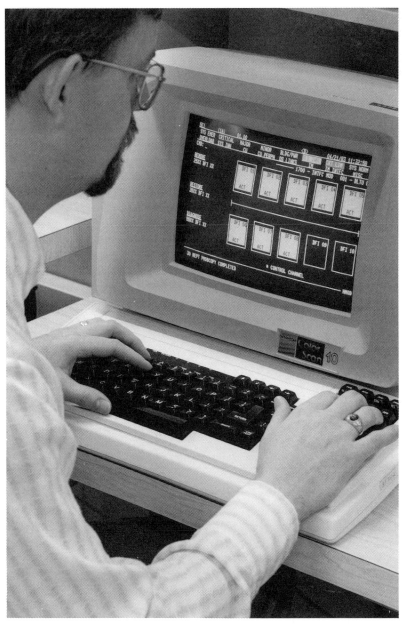

Figure 18.2
Maintenance workstation. The AT&T 5ESS telephone exchange now includes a video monitor that displays colour-coded diagrams, each of which indicates the status of a certain part of the system. A central office technician is checking the status of digital trunks. (*Courtesy of AT&T.*)

358

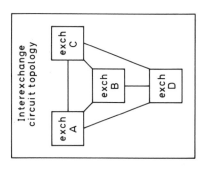

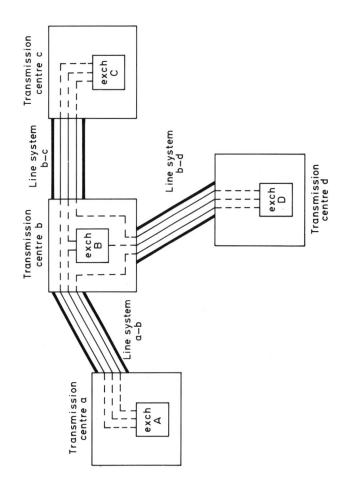

Key:

——— line system (e.g. coaxial cable, microwave system etc.)

— — — individual circuit within line system

——— jumper wires

exch exchange

Figure 18.3
Circuit set-up.

more stringent line conditioning requirements than simple voice grade circuits. The measurements and adjustments ensure that the circuit conforms with the transmission plan. (Chapter 13 refers.) Thus on analogue and mixed analogue/digital line systems and circuits 'line-up' measurements will be made of:

(i)   overall loss in signal strength (in dB);
(ii)   amplitude loss/frequency 'attenuation distortion';
(iii)   group delay (particularly if the circuit is to be used for data);
(iv)   noise, crosstalk, echo etc;
(v)   inter-exchange signalling tests (if appropriate).

Various equipments, including amplifiers, equalizers, filters, and echo controllers are then adjusted to bring the line conditions within the set limits, using measuring equipment as follows:

● signal tone generators (calibrated for precise frequencies and signal strengths);

● calibrated frequency and signal strength detectors;

● noise meters;

● equipment for interexchange signalling or data protocol testing.

First, end-to-end circuit loss is determined by sending a calibrated 800 Hz signal of a known strength, and measuring the received strength of this allows the circuit amplification to be adjusted accordingly. Next, a range of different calibrated signal frequencies across the whole circuit bandwidth is sent, and the received signal strengths are again measured. This allows the frequency distortion equalizers to be adjusted. Then the group delay distortion is corrected using group delay equalizers. This equalization is particularly important for high-speed modem data circuits, and it is achieved by measuring the relative phase of different frequencies relative to the 800 Hz signal. Following this, psophometric noise checks are conducted, together with testing of interexchange signalling systems or of data protocols (e.g. X25 or IBM's SNA) and finally a test call is established.

18.5 HIGH-GRADE DATA CIRCUIT LINE-UP

In high-grade data circuits which use modems over analogue or mixed analogue/digital plant, a number of extra 'line-up' measurements may be necessary as follows:

● weighted noise;

● notched noise;

● impulse noise

● phase and gain hits;

● harmonic disturbance or 'inter-modulation noise';

- frequency shift distortion;

- jitter.

Weighted noise is a measure of the noise in the middle of the channel bandwidth. Noise frequencies in this range are most likely to cause modem errors. Psophometric (European) and C-message (United States) filters are used to measure this type of noise.

Notched noise is measured by applying a pure frequency tone at one end and removing it with a 'notch' filter at the other; the remaining noise is then measured. The notched noise itself arises from the way in which signals have been digitalized or otherwise processed over the course of the link. It is therefore similar to quantization distortion which was discussed in Chapter 5.

Impulse noise is characterized by large 'spikey' waveforms and arises from un-suppressed power surges or mechanical switching noise. It is most common on electromechanical switched networks.

Phase and gain hits are intermittent but only moderate and relatively short (less than 200 ms) disturbances in the phase or amplitude of a signal. Typically less than ten should be recorded in a 15 min test period. Special test equipment is required. More serious gain hits are called 'dropouts'. Gain hits are most troublesome in voice use; phase hits manifest themselves as bit errors in data signals.

Harmonic disturbance may result from the intermodulation of two different signal frequencies F_1 and F_2 when passed through non-linear processing devices. New stray signals of frequencies $F_1 + F_2$, $F_1 - F_2$, $F_1 + 2F_2$ etc. are produced. This type of disturbance is measured by a spectrum analyser.

Frequency shift is also measured by a spectrum analyser. Frequency shift is obviously a problem for a data modem if the frequency received differs to such an extent from that sent as to be misinterpreted. It is most likely to occur when a carrier system or other frequency modulating signal processing has been used.

Jitter, or 'phase jitter' arises when the timing of the pulses on incoming data signals varies slightly, so that the pulse pattern is not quite regular. The effects of jitter can accumulate over a number of regenerated links and result in received bit errors. It can be reduced by reading the incoming data into a store and then reading it back out at an accurate rate, using a highly stable clock controlled by a 'phase-locked' loop circuit.

All of the above parameters should be checked when the line system or circuit is first established. Problems are likely to reflect poorly designed or poorly installed equip-ment and the best remedy is prevention: check the quality of work which is done, since correction circuits are expensive and not entirely effective.

In Chapter 9 we introduced the idea of constellation diagrams for modems. We are now in a position to illustrate their practical use for detecting noise, phase hits and

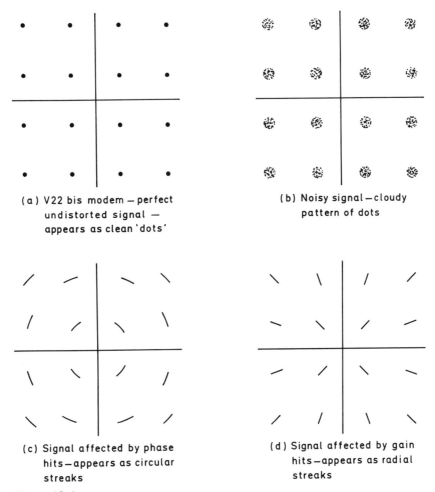

(a) V22 bis modem — perfect
undistorted signal —
appears as clean 'dots'

(b) Noisy signal — cloudy
pattern of dots

(c) Signal affected by phase
hits — appears as circular
streaks

(d) Signal affected by gain
hits — appears as radial
streaks

Figure 18.4
Detecting analogue line disturbances using constellation diagrams.

gain (or 'amplitude') hits on analogue data lines employing modems. Special test equipment may be connected at the receiving end of the line in place of the receiving modem. The test equipment displays on an oscilloscope-like screen the constellation pattern of the received data signal. Disturbances appear as shown in Figure 18.4.

As apparent from the patterns of Figure 18.4, the problem with noise, phase and gain hits is that if they become too great, they result in the incorrect interpretation of the signal—the received bit pattern then includes errors.

18.6 LINING UP DIGITAL CIRCUITS

Digital line systems and their 'tributary' bit streams must conform to a different transmission plan from their analogue equivalents and are therefore lined up in a different manner. The important parameters, as we saw in Chapter 13 are:

- the error rate;
- the network synchronization;
- the quantization (or 'quantizing') distortion.

In practice, the network synchronization (i.e. the jitter and clock accuracy) and the quantization distortion are set by the design of a circuit and can be improved only by redesign. The lining-up process can only address the question of error rate. The accuracy of synchronization depends on the use of highly stable clocking sources and interexchange synchronization links as described in Chapter 13. Quantization distortion is a form of noise, which affects analogue voice and data signals when they are carried over digital media using pulse code modulation (PCM) and other signal processing techniques (Chapter 5 refers). It can be reduced only by redesigning the circuit to avoid multiple signal conversions (e.g. analogue to digital conversion, signal 'compression' etc.). In the case of digital data circuits, these should never be designed to include any form of signal processing since digital data devices are intolerant of the high bit error rates that arise from quantization distortion.

Standard practice for digital circuit line up is to perform a test of digital error performance. This may involve a lengthy stability test over several days or simply be a 'quick-check' (15 min) test. A 'pseudo-random bit pattern generator' (a digital signal generator) is used to provide a test digital signal. During the test the proportion of bit errors (the bit error rate, or BER) is measured, along with the proportion of error free seconds (EFS). Expected values are typically no more than one error in 10^5 or 10^9 for BER and at least 99.5 per cent EFS. The errors, if excessive, may have arisen as the result of any number of different impairments. Some causes can be eliminated easily (e.g. by increasing the transmitted power to reduce the effect of noise). Other causes need more radical circuit checks.

Like their analogue equivalents, digital bit streams are lined up in two stages, first at higher order (e.g. at 140 Mbit/s, if this is the line system bit rate) and then on end-to-end basis for each tributary stream (e.g. 2 Mbit/s, 1.5 Mbit/s). The secondary checking of lower order tributary streams is necessary for the same reasons as in the comparable analogue case exemplified by Figure 18.3.

18.7 PERFORMANCE OBJECTIVES

In recognition of the fact that practical networks can never match idealized performance objectives, CCITT's Recommendation G102 sets out different operating limits for the target 'performance objective', the design 'objective', the 'commissioning objective' (also known as the 'line-up limit') and the 'maintenance limit'.

The 'performance objective' is the ideal performance level for the particular application.

The 'design objective' transfers this objective into a realizable target range, within which economically designed equipment can be expected to operate, given optimum conditions of power supply, temperature, humidity etc.

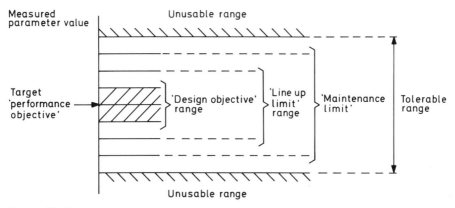

Figure 18.5
Performance objectives: 'line up' and 'maintenance'.

The 'line-up limit' recognizes that the optimum conditions can rarely be achieved in practice, but none the less sets a stringent practical range within which the equipment must operate on first establishment. Any faults identified during line-up, which cause the circuit to operate outside this range, should be cleared and the circuit relined, before taking it into service.

The 'maintenance limit' is the least stringent range of operating conditions but still within a 'tolerable' range as far as the application is concerned. The limit is chosen so that, if exceeded, a fault is considered to exist. The fault should be cleared and the circuit relined-up. As Figure 18.5 shows, each of the last three performance ranges described is a slight relaxation of its predecessor, so giving targets which are achievable in practice. Thus even if a slight degradation in quality occurs after circuit 'line-up', the circuit still operates within the 'maintenance limit' operating range.

A steady deterioration in performance of items in service can be expected as the result of operational use, occasional overload, and general ageing as well as genuine faults. Often the degradation is slow enough to be imperceptible and not to warrant routine checks. For this reason, it is common practice to use automatic alarms to alert maintenance staff to the need for corrective action, so that faults can promptly be eliminated and the equipment returned to its line-up operating range. Two levels of alarm are possible, one at the 'maintenance limit' level, and one at the 'unusable' performance level. Faults on the latter level clearly need more urgent attention than those on the former.

18.8 MAINTENANCE 'ACCESS POINTS'

For the purpose of network lining-up and subsequent maintenance, it is necessary to provide a number of 'test access points' for maintenance. Ideally, 'test access points' are provided at a number of different points in a network to enable easy localization of faults and initial segment-by-segment circuit alignment. A very large number of test points increases the overall network and equipment costs, and may in itself be a source of unreliability. Figure 18.6 shows possible test access arrangements for a

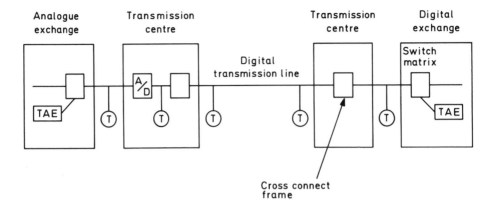

TAE — test access equipment (built into the exchange)

A/D — analogue to digital conversion equipment

(T) — test access point

Figure 18.6
Test access points. TAE, test access equipment (built into the exchange); A/D, analogue to digital conversion equipment; T, test access point.

simple network of two exchanges (one analogue and one digital) interconnected by a digital transmission link. Note how test access points have been provided between all the major items of equipment. This allows the causes of any faults or line-up problems to be quickly narrowed down.

As is typically the case, both exchanges in Figure 18.6 have built-in test access equipment for diagnosing and correcting faults within the exchange itself. In addition, five external test access points enable the exchange's external conditions to be verified and faults in other equipment along the link to be 'localized', 'diagnosed', and corrected.

18.9 LOCALIZING NETWORK FAULTS

In an efficient maintenance organization, faults are discovered before the customer finds them. In this way some of the faults can be corrected even before the customer is aware of them. Means for fault detection include:

- equipment alarms;
- equipment routine testing (so-called 'routining');
- live network monitoring;
- customer fault reporting.

As before, a combination of techniques is usually appropriate. The four are discussed in turn. Equipment alarms are used to indicate an abnormal state (i.e. when the maintenance limit has been exceeded) in an equipment or an environment of crucial importance. (An environment 'fault' might be too high a temperature—computer equipment is prone to failure under such conditions.) Alarms are usually ranked in order of importance, and usually generated by continuous monitoring of some type of 'heartbeat' signal. When the heartbeat stops, it is time for action. For example, the 'pilot' signal on analogue transmission systems is a low level, single tone frequency outside the normal bandwidth range which is continuously transmitted. Absence of the pilot signal at the receiving end is interpreted as a transmission link failure, and it is usually notified to maintenance staff as an audible and/or visible alarm.

Equipment routine testing on the other hand, is suited to any equipment which is fault or wear-prone (e.g. mechanical equipment). It can be carried out manually or by automatic test equipment, the choice depending upon the number and type of devices, their complexity, and the periodicity of tests. For routine network testing, CCITT has described a number of automatic transmission, measuring, and signalling test equipments (ATME) in its 0 series of recommendations.

Live network monitoring helps to pick up transient faults and then avert them; network overload can thus be foreseen, and its adverse effects corrected, as we shall see in Chapter 19. Useful means of monitoring live network performance include call completion rate and congestion statistics, and quality sampling of a small number of connections. For example, a sudden flood of calls or a rapid drop in the call completion rate may indicate the onset of congestion caused by network failure or quality degradation.

Not all faults, however, can be detected other than by the user, and so in the last resort an efficient customer or user fault reporting and handling procedure is paramount.

After detecting a fault, the next step is to localize it and diagnose its nature, all of which may be directly apparent or correctly reported, but usually more diagnosis is required. For example, when a major transmission system fails, a whole host of alarms may go off in the maintenance centre, indicating failure not only of the main transmission system, but also of each of the derived circuits or tributaries. The alarms must be cleared in priority order. Correcting the main transmission fault first may clear the alarms on each individual tributary, but will not do so if lesser problems remain on particular tributaries. Thus in our example, maintenance staff should first attempt to localize the main transmission failure to a given section of the link. This they can do by making use of the various test access points available to them, and by liaison with staff in other maintenance centres through which the link passes. The example is illustrated in Figure 18.7, where the failure in one of the channels in section A–B and the line A–B–C has been detected by maintenance staff in station C. The absence of the main transmission pilot signal in the receive channel at station C has raised an alarm. Telephone liaison with the staff in station B reveals that though this alarm actually notifies the loss of a whole analogue or digital group of circuits, it is evident that this is the likely cause of the problem on the particular A–C channel. If instead the fault had lain on the transmission link between stations B and C, the alarm in station B would not have gone off. Further, if the fault had been in station C's transient channel, the alarms would have gone off at one or both of stations A and B.

Some faults in switched networks can be extremely difficult to trace. A common problem, encountered when trying to locate faults in switched networks is trying to trace which precise links and exchanges were traversed at the time of the fault. It is like

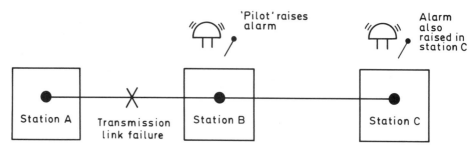

Figure 18.7
Transmission failure and alarm.

knowing your friend is driving between London and Birmingham but not knowing which route he took. You know he has broken down, but where do you look first. Faults of this nature can persist for many months and are finally cleared either as the result of routine maintenance or because routine testing of one of the individual network components revealed the fault.

Not all switched network faults are difficult to trace. For example, a phenomenon known to all experienced maintenance staff is the occurrence of a 'killer trunk'. Imagine that the first choice circuit on the route between the main London-to-Birmingham telephone route has gone faulty. Imagine also that the fault results in any call attempting to use the circuit being immediately failed and released. Affected callers get nothing at all! Not so bad you might think—one circuit faulty in hundreds will not make much difference. But you are wrong—the fault has an incredibly wide-ranging effect, nearly 'killing' *all* the traffic on the route. The reason is that, being the first choice circuit, all calls will attempt to seize it before trying other circuits. But of course, any call that attempts to do so is immediately failed and released. So nearly all calls fail! Hence the name 'killer trunk'. To the experienced maintenance man, however, the killer trunk phenomenon shows up like a sore thumb, because traffic records reveal:

● a huge number of short holding time calls;

● a very low call success (i.e. answering) rate;

● very little overall traffic in erlangs;

● virtually no activity on many circuits.

The condition can be cleared in the first instance simply by 'busying out' the circuit (i.e. making unavailable for use). A repair can then be carried out. The worst effects of a killer trunk can also be prevented by making sure that circuits within the route are chosen randomly—not always scanned in the same order.

Before leaving the subject of network test points and fault localization procedures we should also mention the use of loopback techniques for detecting faults within the 'tail' part of a circuit between the network operator's site and the end-user's terminal. This part of the circuit can be the most inaccessible (e.g. in unmanned premises), or it may be subject to quite adverse conditions in congested ducts, strung between tele-

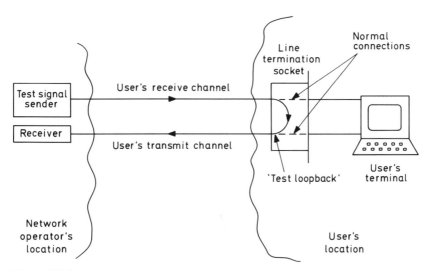

Figure 18.8
Testing a connection tail using a loopback.

graph poles, or even draped across desks. It is just as prone to failure as any other section, and the network operator needs to be able to localize faults here as anywhere else, ideally being able to distinguish such faults from faults in the user's terminal equipment, and preferably without even visiting the user's location. What makes this possible is a circuit 'loopback' or a responding equipment.

A loopback simply 'loops' the user's receive channel directly to the transient channel, so enabling the network operator to send a test signal both ways along the tail and to confirm its correct return. Loopbacks may be either manually operated by the user (on request from the network operator) or invoked by the network operator using a 2713 Hz tone to switch a remotely controlled equipment within the line termination socket at the user's premises. As we see from Figure 18.8, the loopback enables the faults to be localized beyond doubt as being either on the line or in the user's terminal equipment—all without a visit to the user's premises.

Loopback techniques are widely used on point-to-point data connections, where users are intolerant of any significant downtime. PTOs provide them on the tails of their private leased circuits, and other data network providers put them on their equipment circuits. Loopbacks for modems are described in CCITT Recommendation V54.

Another way of testing lines to remote unmanned locations is to use responding equipment which answers calls made to the location, giving standard test signals and other responses in accordance with commands.

18.10 HARDWARE FAULTS

Historically the most common faults have come from mechanical equipment or hardware. Some maintenance organizations are well prepared to deal with component failures, and general wear and tear, by preventive action or by repair, but it must be recognized that the semiconductor age has brought with it much greater levels of reliability and complexity and much greater risk associated with failure. All this has forced change on the handling of software faults.

Nowadays, it is common to design electronic equipment to be tolerant of faults; computer processors and network exchanges have their crucial parts duplicated, the two halves sharing the traffic load and continually monitoring one another for faults. If a fault is detected in either half, then it is shut down and the other half takes over the full traffic load until the faulty half is repaired by maintenance staff, in response to the alarm.

Self-diagnosis software within the computer or exchange is used to localize the fault to a particular circuit board. Correction is achieved merely by sliding the board out and replacing it with a new one. The processor or exchange can then be restored to normal working, and the circuit board can be repaired at leisure or thrown away.

18.11 SOFTWARE FAULTS

Errors in computer programs (or 'software') are becoming the most common cause of faults in telecommunications equipment. Newly developed software is often littered with errors or 'bugs' of an unpredictable nature which come to light in operation, sometimes years later, so that it is difficult to build experience. In addition, since the background of most telecommunications technicians is not in computer programming, the skills for rectification of faults are extremely rare.

Software bugs may take some time to find, and they require the expert attention of specialized 'maintenance support' staff and the original designer, being outside the competence of normal direct maintenance staff.

So, you may ask, how do we keep the service on the exchange running while we sort out the problem? The answer is by initiating a manual or automatic restart procedure which 're-boots' (resets) the whole exchange to allow it to carry on working. This is done by reloading historical data and software, overwriting any corruptions which may have crept in as a result of the software bug. The software and data resident in the exchange processor at the time of failure is downloaded for a later fault diagnosis, which attempts to trace back the processing 'events' leading up to the problem. Meanwhile the exchange is likely to run quite normally for some time, until a similar sequence of events conspires against it again.

Localized restarts of minor functional units are likely to be automatically invoked by the exchange itself in order to minimize the off-air time, but more (e.g. whole exchange) restarts may require manual initiation, in order to prevent an automatic restart from disrupting a large number of unaffected calls at an inopportune moment.

The computerization of many maintenance tasks may have greatly reduced the human numbers required, but the extra skills and knowledge demanded of those staff that remain make them a prized commodity!

BIBLIOGRAPHY

CCITT M-Series recommendations.
CCITT O-Series recommendations.

Clifton, R. H., *Principles of Planned Maintenance*. Edward Arnold, 1974.

Forum 87—5th World Telecommunications Forum. Conference Proceedings. Part 2—Technical Symposium. IV-2 User in Focus—Operations and Management. ITU, Geneva, 1987.

Introduction to Maintenance Planning in Manufacturing Establishments. United Nations Industrial Development Organisation. United Nations, 1975.

Kelly, A. and Harris, M. J., *Management of Industrial Maintenance*. Newnes-Butterworths Management Library, 1978.

Mann, L. Jr, *Maintenance Management*. Lexington Books, 1976.

19

CONTAINING NETWORK OVERLOAD

Network designers like to think that their traffic models are ideally stable, and that their forecasts of future traffic will never be wrong. That was indeed the subject of Chapter 12. In real life, however, such assumptions are unwise, since any one of a number of problems may arise, resulting in network overload, and so congestion. The forecast may under-estimate demand; there may be a short period of extraordinarily high pressure (for example at New Year, Christmas, or any public holiday, or following a natural disaster); or there may be a network link or exchange failure. Monitoring and controlling the network in order to detect and avoid network congestion and the resulting service degradation is a difficult task, but it is being increasingly undertaken by many of the world's telecommunication administrations. This chapter discusses these new 'network management' methods in more detail.

19.1 THE EFFECT OF CONGESTION

Congestion in a telecommunications network manifests itself to customers who are attempting to make calls as a 'network busy' tone, or as a delay in computer or data network response. It annoys the customer, and even worse, the congestion can rapidly increase as the result of customer or equipment repeat attempts. ('Repeat attempts' are further call attempts, made in vain by customers hoping to make a quick connection by immediate re-dialling.) These further call attempts only exacerbate the problem, since they greatly increase the loading on exchange equipment (e.g. exchange processors, number stores, senders and receivers), and they add to the overall traffic volume. The congestion can then spiral out of control. The networks most at risk from congestion are those which have been kept to minimum circuit numbers in order to keep costs down. (Also very much at risk are those networks which employ a high proportion of multi-link overflow routings, since congestion on one link in this type of network rapidly affects other routes, with a consequent 'domino effect').

As a result of practical studies, teletraffic experts have produced models which show that under overload conditions, the effective throughput of a network can actually fall. This makes the overload worse, further reducing throughput, as Figure 19.1 demonstrates.

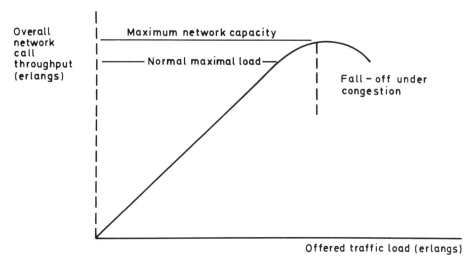

Figure 19.1
The effect of congestion.

Spotting the onset of congestion and taking early appropriate action is crucial to maintaining control of the network. Only by careful control can customer annoyance and unnecessary call failure be minimized.

19.2 NETWORK MONITORING

The most common parameter used to measure network congestion in a circuit switched network (such as a telex or telephony network) is the 'traffic load' (usually measured in erlangs—the number of circuits in use) over a period of time. In a packet network, the length of the delay queue (i.e. the number of packets awaiting transmission) can be used instead. To serve the purposes of network management, the monitoring of these parameters needs to be done frequently, so giving a human 'network manager' a 'real-time' perception of network performance. Other measurements, such as the lost call count (peg count) or machine overload information may also be available—helping quickly to diagnose congestion. But in all cases, care must be taken when interpreting measurements taken over too short a period since short duration measurements or a very small 'sample' of calls can be statistically deceptive. Traffic measurements are not reliable when taken over a time period much shorter than the average call holding time.

Traffic and network performance information nowadays usually comes from computers. Daily, weekly, and monthly usage records can be calculated by computer post-processing of information. The post-processing may only be done well after the day when the traffic itself was recorded—from magnetic tape of information. Such information is clearly of no advantage for recommending immediate alleviation action. The experience of many network operators is that network management monitoring information can be obtained only by setting up a completely distinct computer 'network management system'.

The monitoring part of a network management system relies upon the processing of 'status' information (i.e. information about the current state of the network and its components), output in computer format by stored program control (SPC) exchanges. Commonly this information comprises 'alarms' and other fault messages, but in addition may also include an itemized call record for each call made through the exchange. 'Call records' are merely computer messages from the exchange to the network management system, giving all the details of the call (i.e. number dialled, start time, duration, whether successful or not etc.). From the collation of information relating to all current and very recent calls, a true picture of the current network state may be calculated by the network management system and displayed in some way to a human 'network manager'. Most importantly, the load of traffic and the degree of congestion on each individual route into and out of the exchange can be shown.

If the network manager, in watching this information (or having it 'alarmed' to his or her attention), notices that the congestion level on a particular route, link, or exchange has exceeded a predetermined threshold, or that some other problem has arisen, he or she needs to evaluate the cause of the problem and take corrective action as fast as possible. A quick and correct diagnosis enables the manager to adopt the best available solution.

It is worth reflecting here, that as in road traffic jams and human diseases, the cause of a telecommunication network problem and its symptoms may not both be centred in the same place or region. Causes may lurk a long way from the effect. For illustration, consider the example shown in Figure 19.2.

As a result of failure on the link connecting the tandem exchange T with exchange C, there is growing congestion on the link between exchanges A and T. The congestion is caused by customers at exchange A persisting in vain attempts to make new calls to exchange C. The repeat attempts are resulting in unusually high traffic between exchanges A and T, thereby also affecting other customers at A wishing to connect to B. In a very sophisticated network monitoring scheme, the network manager aims to achieve an instant indication and diagnosis of major problems. This may be easy—for example, as the result of an 'alarm'. In Figure 19.2, the T–C link failure may register an alarm, which will instantly tell the manager the cause of network congestion. In practice, however, real-time network monitoring facilities are only available at a limited number of nodes, so that the network manager may have only an incomplete

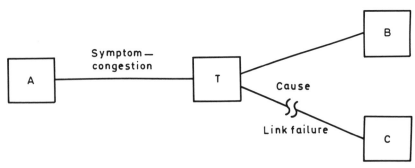

Figure 19.2
Example of symptom and cause of congestion.

knowledge of the network status as a whole. In this case instant diagnosis and resolution of problems will require some guess-work. If in Figure 19.2, exchange A is the only one with network management capabilities (perhaps because it is the only exchange modern enough), then the network manager associated with exchange A could not directly know that link T–C had failed. In this case he has to rely more heavily on 'gut feel' and experience. So what more tools are available to help diagnosis?

The monitoring of a second parameter comes in useful as an aid to diagnosis. The second parameter used is either the 'Answer/Bid Ratio' (ABR) or the 'Answer/Seizure Ratio' (ASR). They are similar. ABR is the percentage ratio of answered calls to a given destination, compared with the number of call attempts ('bits') intended for that particular destination. ASR is nearly the same, but instead is calculated using the number of outgoing seizures from the originating exchange as the denominator. The only reason for using one value rather than the other might be that it is this value that the particular exchange normally measures.

For network management purposes either ABR or ASR should be measured at each originating exchange on a *destination* basis. Let us assume for the remainder of this chapter that we have chosen to measure the ABR by destination as well as each route's traffic intensity in erlangs. Therefore, at exchange A in Figure 19.2, two ABR values are monitored, corresponding to destinations B and C, and the route traffic $A - T$ is also measured. The ABR value is performed at A as follows:

ABR for destination C (measurement at A)
   = number of answered calls to destination C *originated at exchange A during time T*, divided by the total number of calls to destination C originated at exchange A during the same time T

In the example of Figure 19.2 the network manager at A, on noticing that the Erlang threshold has been exceeded on route A–T will also notice that the ABR from A to C has also significantly deteriorated. Meanwhile, calls are still completing from A to exchange B. The network manager at A will not know the root cause of the failure, but this is not necessary. The manager has two courses of action to choose from:

(i) call personnel at exchange C to request further information on the problem, and thus get advice on which network controls to apply or;

(ii) go ahead and implement unilateral network management controls on the basis of the diagnosis to hand.

Which course of action and the exact controls used by the manager will depend upon the individual circumstance.

We have now seen that the monitoring of ABR values is invaluable in diagnosing congestion, but it need not be restricted to being a follow-on action which is considered only after the traffic intensity threshold has been exceeded. There are additional benefits to be gained by monitoring destination ABR values continuously and comparing figures against 'threshold', or 'action required' values. Figure 19.2 is again useful in illustration. Consider the case when a failure of link T–C occurs at night or some other period when demand is so low that because of the low calling rate on the network as a whole, there is no resulting traffic congestion on link A–T. The traffic threshold value in this instance would not be exceeded; and if the network manager was not additionally monitoring ABR on destination C (the value will have dropped to zero completion rate), the problem in completing calls to C might pass unnoticed, and no corrective action would be taken.

Let us not overlook an even simpler method of detecting network problems, which is for all the network transmission and exchange alarms equipment to be brought to the notice of the network manager. Alarms may simply activate bells and lamps, but a sophisticated 'network management system' can have 'equipment alarm' presentation built in, and it can be capable of some automatic problem diagnosis. As with other network status information, alarm information, along with ABRs etc., is most effectively presented to the network manager as a combination of graphical and tabular displays on video terminals. These allow the manager at a glance to gauge overall network status, make a rapid diagnosis of problems, and take corrective or fault-containing action in good time.

19.3 NETWORK MANAGEMENT CONTROLS

With the problem located, what sort of network management control will be appropriate? All network management actions can be classified into one of two categories:

- expansive control actions;
- restrictive control actions.

The correct action to be taken in any individual circumstance needs to be considered in the light of a set of guiding principles, namely:

- use all available equipment to complete calls or deliver data packets;
- give priority to calls or packets, most likely to complete;
- prevent exchange (nodal) congestion and its spread;
- give priority to calls or data packets which can be completed using only a small number of links.

In an expansive action the network manager makes further resources or capacity available for alleviating the congestion, whereas in a restrictive action, having decided that there are insufficient resources within the network as a whole to cope with the demand, the manager can simply fail any new calls coming into the 'hard to reach' (i.e. temporarily congested) destination(s). It makes good sense to fail these calls close to their point of origin, since early rejection of calls frees as many network resources as possible, which can then be put to good use in completing calls between unaffected points of the network.

Expansive control actions

There are many examples of expansive actions. Perhaps the two most worthy of note are:

- Network restoration;

- Temporary alternative re-routing (TAR).

Network restoration, by no means a new technique, has been carried out by many administrations for years; but for some of them the introduction of computerized network management centres has provided an opportunity to apply centrally controlled network restoration for the first time.

Network restoration is made possible by providing more plant in the network than the normal traffic load requires. During times of failure this 'spare' or 'restoration' plant is used to 'stand-in' for the faulty equipment—for example a failed cable or transmission system, or even an entire failed exchange. By restoring service with spare equipment, the faulty line system or exchange can be removed from service and repaired more easily.

Network restoration techniques have historically been applied to transmission links, where it has been usual to plan spare capacity into the network in the form of analogue spare groups, supergroups, hypergroups, or as digital bit-carrying capacity. The spare capacity is built into new transmission systems on a '1 for N' basis; i.e. 1 unit of restoration capacity for every N traffic-carrying units.

The following example shows how 1 for N restoration works. Between two points of a network, A and B, a number of transmission systems are required to carry the traffic. These are to be provided in accordance with a 1-in-4 restoration scheme. One example of how this could be met is with five systems, operated as four fully loaded transmission lines plus a separate spare (which should be kept warm—inactive plant tends not to work when called into action). Automatic 'changeover' equipment can be used to effect instant restoration of any of the other cables, should they fail. An alternative but equally valid 1-in-4 configuration is to load each of the five cables at four fifths capacity under 'normal' conditions. In the former case, the spare cable is used directly as a 'stand-in', should any of the other four fail (Figure 19.3(a)). In the latter case, the failed circuits must be restored in four parts—each of the other cables taking a quarter of the failed portion (as shown in Figure 19.3(b)).

In practice, not all cables are of the same capacity and it is not always practicable or economic to restore cables on a one-for-one basis as shown in Figure 19.3(a). Both methods prevail. Another common practice used for restoration is that of 'triangulation' (concatenating a number of restoration links via third points to enable full restoration). Figure 19.4 illustrates the principle of triangulation. In the simple example shown, a cable exists from exchange A to exchange B, but there is no direct restoration path. Restoration is provided instead by plant which is made available in the triangle of links A–C and C–B. These restoration links are also used individually to restore simpler cable failures, i.e. on the one-link connections such as A–C or B–C.

Because of the scope for triangulation, 'restoration networks' (also called 'protection networks') are often designed on a network-wide basis. This enables overall network costs to be minimized without seriously affecting their resilience to the problems.

The use of idle capacity via third points is the basis of a second expansive action, termed 'temporary alternative re-routing' (TAR). This method is generally invoked only from stored program control (SPC) exchanges, where routing data changes can

(a) Entire spare cable under normal condition

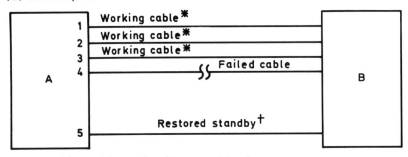

* Each working cable carries its normal load
† Cable 5 'stands-in' to carry the entire load of cable 4
  Normally cable 5 is idle

(b) Spread load under normal condition

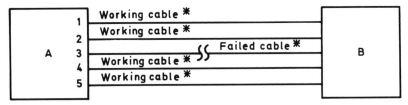

* Each cable normally only 80 per cent loaded
  On failure of any one cable, four 20 per cent components
  are spread over the other four cables

Figure 19.3
Two methods of 1 in 4 restoration.

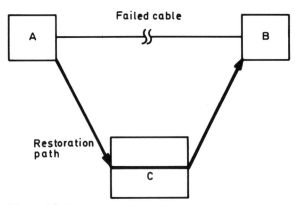

Figure 19.4
Restoration by triangulation.

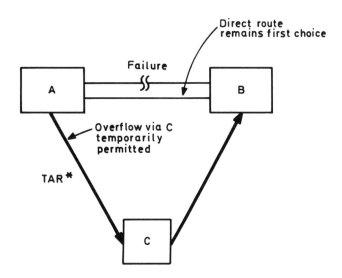

Failure

Direct route
remains first choice

A

B

Overflow via C
temporarily
permitted

TAR*

C

*Transit traffic from A to B via C not normally allowed

Figure 19.5
Temporary alternative re-routing.

be made easily. It involves either routing calls to a particular destination via temporarily different routes, or allowing more route-overflow options than normal. In Figure 19.5, some direct capacity between A and B has failed, resulting in congestion. On noticing this, the network manager has also observed that the routes A–C and C–B are currently lightly loaded (again by chance this traffic has a different busy hour from that of A–B). A temporary overflow route via C would help to relieve congestion from A-to-B. The manager can adapt the normal routing choice list (used at exchange A) to include a temporary overflow via C. This expedient route path is termed a 'temporary alternative route', or TAR.

Restrictive control actions

Unfortunately, there will always be some condition under which no further expansive action is possible (in Figure 19.5, routes A–C and C–B may already be busy with their own direct traffic, or may not be large enough for the extra demand imposed by A–B traffic). In this state, congestion cannot be alleviated. Even worse, all calls made in vain to the problematic destination will be a nuisance to other callers, since these fruitless calls lead to congestion of traffic attempting to reach other destinations. In this case, the best action is to refuse (or at least restrain) calls to the affected destination as near to their points of origination as possible. Perhaps the most flexible of restrictive actions is 'call gapping'.

Under call gapping the traffic demand is 'diluted' at all originating exchanges. A restricted number of call attempts to the affected destination are allowed to pass from each originating exchange into the network as a whole. Within the wider network, this reduces the network overload, relieves congestion of traffic to other destinations,

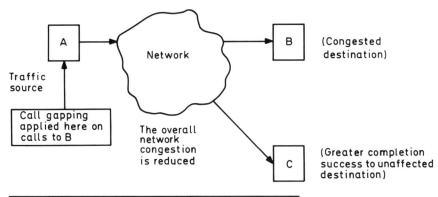

Call gapping type	Effect
1-in-N	Of N calls generated at a particular exchange for the affected destination, only one is allowed to mature into a network call attempt. All others are failed at A regardless
1-in-T	Only 1 call attempt every T seconds is permitted to mature beyond the originating exchange

Figure 19.6
Call gapping.

giving a better chance of completion. There are two principal sub-variants of call gapping, as shown in Figure 19.6. These are the '1-in-N' and '1-in-T' types. The '1-in-N' method allows every Nth call to pass into the network. The remaining proportion of calls, $(N-1)/N$, are failed immediately at their originating exchange (in this case, A). These callers hear network busy tone. The '1-in-T' method, by comparison, performs a similar call dilution by allowing only 1 call every T seconds to mature.

In the first method, the value of N may be varied, to control the level of call dilution (for $N=2$, 50 per cent of calls would be allowed into the network; for $N=3$, 33 per cent etc.). Similarly, in the '1-in-T' method, the value T may be adjusted according to the level of congestion. The greater the value at which T or N are set at any particular instant in time, the greater the diluting effect on calls.

If congestion continues to get worse, value N or T can be increased accordingly. With a very large value of N or T, nearly all calls are blocked at the originating point. The action of complete blocking is quite radical, but none the less is sometimes necessary. This measure may be appropriate following a public disaster (earthquake, riot, major fire etc.). Frequently in these conditions the public are given only one telephone number as a point of enquiry, and inevitably there is an instant flood of calls to the number, few of which can be completed. In this instance, call gapping is a powerful tool for diluting calls, thereby increasing the likelihood of successful call completion for other network users.

19.4 NETWORK MANAGEMENT SYSTEMS

Today it is common for network management computer systems to be developed as an integral part of the modern telephone exchanges which they control. Direct data-link connections between exchange processors and network management systems allow real-time network status information to be presented to the network management systems, and for traffic control signals to be returned. In this way, switch data changes or other network methods can be made quickly. An example might be a switch routing data change to amend routing patterns by the addition of TARs. Another example of a control signal might be an indication to the exchange to perform call gapping on calls to an appropriate destination. Figure 19.7 illustrates the functional architecture of a typical network management system.

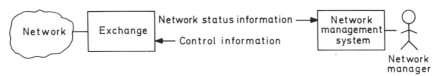

Figure 19.7
A network management system.

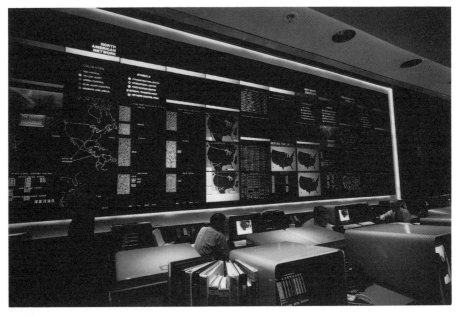

Figure 19.8
Network management centre. AT&T's Network Operations Centre at Bedminster in New Jersey, USA. It controls the AT&T worldwide intelligent network, the most advanced tele-communications network in the world. The centre controls data and voice calls over 2.3 billion circuit miles worldwide, handling more than 75 million calls a day. It is active 24 hours a day, 365 days a year. (*Courtesy of AT&T.*)

Network management systems can be procured either from computer manufacturers or computer software companies, but they are increasingly being offered by switch manufacturers as integral parts of new telephone or data exchanges. As the competition amongst suppliers gets even greater, perhaps we will have to wait only a few years before someone achieves a truly 'self-healing' network—one that remains 'congestion-free' without the continuous assistance of human network managers.

BIBLIOGRAPHY

CCITT Recommendation E410, 'International Network Management—General Information'.
CCITT Recommendation E411, 'International Network Management—Operational Guidance'.
CCITT Recommendation E412, 'Network Management Controls'.
CCITT Recommendation E413, 'International Network Management—Planning'.
CCITT Recommendation E414, 'International Network Management—Organization'.

20

NETWORK ECONOMY MEASURES

People who plan telecommunications networks are always searching for equipment economy measures: reducing the cost of a network for a given information carrying capacity; or, conversely, increasing the information throughput of a fixed network resource. To achieve their aims, they can either maximize the electrical bandwidth available from a given physical transmission path, or—if it is the other kind of economy they want—they can reduce the amount of electrical bandwidth required to carry individual messages or connections. This chapter describes a few of the practical economy measures open to them.

20.1 COST MINIMIZATION

Reducing the cost of equipment required for a given information throughput is important for public and private network operators alike; both will be keen to reduce the quantity and the cost of lineplant and switchgear.

If a given resource, say a transmission link, is already laid on then there is not much to be gained by applying economy measures which have the sole effect of making some of the available capacity redundant. In such circumstances it may be advantageous to 'squeeze' extra capacity from the line, especially if it is nearing its limit. This can be valuable for one of three reasons; first, it enables expenditure on more capacity to be delayed; second, it may be the only practicable means; or third, the cost of duplication may be prohibitive. The first reason might postpone the need for a private network operator to lease more capacity from the PTO (public telecommunication operator). The second case might arise because of a need to make more telephone channels available from a limited radio bandwidth. The third might reflect lack of resources to finance the prohibitive cost of a transatlantic undersea cable.

Earlier chapters in this book have covered one important means of lineplant economy, that of bandwidth multiplexing, either by the (analogue) frequency division multiplex (FDM) method, or the (digital) time division multiplex (TDM) method. This chapter recapitulates briefly these two methods, and goes on to describe some other important techniques including circuit multiplication equipment (CME), statistical multiplexing (used in data networks), speech interpolation, low rate encoding (LRE) and differential adaptive PCM (DPCM and ADPCM).

20.2 FREQUENCY DIVISION MULTIPLEXING (FDM)

Frequency division multiplexing (FDM) provides a means of carrying more than one telecommunications channel over a single physical analogue bearer circuit, as Chapter 3 records. FDM relies upon the carriage of a large electrical bandwidth over the circuit. Bandwidth for individual telecommunications channels is made available by sub-division of the overall bandwidth, much as some main roads are marked out into a number of lanes. Standard large bandwidths are employed over the physical circuit. These are called groups, supergroups, hypergroups etc. They are normally exact integer multiples of a base unit of 4 kHz, which is the nominal bandwidth required for a single telephone circuit.

As we may recall from the example of Figure 20.1, a single 4-wire circuit and a pair of channel translating equipments enable us to derive twelve telephone channels between the endpoints A and B. This compares with the twelve individually wired telephone circuits which might otherwise be required.

In much the same way as 4 kHz bandwidths (individual telephone channels) can be multiplexed by 'CTE' to form an 'FDM group', so FDM groups can be multiplexed by 'group translating equipment' (GTE) to form 'supergroups', and 'supergroups' can be formed into 'hypergroups' by STE.

As we also recall (from Chapter 13), lineplant savings are possible by making up connections out of any number of segments, each comprising a channel or circuit drawn from a number of different cables. This can save the need to lay a new direct cable. The connection from X to Y in Figure 20.2, for example, has been made using two FDM line systems, X–B and B′–Y—without there being any direct wires between the two ends X and Y. Instead all actual wires converge on a single hub-site at B. The use of concatenated (or 'tandem') FDM connections in this way should not be apparent to the end-user provided the interconnections are 4-wire and not 2-wire, since the full bandwidth is carried 'transparently' (i.e. without altering the signal significantly—unlike methods of compression we shall discuss later in the chapter). There is a slight degradation which results from the cumulative effect of repeated multiplexing and demultiplexing, so that the number of tandem sections should be minimized. So,

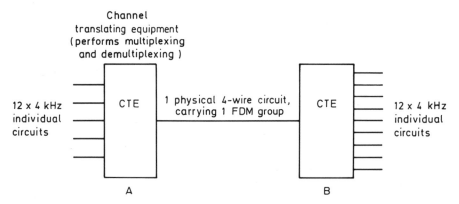

Figure 20.1
Saving lineplant using FDM.

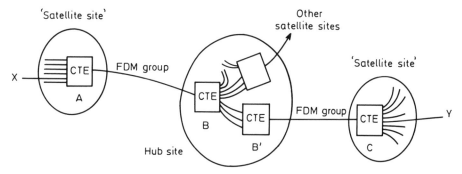

Figure 20.2
Tandem use of FDM systems.

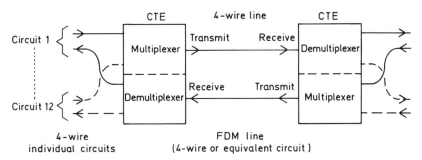

Figure 20.3
An FDM line system in detail.

though 2-wire interconnection between tandem sections is possible it is not recommended because of the nightmare combination of circuit stability and echo problems that it creates (Chapter 13 refers).

In the example shown in Figure 20.2, a connection between users X and Y passes through tandem FDM systems A–B and B' –C. Between A and B it is carried as part of the FDM group. At B it is demultiplexed and remultiplexed (at B') into another FDM group B' –C.

In order that we may later put in context the relative economies of the techniques discussed later in the chapter, it is important to note that FDM systems work in the '4-wire' or 'duplex' mode. By this we mean that simultaneous transmission in both directions is possible at all times. This means that X and Y in Figure 20.2 may talk at precisely the same time, and both messages be conveyed simultaneously. This is possible because a permanent communication path exists in both directions at all times. The diagram of Figure 20.3 illustrates this in detail.

Another method of getting even greater lineplant economy using FDM is to allocate only 3 kHz bandwidth (as opposed to 4 kHz) for each individual speech channel. This was the method used on early transatlantic cables. The method has fallen out of use owing to the bandwidth impairments that result.

20.3 TIME DIVISION MULTIPLEXING

On digital lineplant the FDM technique is not normally applied. In rare cases, however, one of two devices—either a codec (coder/decoder), or a trans-multiplexor (TMUX), is useful as a means of enabling an FDM group or supergroup to be carried on digital plant. The normal multiplexing method for digital signals on digital plant is called time division multiplexing (TDM), as we learned in Chapter 5.

In TDM digital line systems the lowest bit speed (corresponding to a single telephone or data channel) is 64 kbit/s. Thus linesystems operate at bit speeds equal to, or an integer multiple of 64 kbit/s.

Individual channels comprise a continuous series of 8-bit numbers (or octets) corresponding to the signal amplitude (or data signal value) sampled once every 125 μs. The technique of time division multiplexing (TDM) combines channels together by interleaving octets taken from a number of channels in turn. As we recall, the European multiplexing hierarchy for TDM is 64 kbit/s–2 Mbit/s–8 Mbit/s–34 Mbit/s–140 Mbit/s, while the North American standard is 64 kbit/s–1.5 Mbit/s–6 Mbit/s–45 Mbit/s–140 Mbit/s. The equivalent of the CTE used in FDM is called a primary multiplexor (or PMUX), and is illustrated as a reminder in Figure 20.4.

The individual channels of a TDM bit stream are called 'tributaries'. Where the tributary is an analogue signal, such as a speech circuit, the primary multiplexing equipment must first carry out analogue to digital speech encoding before multiplexing the 64 kbit/s tributaries into the 2 Mbit/s stream. The encoding method used for speech signals is called pulse code modulation, and this was also described in Chapter 5.

In common with FDM signals, TDM bit streams operate in a duplex mode and require 4-wire transmission. Also like FDM, individual TDM channels may be concatenated together without significant impairment of the end-to-end signal quality, since the 64 kbit/s or other bit stream is carried 'transparently'.

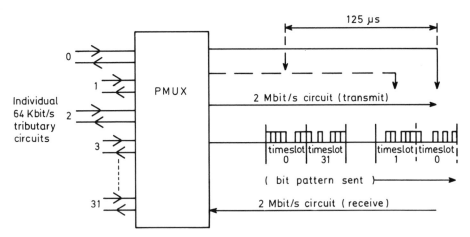

Figure 20.4
Time division multiplex (TDM).

20.4 WAVELENGTH DIVISION MULTIPLEXING

Wavelength Division Multiplexing (WDM) is set to extend still further the bit rate capacity of optical fibre. The technique relies upon the sharing of a single fibre between a number of transmitting lasers of LEDs and receiving LEDs. The different transmitter/receiver pairs are able to share the fibre harmoniously merely by working at different light wavelengths.

20.5 CIRCUIT MULTIPLICATION EQUIPMENT (CME)

'Circuit multiplication equipment', or 'CME', is a term used to describe various types of equipment capable of increasing the number of data or speech circuits that may be derived from a cable or a fixed bandwidth. The term, however, is not usually used to describe *multiplexing* equipment, such as that needed for FDM or TDM.

Circuit multiplication equipments tend to use 'corner cutting' methods as a means of deriving bandwidth economy, by one or a combination of the following practices:

- statistical multiplexing of 'interpolation' of individual channels;

- bandwidth compression of analogue signals, or low bit rate encoding (LRE—of PCM encoded signals);

- data multiplexing;

- data compression.

Different types of circuit multiplication equipment are available, designed either for voice or data network use. Most voice network equipment is described simply as CME, but devices intended for use on data networks include 'statistical multiplexors', 'data multiplexors' and 'data compressors'. All these devices (data and voice) are similar in purpose, the main difference between them being the compression technique used. For example 'statistical multiplexors' employ the interpolation method of bandwidth economy, while voice CME may employ bandwidth compression as well.

20.6 SPEECH INTERPOLATION AND STATISTICAL MULTIPLEXING

We have noted that FDM and TDM transmission systems are designed to allow duplex operation (simultaneous transmission of speech or data in both directions). However, each direction of transmission is probably in use only about 40 per cent of the time. In speech, for example, one or other of the channels is nearly always idle, since both people seldom talk at once, and then probably because the listener wishes to interrupt the speaker. The overall utilization is certainly under 50 per cent, but to make it worse, the speaker leaves gaps between words and between sentences, so that the efficiency is unlikely to top 30–40 per cent. Multiplying the effect, on a total of 60 speech circuits we might expect between 18 and 24 channels to be in use in each direction at any one time! The distinction drawn here between 'circuits' and 'channels' is deliberate. A circuit consists of a two channels—a 'receive' and a 'transmit'. There are 120 channels available in total, 60 in each direction—but only about 40 are in use—between 18 and 24 transmit channels and a similar number of receive channels. Of course there will be short periods when a larger number of channels may be in use, but this is so improbable that we conclude that 80 channels are effectively being wasted. There is considerable scope for economy!

Let us err on the safe side, and allocate 30 channels in each direction (60 in total). This is equivalent to 30 circuits, and can be carried by a single 2 Mbit/s digital line system, rather than the two line systems we would have required for the full 60 circuits. The difference is that we now need a '60 derived circuits on 30 bearers' circuit multiplication device (60/30 CME) capable of squeezing the 60 conversations into the 30 available circuits (60 channels). The first method we discuss for doing this uses speech interpolation, as illustrated in Figure 20.5.

At each end of the transmission link shown in Figure 20.5 a CME terminal equipment is provided. Between the two terminals are 30 speech circuits and a control circuit, each comprising a transmit and a receive channel. The two terminal equipments are normally of identical manufacturer type, and are commissioned and brought into service simultaneously with the 30 bearer circuits. Up to 60 'derived circuits' are connected to the exterior facing side of the CMEs for 'carriage' over the link, but the maximum number of circuits that we may 'derive' will depend upon local traffic characteristics, as we shall see later.

'Interpolation' relies upon interleaving short bursts of conversation from a number of different 'derived' telephone calls on to a single 'bearer' circuit. In order to work well, statistics require the CME to have a large number of bearer circuits available, and to carry a larger number of derived circuits. Thus, for example, during a short burst of conversation in direction A–B, on derived circuit number 1 in Figure 20.5, the CME at the A-end can allocate the use of one of the outgoing bearer channels, say also number 1, for carriage of the burst. The burst might be as short as (or even shorter than) the curtailed phrase 'I've got a...'. At the end of the burst, bearer channel 1 is made idle again. If by chance at this instant a burst of conversation commences

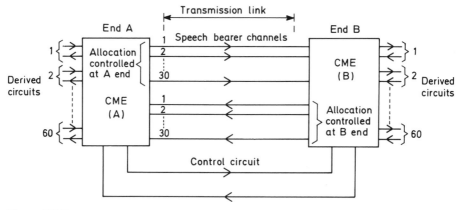

Figure 20.5
60/30 CME designed to use speech interpolation.

on derived circuit number 2, then bearer channel number 1 must also be allocated to carry this subsequent burst. But simultaneously, the A-end talker on derived circuit number 1 may start to talk again. In this instance, the A-end CME must again allocate a bearer circuit to carry the next burst, but since bearer number 1 is not now available, it has to choose a different one, say number 17. This is all right provided the B-end and CME similarly leaps around its incoming bearer channels to reconstruct the conversation.

Constantly, the A-end CME will be allocating and freeing up the bearer channels, in the direction A–B, according to when the A-end person on each of the 60 derived channels, is talking. Similarly, the B end CME will allocate bearers in the direction B–A, for B-end speakers.

If you were to listen in to any of the individual bearer channels, you would hear a train of incessant words and noises, which together would make no sense, since you would be hearing disjointed parts of up to 60 different conversations, as Figure 20.6 shows.

Thus, on bearer number one of Figure 20.6 we might hear the words: 'I've got a ... would you please ... after tea ... thirty-nine tons ... what about ... very well thanks ... six kilometres ... guarantees ... Agreed?'.

The conversations on the bearer channels are decoded by one of two methods. The first method, shown in Figure 20.5, uses a control channel between the two CMEs. This carries information such as 'connect bearer number 1 to derived circuit number 1'; 'connect bearer number 1 to derived circuit number 2', etc. The second method is virtually the same, except that instead of using a dedicated control channel, each burst of speech is carried in a 'packet', the first end of which is coded with some extra control information, which says which conversation it belongs to. The former method is common in CME designed for telephone use, while statistical multiplexors designed for data networks may use the second (packet-switching) technique.

Both statistical multiplexors (for data networks) and CME (for voice networks) use interpolation as a method of lineplant economy. Both types of device only work best when the number of bearer channels is fairly large, and 'gains' of two to three fold are possible (i.e. 2–3 times as many 'derived' circuits as bearers).

Gains of two to three fold (i.e. 2–3 times as many 'derived' circuits as 'bearers') are possible with both statistical multiplexors (for data networks) and CME (for voice application); hardware designs reflect this order of gain. However, although the hardware design of an equipment may suggest the use of a certain number of bearer and derived circuits, only the statistical characteristics of the real traffic can determine how many of each type should actually be connected. In practice few CMEs are wired to use all the bearer circuit and derived circuit ports simultaneously, and some bearer or derived circuit ports (or both) are usually unequipped. Thus the CME in Figure 20.5 could be used as a 54/30 device ($1.8 \times$ gain) or a 60/24 device ($2.5 \times$ gain) or a 48/24 device ($2 \times$ gain). This provides a useful degree of freedom in network planning, but great care is needed in operation if the device is being run either at very high gain ratios (say above $2.5 : 1$) or with a very limited number of bearers (say, less than 15). If, for example you tried to run the device on a 12/6 basis, or alternatively at $5 : 1$ gain (60/12), then severe quality impairments would be likely on the conversations carried. These arise whenever there is no free bearer channel to carry a burst of conversation in a particular conversation. The section of conversation is lost completely, and the

388

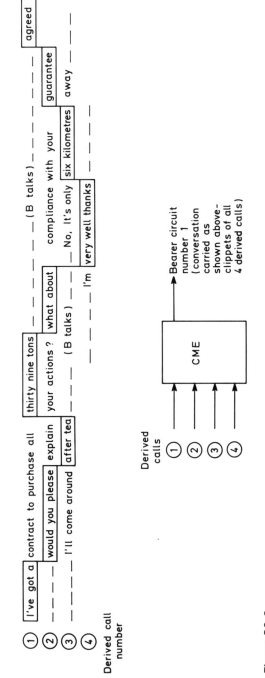

Figure 20.6
Speech interpolation.

listener hears a rather broken-up message. This effect is known as 'clipping', or 'freeze-out'. Later in the chapter methods for controlling it are described.

20.7 ANALOGUE BANDWIDTH COMPRESSION AND LOW RATE ENCODING OF PCM

If, prior to transmission, an analogue signal is passed through a special non-linear electrical circuit to compress its bandwidth, then the saved bandwidth can be used for the carriage of another signal, and an economy can be made. At the receiving end bandwidth expansion will be required to restore the original signal. This method of COMpression and exPANDING, or 'COMPANDING', although quite feasible, is not commonly used on analogue transmission systems as a means of lineplant economy since the benefits are small. The technique is, however, useful as a means of reducing noise interference on the received signal of analogue systems. It is effective as a means of reducing high-frequency noise, since in the expansion stage some of the noise is shifted to a frequency above that audible to the human ear. The technique is the basis of the Dolby noise reduction system, well known amongst audio-cassette tape users.

The real boom in use of bandwidth compression has come with the development of special PCM low rate encoding (LRE) techniques for carrying original analogue signals over digital lineplant. Despite the line bit speeds possible with high-speed electronic

Instant number	Value	Bit string
0	2	00000010
1	6	00000110
2	1	00000001
3	3	00000011
4	1	00000001
5	-1	10000001

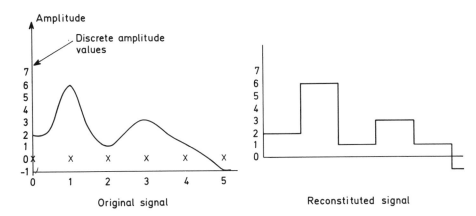

x = Sampling instant

Figure 20.7
Pulse code modulation.

components and optical fibre line systems, many research and development departments are working hard at digital signal bandwidth compression, and rates as low as 8 kbit/s are already quite feasible for carriage of speech (just imagine—eight conversations down a single 64 kbit/s circuit). Similarly 2 Mbit/s video has been achieved (70 Mbit/s or so is required for broadcast standard television).

In the chapter on pulse code modulation, we discussed how an analogue signal could be converted into a digital bit pattern, by 'sampling' the analogue signal at a high frequency to determine the amplitude of the signal. Corresponding to each sample, the amplitude is represented ('quantized') as an integer number and transmitted as a string of binary digits, or bits. Figure 20.7 reminds us of the principle.

The 'normal' sampling frequency for speech is 8000 Hz, and the normal number of quantization levels is 256 (equivalent to 8 bits per 'sample'). Thus the normal telephone bit rate, as we saw in Chapter 5, is 8 kHz $\times$ 8 bits per sample, equals 64 kbit/s, and most digital switching and transmission systems being installed today are based on this rate. However, if we can reduce either the sampling rate or the number of quantization levels without too much affecting the quality of speech, then we can make a direct saving in linespeed. Well, scientists have already found means of encoding speech at much lower rates, including 32 kbit/s, 16 kbit/s and even 8 kbit/s. Indeed, provoked by the need to squeeze more channels out of limited radio bandwidth, 8 kbit/s looks set to be the bit speed used for carrying speech between the base stations and the mobile hand-held telephone units of the latest digital cellular radio telephone systems.

For analogue signals (e.g. video or speech) the difference in quantization values between consecutive amplitude samples is usually small. So even though the amplitude value of a sample could theoretically take any one of 256 (2^8) different values, in practice the sample amplitude rarely differs much from the previous one. *Differential PCM* (a form of LRE) takes advantage of this fact by sending only the difference in amplitude between successive samples, rather than always sending the absolute value of each sample. The example of Figure 20.7 is repeated in the Table 20.1, where the absolute sample values, and the difference between successive samples is shown.

Notice in Table 20.1 how the difference in amplitude values of the consecutive samples can be expressed using only a four-bit string. This is equivalent to a 32 kbit linespeed, and is referred to as 32 kbit DPCM. As with normal PCM encoding, the first bit indicates positive or negative value (0 = positive, 1 = negative), the remaining

Table 20.1
Differential PCM (DPCM).

Instant number	Sample amplitude value	Bit string (normal PCM)	Sample difference	Bit string (32 kbit DPCM)
0	2	00000010	—	—
1	6	00000110	+4	0100
2	1	00000001	−5	1101
3	3	00000011	+2	0010
4	1	00000001	−2	1010
5	−1	00000001	−2	1010

Instant number	Sample amplitude value	Actual sample difference	Bitstring (24 kbit/s D PCM)	Difference actually transmitted
0	2	—	—	—
1	6	+ 4	011	+ 3
2	1	− 5	111	− 3
3	3	+ 2	001	+ 1
4	1	− 2	110	− 2
5	−1	− 2	110	− 2

(a) Sample values

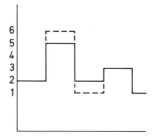

(b) Reconstituted signal

Figure 20.8
Signal distortion resulting from DPCM.

three bits define the differential amplitude. DPCM rates other than 32 kbit/s can be used in our example or any other, but the risk is ever present that if too low a rate is used it may prove to be inadequately responsive to the fluctuation in signal amplitude. Figure 20.8, for instance, repeats our example illustrating the difference values that would be transmitted using a 24 kbit/s DPCM device, using three-bit difference encoding. Also illustrated in Figure 20.8 is the resulting reconstituted signal.

We clearly see that the 24 kbit/s encoding can only cope with a maximum difference between consecutive sample amplitudes of ± 3. In the first part of the time period therefore, where the signal is changing by more than this, the DPCM encoding responds as much as it can, but the lack of adequate responsiveness results in the signal distortion shown in Figure 20.8(b). However, towards the end of the time period, the reconstituted signal recovers to match the original, because the signal fluctuation has reduced.

All differential PCM devices are bound to cause short periods of signal distortion because at least some samples differ by an amount greater than the device is capable of matching, but infrequent distortion in this manner is not critical for certain types of signals. Provided the bit rate of DPCM device is high enough, the subjective quality of the reconstituted signal may be quite acceptable to the viewer or listener. The bit rate is too low if users begin to find the picture or sound quality either 'granular' or 'abrupt'. Normal practice is to ensure that a high proportion of users (say 99 per cent) rate the standard of performance as either 'very good' or better.

Adaptive differential PCM (ADPCM) works in a similar manner to differential PCM. But by a further refinement on the DPCM technique it permits even lower bit speeds. In ADPCM, a 'predictive' algorithm is used to estimate what the next sample value is likely to be. The actual value is then compared with the prediction, and only the bit pattern describing the difference of the two values is transmitted down the line. At the receiving end, the use of the same predictive algorithm together with the correction value received on the line allows the signal to be reconstituted. This technique gives very impressive economies in bit speed provided a suitable predictive algorithm is available for the type of signal to be sent. To date, most research work has concentrated on speech and video based ADPCM algorithms, and in the field ADPCM devices have come to be more common than DPCM devices since comparable signal output performance can be achieved with typically half the bit speed.

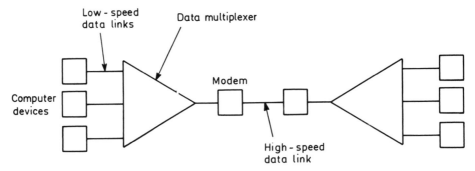

Figure 20.9
Use of data multiplexors.

20.8 DATA MULTIPLEXORS

A simple way of achieving lineplant economy in cases where a number of low-speed data circuits are required between the same two end-points is by the use of data multiplexors. Usually these employ TDM techniques as already discussed, allowing a number of low-speed applications to share the same high-speed link, but achieve lower gains than statistical multiplexors. Figure 20.9 illustrates an example network.

20.9 DATA COMPRESSION

A further way of economizing on data bit rate, and so lineplant, is made possible by the use of data compression techniques. These can be sub-divided into reversible and non-reversible categories. They allow a 33–50 per cent saving in line speed.

The most important example of reversible data compression is called the 'Huffman code'. This works by looking for the most commonly used data characters and using a shorthand code (say only three bits) to represent them, so enabling a saving of five bits from the standard eight-bit ASCII pattern. Unfortunately, to make the short codes for these characters possible, the less frequently used characters have to be coded with more than the normal eight bits, but the relative frequency of use of characters means that, on average, around six bits per character are needed. Irreversible data compression codes (often called 'data compactors') are less frequently used. These in essence corrupt the data signal by removing some of the detailed information content of the message which is adjudged by the transmitting device to be irrelevant. Of course, the information is irrecoverable at the receiving end.

20.10 PRACTICAL USES OF CIRCUIT MULTIPLICATION EQUIPMENT

Different manufacturers produce various types of CME, suitable for slightly different applications. A simple application is the use of a statistical multiplexor or data compressor within a private data network to economize on lineplant between buildings. A typical configuration might be that of a central mainframe computer located in one

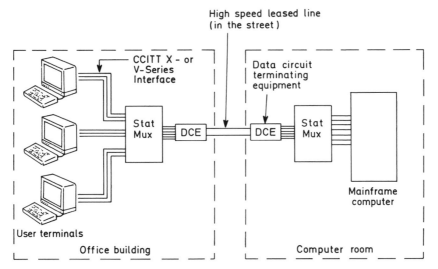

Figure 20.10
Typical use of statistical multiplexors.

building with a number of users' terminals located remotely in other buildings. Figure 20.10 illustrates the use of statistical multiplexors.

Within each remote user's location, a number of terminals have been connected directly to a statistical multiplexor. This employs the interpolation method for allocating the available line bandwidth to the terminals which are active. When data is received from an active terminal it is formatted by the multiplexor into a standard packet format and given a label indicating which terminal it came from. It is then transmitted down the single, high-speed leased line, to the multiplexor at the mainframe computer site. (The DCEs (data-circuit terminating equipment) are merely the line terminating devices required to control the flow of data over the line itself.) The second statistical multiplexor (which could be a built-in part of the computer) dis-assembles all the packets back on to individual circuits, appearing to the computer as if all the terminals were in the same room.

Messages are sent from the computer to the individual terminals in a similar way. A buffer (a sort of 'queue' provided within the statistical multiplex) ensures that no data is lost even in the unlikely event that all the terminals try to communicate with the computer at once. If this should happen, some data is merely delayed and the user may notice a slightly reduced speed of computer response but this is rarely problematic provided the frequency of occurrence is low.

However, if the statistical multiplexor is overloaded with too many terminals, then the buffer may become full, so that the statistical multiplexor has to 'tell' the terminals to stop sending data temporarily. In this case, the user notices a further reduction in the speed of response. Should the condition prevail for long periods it becomes frustrating to the user, and it should be alleviated by reducing the number of terminals connected to the statistical multiplexor. If necessary, another pair of statistical multiplexors and another leased line should be provided.

A larger scale example of CME use might be that by public telephone companies

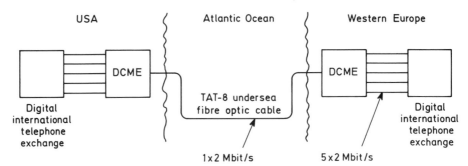

Figure 20.11
Use of DCME on TAT-8.

on an undersea cable. Such use has the advantage of increasing the call carrying capacity of a given cable many times over. Indeed, the digital circuit multiplication equipment (DCME) that has been designed for use on TAT8 (the first transatlantic optical fibre cable, opened in 1988) has the capability of enhancing the call carrying capacity up to five times, from around 8000 actual channels, up to about 40 000 derived ones. This particular DCME uses a combination of 32 kbit/s ADPCM (giving 2 : 1 gain) combined with speech interpolation (giving a further 2.5 : 1 gain). Figure 20.11 shows the configuration in which the devices are used.

Five 2 Mbit/s digital line systems from the international telephone exchange (or exchanges) are connected to the 'derived channel' side of the DCME. Only one 2 Mbit/s line system of the transatlantic fibre system (or other transmission system) is connected to the bearer side. On this 'bearer', timeslot 0 is used in the normal way for synchronization and line system control, and timeslot 16 is used for the circuit allocation control, leaving 30 individual bearer circuits. Meanwhile, on each of the derived 2 Mbit/s systems timeslot 0 is used in the normal manner for synchronization and line system control, and timeslots 1–15 and 17–31 are used as 30 derived speech circuits, so giving up to $5 \times 30 = 150$ derived circuits in total. Timeslot 16 of each desired 2 Mbit/s line system is used for communicating control messages between the telephone exchange itself and the DCME. These messages are necessary to prevent certain degradation conditions. For example the acceptance of new calls can be suspended during periods of 'freeze-out' (as described earlier in the chapter) or during periods of bearer transmission failure. The DCME alerts these conditions to the telephone exchange, which responds by diverting new calls to alternative paths. A suitable control procedure for the purpose is specified in CCITT Recommendation Q50.

20.11 CONSTRAINTS ON THE USE OF CIRCUIT MULTIPLICATION EQUIPMENT

The economies made possible by CME are to some extent achieved by 'cheating', and care is needed in their use. Three constraints need to be kept permanently under review, since the evolution of networks (both in topology and demand terms) can lead to serious impairments caused by the use of CME. The constraints are:

(i) CME derived circuits should only be used for 'approved' applications.

(ii) CME gain ratios should be kept low enough to prevent freeze-out, or suitable control measures should be implemented.

(iii) Circuits comprising tandem links which are all individually circuit multiplexed should be avoided as far as possible.

We explain the reasons in turn. First, particular types of CME are adapted for particular service types. Thus a statistical multiplexor is suitable for data but not voice, and a telephone network DCME is optimized for voice encoded PCM but completely corrupts a digital data signal. These are clear cut cases, but less obvious mismatches can creep in and irritate users. For example, the DCMEs used in public telephone networks must not only perform to a high standard for speech calls, but must also support dial-up voiceband data applications, such as facsimile transmissions. Fortunately, the PTOs have recognized that such data applications are degraded by 32 kbit/s low rate encoding, and have adapted their DCMEs to employ an alternative (40 kbit/s) algorithm on the occasion that a data call is detected. Meanwhile a 24 kbit/s algorithm is used on a simultaneous voice call in order to make the bandwidth available. (This capability is defined in CCITT Recommendation G721.) Less observant network operators might have implemented the DCME and still be receiving the complaints from their facsimile users!

The second constraint is that gain ratios should be kept low enough to avoid the adverse effects of freeze-out. As we have seen, freeze-out (undue contention for 'bearer' circuits) can cause unacceptable delay to users of data statistical multiplexors; it manifests itself in voice CME as unacceptably 'clipped' and abrupt speech. The gain ratio is reduced simply by increasing the number of available bearer circuits or by trimming back on derived circuit numbers.

Finally, take a warning about the ill-effects of the tandem use of CME. In data applications, tandem links of statistical multiplexors can lead to significant transmission delays. Worse still, the use of tandem CME in voice networks can cause the signal to become unintelligible. All this may seem obvious, but it can be difficult to foresee some of the permutations that can arise, particularly in switched networks! As we saw in Chapter 13, each ADPCM encoding introduces about three quantization distortion units. Within an overall budget of 14 qdus you cannot afford to be liberal, so be careful when designing networks like Figure 20.2 using CME!

BIBLIOGRAPHY

Cappellini, V., *Data Compression and Error Control Techniques with Applications*. Academic Press, 1985.

CCITT Recommendation G721, '32 kbit/s Adaptive Differential Pulse Code Modulation (ADPCM)'.

Held, G., *Data Compression—Techniques and Applications, Hardware and Software Considerations*, 2nd edn. John Wiley & Sons, 1987.

Huffman, D. A., *A Method for the Construction of Minimum Redundancy Codes*. Proceedings IRE 40-1098, 1952.

Saito, S. and Nakata, K., *Fundamentals of Speech Signal Processing*. Academic Press, 1985.

Smith, D. R., *Digital Transmission Systems* (Good review of ADPCM). Van Nostrand Reinhold/Lifetime Learning Publications, 1985.

Tugal, D. and Tugal, O., *Data Transmission—Analysis, Design, Applications*. McGraw-Hill, 1982.

Wade, J. G., *Signal Coding and Processing—An Introduction Based on Video Systems*. Ellis Horwood, 1987.

21

TECHNICAL STANDARDS FOR NETWORKS

In the past, telecommunications networks have been evolved in the minds of their designers to meet well defined but changing user demands. These broadly innovative influences are bound to persist and users can look forward to ever more sophisticated telecommunications services in the future. It was only about twenty-five years ago that customer-dialled international telephone calls first became possible, and in those days such calls were for the rich alone. Nowadays the advance of technology has made them so much cheaper that everyone takes them for granted. In the late 1960s everyone marvelled at the first live satellite broadcasts; now there are so many satellites in space that many of them no longer have names, only longitudinal geographic location references. Every day the world's communicators move larger and larger volumes of video, text, speech, and data information around the globe, and all without a second thought.

The revolution in communications that we have seen, and the ever-widening scope for information transfer, has only come about through an almost fanatical emphasis upon the international compatibility of telecommunications networks and equipment, underpinned by worldwide agreement on the technical standards without which interconnection between networks in different countries and the interworking of different network types would be impossible. This chapter discusses the various bodies involved in setting these standards. It outlines in particular the recommendations of CCITT (International Telegraph and Telephone Consultative Committee)—the world's most authoritative body on telecommunications technical standards.

21.1 THE NEED FOR STANDARDS

Whenever two or more pieces of equipment, built by different people or different manufacturing companies are called on to work harmoniously together within the same network, there is a need for technical standards. The 'standard' defines comprehensively, the interface, or set of interfaces to be used between various equipments. It describes the functions that each of the equipments are required to perform, and it ensures that the signals passed between the equipments are fit for their purpose, and unambiguous.

Figure 21.1
Can you hear me? Drawing by Patrick Wright. (*Courtesy of M. P. Clark.*)

Only a network designed and built in isolation could exist without carefully defined and documented technical standards for the interfaces. Such networks are extremely rare, for even when manufacturers develop their own special interfaces for interconnecting two pieces of their own equipment, these interfaces usually conform to their own documented 'proprietary standards'. An example of a proprietary standard is the 'systems network architecture' (SNA) which can be used as the basis for network design (called an architecture) interconnecting computer systems. SNA is similar to CCITT's (or ISO's) OSI model, but was developed by the IBM corporation and was in use by their customers before the CCITT standards had been agreed.

Most networks are made up of equipment supplied by many different manufacturers, who usually co-operate with the network operators and other interested parties to define the standards necessary to ensure correct interworking of equipment. The application or potential of any particular 'standard' is usually the determining factor in how many parties need to be involved in developing and documenting it. Consequently many slightly different groupings of manufacturers, network operators, users, and regulatory bodies, both national and international, have sprung up over the years to address the development of various different sets of standards. Each grouping (often called a 'standards body') tends to specialize in the interfaces required between particular types of equipment (say computers, or facsimile machines, or 'customer premises equipment (CPE)' or safety). The remainder of the chapter describes a number of the most prominent ones.

21.2 WORLDWIDE INTERNATIONAL STANDARDS ORGANIZATIONS

International network interconnections have always presented something of a challenge for the engineers developing them. Historically, international telecommunications between different countries have been provided by means of gateway switches, which have provided a means for interfacing and interworking a common international standard with the plethora of slightly differing national networks. International standards have thus provided the foundation of the recent boom in international communication. And the boom is likely to continue as long as each national network evolves slowly towards the international standard, leading ultimately to a homogeneous worldwide network. But this is only one of the benefits to be derived from international standards; no less important, with users of even the smallest networks insisting on conformance, terminal equipment manufacturers will be obliged to produce devices of worldwide compatibility. Thus even the interface between a personal computer and its associated printer will conform to an international standard eventually, with the happy result that there will be more choice to satisfy the customer, and larger markets to entice the manufacturer.

The most important worldwide standards organizations in telecommunications are the International Organization for Standardization (ISO) (colloquially the 'International Standards Organisation') and the International Telecommunications Union (ITU). A third body, but only important in the agreement of standards for satellite working, is the International Telecommunications Satellite Organization, called 'INTELSAT'.

International Organization for Standardization (ISO) The ISO is a 'voluntary' organization, composed of and financed by the national standards organizations (e.g. British Standards Institute, American National Standards Institute etc.) of each of the member countries. Like each of its component national organizations, ISO lays down standards for practically every conceivable item, not only telecommunications, but also, the colour scheme for electrical wiring, and even the standard sizes for paper (e.g. A3, A4 etc.). The organization has a number of sub-committees, but in the main it does not produce standards from scratch. It tends merely to 'brush-up' standards submitted by its member organizations. In this way the local area network standard 'ISO 8802' started life as 'IEEE802', generated initially by the American Institution of Electrical and Electronic Engineers (IEEE).

Address
International Organization for Standardization
Rue de Varembe, 1
Case Postale 56
CH—1211
Geneve 20
Switzerland
Tel: (41) 22 734 12 40

International Telecommunications Union (ITU) The ITU is an agency of the United Nations, responsible for overseeing all aspects of telecommunications. Organizationally, the ITU is split into five permanent parts, all of which are head-quartered in Geneva. These are shown in Figure 21.2. The General Secretariat is responsible for overall administration, finance, and publication of regulations, journals and technical 'recommendations' (their name for 'standards'). The IFRB (or International Frequency Registration Board) serve as 'custodians of an international public trust', regulating the assignment of radio frequencies throughout the world, and so preventing interference between radio 'stations'. CCITT and CCIR, the third and fourth parts of the ITU are 'consultative committees' which generate technical standards, respectively for telephone/telegraph networks (including data networks) and for radio communications. CCITT stands (in French) for 'International Telegraph and Telephone Consultative Committee', and CCIR (in French) for 'International Radiocommunication Consultative Committee'. The fifth part of the ITU, set up by the plenipotentiary meeting in June 1989 is the BDT (Telecommunications Development Bureau). BDT will have the same status as CCITT and CCIR and its aim is to secure 'technical co-operation' and raise finance to help the less-industrialized countries develop their telecommunications networks. More than 150 countries are members of the ITU and participate in the work of the consultative committees. Delegates to the individual 'study groups' meetings comprise not only the main public network operators, but also scientific and industrial organizations, private companies and equipment manufacturers. In the main, the study groups generate 'recommendations' from scratch, based upon the 'contributions' of delegates.

The ITU administers itself by means of large plenipotentiary meetings held every four years. At each meeting the constitution and structure is reviewed. In Nice, June 1989 a new constitution and convention was agreed. It was also at this meeting, amongst great controversy, that the BDT was set up.

INTERNATIONAL TELECOMMUNICATIONS UNION (ITU)				
General Secretariat Administration Finance	IFRB International (Radio) Frequency Registration Board	CCIR Consultative Committee for International Radiocommunication	CCITT Consultative Committee for International Telephones & Telegraphs	BDT Telecommunications Development Bureau

Figure 21.2
Organization of the ITU.

Address
International Telecommunications Union
Place de Nations
CH—1211
Geneve 20
Switzerland
(Same address for CCITT and CCIR)
Tel: (41) 22 730 51 11

International Electrotechnical Commission (IEC) Also known as the Commission Electrotechnique Internationale (CEI), IEC exists to promote international co-operation on electrical and electronic standards. It works in a similar manner to ISO, gaining international consensus of opinion and issueing standards in nine main fields, the most important to us being those affecting telecommunications and electronic components and those on telecommunications equipment and information technology. IEC has equal status and works in close co-operation with ISO, whose prime interest is in the non-electrical fields.

Address
International Electrotechnical Commission
Rue de Varembe, 3
CH—1211
Geneve 20
Switzerland
Tel: (41) 22 734 01 50

INTELSAT—The INternational TELecommunications SATellite organization is run by subscription between the world's main international public network operators. Its prime purpose is to develop, procure, and operate satellite services between countries. In support of these aims it develops its own standards specifically for the satellite service (e.g. from satellite earth station to satellite in space, or for 'multiple access' operation (Chapter 13 refers)).

Address
INTELSAT
3400 International Drive NW
Washington DC 20008
United States of America
Tel: (1)-202-944-6800

21.3 REGIONAL AND NATIONAL STANDARDS ORGANIZATIONS

A number of regional and national standards organizations are also in existence. In some cases these duplicate the international standards, and may even be at variance with them, a situation which is bound to persist, because it takes an unconscionable time to agree on an international standard. However, some users cannot wait and feel themselves obliged to develop their own 'interim' standards, and these too tend to persist even after the emergence of the world standard, because of the investment tied up in them. Notable regional and national organizations involved in such standards are:

Comité Européen de Normalisation (CEN) CEN is the European equivalent of ISO. Constituted of member national standards bodies, it generates European norms (ENs), but unlike ISO standards ENs are mandatory standards in the signatory countries to CEN. CENELEC, a related organization of CEN, is the European equivalent of IEC, generating European electrotechnical standards. Both share the same address.

Address
Comité Européen de Normalisation
Rue Brederode, 2
BTE 5-1000
Brussels
Belgium
Tel: (32) 2 519 6811

European Conference for Posts and Telecommunications (CEPT) The Telecommunications Commission (T Com) of CEPT was, until 1988, prominent in much the same way as CCITT, in the development of European telephone, telegraph and data networks standards, and many CEPT recommendations have been contributed to CCITT, and have formed the basis of CCITT recommendations. However, in 1988, prompted by the desires of the European Economic Community (EEC) to develop rapidly the network interfaces needed for the establishment of a Pan-European network infrastructure, CEPT elected to hand over all its technical standards development work to a new permanently staffed organization called ETSI (European Telecommunications Standards Institute). Nevertheless, CEPT lives on, as an organization of European public telecommunications operators discussing mutual strategy and planning.

Address
Office de Liaison CEPT
Case Postale 1283
CH—3001
Berne
Switzerland
Tel: (41) 31 62 20 80

European Telecommunications Standards Institute (ETSI) Funded by EEC money, ETSI was founded at Nice in France in 1988 to develop the network interfaces and other technical standards necessary for a homogeneous, and Pan-European telecommunications network. It has over 160 members from more than 20 countries in the CEPT and European Free Trade Association (EFTA), as well as EEC members. The technical work is undertaken by full-time staff, seconded from public network operators, manufacturers, trade associations, users, research bodies and national standards bodies. In addition a number of special voluntary working parties are established to deal with priority items.

ETSI is composed of a technical assembly and a permanent secretariat, but the standards development work is carried out by working parties and project teams. Technical documents produced by ETSI may become one of three types of document as listed below:

- EN or ENV (European Norm) Standards. To become ENs or temporary ENs (ENVs), the standard needs to be put up for formal adoption by CEN/CENELEC.

- ETS—(European Telecommunication Standards). ETSs are voluntarily accepted standards and guidelines.

- NETs—(Normes Européennes de Télécommunications). These are the most important standards, since they are mandatory across Europe—used for terminal equipment approval.

Address
European Telecommunications Standards Institute
Espace Beethoven
Sophia—Antipolis
BP62
F-06561-Valbonne Cedex
France
Tel: (33) 92 94 42 00

European Computer Manufacturers Associations (ECMA) A consortium of European computer manufacturers, working on a voluntary basis to develop computer interconnection standards. ECMA have often debated standards within Europe before contributing to CCITT. An example is the OSI transport protocol, ECMA-72 and ISO 8072. Another commonly used ECMA standard is ECMA-80, which is a standard for a CSMA/CD type local area network (LAN) operating over coaxial cable.

Address
European Computer Manufacturers Association
Rue du Rhône, 114
CH—1204
Geneve
Switzerland
Tel: (41) 22 735 36 34

American National Standards Institute (ANSI)   ANSI is a voluntary national standards organization of the United States of America, and is a member of ISO. It is the United States' clearing house for standards—generally adopting standards proposed by smaller companies and organizations within the United States. Notable achievements of ANSI have been the computer character code ASCII (American Standard Code for Information Interchange) and the computer programming languages, COBOL and FORTRAN. ASCII was the forerunner to CCITT's international alphabet number 5 (IA5—CCITT Recommendation T50) which defines the alphabet in which computers 'talk' to one another.

Address
American National Standards Institute
1430 Broadway
New York
NY 10018
United States of America
Tel: (1)-212-354-3300

Electronic Industries Associations (EIA)   EIA is a national trade association developing electrical standards primarily for North America. Notable achievements of EIA have been the RS232 and RS449 interfaces used between computer equipment.

Address
Electronic Industries Association
2001 Eye Street NW
Washington DC 20006
United States of America
Tel: (1)-202-457-4900

Institute of Electrical and Electronics Engineers (IEEE)   This is a professional institution for worldwide electrical engineers based in the United States. It lays down professional and technical standards and codes of conduct. A notable achievement of IEEE was the IEEE Local Area Network (LAN) standards in the 802 series, first published in 1983. These are now used throughout the world as the standards for 'Local Area Networks (LANs)'. They have subsequently been adopted by the International Organization for Standardization as Standard ISO 8802.

Address
Institution of Electrical and Electronic Engineers
345 East 47th Street
New York
NY 10017
United States of America
Tel: (1)-212-644-7910

Exchange Carriers Standards Association (United States)—commonly called the 'T1 Committee' A leading body in agreeing standards for the deregulated networks of the United States, the T1 committee is open to any subscribing members, but these are mainly United States exchange carriers. Standards under the auspice of T1 include most United States PTO (or 'exchange carrier') networking standards (e.g. T1-ISUP, the T1 version of the number 7 integrated services digital network user part signalling described in Chapter 28. T1-ISUP is used mainly in the United States). The standards work itself is carried out by a number of sub-committees called variously T1-D1, T1-X1 etc.

Address
Exchange Carriers Standards Association
5430 Grosvenor Lane
3200 Bethesda
Maryland 20814-2122
United States of America
Tel: (1)-301-564-4504

The Society of Manufacturing Engineering (SME) This is an American society formed of subscribing manufacturing organizations. It is responsible for the early development work on the Manufacturing Automation Protocol (MAP) and the Technical and Office Protocol (TOP).

Address
Society of Manufacturing Engineers
PO Box 930
1 SME Drive
Dearborn
Michigan 48121
United States of America
Tel: (1)-313-271-1500

The Telecommunication Technology Committee (Japan) (TTC) This body (set up in Japan at the time of liberalisation of telecommunications network services) is responsible for laying down network interface standards. It is composed primarily of leading Japanese scientists and engineers from network operators and equipment manufacturers. In the same way that T1 and ETSI set the pace for telecommunications standards in North America and Europe respectively, so TTC sets the *de-facto* standards for the South East Asian region.

Address
The Telecommunication Technology Committee
Second Floor
Nishi-Schinbashi Abe Building
3-12-10 Nishi-Shinbashi
Minato-Ku
Tokyo 105
Tel: (81) 3 432 1551

British Standards Institution (BSI) As an example of the many other national standards organizations, the British Standards Institution (BSI) produces telecommunications and other standards and is a member of ISO.

Address
British Standards Institution
Linford Wood
Milton Keynes
MK14 6LE
United Kingdom

21.4 REGULATORY STANDARDS ORGANIZATIONS

Since many of the world's governments have recently become keen to deregulate their telecommunications markets, allowing free competition particularly between manufacturers of 'customers premises equipment', there has been an increasing need for certain network interface standards to be mandated by law, and 'policed' by a regulating body. (Customer premises equipment, or CPE for short, is the general name applied to any type of equipment connected to the public network. Thus telephones, computer terminals and other similar devices are all 'CPE'.) The use of a mandated standard for the interface between network and CPE ensures the protection of the network operator's staff and equipment, meanwhile also assuring CPE manufacturers that their equipment operates satisfactorily over the network. The following organizations are active in laying down mandatory technical standards.

European Telecommunication Standards Institute (ETSI) Already mentioned in this chapter, one of the first jobs of ETSI, as part of the programme for a homogeneous pan-European telecommunications network, will be to lay out the CPE-to-network interfaces that will become mandatory in each of the EEC member states and other signatories of CEPT's memorandum of understanding. The standards will be known as NETs ('Normes Européennes de Télécommunications'—French for 'European Telecommunications Standards'). Once a NET has been adopted, both public network operators and CPE manufacturers will have to prove that their equipment conforms to the NET interface, before legal approval is given either for the launch of a new network service or for the sale of CPE. The NET programme, however, is significant in that CPE approval and certification (the 'green dot') will be valid Europe-wide (in any one of the signatory countries). Table 21.1 lists the NETs which in 1989 were either approved or under study.

Table 21.1
Mandatory European telecommunications standards.

Normes Européennes de Télécommunication (NETs)	
Access Group	
NET 1	X21
NET 2	X25
NET 3	ISDN basic rate interface (Chapter 24)
NET 4	PSTN normal telephone access
NET 5	ISDN primary rate interface
NET 6	X32
NET 7	ISDN terminal adapter
Mobile Group	
NET 10	Cellular access
NET 11	Telephony terminal
Modems Group	
NET 20	General modem
NET 21	V21 modem
NET 22	V22 modem
NET 23	V22 bis modem
NET 24	V23 modem
NET 25	V32 modem
Terminals Group	
NET 30	Group 3 facsimile terminal
NET 31	Group 4 facsimile terminal
NET 32	Teletex terminal
NET 33	Telephony terminal

Federal Communications Commission (FCC) The FCC is the regulatory body in the United States, reporting directly to Congress, which lays down the code of practice to which network operators and equipment manufacturers must conform.

> *Address*
> Federal Communications Commission
> 1919 M Street NW
> Washington DC 20554
> United States of America
> Tel: (1) 202-632-7000

Office of Telecommunications (OFTEL) OFTEL is the regulatory body for telecommunications within the United Kingdom, overseeing the activities of all public and private telecommunications operators and users. Part of the UK government's Department of Trade and Industry, perhaps its most important publications are the *Branch Systems General Licence (BSGL)* and the *Network Code of Practice*, which together govern the use and connection of public and private networks in UK.

Address
OFTEL
Atlantic House
Holborn Viaduct
London
EC1N 2HQ
United Kingdom
Tel: (44)-71-353 4020

British Approvals Board for Telecommunications (BABT) BABT is the independent body in the United Kingdom which undertakes customer premises equipment (CPE) approvals, confirming that submitted equipment conforms to the national network interface. With the emergence of the European 'NETs' programme, BABT will also be the UK's main registered body for approvals which will apply throughout the European Community and other NET signatory countries.

Address
British Approvals Board for Telecommunications
Claremont House
34 Molesey Road
Hersham
Walton-on-Thames
Surrey
KT12 4RQ
Tel: (44) 932 222289

21.5 PROPRIETARY STANDARDS

Despite the phenomenal number of standards organizations around the world, a few of which have been mentioned, some companies none the less feel that they can steal a march on their competitors by developing their own 'proprietary' (as opposed to 'open') technical standards. By this we mean that the standards are not available for free use by any manufacturer; either they may not be published at all, or alternatively a 'licence fee' is demanded when the standards are used in the design of new equipment. It is usually only large and powerful organizations who can afford to retain their standards as 'proprietary'. The markets such companies can command for their proprietary technology make it worth while to protect the technical rights. Other companies may find it lucrative to 'buy-into' these standards, in order that they in turn may sell compatible 'add-on' equipment, but the lead company retains an edge. Examples of companies who have been successful in retaining control of proprietary standards are AT&T and IBM.

American Telephone and Telegraph (AT&T) A major telecommunications carrier as well as computer and exchange manufacturer in the United States, AT&T draws its technical strength from its famous research arm, Bell Laboratories. An example of one of AT&T's proprietary standards is the 'UNIX' operating system, used as the operating software in some computer systems.

United States address for all documentation
AT&T Documentation Centre
AT&T Technologies
PO Box 19901
Indianapolis
Indiana 46219
United States of America
Tel: (1)-800-432-6600
    (1)-317-352-8557

European address for Unix documentation
AT&T Unix Europe Limited
International House
Ealing Broadway
London W5 5DB
United Kingdom
Tel: (44) 1-567 7711

Bell Communications Research (Bellcore) Bellcore is the jointly owned research arm of the seven United States Regional Bell Operating Companies (RBOCs). It was formed partly with staff from AT&T's Bell Laboratories at the time of AT&T divestiture in the mid 1980s. Bellcore has gained considerable respect for its networking standards called 'Technical References (TRs)—in particular in the field of intelligent networks and services (Chapter 26). Preliminary versions of TRs are called TAs (Technical Advisories).

Address
Customer Service
Bellcore
1A-234
60 New England Avenue
Piscataway
New Jersey 08854
United States of America
Tel: (1)-201-699-5800

British Telecom (BT) The largest public telecommunications operator in the United Kingdom, and descended from the state-owned 'Post Office Telecommunications' and formerly GPO (General Post Office). British Telecom remains a predominant force in UK network standards setting. The standards are nowadays catalogued as British Telecom Network Requirements (or BTNRs). Examples include BTNR 190 (DPNSS—Digital Private Network Signalling System).

Address
BTNR Coordinator
Network Systems, Engineering and Technical Department
Room 445
Williams National House
11–13 Holborn Viaduct
London EC1A 2AT

International Business Machines (*IBM*) Perhaps the world's most important computer manufacturer, it was IBM who gave us the standard hardware and bus configuration for many mainframe and personal computers and the systems network architecture (SNA) for communicating data between interconnected computers.

UK address
IBM Technical Publication Centre
PO Box 117
Mountbatten House
Basing View
Basingstoke
Hampshire RG21 1EJ
United Kingdom
Tel: (44) 256 478166

European address
IBM Software & Publications Centre (SPC)
IBM Denmark A/S
Sortemosevej 21
DK 3450
Alleroed
Denmark
Tel: (45) 2 93 55 11

United States address
IBM Mechanicsburg Distribution Centre
180 Kost Road
Mechanicsburg
Pennsylvania 17055-0786
United States
Tel: (1)-717-691-2055

21.6 THE STRUCTURE AND CONTENT OF CCITT RECOMMENDATIONS

Finally, we take a look into the structure and content of the 'recommendations' of the world's leading authority on networking standards, CCITT. CCITT 'recommendations' are approved every four years, at a plenary assembly of all the member organizations. This forms the culmination of four years development work, improving existing recommendations, establishing new, and deleting obsolete ones. The recommendations are then published in a series of coloured volumes—a process taking an entire year since all the transcripts have to be translated and published in all the working languages. Each four-yearly issue is in a single colour, but the colour of each subsequent issue is changed. Together, the whole set of volumes is referred to colloquially as 'The Blue Book' (1988), or 'The Red Book' (1984), according to the colour of the year. Thus:

- The *yellow* book was approved in 1980.
- The *red* book was approved in 1984.
- The *blue* book was approved in 1988.

Each coloured book is an improvement over its predecessor, but there is no guarantee of compatibility between equipments which have been designed to different coloured versions of the same CCITT recommendation. Thus, for example, there is no

Table 21.2
CCITT series of recommendations.

Recommendation Series	Subject coverage
A	Organization work of the CCITT
B	Means of expression (Terms etc.)
C	General telecommunication statistics
D	General tariff principles, charging and accounting
E	International telephone service operation
F	Telegraph services—operations and quality of service
G	General characteristics of international telephone connections and circuits
H	Line transmission of non-telephone signals (TV etc.)
I	Integrated services digital networks (ISDN)
J	Sound programme and television transmissions
K	Construction, installation and protection of cables
L	and other outside plant
M	General maintenance principles
N	Maintenance of sound programme and TV transmission circuits
O	Specifications of measuring equipment
P	Telephone transmission quality
Q	General recommendations on telephone switching and signalling
R	Telegraph transmission
S	Alphabetical telegraph terminal equipment
T	Terminal equipment and protocols for telematic services
U	Telegraph switching
V	Data communication over the telephone network (modems etc.)
X	Data communications networks
Z	Specification description language (SDL) and man machine interface (MMI)

guarantee that CCITT 7 signalling (1988 or blue book version) is compatible with CCITT 7 signalling (1984 or red book version).

Each 'book' of recommendations, however, is usually cohesive and compatible within itself and covers a wide range of subjects. Table 21.2 provides an overview of the subject areas, covered by the various 'series' of recommendations, which go to make up each book. Recommendations are usually known by their series letter and a number; X25 for example is a particular type of 'data communications network interface' in the X series of recommendations, in fact the one used in packet-switched networks.

BIBLIOGRAPHY

BSI Standards Catalogue, 1988.

CCIR 1986. ITU, Geneva.

CCITT Red Book (1984). Blue Book (1988). ITU, Geneva.

Folts, H. C. (ed.), *McGraw-Hill's Compilation of Data Communications Standards*, Edition III, Volumes 1–3. Mcgraw-Hill, April 1986.

Foreign National, International and US Industry Standards. Information Handling Services/ VSMF, 1986.

Franco, G. L., *World Telecommunications—Ways and Means to Global Integration*. Le Monde Economique, 1987.

ISO Catalogue, 1988.

Omnicom Index of Standards. McGraw-Hill, 1988.

Quick Reference to IEEE Standards. IEEE, 1986.

Stokes, A. V., *OSI Standards and Acronyms*, 2nd edn. Blenheim Online Publications, 1988.

The User View of Communications Standards—Some Suggestions for the Selection and Adoption of Communications Standards, for Users of Information Technology. Information Technology Standards Committee for Private Sector Users, 1st edn, Jan. 1983.

PART 3

SETTING UP
NETWORKS

22

BUILDING, EXTENDING AND REPLACING NETWORKS

It is rare to come across a telecommunications network that is not in a state of continuous evolution. At its simplest, a network could be expanding in order just to cope with increased demand. But in addition the network may be expected at the same time to provide increasingly sophisticated telecommunications services. For example, since the mid 1980s, many of the world's public telecommunications operators (PTOs) have responded to customer demand by the introduction of a 'freephone service' offering the facility for call recipients to pay for the calls.

As an introduction to the subject of network evolution, which the operator must be able to cope with, this chapter describes how networks may be built or extended to meet capacity and service needs. Particular attention is paid to the design factors inherent in the equipment ordering process—these are crucial to the success of a network evolution plan. In addition, and because it is not always possible to achieve the evolution without entire replacement of the network and equipment, the chapter describes various methods by which network modernization can be achieved without a major disturbance or interruption to the established service.

22.1 MATCHING NETWORK CAPACITY TO FORECAST DEMAND

No matter what type of telecommunications service it provides, network costs are minimized by matching capacity to the demand. Both over-provision and under-provision increase costs. When equipment is provided before it is actually needed, higher capital expenditure is incurred at an earlier date. In addition, there are higher running costs associated with equipment maintenance, accommodation, staffing etc. What is more, when the equipment finally comes into use it may already be obsolescent and in need of early replacement. Under-provision, which some network operators have to put up with through lack of capital, also is expensive, because of the work needed to relieve congestion and to maintain overloaded equipment. Further, higher costs are incurred in rearranging the network to incorporate new exchanges, as a result of the lack of network 'manoeuvring room'.

Figure 22.1 illustrates a graph of forecast demand, and shows a typical network operator's equipment provision schedule, in which 'steps' of extra capacity are provided to meet the projected demand. The demand may represent the total (local and long distance) traffic originated in a given area, in which case each 'step' on the capacity profile could represent the provision of a local exchange extension, or of a new local exchange. Alternatively, the demand curve could correspond to a particular type of traffic. One example might be the 'trunk' or international traffic generated by a number of local exchanges, covering a wide area. The steps of capacity shown in Figure 22.1 might then correspond to new trunk or international exchanges. Another example might be a case where the demand predicted in Figure 22.1 is the forecast demand for a new telecommunications service, for which some special new equipment will be required. The steps then relate to the necessary provision schedule of that equipment.

It is seldom possible to match the fitted capacity exactly to the demand, since the practicalities of exchange provision or extension usually allow capacity to be added only in fairly large chunks. It is therefore normal to keep the 'steps' of the fitted capacity graph above the demand, so as to ensure that the new equipment is on hand before the demand seems likely to exceed capacity. This reduces the risk of under-capacity. Where a network has slipped into under-capacity, new exchanges often become congested on the day they are opened—caused by the fact that forecasters tend to underestimate the amount of suppressed traffic in congested networks.

The forecast of demand is worked out as described in Chapter 11 on forecasting. For purposes of exchange provision, the demand is normally quoted in terms of the

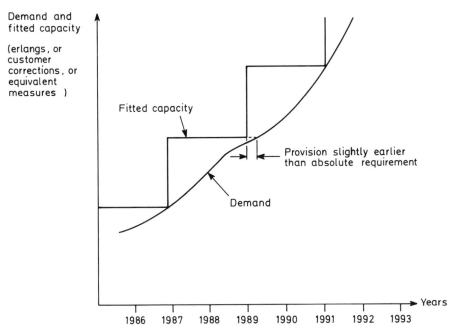

Figure 22.1
Forecast demand and fitted capacity.

highest instantaneous network throughput that is required. In circuit-switched terms, the important parameters are the traffic intensity (i.e. the busy hour traffic in erlangs) and the number of customer connections required. In packet-switched networks, the equivalent of the traffic intensity is the number of 'segments' of data to be carried per hour.

Forecasts form a direct tool for determining the forward network provisioning schedules, but in order to be useful the parameters to be forecast need first to be thought out carefully. Before making any forecast, the network planner has to determine the geographical area which the traffic forecast is intended to cover, and the type or destination of traffic (local, trunk, international) which any new exchanges will carry. In other words, he must have some idea what type of network topology will be employed, and how particular services will be supported. This depends upon a number of factors:

● terrain conditions;

● density of telephone customers;

● volume of traffic generated by each customer;

● volume and proportion of traffic to other local, trunk or international destinations;

● the established network (if any), its capability for extension, its suitability for the support of any new services (if required), its state of repair, and its degree of obsolescence;

● the reliability, optimum size, and service capabilities of contemporary equipment which might be used to extend or replace the established network.

Designing a new network from scratch, to serve an area which has previously been ignored, has the advantage of allowing a 'clean slate' approach, but it is likely to be constrained by the amount of capital available. This may well limit the initial network design to choosing the optimum locations for such exchanges as can be purchased, and of deciding upon each exchange's service or 'catchment' area. This task is best carried out by use of a map of the area to be served, marking it up to show the density of traffic likely to be originated and terminated in each spot. In areas generating low volumes of traffic (e.g. rural areas) it may prove economic to provide a medium size exchange to cover a very wide area, but in extreme conditions it is usually efficient to employ a large number of smaller exchanges. The two diagrams in Figure 22.2 demonstrate the trade-off exchange equipment against lineplant. In Figure 22.2(a) only one exchange is used, and each customer is connected to it by direct lines. Figure 22.2(b) shows an alternative network using smaller exchanges and it incurs greater cost in switching equipment, but the overall lineplant requirements and costs are reduced.

Option (a) will generally be appropriate for urban areas, and areas where lineplant can be provided relatively easily and relatively cheaply. Option (b) might be more appropriate as a means of serving 'pockets' of remote network customers, particularly where the terrain makes the provision of lineplant difficult. An interesting feature of option (b) is the emergence of a hierarchical network structure, where the central exchange has taken on the role of transit switching between all the other exchanges. This typical occurrence in real networks was discussed more fully in Chapter 12.

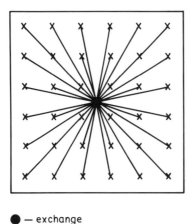 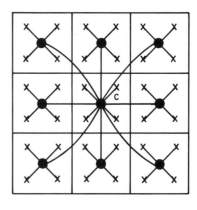

● — exchange

X — customer station

C — central exchange ; also carries
transit (or 'trunk') traffic.

(a) One large exchange
covering large area

(b) Nine smaller exchanges
covering the same area

Figure 22.2
The lineplant versus switch equipment trade-off. •, exchange; ×, customer station;
C, central exchange; also carries transit (or 'trunk') traffic.

For both networks shown in Figure 22.2, it is necessary for the network operator
to forecast the future demand from customers, not only in terms of the traffic in
erlangs (or equivalent), but also in terms of the total number of customers demanding
an exchange connection. Shortage of either type of capacity may lead to customer
dissatisfaction and loss of business. Shortage of customer connections means that new
customers cannot be connected. Shortage of busy hour throughput (i.e. erlang)
capacity means that the needs of the established customers are not met. Graphs may
thus be drawn, similar to Figure 22.1, for both fitted traffic-carrying capacity in
erlangs and the number of customer connections. Such a graph enables the network
planner to decide upon the dates and sizes of future exchange extensions, or new
exchange provisions. In this way the capacity may be matched to demand.

If the rate of growth in demand is slow, then small exchange extensions may be a
good way of increasing the capacity. However, in cases where growth is very rapid,
it may be more appropriate to provide completely new exchanges, with large capac-
ities. In addition, during rapid growth, the provision of entirely new exchanges presents an ideal opportunity for adjusting the network topology, perhaps splitting an
exchange area into two parts, with one area to be served solely by the new exchange
while the old exchange continues to serve the remaining area. Figure 22.3 illustrates
the splitting up of an established local exchange area into two parts. Exchange A is
the established exchange. Forecast growth in connections at a new town in the east
of the catchment area will exhaust the capacity of exchange A, and so it has been
decided to locate a new exchange in the new town (at B) and split the exchange area
accordingly. This split means that the new exchange will cater for any further growth
in the region around the new town, while the offload of traffic from exchange A on

to the new exchange, will also allow for further growth in traffic in the western area. This traffic will be served by exchange A. The decision to split the area in this case is quite straightforward, since a large quantity of new local line wiring will be needed and can be established at the new exchange B. However, in the case where all the wiring is already centred an exchange A, the economics would likely have forced an extension there.

Going back to Figure 22.2, option (b) can be used to demonstrate a further point. A forecast of the exchange connections and busy hour erlangs must be drawn up for each of the exchanges shown in option (b), in the same way as already described. But in addition, in the case of the central exchange C, a forecast is also required for the likely 'transit' (or trunk) traffic to be carried by this exchange. This is the traffic between different outlying small exchanges which is routed via exchange C. In practice, a forecast will also be required for the traffic which exchange C will have to carry between any of the exchanges shown and to-and-from other geographic areas, not

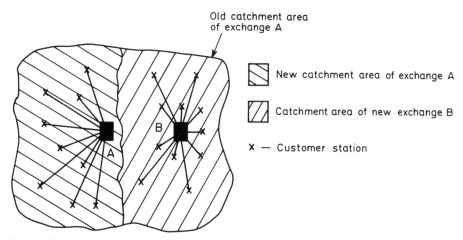

Figure 22.3
Splitting a local exchange area.

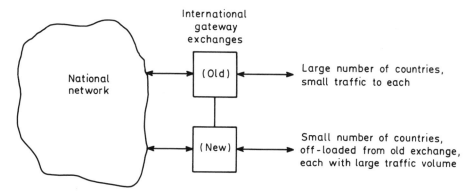

Figure 22.4
Reconfiguring international exchanges.

illustrated in Figure 22.2. In effect, just as local exchanges may be extended or may have their catchment areas adjusted, so may trunk and international exchanges (like exchange C).

Take as an example a single international exchange which is used initially to serve the international traffic needs of an entire country. Traffic growth may be such as to exhaust the capacity of the exchange, necessitating the provision of a new one. One way of reconfiguring the network in these circumstances would be to offload some of the overseas routes, corresponding to those overseas countries which have the largest volumes of traffic. The established exchange will thereby gain growth capacity (resulting from the offload), while the new exchange will serve and provide growth capacity for a small number of heavy traffic routes. Figure 22.4 illustrates this example. Similar network reconfigurations to those shown in Figures 22.3 and 22.4 might also be appropriate for trunk or other transit exchanges.

22.2 OTHER FACTORS AFFECTING THE NEED FOR NEW EXCHANGES

Apart from a straightforward increase in demand, there are a number of other factors which a network operating company may take into account when deciding upon the best time to replace exchanges. These are:

- the age and lifetime of the existing equipment;
- the obsolescence (or continuing usefulness) of the existing equipment;
- the comparative running costs of existing and replacement equipment;
- the new service demands of customers.

Historically, electromechanical telephone exchanges were planned and operated on the basis of a 20 or 25 year lifetime. Over this period, the basic technology did not change very much, and the eventual need for replacement was governed by the increasing wear, and the consequent increase in running costs—associated with maintenance, spares, and labour. Modern computer controlled equipment has a much shorter lifetime, perhaps as little as from five to seven years. The problem is not the reliability and wear of the electronic components, but the increasing rate of change of technology, and customer service expectations. Equipment lifetimes are becoming shorter, as the result of more rapid equipment obsolescence. An exchange might have plenty of copper wire termination ports but these cannot be used to terminate fibre lineplant unless converters are installed. This reduces the useful capacity of the exchange. In a similar way, older signalling system ports (for example pre-digital ones) become obsolescent and also reduce exchange capacity. Finally, exchange software needs occasional upgrading to provide either the capability for new services or the updating of maintenance capabilities. Without upgrade, certain categories of traffic may have to be diverted to newer exchanges for processing.

With older exchanges the running costs alone may render them uneconomic. Recently many of the world's public telecommunications operators have been modernizing their networks prior to the originally intended life-end of the equipment. The modernization is carried out in order to reduce ongoing running costs, particularly in manpower terms brought about by the use of digital (as opposed to analogue) transmission and stored program control (SPC) exchanges.

22.3 FACTORS IN DETERMINING AN EXCHANGE PROVISION PROGRAMME

Figure 22.5 shows a similar demand curve to that in Figure 22.1, but in this case the diagram takes account not only of the need to match capacity to the forecast growth in demand, but also to the decay in the effective capacity of existing equipment. The decline in effective capacity comes about because of the gradual obsolescence of equipment and the eventual need for replacement of life-expired exchanges.

Over the period up until 1990 the forecast is that the effective capacity of the existing system in Figure 22.5 will diminish due to the redundancy of equipment, in itself the result of obsolescence. Over this period the new exchanges, (1), (2) and (3) are required to handle the forecast growth in traffic, and to make good the shortfall which would otherwise arise as a result of the reducing effective capacity of existing equipment. The diagram shows the planned 'changeover' replacement of exchange (1) after a seven-year lifetime. Diagrams similar to that shown in Figure 22.5 are an invaluable tool in determining future equipment provision programmes. The example shown in Figure 22.5 demands a forward provision programme as shown in Table 22.1. Any provision programme is critically dependent upon the closure dates chosen for the older units, and also upon the installed size of the new units.

Counteracting some of the disadvantages of over-provision (discussed earlier in the chapter), the provision of larger units in advance of demand reduces the frequency of exchange extensions and new exchange provisions, and it brings benefits in the way of reducing and simplifying the installation work programme. The use of larger exchanges also tends to simplify the job of interexchange network and traffic design (for example in Figure 22.2 no interexchange network is required for option (a), while

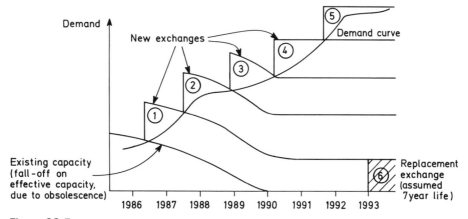

Figure 22.5
An exchange provision programme.

Table 22.1
Exchange provision programme.

New exchange no.	Date of introduction
1	1986
2	1987
3	1988
4	1990
5	1991
6	1993

a relatively complicated network is required to interconnect the nine exchanges of option (b)). On the other hand, the use of a larger number of small units may be the best answer in difficult terrain, or in cases where there is an established workforce, network or building already available. In the end the overall exchange provision is a compromise between a number of constraints.

22.4 DETERMINING A STRATEGY FOR NETWORK EVOLUTION

The efficiency and reliability of a network ultimately depends upon the expertise of its designer, and there is no substitute for experience in developing network evolution schemes tailored to particular circumstances. None the less, it is helpful to have a system for selecting the best available evolution scheme when there is more than one to choose from. A systematic approach also guards against the accidental oversight of crucial considerations. The factors listed below set out a framework for determining and evaluating alternative network evolution schemes, and by using them, the network planner can devise a forward strategy for evolution and development.

Capacity exhaustion date This provides the 'backstop' date, by which new capacity must have been provided. Some evolution schemes may be precluded if they do not provide extra capacity early enough.

Short-term network constraints It may be impossible to avoid provision of further, already obsolescent, equipment in order to meet a short-term peak in requirements (for example, a new international exchange may need to include a few lines using an archaic signalling system, to meet growth to a particular country, prior to modernization of equipment in that country). It is no good buying equipment so modern that it will not interwork with the existing network. Some evolution schemes or particular types of equipment may therefore be precluded.

Objective long-term network topology Over the lifetime of any exchange, the network will constantly be evolving towards the goal of the long-term strategy. A good network planner will optimize the procurement date and design of the exchange to maximize the value of the exchange throughout its life, and ensure its compatibility with the long-term objectives.

Evaluation of overall costs The total lifetime costs of an exchange (or of a complete exchange provision programme) include not only the capital costs associated with the provision of the equipment, the network, the accommodation, the peripherals, the training, the documentation and the spares, but also the ongoing running costs associated with equipment maintenance, rates, bills, electricity, staff costs etc. While one type of equipment may be cheaper to buy initially, it may be significantly more expensive in running costs. A 'Discounted Cash Flow' (DCF) analysis, to determine the 'Net Present Value' (NPV) (also called the 'present worth') costs of the exchange, or of the entire provision programme is often carried out. The example in Table 22.2 compares the net present value of providing a single large exchange of one equipment type, against an alternative strategy of purchasing a different manufacturer's exchange equipment, which allows some of the capital outlay to be deferred by phasing the overall provision capacity. The equipment capital costs shown in Table 22.2 compare the provision of a large unit costing £10m in one go, with the provision of an initially smaller unit which is extended twice, and costs £12m in total. The running costs (maintenance and manpower) are assumed to be 5 per cent per annum of the equipment value (a typical assumption). The accommodation costs are common to both cases, and include £1m initial purchase and fitting out of the building, plus £0.1m per annum rent and rates etc. It is interesting to see from the full-life cost analysis shown in Table 22.2 how, despite the fact that the overall expenditure is cheaper for Option 1 than for Option 2 (£15.2m as against £16.7m), it is none the less cheaper in present value terms to adopt Option 2 (£13.45m compared with £14.13m). The reduced

Table 22.2
Discounted cash flow analysis.

				Year				
Option 1 (large exchange)	1987 (£m)	1988 (£m)	1989 (£m)	1990 (£m)	1991 (£m)	1992 (£m)	1993 (£m)	Total (£m)
Equipment capital costs	10	—	—	—	—	—	—	10
Running costs	0.5	0.5	0.5	0.5	0.5	0.5	0.5	3.5
Other costs (accommodation etc.)	1.1	0.1	0.1	0.1	0.1	0.1	0.1	1.7
Total cost in year	11.6	0.6	0.6	0.6	0.6	0.6	0.6	15.2
Present value of cost (10% discount)	11.6	0.54	0.54	0.49	0.44	0.39	0.35	14.13

				Year				
Option 2 (large exchange)	1987 (£m)	1988 (£m)	1989 (£m)	1990 (£m)	1991 (£m)	1992 (£m)	1993 (£m)	Total (£m)
Equipment capital costs	4	—	4	—	4	—	—	12
Running costs	0.2	0.2	0.4	0.4	0.6	0.6	0.6	3
Other costs (accommodation etc.)	1.1	0.1	0.1	0.1	0.1	0.1	0.1	1.7
Total cost in year	5.3	0.3	4.5	0.5	4.7	0.7	0.7	16.7
Present value of cost (10% discount)	5.3	0.27	3.65	0.36	3.08	0.41	0.37	13.45

present value results from the deferred outlay. In the early years, more money is therefore 'left in the bank' (we may think of it as earning interest).

From Table 22.2 the planner could equally decide that £0.68m was cheap insurance against unplanned demand. In either case a full cost analysis would also include the cost of transmission plant provisions and rearrangements—these can be much greater than the switching costs.

In a private network, where line capacity is leased from the public telecommunications network operator, it is easy to match lineplant to demand—pay as you need.

The profile of installation work The profile of installation work may be an important factor, since it may be impossible for the chosen manufacturer to deliver equipment at a very fast rate (for example, a rapid and large-scale network modernization may require very large resources beyond the means of some suppliers). It may also prove impossible for the installation workforce to match a stop–start programme. It is much better, once an installation team has been established, to maintain as steady a workload as possible.

Recruitment of maintenance staff Consideration should be given to the maintenance of both old and new exchanges. Transferring and retraining staff from old units on to new ones may help reduce recruitment difficulties, but the ability to do so will depend upon the relative geographic locations of the two units and the skills match of the staff involved.

Associated network changes The introduction of a new exchange to support a new service, or as part of a technology modernization strategy, may have to be co-ordinated with other network rearrangements, or with public announcements.

Ongoing network administration One provision strategy may lead to much easier ongoing network administration than another. Perhaps one strategy allows 'centralized maintenance' (all maintenance staff performing remote maintenance from a single central location). One scheme may ease the task of monitoring ongoing network congestion and quality performance. Another might ease the task of network design, reducing the planning manpower that is needed. Finally one strategy might leave the traffic less prone to network failure, so that less traffic would be lost in the event of a single exchange of transmission link failure.

Available exchange size Exchange costs are often 'front loaded'. (By this expression we refer to the fact that you have to pay for a building, a maintenance staff and a central processor for an SPC exchange, no matter what its size.) Per unit of capacity, larger exchanges therefore tend to be cheaper.

Commercial conditions for equipment provision Open tenders for equipment, sent to a number of potential suppliers will often attract lower prices, since competing equipment providers are eager to gain maximum business. In order to allow open tendering, the requirements for the equipment should be kept as open as possible. Non-competitive tenders, for the extension of existing equipment by the original supplier, should be avoided where possible. (Where an extension is anticipated, it is better to make this part of the initial tender and contract rather than negotiate it after the first installation has been made.) When one supplier is ruled out, either by the stringency of the requirements or because of equipment being extended, there is little incentive for this supplier to offer competitive prices. Some network operating companies deliberately buy equipment of two compatible makes. Even though in the very short term this is clearly not the cheapest option, it protects them from becoming wholly dependent on either of the individual manufacturers. (We discuss the commercial considerations for contract placement in the next chapter (Chapter 23) on 'selecting and procuring equipment.')

Equipment provision lead times   There is a finite period of time required to develop new software for exchanges, to manufacture the hardware, and to install and test the equipment. This time is called the 'equipment lead-time', and must be taken into account when determining the exchange (or other equipment) provision programme. Indeed, some manufacturers may rule themselves out of possible consideration for a contract because the lead-time they are able to tender is not short enough. Insufficient respect for the lead-time, or over-optimistic estimation of development timescales leads to delayed opening dates, and almost inevitably to network congestion or some other problem.

Suppressed demand   A factor sometimes worthy of consideration, when injecting new capacity into congested networks, is the quantity and likely effect of any suppressed traffic demand. In particular, will an unexpected heavy load present any problems to the new exchange, and have sufficient tests been planned?

22.5 COMPARISON OF STRATEGY OPTIONS

When planning the evolution of networks, it is usually best to compare a number of different provision strategies, evaluating the differences (as outlined in the previous section) between perhaps:

● a strategy relying on extension of existing exchanges;

● a strategy adding new exchanges;

● a strategy replacing all exchanges;

● a long-term strategy for a small number of large exchanges;

● a long-term strategy for a large number of small, geographically dispersed exchanges.

The options used in the evaluation should be restricted to a small number of practical options (say three or four) and compared on as many factors as possible. Always, a complete long-term provision programme should be worked out, even though it may be necessary, in the short term, only to commit expenditure to a single new exchange or transmission line system. This ensures that consideration has been given to how the new item of equipment will operate in the prevailing network throughout its life. The analysis will finally lead to a decision about the exchange size, location, in-service date, service requirements, and possibly which manufacturer is to be asked to provide it. Similar analysis can be used to determine the required-by date and capacity for a new line system.

22.6 EXCHANGE DESIGN AND SPECIFICATION

Having decided upon the size, location, and in-service date for a new exchange or exchange extension, next comes the task of mapping out the design of the exchange itself. This is usually conducted in two phases. First an outline design sets out the functions and overall requirements, leading to an ultimate and more detailed specification. The specification must list the exact functions to be provided, defining the required software capabilities and performance requirements. It must also include the signalling systems to be provided, the routing and number analysis capabilities required, the billing and charging features needed, and the expected maintenance and network management features etc. These are all features without which the exchange could not operate, either in isolation or within the network. In addition, the specification needs to list any other interfaces to be provided so that the exchange can be interconnected to, and interwork with, any peripheral equipment. Such 'peripheral' equipment might include echo suppressors, circuit multiplication equipment, statistics post-processing computers (for the processing of accounting, traffic recording, and billing records), recorded announcement machines, network management systems, network configuration databases etc.

Each interface needs to be clearly and unambiguously defined, in order that the exchange interworks correctly with the peripheral equipment. The likelihood of incompatibility is increased when the two pieces of equipment are provided by different manufacturers, and for this reason, standard interfaces should be used as far as possible. The use of standard interfaces also tends to reduce the development costs incurred by the manufacturer, since these interfaces are likely to have been developed already. A lower equipment price therefore pertains (as compared with an equivalent exchange for which special development has to be undertaken). Figure 22.6 illustrates a possible exchange configuration for a modern day, digital and stored program control (SPC) exchange.

In Figure 22.6, the exchange is shown surrounded by a number of different types of equipment, each with its own particular interface. There are computer systems, one performing the 'network management' functions of network status monitoring and network control (as already described in Chapter 19); the other is a database, with an up-do-date record of the network configuration surrounding the exchange. Both systems are connected electronically to the exchange in order to be kept automatically up to date with any changes in the network. Two X25 datalinks are being employed for the interconnection, but in a few years a more comprehensive technical standard is expected to be available for this purpose. Developed by CCITT, the interface will be covered by the recommendations on what is called the 'Telecommunications Management Network' (TMN). TMN will be an all-purpose interface standard for connecting peripheral computer systems into exchanges. Especially it will cover the needs of operation and maintenance, network management, and network performance

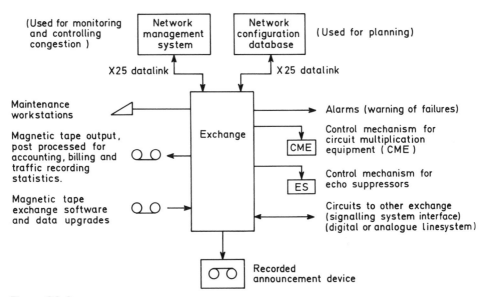

Figure 22.6
Some typical exchange interfaces and peripherals.

monitoring. An alternative interface which might be used is that being developed by the 'OSI Network Management Forum'.

The exchange in Figure 22.6 is shown with magnetic tape interfaces for downloading accounting, billing, and traffic statistics, and uploading software and exchange data. The important 'interface' in this instance is the format and informational content of the data on the tape. The peripheral computer needs sufficient information in order to carry out any necessary accounting, billing or traffic recording functions, and it needs to understand the format in which this information is passed out by the exchange, in order that the data may be correctly interpreted.

Another couple of interfaces shown on Figure 22.6 are the control mechanisms for circuit multiplication equipment and for echo suppressors. These can be relatively simple, 'hard-wired' electrical leads, indicating simple on/off controls, or they may be more sophisticated interfaces. For example, in 1988 CCITT recommended a special control interface for use between digital exchanges and DCME.

The final interfaces shown in Figure 22.6 are those intended for maintenance workstations, for recorded announcement devices, and for reporting equipment alarms. In the case where the peripheral devices required are provided by the exchange manufacturer as part of the overall contract for the exchange, it may not be important for the network operator to specify the interface in great technical detail. The exchange manufacturer is then free to use a 'proprietary' interface, probably specially designed for the purpose. The benefit to the network operator might be simpler working practices at the maintenance workstation, or perhaps more informative alarms or recorded announcements.

Figure 22.6 is not meant to illustrate a comprehensive or exclusive set of interfaces; there are other interfaces which are appropriate, depending upon the type and intended use of the exchange.

Just like exchanges, line terminating equipment, cross-connect frames, and multi-plexors require maintenance and alarms, and also like exchanges, they are under pro-gressive development so as to be capable of being remotely maintained or reconfigured using datalink interfaces. For these reasons, large portions of the specifications may be the same or similar.

Exchange specifications are often produced in two stages. An initial outline design lays out the network topology and describes the overall 'functional' requirements of the exchange, and it leads on to the lengthier detailed specification. The next chapter covers in some detail the points to be considered when preparing a full specification. The remainder of this chapter concentrates on the important considerations and tactics appropriate to the outline design stage.

22.7 OUTLINE CIRCUIT-SWITCHED DESIGN: CIRCUIT NUMBERS AND TRAFFIC BALANCE

In designing an exchange it is crucial to order the right number of circuits and the correct 'mix' of different signalling system termination types. If there are not enough terminations of a particular kind, or too few 'outgoing' or 'incoming' circuits, then the exchange cannot be fully loaded and its effective traffic-carrying capacity is reduced. Figure 22.7 shows two scenarios in which inappropriate circuit and signalling mixes have led to premature exchange exhaustion.

In Figure 22.7(a), with the circuits that have been ordered, the exchange cannot be supplied with enough incoming traffic, since the incoming circuits have been exhausted. The spare outgoing circuit terminations are wasted, and the effective exchange capacity is correspondingly reduced. A similar wastage could also have resulted from a shortage of outgoing circuits and an excess of incoming ones. In this case, incoming circuits would have to be left spare, in order to prevent the exchange running into congestion. The best way to alleviate either of these conditions is to carry out a small exchange extension (if possible) of incoming or outgoing circuits as appropriate.

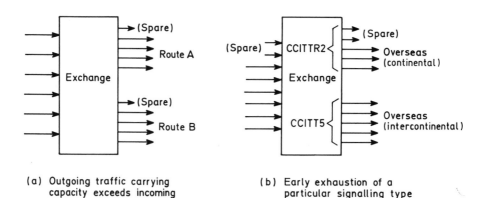

(a) Outgoing traffic carrying capacity exceeds incoming

(b) Early exhaustion of a particular signalling type

Figure 22.7
Poor circuit mix causing premature exchange capacity exhaustion.

In Figure 22.7(b), a slightly less obvious traffic imbalance has crept in. Illustrated is an international exchange, drawing incoming traffic from a national network, and connecting it onward to both continental and intercontinental overseas destinations. For continental destinations, CCITTR 2 signalling is used; for intercontinental destinations, the signalling system is CCITT 5 (see Chapter 7). The diagram illustrates the exhaustion of CCITT 5 signalling terminations. Although overall there are still incoming and outgoing circuits available (incoming from national, and outgoing in CCITTR 2), any further loading of the exchange will lead to congestion on the CCITT 5 routes. This premature exhaustion comes about because when further incoming circuits are connected from the national network they feed a mix of further traffic to both the intercontinental (CCITT 5) and continental (R2) destinations. But no further CCITT 5 circuit termination can be made available. The imbalance is therefore in the relative quantities of CCITT 5 and R2 signalling terminations, and the actual ratio of signalling termination types should match the balance of intercontinental to continental traffic. Two methods are possible for alleviating the type of premature exhaustion shown in Figure 22.7(b): one is to order an exchange extension to correct the CCITT 5 shortfall, and the other is to presort the traffic (at a prior exchange within the national network), so that a disproportionate amount of continental traffic is sent to this particular international exchange. This allows the spare incoming and outgoing-R2 terminations to be used, without adding traffic to cause congestion on the intercontinental (CCITT 5) routes. Of course, in this circumstance, some other international exchange will have to carry a disproportionately large share of intercontinental (CCITT 5) traffic.

The following procedure will help to determine an appropriate exchange termination mix, and to avoid such traffic imbalances. It should be amplified with care, and due consideration should also be given to studying the whole range of expected network configurations in which the exchange will be used throughout its life. For each configuration:

(i)   Calculate the circuit number, signalling system types, and traffic direction requirements to meet the busy-hour requirements of each individual route connected to the exchange.

(ii) Add up the total incoming and outgoing traffic in erlangs, and compare these totals with the total incoming and outgoing circuit numbers. (Bothway circuits, if any, should be considered as half incoming, half outgoing.) Ensure that the numbers are balanced, or within about 10 per cent of one another, but do not arbitrarily re-adjust the numbers to give a closer match. If an imbalance exists, go back and check it. One effect which can lead to apparent imbalance is that of non-coincident route busy hours. Figure 22.8 shows an example in which the optimum circuit numbers may at first glance appear misaligned, but in fact are not. Outgoing Route A carries mainly business traffic, predominantly during the daytime. The busy hour itself is at 9 am, when the traffic is 5 erlangs (requiring 11 circuits on 1 per cent tables). Route B by contrast carries mainly residential traffic, with a busy hour traffic load of 4 erlangs at 8 pm (requiring 10 circuits). Route C is the incoming route, serving both outgoing routes; its busy hour traffic is 7 erlangs, at 7 pm (requiring 14 circuits). Summing up the route busy hour traffic values, and incoming/outgoing circuits, there appears to be an imbalance, as demonstrated by Table 22.3. However, as the analysis above shows, the imbalance of circuits is relatively extreme (being 50 per cent extra on the outgoing side). Large real exchanges rarely show an imbalance greater than about 10 per cent.

(iii) Other traffic balancing studies may be appropriate. A useful one to perform on international exchanges is of national 'side' traffic to international 'side' traffic. Similar balancing of local 'side' to trunk 'side' traffic could be done for trunk (toll) exchanges.

(iv) Finally, it is worth calculating the peak cross-exchange traffic and comparing this with the sum of the individual route busy hours. The maximum traffic across the exchange (the cross-exchange traffic) will be less than the sum of all the route busy hours, unless all the route busy hours are coincident in time. This gives a plausibility check on the exchange traffic design. The nominal value of the cross-exchange traffic is also invaluable to the exchange manufacturer, since it is the maximum call throughput rate which the exchange common equipment (for example, the central processor) must be capable of handling.

(v) Should any of the traffic balances highlight unexplainable imbalances, then the original design should be checked.

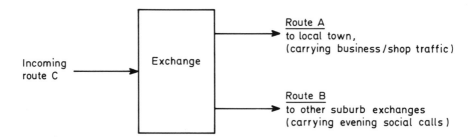

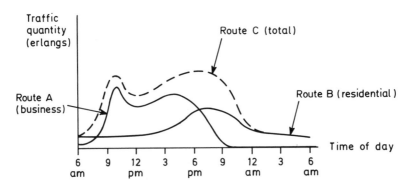

Figure 22.8
Traffic balance and non-coincident route busy hours.

Table 22.3
Apparent traffic and circuit imbalance.

	Total incoming	Total outgoing
Route busy-hour traffic in erlangs	7	$5 + 4 = 9$
Appropriate number of circuits	14	$11 + 10 = 21$

22.8 OUTLINE DESIGN OF OTHER TYPES OF NETWORK

Similar traffic balance considerations to those in the previous section should be applied when designing networks other than circuit-switched ones. For example, in a traffic balance of a packet-switched exchange the overall packet volume and rate carried by each route and the exchange as a whole would be taken into account.

22.9 THE EFFECT OF LOW CIRCUIT INFILL ON EXCHANGE AND LINEPLANT PLANNING

One final trap for the unwary exchange or lineplant planner, is the effect of low circuit infill. The problem arises when circuits are multiplexed together within the exchange, either on a higher order analogue FDM system, or as a 2 Mbit/s (or 1.5 Mbits) higher bit rate digital line system. In such a case, the whole analogue or digital 'group' must be provided between the two end-points, even if the circuit demand is not sufficient to need all the circuits in the 'group'. (Twelve circuits for a normal analogue group, 30 circuits for a 2 Mbit/s digital line system, 24 circuits for 1.5 Mbit/s.) The effect is known as low circuit 'infill'. In cases where the circuit 'infill' is lower than 100 per cent, extra lineplant and exchange termination capacity may be required.

Figure 22.9 shows an example in which a total of 20 circuits is required between exchange A, and two other exchanges, B and C. The terminations available on all the exchanges are standard 2 Mbit/s (European) digital terminations, and the only lineplant is also 2 Mbit/s. In result, two 2 Mbit/s digital line systems must be used, together with two 2 Mbit/s exchange terminations at exchange A. In other words, unless instead the planner reverts to transit routes to access the destinations, the equivalent of 60 direct circuits must be provided, with an effective circuit infill of 10 circuits within each 2 Mbit/s line system, i.e. 33 per cent infill.

Poor infills can also occur in higher order line systems. For instance, within an analogue supergroup it may be that only four FDM groups are used. (To provide the circuits using a supergroup may still be cheaper than providing separate lines for four individual analogue groups.) A similar example in digital practice might be the use of only three 2 Mbit/s 'tributaries' of an 8 Mbit/s digital line system, or the use of only two 34 Mbit/s 'tributaries' within a 140 Mbit/s digital line system etc. So, as with

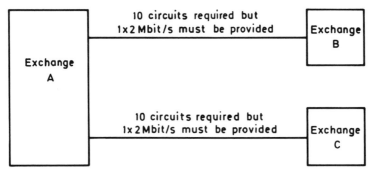

Figure 22.9
The effect of poor circuit infill.

exchange terminations, extra lineplant capacity must be provided on all occasions when poor circuit infills are encountered.

22.10 FUNCTIONAL REQUIREMENTS OF EXCHANGES OR LINE SYSTEMS

We have now discussed in some detail the traffic flow design of exchanges, but what other functional characteristics will be required? The answer depends upon the expected performance of the switch. The next chapter on equipment procurement, provides a rough checklist of items which may be appropriate to the final specification.

22.11 METHODS OF NETWORK OR EXCHANGE MODERNIZATION

Finally, before going on in the next chapter to discuss the production of detailed equipment specification, it is worth considering various tactical methods for network or exchange modernization. The earlier part of this chapter considered how to determine the provision programme for exchanges or new line systems, and talked about the need for new equipment—to meet growth or extension in demand, for equipment upgrade or replacement of life-expired equipment. What were not covered in any detail were the tactical methods that might be employed for network modernization (or replacement). Basically, two methods are available for either exchange or lineplant modernization; they are the 'integration' method and the 'overlay' method.

Integration   A new exchange or line system is provided in a manner fully integrated with the existing network. A typical example might be a one-for-one or one-for-many replacement of older equipment. Figure 22.10(a) shows a one-for-many replacement of four small analogue local exchanges with a single larger modern exchange. Figure 22.10(b) shows a one-for-one transfer of an analogue line system on to a digital replacement.

The integration method of replacement is generally used where the old and new technologies are expected to coexist for a long time, as part of a gradual moderniza-

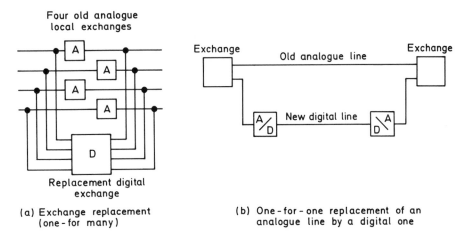

Figure 22.10
Methods of equipment replacement.

(a) Exchange replacement (one-for many)

(b) One-for-one replacement of an analogue line by a digital one

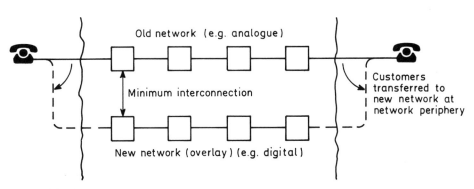

Figure 22.11
Network overlay.

tion regime. The changeover of individual exchanges or line systems may be carried out slowly, circuit by circuit, or by a special rapid wiring 'changeover' device.

Network overlay When network operators are embarking on more radical modernization, perhaps undertaking a heavy capital investment programme to replace the network very quickly with modern equipment, an overlay network approach may be the most economic method. Network overlay is also a good method of introducing networks for new or special services. In the 'overlay' method, the new equipment is linked together as an entirely separate (or 'overlaid') network. Customers are then transferred from one network to the other at the customer-to-network interface on the periphery of the network. Figure 22.11 demonstrates the principle.

The advantage of using a network overlay is the much reduced number of interconnections required between old and new networks. Interconnection is often still needed in order that customers of 'old' and 'new' networks may intercommunicate, but the extent, and therefore the cost of large-scale 'interworking' between the two networks can be contained. For example, the amount of analogue-to-digital conversion equipment between an old analogue network and a new digital overlay network can be minimized. In addition, the man-effort associated with the extensive network reconfiguration of interworked networks can be largely contained.

BIBLIOGRAPHY

Cassimatis, P., *A concise Introduction to Engineering Economics*. Unwin Hyman, 1988.

Courville, L., de Fontenay, A. and Dobell, R., *Economic Analysis of Telecommunications—Theory and Applications*. North Holland, 1983.

Flood, J. E., *Telecommunications Networks*. Peter Peregrinus (for IEE), 1975.

Mayer, R. R., *Production and Operations Management*, 4th edn. McGraw-Hill, 1982.

Morgan, T. J., *Telecommunications Economics*, 2nd edn. Technicopy, 1976.

Sprague, J. C. and Whittaker, J. D., *Economic Analysis for Engineers and Managers*. Prentice-Hall, 1986.

23

SELECTING AND PROCURING EQUIPMENT

The previous chapter described the factors in outline equipment design which are crucial to the success of the network evolution plan. Having a good plan is one thing; executing it is another, and in this respect the ordering and delivery of the equipment should receive considerable attention. This is an activity often referred to as 'procurement', and it can be carried out very successfully by highly mechanistic management methods. Various 'project management' techniques are available which tackle network design as if it were the input of a procurement 'process', much as 'raw material' is the input to a manufacturing 'process'. In production-line fashion, project management techniques lead the project through its design and checking stages, and on to ordering, installation, and testing. But in the same way as poor raw material has an effect right through to the end product of a manufacturing chain, so the defects of a poor network design cannot be made good during implementation. This chapter describes a typical methodology for selecting and procuring equipment, starting from the outline equipment design.

23.1 TENDERING FOR EQUIPMENT

Except for some of the largest corporations, very few companies that operate networks also manufacture the 'kit' that goes to make them up; everything from personal computer links upwards has to be bought from other manufacturers. For small pieces of equipment like modems the purchase may be straightforward, since a range of items meeting a standard specification (say, a CCITT V-series recommendation) is available from a number of manufacturers. The purchaser then has time to concentrate on the finer differences (size, price, reliability, ease of maintenance etc.) of the various equipments, and to choose what suits him best. More costly equipment, like major transmission systems or large exchanges, can seldom be bought 'off-the-shelf'. Instead, the differing circumstance of each network and application means that a considerable amount of 'adaptive engineering' is required in each case, to modify or upgrade the manufacturer's own 'basic' equipment design. Selecting suitable equipment is then much more difficult, since at the time of order, it may be that none has yet been developed. Under these circumstances, the normal method of equipment selection is by 'invitation to tender', and procurement follows after a contract has been placed with the manufacturer who has 'tendered' a price and a technical conformance nearest to the specification.

As well as a technical specification, an invitation to tender includes a number of commercial and other contract conditions. The specification itself either lays out in precise detail the individual electrical components and connections to be made, or it is a 'functional specification' which lays out only what general functions must be performed and what external interfaces are required. 'Functional specifications' are preferred by tenderers and manufacturers alike because they place fewer constraints on the internal design of the equipment. They permit a wider range of manufacturers to consider adaptation of their products to meet the external interface requirements. For the purchaser, the larger the number of equipment manufacturers kept in competition and providing 'standard' equipment the lower the tendered prices will be. For the manufacturer the benefit is the larger market.

Each 'tender' sets out the degree to which the equipment can be adapted to conform with the specification, the degree to which the commercial conditions are considered acceptable, and the quoted price. Even if none of the tenders meet the specifications and commercial conditions in every respect, purchasers can at least gain an understanding of each piece of equipment and decide upon their own trade-offs between various factors, which are:

(i)   price and conditions for payment;

(ii)   amount of further technical development required by an existing product, and how much confidence there is that it can be achieved;

(iii) ease of equipment maintenance;

(iv)   time required for installation, and whether the in-service date is early enough;

(v)   quality of product and installation;

(vi)   modularity of kit; the ease with which parts may be replaced or with which a subsequent equipment extension may be carried out;

(viii) warranty and after-sales service;

(ix)   the area and the standard of accommodation required;

(x)   robustness and reliability of equipment;

(xi)   political factors (nationality of manufacturer etc.);

(xii) software licence and conditions.

The process of invitation-to-tender and response is illustrated in Figure 23.1 and is much the same, no matter what type of equipment is being purchased. The length of each of the working documents (specification, tender etc.) depends upon the complexity of the requirement and the confidence gained from previous experience, but the contract itself needs only to be a short document committing the manufacturer to conform with the tender response. In the event of a failure to conform with the requirements of the contract, a number of penal clauses usually cover each of the possible eventualities. For example, failure on the part of the manufacturer to develop a given function may result in no payment, while failure on the part of the purchaser to pay a given instalment, or to provide some detailed information on a given date may nullify the contractual obligation placed on the supplier.

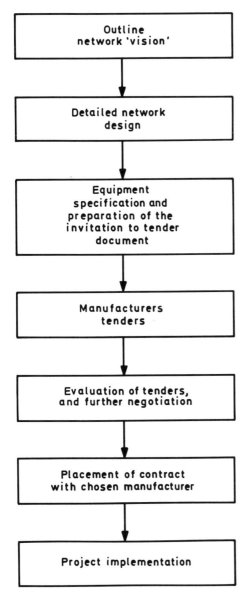

Figure 23.1
Procuring equipment.

23.2 PROJECT MANAGEMENT

Even before the equipment goes out to tender, the job of project management must have commenced, to co-ordinate all the separate aspects associated with the provision. Accommodation will be needed (some to a clinical standard)—built or refurbished by a separately contracted company; power and standby generator equipment may need to be purchased and installed; and all sorts of changes in the network may have to be undertaken.

During tendering, a sound project management keeps each task on schedule, extracting prompt replies from manufacturers, insisting on rigorous evaluation and timely contract placement. At the next stage the job becomes one of overseeing the manufacturer's work and quality, and of co-ordinating the supporting in-house or other contractors' activities.

Later on in the process, prior to the 'contract completion date', the purchaser should test the equipment, especially any new developments, to check conformance with the specification before final acceptance and payment.

The key to successful project management is early project planning. This must be thorough and realistic. Forgetting some activities, or programming too much or too little time for other activities, can seriously upset the smooth running of the entire project.

It is valuable during the project planning stage, to undertake a 'critical path analysis'. This sets out the relationship between each of the sub-projects, laying out the full sequence of 'events' (or sub-projects) and the dependency of any individual 'event' start date upon the completion of an earlier 'event'. For example, it would not be possible to fit out a building before it had been built, but it might not matter whether the electrical wiring or the plumbing were to be carried out first. This consecutive sequence of interdependent sub-projects determines the earliest date at which the project could be completed. If any of the sub-projects where to take longer than scheduled, then the earliest possible project completion date would slip by at least the same number of days. Figure 23.2 illustrates a simple critical path analysis project schedule.

The exchange installation project depicted in Figure 23.2 comprises ten sub-projects or 'events'. Four of these 'events' make up the 'critical path' of the project. These are the building and fitting out of the accommodation, followed by the exchange installation, the exchange testing, and the wiring-up of the network. None of these activities can commence before the previous one has been completed, so that the consecutive time period required for their completion determines the minimum overall project length. Slippage in any of these four sub-projects delays the project's earliest completion date (i.e. the exchange opening).

In contrast, the other six events do not lie on the critical path. This means that a degree of slack time is available. For example, a total of 29 months is available to recruit and train maintenance staff, but only ten months is required. These tasks could therefore be delayed for 19 months after the project start date without affecting the project end date. In fact, the only constraint in this case is that the training of staff cannot commence before January 1991, after the exchange installation, otherwise there is no equipment to be trained on. This relationship is shown by the dotted line arrow on Figure 23.2, indicating a 'dependent event'. The use of a dotted line arrow

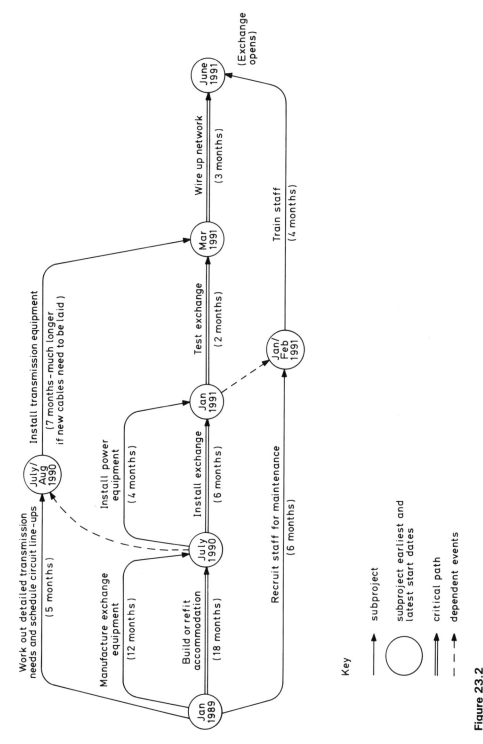

Figure 23.2
Simple critical path analysis for exchange installation project.

indicates that it is only a 'one-way dependency', rather than two-way. In other words, the exchange must be installed before training can commence, but in our example it is not the case that the maintenance staff need to have been recruited before exchange testing can commence.

Other events which are not time critical are the ordering of circuits and the installation of transmission equipment. The ordering must precede the installation, but the ten months needed for the work is far less than the 26 months available. The transmission equipment installation cannot commence until July 1990 at the earliest, after the accommodation has been prepared.

Not only does a critical path analysis provide a checklist of all the tasks to be undertaken, it provides a straightforward and mechanistic method for marshalling resources and ensuring the timely completion of critical path events. Further, it allows the scheduling of non-critical events in the most efficient and convenient manner possible within the time-scale. For example, staff training in Figure 23.2 could start any time between January and February 1991, according to convenience.

23.3 PROCUREMENT POLICY

Before purchasing items of equipment, it is prudent for any company to determine a proper procurement policy. A procurement policy protects the interest of the purchaser. It is no good always buying the cheapest equipment if this means the eventual acquisition of a diverse range of slightly incompatible equipment, each piece requiring different spares and differently trained maintenance staff, and with insoluble compatibility problems should the equipments ever have to be integrated. On the other hand, if a company always purchases equipment from a single supplier, then dependency creeps in; the supplier's prices may escalate to give a better profit margin; worse still the supplier may go out of business or become unreliable or un-cooperative. Using even a small range of different suppliers helps to maintain competition between them. An example of a complete equipment purchasing policy might be that 'new equipment purchased should be compatible with an IBM equipment'.

At the time of tendering for equipment, due account should be taken of the company procurement policy. It makes poor sense to write a specification which precludes all but one supplier, when the company procurement policy is to promote competitive tendering, with a long-term objective of maintaining at least two suppliers. Overworded and complicated specifications tend to have the effect of precluding suppliers—so that in general it is best to try to keep conditions of tender and contract as simple as possible.

23.4 PLANNING DOCUMENTATION

Before commencing any major project, whether it involves procurement or not, the network operator must determine the long-term network strategy or 'vision' of which the project forms a part. The strategy should set out the long-term objectives for network development, and against the backdrop of short-term constraints it should map out a series of projects to achieve the long-term goal. Typically, the strategic

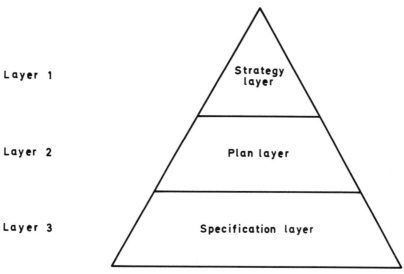

Figure 23.3
A layered documentation structure.

goal may be reached in a number of different ways, and the analysis leading up to a recommendation of a particular plan should consider and reject a number of alternative options.

Having laid out the long-term strategy and mapped it into a series of component projects, the planning, co-ordination, and documentation of each individual project must commence. Any project is best documented in a controlled and layered structure so as to ensure easy reference to detailed information while retaining consistency and oversight across the whole job.

Figure 23.3 shows a schematic documentation structure of three layers, oriented in a triangular fashion. At the top or 'strategy' layer the documentation is light, laying out the overall objectives, the end goal, the resources to be used, and in outline, the major component projects of the strategy. At the second or 'plan' layer a number of separate planning documents exist, each comprehensively describing an individual project. The plan has to cover everything: service and project time-scales, staff responsibilities, network topology, maintenance procedures, customer service arrangements, accommodation requirements, sales, procurement, customer billing, training and all other conceivable aspects. Ideally, the document refers to the strategy that it supports, and also explains the existence and interrelationship of any more detailed specifications which exist to support it. The third and last layer of documents, 'specifications' covers the detailed technical definitions and procedures.

Specification documents are the meat of the project. It is documents at this level of detail that are tendered to prospective manufacturers and are subsequently used for definitive reference. That is not to say that layer 1 and 2 documents are any less important, since they hold the entire project and strategy together.

Given a proper document structure, a document registration procedure should be set up to ensure proper quality inspection of documents. Documents should be checked for accuracy, comprehensiveness, and clarity by persons other than their original authors. To ensure that there is no confusion over discrepancies between differently updated or amended versions of the same document, a proper re-issuing and change-control procedure should be adopted.

23.5 THE TENDER DOCUMENT

The tender document, or 'invitation to tender' describes the equipment to be supplied. It is prepared by the network operator whenever a major item of equipment is to be procured. It describes what technical functions the equipment is expected to perform and it includes any commercial conditions which apply to the supply contract. Once prepared, the tender document is sent to one or more prospective suppliers, who consider the requirements and send back formal replies, stating what they are able to offer and the price to be charged if their offer is taken up.

The tender document and the chosen manufacturer's response to it form the documentation of contract. Any subsequent contractual dispute between the network operator and the supplier is resolved by reference to this documentation. From the network operator's point of view, it is important to ensure that the tender document makes the obligations on any contracted supplier absolutely clear. The supplier on his side aims to make it clear in the response, which of the specified items can be supplied and which are not included in the offer. The most harmonious business relationships, and the most successful projects are based on accurate understanding. In the remainder of the chapter we describe items that should to be included in the tender document.

Tender document

Commercial conditions of contract

In the opening section of the tender document a general introduction should be given, explaining briefly the type of equipment required and in what time-scales. This prevents any 'no-hope' supplier from wasting further time by analysing all the detail. The commercial conditions of contract should go on to explain in detail what items of work are the obligation of the supplier, the time-scales for delivery of each item, and the terms on which payment is to be made (for example, installments as each part is delivered, or 'only when the whole job is done'). The commercial conditions also need to lay out the penalties for failure to conform with the requirements of contract. These penalties sometimes include an obligation on the supplier to correct errors and may call for a sum of money to be paid to the network operator as compensation for 'liquidated damage'.

Other general conditions of contract

The general conditions of contract may clarify any other legal or general constraints which the network operator might wish to impose. This section might cover the ownership of any document copyright, the ownership of 'patent rights' and 'intellectual property rights (IPRs)' of any new items that may be developed, and may cover general conditions pertaining to the disclosure of information. The section may constrain the supplier not to make press releases and should cover the procedure for resolution of contract disputes. It could cover both parties' rights to terminate the contract, and (if relevant) the privilege of one party in the event of the other's bankruptcy. Finally, the section may clarify how particular legal constraints on one of the parties affects the contract. For example, let us imagine that the network operator is obliged by law not to release information to a particular third party. This information may, however, be required by the supplier in order that the contract can be fulfilled. In such an instance the supplier should be obliged by contract not to release the information to the third party. This type of situation can easily arise. Imagine a telephone company, A, buying an exchange from company B, where company B is both a telephone network operating company and an exchange supplier. Now company A may need to reveal to its exchange supplier details of its network configuration, and this may include details pertinent to network business held jointly with one of company B's network operating competitors. In this instance company A needs to release details to the exchange supplier part of company B but with the proviso that this information is not passed back to company B's network administrators, who might gain unfair competitive advantage from it.

Technical conditions

The technical part of the specification sets out the type of equipment required and its detailed functions. Once upon a time many of the large public telecommunications operators (PTOs) had a hand in designing the equipment. In those days the specification was at a level of detail of an actual equipment design, almost a detailed circuit and component design, so that contracts were principally for manufacture. Nowadays, however, it is more common for purchasers (including PTOs) to provide a 'functional specification', laying out no more than the broad *functions* the equipment is expected to perform, together with the details of any external interfaces enabling the new device to interwork correctly with other network components. As we have already noted, the use of a functional specification as opposed to an 'equipment design specification' promotes competition amongst prospective suppliers, with a long-term benefit of lower costs and higher quality. Items which may be pertinent in the specification are shown here and are explained below:

● Accommodation and environment

● Network configuration and topology

● Operation and maintenance

● Equipment size and performance

● Functions

- Control interfaces
- Power supply

(i) The *accommodation* and *environment* in which the equipment is to be installed. Any constraining floor layouts, temperature considerations or other adverse conditions (e.g. siting of equipment in a manhole).

(ii) The *network configuration and topology* in which the equipment is to work, including the specification of any interfaces to other network components.

(iii) The *operation and maintenance* requirements. Whether remote control or monitoring of equipment is required. What test points and test equipment is required. Whether internal diagnostic programmes are required. What failure alarms are required.

(iv) The *equipment size* and expected *performance*. For an exchange, the total daily traffic throughput in calls, and the peak hour traffic intensity (number of simultaneous calls) should be quoted, together with performance expectations under overload. The number of busy hour call attempts (BHCA—for circuit-switched networks) or packets (packet-switched) may be important to the dimensioning of the exchange processing capacity, and the total number of daily call attempts may govern how much storage capacity is required for any statistical information which may be required from the exchange. For a transmission system, the bandwidth of the system or the bit rate needs to be quoted, along with its jitter performance, sensitivity to noise, and other relevant information.

(v) The *functions* of the equipment. For an exchange this includes defining the call control and call routing capabilities that are required, together with the method of switching, the method of changing routing data held in the exchange, and the signalling systems to be provided. In addition recorded announcements may be required, along with particular methods or formats for output of traffic recording, call charging, and accounting information. Transmission equipment may need to provide maintenance staff with some means of remote monitoring failures, and of reconfiguring equipment. An automatic switchover to an alternative line system may be required.

(vi) The *control* interfaces. It is increasingly common for network operators to expect to be able to control, monitor and reconfigure the equipment within their network from a single computerized 'control centre'. Any equipment destined for such a network requires the necessary computer interfaces to be built in. These must be covered by the specification. Conversely, the equipment supplied may be controlling other devices, and so it too must conform to a defined control interface. (An exchange, for example, may have to control external echo suppressors or circuit multiplication equipment, and a computer may be required to monitor and control any of a whole range of devices.)

(vii) The type of *power supply*. The supplier may have to provide equipment for working at a given voltage. Alternatively, the power equipment (including batteries and back-up generators) may have to be provided by the supplier.

Often electronic equipment demands 'uninterruptible power supplies (UPS)', since computer data can be lost as the result of instantaneous breaks in power.

Reliability conditions

Reliability expectations should be quoted as a series of measurable parameters, quoted with reasonable target values. Measures such as 'the number of separate instances of failure, measured month on month, shall not exceed...' are much better than more vague parameters, e.g. 'the mean time between failures (MTBF) shall be greater than 3 years'. The problem with the latter statistic is that the long measurement period that is involved makes it virtually unenforceable. Obligations of the supplier on failing to meet the reliability conditions should also be laid out. For systems like public telephone exchanges or high grade computer control systems, where no break in service is acceptable to customers, availability values as high as (or even better than) 99.9 per cent are common. Other considerations of reliability are the conditions and prices for the supply of spares and for after-sales support, ideally lasting for the lifetime of the equipment. The network operator does not want to be left in the position, a short time after equipment purchase, where the supplier changes the product range, and stops producing spares for the now obsolescent equipment.

Project management conditions

The tender document should lay out any special procedures for project management. The purchaser for instance may make conditions on work practices of the supplier, constraining noise, mess, access to site, health and safety. In addition it may be necessary to clarify responsibilities: whether for instance the supplier or the purchaser is to wire-up the equipment on site. Particular installation work practices may have to be adopted. Also, for a large and lengthy project, the purchaser may wish the supplier to build check-points into the development and installation, in order that the rate of progress through the project can be fairly closely monitored.

Quality assurance

The tender document should set down the particular measures that will be applied by the purchaser during 'acceptance testing' to determine whether the equipment is fit-for-purpose. Usually a fair proportion of the monies are withheld until these tests have been completed. The tender document may also mandate the supplier to carry out internal inspections at various stages of development and manufacture, and may demand adherence to a quality work practice methodology such as the British Standard BS5750.

Documentation and training

In some cases, particularly when an 'open tender' is undertaken between a number of competing suppliers, the purchaser may end up choosing a type of equipment that is new to the network. In this case the purchaser requires training for the maintenance and administration staff, and full documentation for later reference. This should be stated in the tender document, so that the price for it gets full consideration. Alternatively, it should be the subject of a separate contract.

Definitions

Finally, it is usual to include a glossary of definitions within the tender document, as a ready reference for terms of jargon or unfamiliar abbreviations. This is merely good documentation practice. The following list illustrates the parts of a complete tender document, but of course the size of it and the omission of any irrelevant parts are purely at the purchaser's discretion.

- Commercial conditions
- Other general conditions
- Technical conditions
- Reliability conditions
- Project management conditions
- Quality assurance
- Documentation and training
- Definitions and glossary

23.6 SUMMARY

In summary, the procurement of equipment can be a relatively straightforward and mechanistic process provided the necessary planning is undertaken at an early date, and provided care is taken in preparing the documentation. The commonest causes of project failure stem from faults in the very early stages. The cost of rectification can be enormous.

BIBLIOGRAPHY

British Standard BS5750 'Quality Systems'. Part 1—Specification for design, manufacture and installation. Part 2—Specification for manufacture and installation. Part 3—Specification for final inspection and test. Part 4—Guide to the use of part 1. Part 5—Guide to the use of part 2. Part 6—Guide to the use of part 3. (BSI Handbook 22, 'Quality Assurance'). BSI.
Burbridge, R. N. G., *Perspectives on Project Management*. Peter Peregrinus (for IEE), 1988.

Camrass, R., *Buying a PABX—How to make a Sensible Choice*. Oyez Scientific & Technical Services, 1983.

Egan, M., *Communications—The most comprehensive guide to the UK Telecommunications Industry (Addresses)*. Macmillan, 1986.

Graither, N., *Production and Operations Management*, 2nd edn. The Dryden Press, 1984.

Gumhalter, H., *Power Supply Systems in Communications Engineering*. Siemens Aktiengesellschaft/John Wiley & Sons, 1984.

Henkin, H., *Drafting Engineering Contracts*. Elsevier Applied Science, 1988.

ISO 9000-series, Standards on Quality Systems.

Lockyer, K., *Critical Path Analysis—and other Project Network Techniques*, 4th edn. Pitman 1984.

Rodgers, D., *PABX Purchase and Installation*. Telecommunications Press, London, 1985.

Part 4

EMERGING
TECHNOLOGIES

24

INTEGRATED SERVICES DIGITAL NETWORK (ISDN)

Many public telecommunications operators throughout the world are committed to modernizing their networks by introducing digital transmission and digital switching into the public-switched telephone network (PSTN). At the same time, many are at various stages of research or development of an integrated services digital network (ISDN) which will allow customers access to a variety of services while reducing the cost of provision of these services both to the administration and to the customer. So what is an ISDN? What is its value? This chapter answers these questions, and goes on to explain the technical detail of how ISDN operates and the user benefits that can be gained.

24.1 THE CONCEPT OF ISDN

The internationally agreed definition of ISDN is a 'network evolved from the telephony integrated digital network (IDN) that provides end-to-end digital connectivity to support a wide range of services, including voice and non-voice services, to which the users have access by a limited set of standard multi-purpose customer interfaces'.

The above definition by CCITT makes a number of points. First, that ISDN requires a digital network. Second, that this digital network is not one between exchanges, but it extends to the customers. Third, it provides not only telephony, but a variety of services. Fourth, a customer does not require a separate interface for each service provided by ISDN, but can use all services via one access point or, at worst, via 'a limited set of standard multi-purpose customer interfaces'. Figure 24.1 illustrates these principles.

Over the last 20 years there has been development of a number of networks dedicated to the transmission of data. These include packet-switched networks, circuit-switched data networks and leased line networks. In addition to these, there is the much older telex network. The provision of these independent networks has made

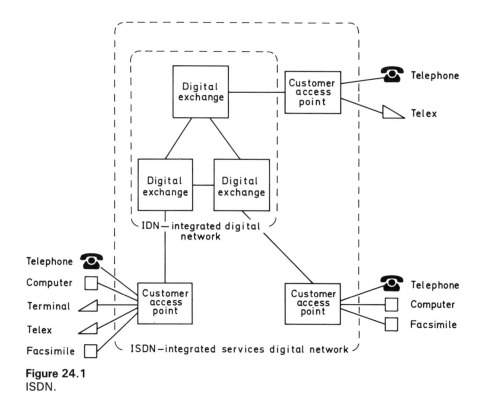

Figure 24.1
ISDN.

each of them relatively expensive, and a great deal of research has been aimed at integrating these networks with the PSTN (public-switched telephone network).

ISDN (Integrated Services Digital Network) will remove the need for telecommunications customers to have separate physical links to each service network. Under ISDN, a single physical connection (or rather one of two types) is provided to a customer's premises and a wide range of services can be made available from it. Further, if required, the services may be used simultaneously, because access is not restricted to each individual service in turn. But as well as greatly enhancing the established services, and reducing their cost of provision, ISDN will introduce a powerful range of new ones.

The services of ISDN are classified into three types, 'bearer services', 'supplementary services' and 'teleservices'. An introduction to each of these precedes our technical and business benefits discussion of ISDN, which takes up the remainder of this chapter.

24.2 'BEARER', 'SUPPLEMENTARY' AND 'TELE' SERVICES

The new ISDN interfaces enable the integration of telecommunications services or 'teleservices' by breaking them down into smaller component parts, called 'bearer' and 'supplementary' services. A 'teleservice' comprises all elements necessary to support an end user's 'particular purpose or application'. Thus examples of teleservices are 'telephony', 'facsimile', 'videotelephone' etc. The specification of a teleservice usually calls upon 'bearer services' and 'supplementary services' but may include additional information specific to the particular application. Thus the 'Group 4' facsimile teleservice specification calls up a 64 kbit/s 'bearer' capability and the CCITT T90 recommendation on image-encoding.

A bearer service is a simple information carriage service, at one of a number of available bandwidths (or rather bit-rates). At its simplest, this might be a normal switched speech (public-switched telephone network) service. At its most advanced, it could be a high-speed, 64 kbit/s data service, capable of carrying a switched 'picture phone' or slow-speed video service. The range of bandwidths available also gives scope for carriage of today's voice-band data (e.g. facsimile), telex and packet-switching services.

'Supplementary services' are so-called because, on their own, they have no purpose or value. They are always provided in conjunction with one of the bearer services, to which they add a 'supplementary' value. Examples of supplementary services are 'call diversion' (redirecting incoming calls to another number) and 'calling-line identity' (the identification of the caller to the called party before or during answer). Another example is the 'ring back when free' service, which saves callers from futile repeat attempts when the called party is busy, by leaving the network to establish the call as soon as both parties are free. In all of these examples, the associated bearer service is 'speech'. In some networks, some supplementary services are available even in the PSTN, but for many more ISDN's supplementary services will add a powerful new dimension.

Summarizing, examples of the three types of services included as component parts of ISDN are:

1. Bearer services

Speech	at an appropriate bit-rate.
3.1 kHz	the bearer service required for voiceband data—i.e., dial-up modems and analogue facsimile machines.
64 kbit/s	high-speed data.

2. Supplementary services
   call diversion;
   calling line identity;
   ring back when free.

3. Teleservices
   facsimile;
   telephony;
   videotelephone.

24.3 ISDN INTERFACES AND END-USER APPLICATIONS

In order to use the ISDN, customers either have to re-equip themselves with PBXs and user terminals designed with one of the new ISDN interfaces (of which there are currently two as explained below), or use 'terminal adapters' in conjunction with existing terminals. A simple terminal might be an ISDN telephone, or an ISDN telex terminal. A more advanced device might be a personal computer with sophisticated file-transfer capabilities, or a 'group 4' facsimile machine offering high-speed facsimile service of almost photocopier quality.

Worldwide standards for the new ISDN network interfaces are set by the I (ISDN), G and Q series of CCITT recommendations. Two types of customer-to-network interface are specified, these being the 'basic rate' and 'primary rate' interfaces, sometimes shortened to BRI and PRI.

Both basic and primary rate interfaces comprise a number of B ('bearer' or 'user information') channels plus a D ('data', but best thought of as a 'signalling') channel. The B channels carry the bearer services while the D channel enables the customer to signal to the network how each bearer channel is to be assigned and used at one time (for example, 'connect bearer (B) channel number 1 to Mr A for telephone service').

24.4 BASIC RATE INTERFACE (BRI)

The BRI (or BRA—basic rate access) comprises three channels, so-called '2B + D', and is the interface that is used to connect most end-user terminal equipments, either directly to the public network or via a company private branch exchange (PBX). The bandwidth itself can be made available to the terminal in a number of different ways, using any of the 'S', 'T', or 'U' variants of the interface, as the diagram of Figure 24.2 shows. Each of the variants S, T, and U are 'basic rate interfaces' but they differ in their physical realization.

The *T interface* provides for a connection point between new ISDN terminal equipment (TE) and the ISDN network terminating equipment (NT1). The NT1, provided as part of the network operator responsibility in Europe, is merely an exchange line terminating device.

The *U interface*, often known as the 'wires only' interface, is the 2-wire line, 2B + Q, interface, up to about 7 km in length, incoming from the public exchange, and designed to work as far as possible on existing 2-wire copper pairs. Its definition came about because of regulatory requirements in the United States which debar PTOs from offering the S/T interface which the European PTTs intend to offer, because in the United states the NT1 is deemed to be 'customer premises equipment'. The U interface will also be important for cellular radio based ISDNs.

The *S interface* is physically and electrically identical to the T interface, but for the academics amongst us it is the name we give to the interface after it has passed through a call routing equipment (NT2). An example of an NT2 is an ISDN PBX. Because S and T interface are the same, some people refer to the 'S/T interface', or sometimes the 'S_0 interface'.

The final interface shown in Figure 24.1 is the *R interface*. This is the general nomenclature used to represent any type of existing telecommunication interface (such as X25, RS232C or X21, as we discussed in Chapter 9). By using a terminal adapter (TA) we make possible the use of the ISDN for carrying information between existing data, and telephone terminals—thus easing the user's problem of deciding when to change over from older style networks (e.g. the telephone or public packet data network).

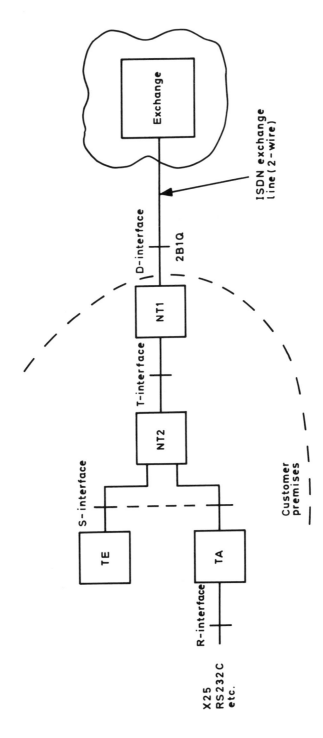

Figure 24.2
The basic rate interface (BRI). NT1 = Network Terminating Equipment 1; NT2 = Network Terminating Equipment 2; TE = Terminal Equipment (ISDN terminal); TA = Terminal Adaptor.

24.5 THE S/T INTERFACE SPECIFICATION

The S/T interface is a 4-wire or equivalent interface, allowing 192 kbit/s data carriage in both directions. The 192 kbit/s comprises 144 kbit/s of 'user information' (data supplied by the terminal), together with an overhead of 48 kbit/s which is needed to administer and control the interface, to control the 'passive bus', for example as described below.

The 144 kbit/s is used by the terminal as two B channels (both 64 kbit/s) and one D channel (16 kbit/s). Either or both of the B channels may be in use at any one time, and when used simultaneously are not constrained to be connected both to the same destination.

The passive bus is a feature of the S/T interface that enables up to eight terminal devices to be connected simultaneously to the same basic-rate interface. It is shown in Figure 24.3.

The capability for multiple-terminal connection to the BRI in this manner is crucial to the support of incoming calls since, prior to the receipt of an incoming call, it is not possible to know which device (e.g. facsimile machine, telex machine, telephone) should be connected to the line, because the type of the sending device is not known. A procedure called 'terminal compatibility checking' which is part of the D-channel signalling ensures that the call is answered only by the correct terminal device (i.e. the one of up to eight connected to the bus which is compatible with the sending device).

The S/T interface and the passive bus are intended for within-building use on customer premises. They are defined by CCITT Recommendation I420. They are provided directly from the NT1. On the 'terminal-side' of the NT1, there may be one of three wiring configurations:

(i)   a short and full facility passive bus up to 150 m in length;

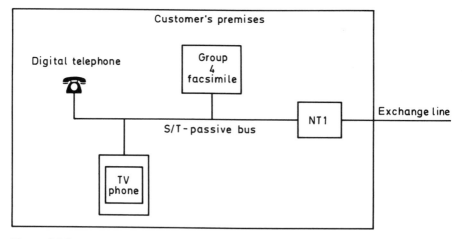

Figure 24.3
Passive bus at a customer's premises.

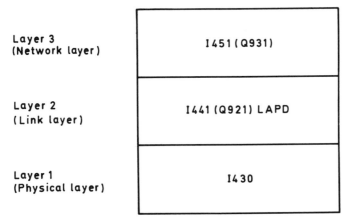

Figure 24.4
Protocol stack in the ISDN D channel.

(ii) an extended passive bus up to 500 m in length;

(iii) a single point access line up to 1000 m in length.

Protocols defined for use over the basic rate interface are defined by CCITT Recommendation I420. (The European version of it is called 'NET3'.) The B channels are left as clear channels for users, while a three-layer protocol stack, conforming to the first three layers of the OSI model, is tightly defined for the D channel, as Figure 24.4 shows.

The physical layer protocol (CCITT Recommendation 1430) provides a 'contention resolution scheme' to ensure that messages from different terminals on the passive bus do not collide (Recall the CSMA/CD technique for LANs, discussed in chapter 9.) At the second layer (I441 or Q921) the protocol is called LAPD (Link Access Procedure in the D-channel). Like HDLC, the procedure controls the data flow over the D channel, ensuring that the data buffers are not overfilled and that the synchronization of data is maintained. It includes some degree of data error checking. In particular, LAPD provides for a number of connections to be simultaneously maintained with the multiple terminals on the passive bus. Finally, the network layer protocol (layer 3 protocol) is defined by CCITT Recommendation I451 (and duplicated in Recommendation Q931). This protocol allows the transfer of dial-up signalling information and for the control of B channel connections. It allows the user's terminal to negotiate with the network for the setting up of an appropriate terminal device at the destination end. It is this protocol that includes the 'terminal compatibility' checking procedure.

24.6 USE OF THE BASIC RATE INTERFACE

Given the small number of channels, the BRI is well suited for digital telephone customers, computer networks, small-business users and sophisticated residential customers. (An example might be a businessman accessing his company computers from home.) *Circuit-switched calls* at any of the three bit rates corresponding to speech, voice-band data (e.g. facsimile) or 64 kbit/s data (e.g. inter-computer links) are set up over one of the B channels, using the D channel for signalling the request and managing the connection. The ISDN exchange responds to such requests by extending the relevant B channel using a circuit-switched inter-exchange connection through to the destination, as indicated by the dialled ISDN number. (The ISDN number is similar to the PSTN number—see Chapter 14.) However, although the customer-to-customer interface will use a B channel with 64 kbit/s capability, not all of this bit rate will necessarily be extended. Instead, within the interexchange part of the ISDN, a connection with a bit rate appropriate to the requested bearer service is provided. Thus in the case of a 64 kbit/s bearer service, a 'clear' digital 64 kbit/s path will be chosen. However, when only 'speech' bearer service is requested, the exchange may select either a standard digital telephone channel of 64 kbit/s or may choose instead a lower bit rate connection, routed via circuit multiplication equipment

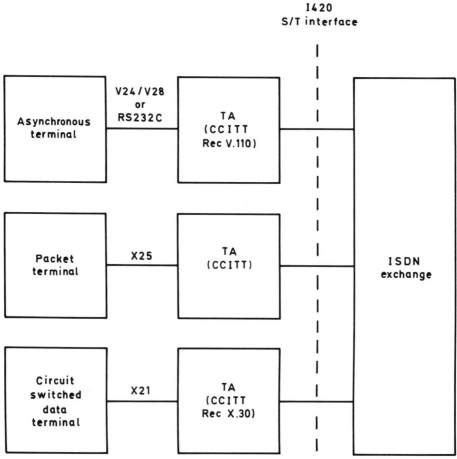

Figure 24.5
Common ISDN terminal adapters.

(CME—see Chapter 20) or perhaps even over analogue transmission plant. Similarly, for 3.1 kHz bearer service, a connection of at least 3.1 kHz bandwidth will be selected.

Calls connected over ISDN in a circuit-switched manner include telephone calls, telex calls and 64 kbit/s circuit-switched data calls. *Packet-switched* connections may also be established over ISDN, either using the B channel or the D channel. In the B channel mode, a circuit-switched (B channel) connection is first established to a nearby PAD (packet assembler/dissembler) on the packet network, and then X25 protocols are used during the 'conversation' phase of the call. However, a more efficient method of packet-switched connection is to use the D channel. In this instance the user's data packets are interleaved with other signalling messages on the D-channel, and are transferred between ISDN exchanges either using the CCITT 7 inter-exchange signalling network or via the established X25 packet network.

24.7 ISDN TERMINALS

Terminals used in conjunction with the BRI must conform with the new I420 S/T interface. This interface alone is not sufficient to ensure the correct operation of

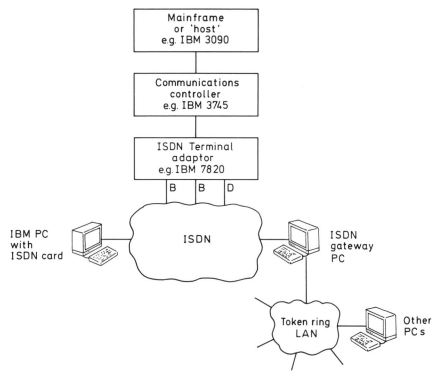

Figure 24.6
ISDN in an IBM computer network.

Figure 24.7
ISDN terminal in use. ISDN telephones on executive secretaries' desks at Bell Laboratories in Indian Hill, Illinois, USA. These 40-button digital telephones provide such features as calling number identification, displayed on the liquid crystal display. (*Courtesy of AT&T.*)

terminals for particular applications, since it only ensures correct establishment of the connection between like terminals. In addition, higher layer protocols (corresponding to the higher application layers of the OSI model) must be defined so that the two end terminals can interpret the data they are sending between one another. For example, two TV phones operating over the ISDN must use the same picture coding technique (OSI layers 6 and 7) if the picture is to be transferred successfully. Unfortunately the agreement of such standards (spanning all conceivable applications—particularly those using the full capabilities of the 64 kbit/s data rate) has been slow, and to date few applications other than reliable old telephony (using A-law or Mu-law PCM) and 'group 4 facsimile' (using CCITT Recommendation T90 code) are standardized. Others likely to be available soon include 64 kbit/s TV phone (videotelephone) and 64 kbit/s personal computer file transfer protocols.

Developers of the ISDN recognized early on that the development of, and investment in, new terminals would be a constraining influence on ISDN take up, and so a range of 'terminal adapters (TAs)' have been defined in CCITT recommendations. These convert established interfaces into a suitable format for connection to the ISDN S/T interface.

The diagram in Figure 24.5 shows some of the possible terminal adapter configurations, giving the CCITT recommendation numbers pertaining to them. Figure 24.6 illustrates a typical mainframe and personal computer network of the future employing IBM equipment, the ISDN and terminal adapters. (For a recall of the IBM network architecture return to Chapter 9.)

24.8 PRIMARY RATE INTERFACE

The PRI (also the PRA, or Primary Rate Access) is a 4-wire interface optimized for connection between the public network and customer PBXs, and is best suited to the needs of medium to large customers. The format of the European PRI is 30B + D (and is specified in 'NET5'); i.e. 30 separate 64 kbit/s B channels, plus one 64 kbit/s D channel. (Note that the D channel is given greater capacity than in the BRI, commensurate with the larger number of channels to be controlled. However both PRI and BRI use a similar protocol stack.) The PRI format has been carefully designed to conform with the standard 32-channel, 2048 kbit/s pulse-code-modulation transmission format specified in CCITT Recommendation G703. In order also to conform with the CCITT 24-channel PCM standard (1544 kbit/s), an alternative primary rate interface of 23B + D is also defined. This will be the one used in North America and Japan.

The primary rate interface is covered in CCITT Recommendation I421 but is not yet fully defined. Early national versions are more fully defined and specified by national standards bodies. The PRI is illustrated in Figure 24.8, where the PBX is providing a concentration function between a number of end users individual BRIs and the network PRI.

For economic reasons, the exact traffic value at which a customer should subscribe to PRI service as opposed to multiple BRI service depends on the pricing policy of the local administration. Typically, PRI should be worthwhile at values above about 10 erlangs (10 simultaneous calls—say 20 exchange lines or even less).

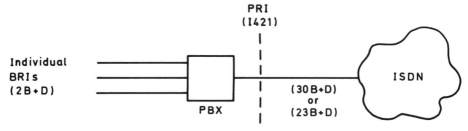

Figure 24.8
Primary rate interface (PRI).

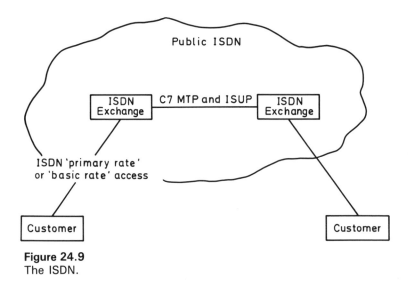

Figure 24.9
The ISDN.

24.9 THE PUBLIC NETWORK AND ISDN

A radical new style of customer/network interface as implied by ISDN must be accompanied by equally radical changes to the public network. The exchanges must be fully digital, switching individual channels at 64 kbit/s, and must support a common channel signalling protocol between exchanges. The CCITT 7 Integrated Services User Part described in Chapter 28 (C7 ISUP: Recommendations Q761–Q764), is intended for the support of ISDN. This must be used with the CCITT 7 'Message Transfer Part' (MTP) and a network of digital transmission links illustrated in Figure 24.9.

An alternative signalling system to C7 ISUP which may be used as an interim measure is the TUP+ (Telephony user part enhanced) signalling system, specified by CEPT/ETSI particularly for use in Europe. TUP+ provides for restricted ISDN capabilities that will be superseded by development of the ISUP. None the less it is expected to be the main intra-European signalling system for ISDN in the early 1990s.

24.10 DEPLOYMENT OF ISDN

In any modern network, there needs to be a staged plan for the changeover from analogue to digital communications and a further plan for the introduction of ISDN. Since ISDN requires a digital network, and since it will be paid for mainly by the business community, it may be possible for a network operator to arrange for the first digital exchanges, with digital transmission links interconnecting them, to be introduced in business areas. Then, ISDN can be provided at marginal cost. Thus, while the changeover takes place in the rest of the country, the business community is already enjoying most of the benefits of ISDN, and the national economy is enjoying the investment which it generates.

It must be realized that the largest business investors are the multinational companies. While they require good national telecommunications, technology is now such that they are coming to expect the same facilities internationally. The exchange of data between computers in different countries is already commonplace, and facilities such as international videoconferencing and fast facsimile will rapidly become requirements of many large companies.

A national ISDN is not enough, and a PTO must be confident that it can interface with ISDNs in other countries. How can this be guaranteed when telecommunication system manufacturers in various parts of the world are independently developing their products? The answer is that all systems should work to the international standards recommended by the International Telegraph and Telephone Consultative Committee (CCITT). The first ISDN standards were documented in CCITT's 'Red Book' in October 1984. They were updated in the 'Blue Book' of 1988, and intense effort is continuing to ensure that different countries do not install incompatible systems. Even so, there are still significant differences between the various early national ISDNs. This may mean that terminal equipment compatible with one network is incompatible with that provided in another.

Given the fact that CCITT standards for ISDN are still not finalized, there are a number of reasons why a PTO may have elected to develop its own pre-standard national ISDN:

(i) First, it gives a PTO an opportunity to gain experience with the new technology and simultaneously start to recoup some of the huge development costs already incurred in ISDN development and the installed digital network.

(ii) Second, it provides an interim solution to the high-speed digital connection services that some major customers have been demanding, and helps to stimulate end-user terminal manufacturers to invent and develop new applications that may have not previously been possible.

(iii) Third, it breaks the deadlock between CPE manufacturers and network operators, both waiting in a 'chicken-and-egg' fashion for the other to make the first ISDN move. After all, ISDN terminals are of no value without an ISDN, and an ISDN is no use without terminals.

The appearance of CCITT's 1988 recommendations, including stable protocols for basic call control in both the Q931 (D channel) and the ISUP (inter-exchange) signalling systems, means that we can expect widescale ISDNs with international interconnections to start to appear in the early 1990s. The interest and intervention of some governments, who are eager to ensure the existence of good telecommunications networks, including ISDNs, will see to that! Thus the European Economic Community (EEC) aims to create a pan-European ISDN by 1992—in time for its Single European Market (SEM) due in December 1992.

24.11 THE MARKETING OF ISDN AND THE EARLY USER BENEFITS

The early years of ISDN are likely to see marketing stress and user benefit in two main areas:

(i) The quality, cost, and 'frills' benefits made possible for telephone users of the ISDN.

(ii) The new high-speed data (64 kbit/s) applications (e.g. high speed and high quality facsimile, fast computer file transfer, videotelephones etc.).

The quality improvements to ordinary telephone calls result from the introduction of digital transmission to the local access line, historically a major source of signal 'noise' (e.g. 'fried eggs' noise). Cost benefits are likely to accrue to PBX users of the primary rate interface (PRI), since the PTO can pass on the cost advantage of getting 30 telephone circuits (Europe) (23 circuits in North America and Japan) from a single 4-wire access line.

Thus a cost saving should be possible by turning 30 (or 23) analogue exchange lines over to ISDN primary rate interface. A problem facing some PTOs, however, is that 1.5 Mbit/s or 2 Mbit/s digital access lines may already be available to customer PBXs, and the extra cost of ISDN signalling upgrade may not be perceived by the customer to be justified by the additional service benefits of ISDN.

ISDN service benefits form the central theme of many PTOs marketing campaigns for ISDN. These will accrue from the new supplementary services as they become available—e.g. 'ring back when free', 'calling line identification', 'call redirection'. The latter may be particularly useful to businesses receiving calls from customers, since it could be used for automatic presentation of the customer's details on a computer screen—ready for quick reference. A new interface, designed to use ISDN-like signalling and called ASAI (adjunct/switch application interface), will enable adjunct devices (e.g. computers) attached to ISDN exchanges to control or respond to supplementary service and other ISDN signals. Thus an ISDN PBX receiving a call and an associated calling line identification signal, could trigger action from an adjunct computer connected to the PBX. The call receiver might then be provided with a range of information on his computer screen (e.g. customer name and address)—simultaneously being presented with the call. Alternatively, the adjunct computer might take charge of the call—instructing the PBX to divert the call or take some other action.

New ISDN data applications can be targeted and promoted in an entirely different fashion, aimed at customers' needs that had not previously been met. The explosion of interest in facsimile, for example, and the continuing growth in courier and other forms of document transfer suggest that the new high-quality 'group 4' facsimile machines will enter a fertile market. Another likely early use is full motion but low quality videoconferencing and picture phone devices.

Perhaps the simplest, and single most beneficial effect of ISDN will be the common specification for worldwide network access. Ultimately this will mean that any I420 or I421 standard device can be connected to any ISDN anywhere in the world. On the other hand, slight local variants have always thwarted these ideals in the past and the cynical amongst us might say it will happen again. It is clear that ISDN needs the wholehearted commitment of network operators, terminal manufacturers and users alike—without it ISDN will remain only a subject of marketing 'hype' for years to come.

A company's ISDN portfolio will depend on its assessment of its important markets and applications and upon its networking constraints. Thus some have decided to push the 'improved telephone' aspect of the primary rate interface first, while others have concentrated on the new high-speed data applications.

Given the difficulty in matching up and interconnecting early ISDNs, a typical evolution path is likely to be: first—an unsophisticated 64 kbit/s switched service, followed later by the supplementary service benefits of fully-developed ISDN signalling. Finally, consistent international services will be possible.

24.12 NETWORK INTERWORKING

Network operators need not switch instantly from their current networks to an entirely new ISDN. The provision of network interworking avoids this need. Many administrations are providing early ISDN services simply by deploying a small overlay network, serving only a restricted number of ISDN customers, but allowing cross-connection to customers of existing telephone and data networks. This has the benefit of constraining the amount of early investment, but has the disadvantage of having to interwork with a number of separate existing networks, each with its own protocols (as in Figure 24.10). Such an overlay network allows an ISDN customer to communicate with customers on all the other networks, but it may have severe service constraints.

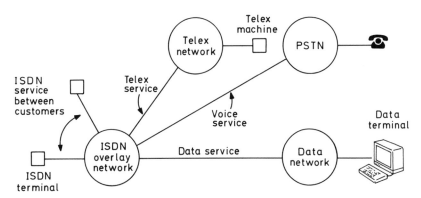

Figure 24.10
ISDN interworking.

24.13 COMPANIES' PRIVATE ISDNS

ISDN in the private network arena will be as important as public ISDN. After all, it is the needs of the major telecommunications users that it aims to serve. Companies thus face a major strategic decision on when to start their (expensive) network evolution to ISDN and what technical standards to adopt. The decision needs to be tempered by the geographical locations in which they operate and the demands of the telecommunications services they use.

World standards for private ISDN are lagging behind those already developed for the public network, and a range of 'interim' and 'proprietary' standards face the private network planner in 1990.

The standards bodies have just commenced work on an adapted version of the PRI for use on private network connections between PBXs. The interface is known generically as the 'Q interface', and is illustrated by Figure 24.11.

Work commenced on the Q interface in 1988, and the signalling system to be used over it has come to be called Q-sig. In its earlier absence, the established *de-facto* standard for inter-PBX ISDN signalling is the 'Digital Private Network Signalling System' (DPNSS), which was originally defined by PBX manufacturers and network operators led by British Telecom in the United Kingdom.

Another interface of importance to private ISDN operators will be that enabling a mainframe computer to be connected to an ISPBX for high-speed data transfer over

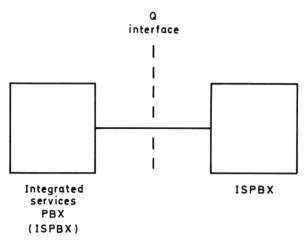

Figure 24.11
The Q interface between ISPBXs.

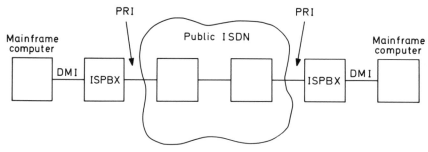

Figure 24.12
AT&T's digital multiplex interface (DMI).

the ISDN. This arena also is the realm only of proprietary standards at the moment, and AT&T's Digital Multiplex Interface (DMI) provides one solution, as shown in Figure 24.12.

Finally, we should not leave the subject of private companies' ISDNs without at least mentioning the scope for direct connection of company telephone exchanges with the public network using signalling system 7. Such access is likely to be available at least in the United States. The prime benefit is the increased network control capabilities of CCS7 (common channel signalling system 7) when compared with the PRI.

24.14 'BROADBAND' SERVICES OVER ISDN

Despite the fact that 64 kbit/s is a high bit rate compared with those possible over many established data networks, it is none the less inadequate for many applications of very fast inter-computer communication or video image transfer. In this context, the type of ISDN we have described in this chapter is sometimes called 'Narrowband ISDN'. The emergence of new broadband ISDNs promises much greater bandwidths as Chapter 29 reveals. However, in the shorter term, an adaptation of narrowband ISDN gives scope for greater bit rates than 64 kbit/s simply by combining an integral number, n, of individual 64 kbit/s channels as '$n \times 64$ kbit/s' bit rate path (i.e. 128 kbit/s, 192 kbit/s, 256 kbit/s etc). This is not quite as easy as it sounds. In concept it may seem simple to take consecutive bytes of a 128 kbit/s bit stream and feed them alternately on to two independent 64 kbit/s paths (a sort-of 'reverse-TDM' process). In practice, having done so, it is much more difficult at the receiving end to ensure that the bytes carried on the two separate paths go back into a single string in the correct sequence. We have to 'maintain bit sequence integrity'. Figure 24.13 attempts to illustrate the problem.

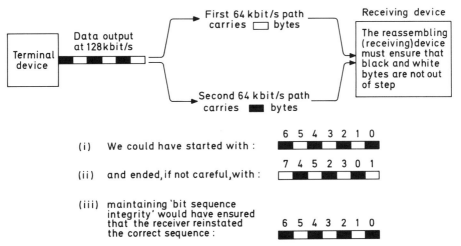

Figure 24.13
Maintaining bit sequence integrity.

As we continue down the evolution path of ISDN, the signalling techniques protocols, and new terminal applications will be the only limiting factors in determining its success and its rate of take-up.

BIBLIOGRAPHY

Bocker, P., *ISDN—The integrated Services Digital Networks: Concepts, Methods, Systems.* Springer-Verlag, 1988.

CCITT I-Series Recommendations.

CCITT Basic Rate Interface (BRI) Recommendation I420.

CCITT Primary Rate Interface (PRI) Recommendation I421.

CCITT 7, ISUP (Integrated Services User Part) Signalling System. CCITT Recs Q760–9.

CEPT TUP + number seven signalling system, CEPT T/SPS 43-02.

DASS2 (Digital Access Signalling System)—British Telecom Network Requirement (BTNR) no. 191.

DPNSS (Digital Private Network Signalling System)—British Telecom Network Requirement (BTNR) no:190.

Franco., G. L., World Communications—ways and means to global Integration. Le Monde Economique, 1987.

Griffiths, J. M. (ed.), *ISDN Explained: Worldwide Network and Applications Technology.* John Wiley, 1990.

Heldman, R. K., *ISDN in the Information Marketplace.* TAB Books, 1988.

Kitahara, Y., *Information Network System—telecommunications in the Twenty-first Century.* Heinemann, 1983.

Network Interface Description for [USA] *ISDN Customer Access* [U reference point]. Bellcore TR-TSY-000776 Issue 1. August 1989.

NET3—European Basic Rate Interface for ISDN (produced by ETSI—see Chapter 21).

NET5—European Primary rate Interface for ISDN (produced by ETSI—see Chapter 21).

Primary rate Interface (PRI) Capabilities and features. AT&T TR41459, 1989.

Ronayne, J., *The Integrated Services Digital Networks: From Concept to Application.* Pitman, 1987.

Rutkowski, A. M., *Integrated Services Digital Networks* Artech House, 1985.

Sarch R. (ed.), *Integrating Voice and data.* McGraw-Hill, 1987.

Stallings, W., Tutorial—Integrated Services Digital Networks (ISDN). IEEE/Computer Society Press, 1985.

25

THE MESSAGE HANDLING SYSTEM (MHS)

The message handling system is a concept that will ultimately lead to the interconnectivity of all different types of message conveying systems—e.g. telephone, telex, facsimile, electronic mail etc.

To commence the long road towards development of all the necessary technical interfaces that will be needed, MHS sets out a simple model of basic interconnection between systems. As it does so it defines a new dictionary of standard terms and jargon to describe the various phases of a communication. In places it may even seem trivial, but the careful definition is worthwhile if it will solve today's incompatibilities. This chapter seeks to explain the model, unravel the jargon and describe the initiatives that will result.

25.1 BACKGROUND

It is commonplace today for a wide variety of telecommunications technologies to be provided in the same office. Often a number of technologies provide similar but separate means for conveying identical information, but many users still employ more than one of the methods. Telex, facsimile and electronic mail all provide means of transferring text or documents. One or other of the different devices may be favoured for conveying information to a given destination, depending upon the facilities available at the far end, for despite the commonality of each of the different methods of document transfer, they are mutually incompatible and will not interwork.

MHS defines a simple model of the procedure needed for documents to be passed electronically in a 'store-and-retrieve' fashion between a wide range of different types and makes of office business machines (e.g. word processor, facsimile machine, teletype—maybe even the telephone as well). In time, the model will lead on to the development of communication standards, interfaces and equipment needed in order that machines of dissimilar types may intercommunicate.

25.2 THE CONCEPT OF MHS

The underlying assumption on which MHS is founded is that during any period of communication only the information content of a message is important. The corollary is that the means of message conveyance is immaterial, and the format of the information (e.g. facsimile, telex, electronic mail etc.) may be changed between the originating and destination stations. Thus the originator need not worry about whether a compatible facsimile or telex machine is available at the destination, since MHS will provide conversion to the required format.

CCITT's X400 Recommendation defines a model of a 'message handling (MH) service', provided by means of a 'message handling system' (MHS). The MH Service allows the conveyance of messages across a network on a 'store-and-forward' or 'store-and-retrieve' basis. Thus information between sender and recipient can be sent without interrupting the recipient from his current activities: the message is dealt with when it is convenient. This is important in computing activities if the receiving machine is already busy. It is also important for human recipients who might be away from their desk. Much like a postal system, messages may be 'posted' into the system at any time of day and will be delivered to the recipient's 'mailbox'. The recipient is free to sort through the incoming 'mail' at any time of day, and as soon as an item is accepted by the designated recipient, automatic confirmation of receipt can be despatched back to the sender. This eliminates any possible concern about whether the recipient has seen a given item.

25.3 THE MHS MODEL

The MHS model provides a design aid to the development of networks providing message handling services. It is based upon the principles of the International Organization for Standardization's 'Open Systems Interconnection' (OSI) model (as described in Chapter 9). Like OSI, MHS is a layered model with general purpose application. Unlike the OSI model, only 'high level' functions are described in the MHS model. In fact, the MHS model is functionally equivalent to the OSI layer 7, or 'application' layer.

Any realization of a higher layer protocol of the OSI model relies upon implementation also of the underlying layers. The message handling system is no exception. The MHS defines the format of messages that may be sent between end-points and the procedure for doing so. It does not state any particular physical or electrical means of conveyance. Thus any information which may be represented in binary code may be conveyed on a MHS. Equally, any network capable of carrying the information and conforming to OSI layers 1–6 will do. The network must:

(i)   establish the transport service (OSI layers 1–4). Suitable protocols in conjunction with any of the telephone, packet-switched data, LAN or leased line networks will do;

(ii) establish a 'session' for 'reliable transfer' of messages (OSI layer 5);

(iii) but MHS uses a null 'presentation' (data syntax) layer (OSI layer 6).

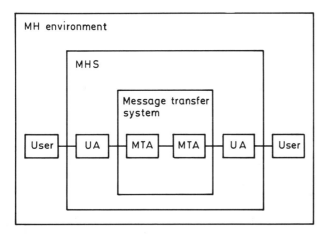

Figure 25.1
Functional model of MHS. MH, message handling; UA, user agent; MTA, message transfer agent. (*Courtesy CCITT—derived from Figure 1/X400.*)

Figure 25.1 shows how MHS operates to provide a service to users within a 'message handling environment'. All the functions within this environment need to be standardized. For this purpose the model defines a number of terms, as follows.

The two basic functions of MHS are the 'user agent' (UA) and the 'message transfer agent' (MTA). These are not usually distinct pieces of equipment in their own right, but a number of these functions may be contained within a single physical piece of equipment. The user agent (UA) function is undertaken by a personal computer (or similar piece of office equipment) which converts the user's message into a standard form, suitable for transmission. Having done the conversion, the user agent 'submits' the message to the local 'message transfer agent' (MTA). The message is conveyed via a number of MTAs, collectively called the 'message transfer system', to the recipient's MTA. This is called the 'message transfer function'. Each MTA is usually a mainframe computer or other device capable of storing and forwarding information. Interconnection of the UA to MTA and between the MTAs may use any type of network which provides the functions of OSI layers 1–6.

The two services provided by message handling systems are called the 'inter-personal message services' (IPMS) and the 'message transfer service' (MTS). These correspond to the services provided by the user agent and the message transfer agent respectively, as Figure 25.2 demonstrates.

Figure 25.2 shows three different implementations of user agent (UA). This underlines the point that the user agent is only a conceptual function, which can be implemented in a number of ways provided it interacts with the MTA function in the standard fashion. The three different implementations of user agents shown in Figure 25.2 are:

(i) A user agent within a computer or word processor. The IPMS provided by this machine is utilized by a dumb (asynchronous) computer terminal.

(ii) An intelligent terminal, within which the user agent function is contained.

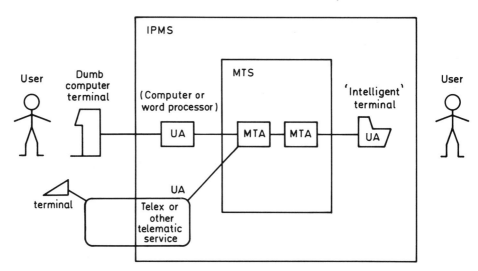

Figure 25.2
Message transfer (MTS) and interpersonal message service (IPMS).

(iii) An interworking device, interfacing an existing telematic (text based) network (telex or facsimile etc.) to one of the MTAs.

The end-users of a MHS may be either persons or computer applications. Users may be either 'originators' or 'recipients' of messages. Messages are carried between originator and recipient in a sealed electronic 'envelope'. Just as with the postal equivalent, the envelope provides a means of indicating the name of the recipient, and of making sure it reaches the right address with the contents intact. In addition further information may be added to the envelope for special delivery needs. For example, a confidential marking could be added to ensure that only the named addressee may read the information.

The basic structure of all messages in X400 is thus an 'envelope', with a 'content' of information. The 'content' is the information sent by an originating user agent (UA) to one or more recipient user agents (UAs). This simple message structure is shown in Figure 25.3.

MHS defines three different types of envelope (although all three are very similar in structure). These are called the 'submission envelope', the 'relaying envelope', and the 'delivery envelope'. The submission envelope contains the information prepared by the UA, and is labelled with the information that the MTS will need to convey the message, including the address and any markings such as 'confirmed delivery' and 'priority'. This envelope is 'submitted' by the user agent (UA) to the message transfer agent (MTA). Upon reaching the MTA, the content of the message is transferred to a 'relaying envelope' for relay through the message transfer system to the destination MTA. Finally the message is delivered to the recipient's user agent in a delivery envelope.

In reality, the 'envelopes' are simply some extra data, sent in a standard format, as a sort of package around the user's real message. All three types of envelope are very similar in structure, though slightly different information is pertinent in each case.

(Content)   (Envelope)

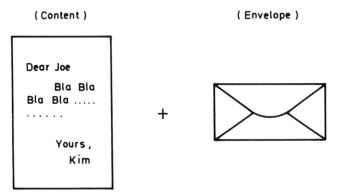

Figure 25.3
Basic message structure in MHS. (*Courtesy of CCITT—derived from Figure 2/X400.*)

25.4 LAYERED REPRESENTATION OF MHS

As with the OSI reference model, it is easy to conceive MHS as consisting of functional layers for which:

● Similar functions are grouped in the same layers.

● Interactions between layers are minimal.

● Interactions within the layers are supported by one or more distinct 'peer protocols'.

The principles of layered protocols were covered in Chapter 9, where the OSI model and 'peer protocols' are described. In particular, MHS comprises two functional layers, as shown in Figure 25.4, and since MHS itself resides at layer 7 of the OSI model, these two layers are sub-functional layers of OSI layer 7.

A number of new terminologies are introduced by Figure 25.4 and need explanation. The two layers of MHS are the user agent layer (UAL) and the message transfer layer (MTL). These are the layers in which the user agents and message transfer agents operate.

The user agent layer comprises 'user agent entities' (UAEs) and an interacting protocol, called 'Pc'. The protocol 'Pc' governs the content, format, syntax and semantics to be used when transferring information between the two user agents. (Pc stands for 'content protocol'.) A particular type of content protocol is called 'P2', and is also known as the 'interpersonal messaging protocol'. It is covered by CCITT's Recommendation X420. The 'user agent entity' is the capability, within the user agent which is required to convert the base information into the form demanded by Pc (or P2).

The message transfer layer (MTL) comprises 'submission and delivery entities' (SDEs), 'message transfer agent entities' (MTAEs) and interacting protocols called 'message transfer protocol' (or 'P1'), and the 'submission and delivery protocol' (or 'P3').

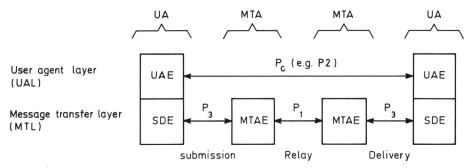

Figure 25.4
Layered representation of MHS. (*Courtesy of CCITT—derived from Figures 12 and 13|X400.*)

The submission and delivery entity (SDE) is the part of the user agent that places the information (or 'content') provided by the user agent entity (UAE) into the 'submission envelope'. The envelope is then addressed using the protocol P3. The 'message transfer agent entity' (MTAE) is the functional part of the 'message transfer agent' (MTA) which relays the message in the 'relaying envelope'. The 'message transfer protocol', P1, is used to perform the message relay. Finally the message is delivered to the destination SDE in the 'delivery envelope', using protocol P3. The full messaging sequence is thus:

(i)   The user agent entity (UAE), which is a functional part of the user agent (UA), assists the user to compose the 'content' of the message in P2 (or some other Pc) format.

(ii) The submission and delivery entity (SDE), also a functional part of the user agent, envelopes and addresses the message, and then 'submits' it to the message transfer agent entity (MTAE) using P3 protocol.

(iii) The message transfer agent entity (MTAE), which is a functional part of the message transfer agent, re-envelopes the message in a P1 'relaying envelope', and relays it in a store and forward fashion via the MTAEs of other MTAs, until reaching the destination MTA. For the relay, protocol P1 is used.

(iv) The destination MTAE returns the message to its P3 envelope and delivers it to the SDE of the destination UA, which removes the envelope.

(v)   The UAE part of the destination UA reformats the information to the original style and conveys it to the user.

As an aside, it is worth noting at this point that MHS is still continuing to be developed by CCITT, and whereas CCITT's 1984 version contained a version of P3 protocol which was a store-and-forward protocol, this had the weakness that it had no way of preventing an MTA from attempting to forward a message to a UA, such as a personal computer, which might have been switched off at the time. In that case the message would be lost. The more recent 1988 version of the P3 protocol has been further developed. P3 + (as it is known) is a 'store-and-retrieve' protocol. In this version, the message is retained by the destination MTA until the recipient user agent chooses to retrieve it. P3+ appears as part of CCITT's 1988 recommendations.

25.5 THE STRUCTURE OF MHS MESSAGES AND MHS ADDRESSES

Let us now consider the structure of MHS messages and the structure of messages which are used to support the interpersonal message service (IPMS). Figure 25.5 shows the imaginary nested structure of a message, sent in reality as a long stream of binary coded data.

The structure is apparent from Figure 25.5, although some other parts of the message are also shown. The content may be either user information, such as text, facsimile diagrams etc., formatted in the P2-style as 'heading' and 'body', or it can take the form of a status report. A status report is a special type of message required for the correct functioning of UAEs. A status report may be used for instance, to convey 'receipt confirmation messages'. The 'heading' of the message characterizes the user information with elements such as 'subject', 'to', and 'copy'. The body is the text and diagrams. The content of the message must be formatted using a content protocol, Pc. One type of content protocol already discussed is the P2 'interpersonal messaging protocol'. If required, however, a different type of content protocol can be specially developed for a given application.

The envelope of a message carries the address of the recipient. CCITT Recommendation X400 suggests that the address should either be:

- A unique identifier of entirely numeric characters. This is called a 'primitive name', and indicates the destination user agent in terms of its country, its 'management

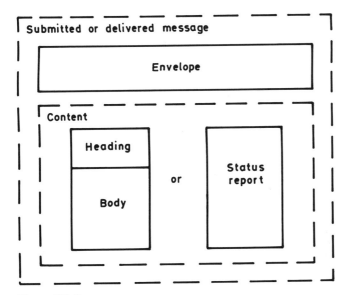

Figure 25.5
Message structure for IPMS. (*Courtesy of CCITT—derived from Figure 2/X420.*)

domain' (explained below), and a unique customer number. Alternatively the address may be:

● A unique name-style identifier, correctly called a 'descriptive name', comprising a personal name, an organization name and the name of an organizational unit.

All addresses in MHS must be unique, and denote exactly one user. An example of a primitive name might be a number such as 12345926 etc. By contrast an example of a descriptive name is:

'The chairman of the XYZ company'

The scope for using descriptive names is helpful to human users, who no longer have to remember arbitrary numbers for calling their friends or close associates. However, the name must be sufficient for the message handling system to recognize the user uniquely. A name such as:

'the floor cleaner of the XYZ company'

will not do as a descriptive name, unless the XYZ company only has one floor cleaner.

25.6 MHS MANAGEMENT DOMAINS

A management domain might be a complete MHS network, or it may comprise a small number of MTAs owned and run by a single organization. A 'management domain' might be a large company's private MHS network, or a network run by the public telephone company. The two types of domain are called, respectively, a private management domain (PRMD) and an 'administration management domain' (ADMD). Figure 25.6 illustrates the management domains concept, and shows the interconnection of a number of different domains. The use of management domains allows the addresses in each domain to be allocated independently of addresses in other domains. Thus the same address 'The chairman' may be used to apply to two different people, being the individual users of separate companies' MHS networks.

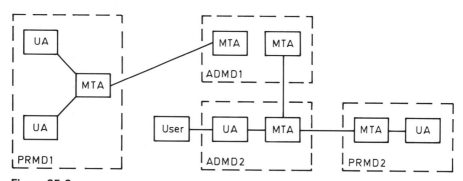

Figure 25.6
Administration and private management domains.

Figure 25.6 reflects the different examples encountered in real life, where some public telecommunications operators may provide only message transfer (i.e. network) facilities while others provide customer premises equipment as well. Interconnection between all the management domains is by standard P1 (message transfer) protocol.

25.7 MHS AND THE OSI DIRECTORY SERVICE

In order that the MHS may deliver messages to the correct recipients, it needs to:

- validate the address names of originators and recipients (originator/recipient, or 'O/R name'), then

- determine the necessary call routing, and finally,

- determine whether data format, or code conversion is required.

A large quantity of reference information, giving details and addresses of all the users, is required so that first, the sender can address the message correctly to the intended recipient and second, so that the UA and MTA can determine where this person is and so deal with the message accordingly. The information needed is drawn from the 'OSI directory service' which is prescribed by CCITT recommendations in the X500 series. These recommendations lay out a standard structure in which information may be held in a database, and they provide for a standard method of enquiry. Besides determining the correct routing paths for MHS messages, the service can also be used to answer users' directory enquiries. A widely held view is that large-scale corporate and international interconnection of electronic mail and other messaging devices cannot take place using the X400 standard until the X500 directory service has been fully defined and realized.

25.8 MESSAGE CONVERSION AND CONVEYANCE USING MHS

MHS is intended to recognize all of the standard binary codes (e.g. telex, facsimile etc.) and provide for translation between them by interaction with the OSI presentation layer (layer 6). In the case of a telex machine using the message conversion capabilities of a message handling system to send a message to a facsimile machine, MHS would convert from telex characters into a facsimile message format. In time, some people hope that even speech to text conversion may be possible. So far, however, not all translation functions are understood and further technical development will be required before we have a system which provides for all types of translation.

The standard codes understood by MHS include:

- text in IA5 computer code (international alphabet no. 5);

- telex (international alphabet no. 2, or IA2);

- facsimile;

- digitalized voice;

- word processor formats.

There is no restriction on the number of different types of code which may be used in a single message. Thus messages of mixed text, facsimile diagrams, and even voice can be sent, although there may of course be a limitation on what the receiving device will comprehend.

Early application of MHS by companies and other users is for conveying documents in electronic form. A prime use of the standard X400 protocols will be to interconnect today's incompatible electronic mail systems. In addition MHS has provided the foundation for a new era of paperless trading between companies, called EDI (electronic data interchange). In this environment it is invaluable for large scale distribution of documents, for rapid confirmation of receipt, and for delivery of messages in electronic form, allowing subsequent processing if necessary.

25.9 SETTING UP A MESSAGE HANDLING SYSTEM

Figure 25.7 illustrates a practical implementation of a message handling system, showing the items of equipment that the user may be familiar with.

Figure 25.7 shows a network comprising a personal computer, a mainframe computer and another mainframe computer. The 'message handling system' is only a small function, but one which each of the computers must be capable of performing. The MHS itself is a small amount of software in each computer—and so invisible to the user.

The personal computer on the left of Figure 25.7 has within it communications software which performs a user agent function, as well as the functions of OSI layers 1–6. The physical network connecting it to the message transfer agent (MTA), which resides in the mainframe computer in the centre of the figure, is a packet-switched network. In the jargon of MHS the personal computer is said to be an 'S1-system', since it comprises only a user agent entity (UAE) and a submission and delivery entity (SDE) but no message transfer agent entity.

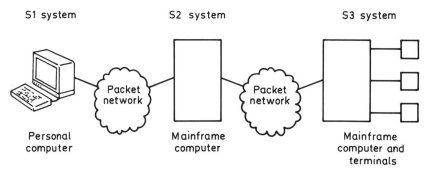

Figure 25.7
Physical example of MHS realization.

In contrast, the mainframe computer shown in the centre of Figure 25.7 is the main 'databank' of the MHS system. It comprises only a message transfer agent entity (MTAE) together with communications software supporting the functions of OSI layers 1–6. The MTAE performs the function of relaying messages between user agents (UAs), and when necessary it stores messages for later retrieval by recipient UAs. Since this mainframe computer contains a MTAE, it is called, in MHS jargon, an 'S2-system'. The computer may perform other normal data-processing functions in addition to its MHS communication functions, although many companies may choose to devote a mainframe computer for use as a message handling system (S2-system). Finally, the computer shown on the right-hand side of Figure 25.7 comprises both a user agent entity (UAE) and a message transfer agent entity (MTAE). This is called an S3-system. Mainframe computers dedicated to use as MHS S3 systems are also likely to become common.

To sum up, we have distinguished three types of systems as realizations of MHS:

● Systems containing only user agent (UA) functions (i.e. UAE and SDE). These are termed 'S1 systems';

● Systems containing only message transfer agent (MTA) functions (i.e. MTAE), termed 'S2 systems';

● Systems containing both UA and MTA functions (i.e. UAE and MTAE), termed 'S3 systems'.

The diagram in Figure 25.8 shows the functional structure of the equipment of Figure 25.7. It illustrates the 'stack' of OSI protocols and functions required within the actual

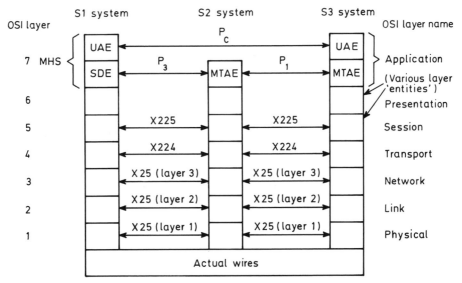

Figure 25.8
Packet based realization of MHS.

equipment, in order to realize MHS in practice. The diagram shows how the 'protocols' at each of the OSI layers are communicating between S1, S2 and S3 systems on a 'peer' level. A packet network based MHS is shown, using the CCITT X25 packet protocol at layers 1–3. Meanwhile layers 4 and 5 protocols being used are those defined in CCITT Recommendations X224 and X225. The presentation layer has a null protocol, and then P1, P3 and Pc are used at layer 7. Layers 1–3 might equally have been based on a LAN (local area network), or a telephone network, and in these cases the X25 packet protocols would have been unsuitable.

25.10 SUMMARY

The message handling system X400 will provide for users an extremely flexible way of interconnecting different types of office equipment. It operates in a store-and-retrieve fashion and will be an important tool in computer communication, particularly for:

● Passing electronic mail and documents between proprietary computer systems;

● Data file transfer between computers separated by time zones and geography —using the public packet network.

The usefulness of the X400 service will depend upon the availability of directory information in order that messages can be delivered to the correct recipient. This will govern the rate at which today's different electronic mail networks may be linked together. In essence this means that many potential users must wait for full development of the new X500 directory service.

Ultimately, X400 will provide a most powerful means of interconnection, allowing paperless interaction not only between different equipment but also between different companies. It heralds the first real opportunity for paperless trade based on electronic data interchange (EDI), and it may eventually lead to a genuinely 'paperless office'.

BIBLIOGRAPHY

CCITT Recommendation X400, 'Message Handling Systems: System Model Service Elements'.
Electronic Message Systems 87. Proceedings of the International Conference held in London, November 1987. Online International Ltd, 1987.
Roberts, S. and Hay, T., *Electronic Message Systems and Services.* Comm. Ed. Books, 1989.
Vervest, P., *Electronic Mail and Message Handling.* Pinter, 1985.
Vervest, P., *Innovation in Electronic Mail.* Eburon, 1986.

26

INTELLIGENT NETWORKS AND SERVICES

By storing a massive 'memory' of customer and service information in a network, and referring to it while setting up calls, and as a historical record of network use, a phenomenal new range of services becomes possible. The effect is almost as if the network had some degree of 'intelligent' power of thought. This chapter commences by describing the 'intelligent networks' as a concept, and then goes on to give examples of the new services we can expect from it.

26.1 THE CONCEPT OF INTELLIGENT NETWORKS

The concept and development of 'intelligent networks' (IN) originated from North America. The forerunner was AT&T's database 800 service, and AT&T continues to be a key player. More recently, widely publicized work has also originated from the RBOCs (the American Regional Bell Operating Companies—or local telephone companies), in conjunction with their jointly funded research arm, Bellcore.

The concept is based on the premise that all services can be broken down into elemental capabilities called 'functional components'. For example, a simple service may include providing dial tone, collecting digits, performing number translation, switching the connection, and charging at an appropriate rate. If we now were to examine a second service, then we would find that some of the functional components used in that service would overlap those already identified in the first. If a comprehensive set of these functional components could be implemented at every exchange (or so-called 'service switching point (SSP)') and if a suitable means of controlling the exchanges, from new powerful and remote computers called 'service control points (or SCPs)' could be found then new and much more powerful services could be implemented simply by writing software (a 'service script') for the SCP—enabling it to manipulate the SSP(s).

26.2 INTELLIGENT NETWORK ARCHITECTURE

In the past each exchange had at least a small amount of 'intelligence' comprising software programs and related data for call routing and service control of 'basic' telephone services. But when we speak of an 'intelligent network' we mean a network equipped with a much larger information reference store and with software capable of controlling far more powerful services.

The 'intelligence' can be added to the network on either a 'distributed' or a 'centralized' basis, according to the circumstances of the established network, the equipment to be used, and the service to be provided. Here we compare and contrast the two architectures as a means of illustrating the scope of possibilities.

In a network employing *distributed intelligence* the information required for advanced call routing and service control is spread over a large number of sites or exchanges. Each exchange stores a large bank of information necessary for set up and control of the wide range of services it is expected to offer. This will include a store of customer-specific data (the information pertinent to a given customer's network), as well as some 'service logic' to tell the exchange exactly how each sophisticated service works, and the procedure for setting up calls. This sort of intelligent network could be created by continual enhancement of today's exchanges—progressively adding software and hardware to cope with new service needs. The advantage of such an approach (storing information at a large number of exchanges) is that the service becomes available at all existing exchanges and the call handling capacity is large. The disadvantages are that the exchange software becomes very complex and the job of keeping all the exchanges' software up to date is unmanageable. Not only that, but the software and data duplication increases the risk of inconsistencies and may affect the smooth running of both the service and the network as a whole. (For example, two exchanges may hold conflicting data because they were updated by different people at different times).

In a *centralized intelligence* network architecture, data and service logic supporting advanced network services is kept in a single, or small number of central network computers called 'service control points' (SCPs).

Figure 26.1 illustrates what is fast becoming the standard 'intelligent network' architecture. In the lowest tier of a centralized intelligent network are a number of 'service switching points' (or SSPs). These are really just telephone exchanges which have been developed to include a 'new intelligence network' interface. The interface allows the exchange to refer the call control of advanced service calls to the service control point (SCP)—allowing the SCP to manipulate subsequent actions of the exchange.

The service control point (SCP) is likely to be a specialized computer, distinct and maybe remote from the exchange, but connected by a CCITT 7 signalling link. It holds the 'knowledge' as to how a service works and the customer specific data. In response to an exchange request to deal with an 'advanced service' call attempt, the SCP sends a sequence of 'primitive' commands to the exchange, using CCITT 7 signalling. The commands direct the exchange to perform the necessary sequence of simple switching actions which combine to appear as a more complex service offering (e.g. 'collect digits', 'connect switch path'), thus the call control is carried out by the SCP rather than the exchange (SSP). But the connection itself never passes through the SCP. Connections are always only switched by exchanges (the SSPs).

During call set-up, call processing is suspended in order that the SSP may refer to SCP. The reference may reveal whether a given caller is permitted to be connected to a given number. Alternatively the interruption may allow a credit card to be validated before accepting call charges or may allow a freephone number to be interpreted into its real 'telephone number' address.

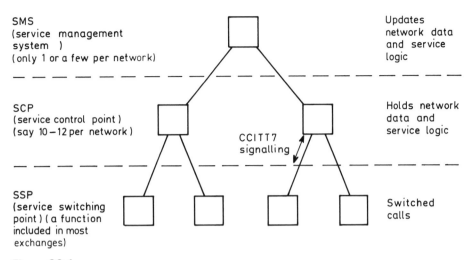

SMS
(service management
system)
(only 1 or a few per network)

Updates
network data
and service
logic

SCP
(service control point)
(say 10–12 per network)

CCITT 7
signalling

Holds network
data and
service logic

SSP
(service switching
point) (a function
included in most
exchanges)

Switched
calls

Figure 26.1
Intelligent network architecture (centralized intelligence).

The referral from SSP to SCP is carried out using CCITT 7 signalling. A message is sent to the SCP containing the dialled number and any other known information about the called or calling party. The SCP interprets the call request using the received information and its own store of data, and then returns the sequence of commands back to the SSP, again using CCITT 7 signalling. The specially developed 'user parts' of CCITT 7 signalling which enable this interaction are called the 'signalling connection and control part' (SCCP) and the 'transaction capability application part' (TCAP). These are described in more detail in Chapter 28.

The number of SCPs deployed in any given 'intelligent' network depends upon a number of factors, including the complexity of the service logic required to support the advanced services, and the traffic demand for them. One option is to allocate one SCP for each individual advanced service, but for a large number of services we would need a large number of SCPs. Some experts therefore favour SCPs which are capable of handling a number of different services, so that the number of SCPs in a network can be kept down to a handful—with corresponding cost savings. In this case each will cater for a number of services, but each service will be duplicated over more than one SCP to prevent total loss of the service in the event of an SCP computer failure.

Only full realization of the architecture will answer our curiosity about the number of SCPs that will be needed in a network based on the Bellcore architecture. In the meantime, the best gauge is given by the intelligent architecture used to support early AT&T services (800, credit card etc). In this proprietary architecture there already exist many NCPs (Network Control Points—the equivalent of SCPs).

Returning to Figure 26.1 we see that the top of the intelligent network hierarchy is dominated by the 'service management system' (SMS). This is another computer system connected to all SCPs and used to update them. It ensures that the data held in all SCPs is comprehensive and consistent. The fact that only one, or a small number of, SMS exists makes the manual task of administering data held within the network a great deal easier. The service risk of running only a single SMS is not material because it is only an 'updating machine'. The service availability is affected mainly by the reliability of the SCPs and SSPs.

26.3   BENEFITS OF INTELLIGENT NETWORKS

The overall benefits of 'intelligent networks', apart from the eased storage and updating of network data are:

- More rapid introduction of new network services across a wide geographic area and with only minimal impact on existing network and switching equipment. (The introduction of a new service requires new service and data to be loaded into the SCP by the SMS. In most cases the SSP will not need development, needing only to respond to the usual SCP commands.)

- Reduced cost of enhancing services (simply by upgrading SCP programs via the SMS).

- The ability for rapid reconfiguration of services, allowing continual returning to meet changing market needs. (The use of a single SMS means that the job of co-ordinating network upgrades is largely eliminated.)

- The ability to give the limited customer control and management facilities if required. (By providing special customer terminals connected to the SMS, customers could be authorized to make some changes specific to their own networks.)

26.4   INTELLIGENT NETWORK SERVICES

We now go on to review a number of the initial services made possible by 'intelligent networks'.

At present many PTOs' 'intelligent networks' and services operations are based on 'proprietary' or 'interim' standards. Most consist of overlay networks made up of a single manufacturer's type of exchange and a single computer type for the SCP or SCPs. However, as world standards develop we can expect the manufacturer's equipment (exchanges and SCPs) to become more compatible and interchangeable, and as a result the services offered by all PTOs will become more co-ordinated.

Calling card

Using 'network intelligence' to validate 'calling card' or 'credit card' account numbers and as a historical record of transactions, an alternative means of paying for calls is possible. Calls made from any telephone can be charged to a special 'calling card' account. The bill can then be sent to the card-holder's address. Maybe such a service will make obsolete the familiar public payphones, or at least the ones for which you need handfuls of coins to operate.

There are three ways of initiating a call. In one, the caller tells the operator the card number and the 'personal identification number' (PIN). The operator types this infor-

Figure 26.2
Credit card telephone. Telephone specially designed to allow payment for public telephone calls by credit or calling card. (*courtesy of British Telecom.*)

mation into a computer which interrogates the SCP to check that the card is valid, and subsequently charges the cost of the call to the appropriate account. An alternative is an automatic version relying on the customer being prompted to dial in his card account number and PIN using a DTMF telephone. Finally, a specially designed telephone with a 'card-wipe' system might also be available. In this instance, a magnetic strip on the reverse of the card is 'wiped' through a narrow channel on the telephone the telephone 'reads' the magnetic strip to derive the 'calling card' (or standard credit card) type and number, and automatically validates the card, notes its expiry date and other details by using the network intelligence in the same way as above. If the card is not valid, or if the caller dials in the wrong PIN then the call is not permitted.

The beauty of using central intelligence of the SCP to validate cards is the scope it gives for tailoring calling capabilities of the card to its owner's needs. A student's parents can give their son a calling card with which he can only 'call home'. Other calls are at the student's expense. Similarly, a company representative can be given the means to call his office.

Calling card service is growing in popularity in countries where it is already available, and most major PTOs are planning to introduce it. Figure 26.2 illustrates an example of a telephone designed especially for automatic card validation.

Freephone service (or 800 service)

'Freephone' service (or 800-service) is available in a number of countries. In the UK, callers who dial a number in the 0800 range, and in the US callers who dial a 1-800 range number have those calls completed entirely free of charge. The call charge is paid by the recipient of the call.

Freephone service gives companies a way of persuading people to call them. A company may wish to promote calls to follow up an advertisement campaign, or to allow customers to call the service department, or maybe to allow their travelling representatives to call the office.

Network intelligence plays two key roles in support of the freephone service. First,

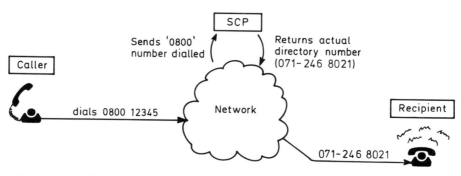

Figure 26.3
Automatic freephone service.

the 0800 number dialled (say, 0800 12345) must be converted into the receiving company's actual number, say 071-246 8021, otherwise the normal telephone network will be incapable of completing the call. The second role is to record the total number and duration of calls made, in order that the call recipient can be charged in due course.

Figure 26.3 shows a diagram of automatic freephone service. The caller has dialled the number of 0800 12345 into the network. The SSP sends the number to the SCP, which returns the normal telephone number to the network (to allow routing), and records the time of day, call origin and call duration, in order that the recipient (071-246 8021) may be charged for the call by normal quarterly account.

900 service

The United States '900 service' uses a similar intelligent network to that of the 800 service, but rather than calls being free to callers they are charged at a premium. The premium charge covers not only the cost of the call itself, but also the cost of 'value-added' information provided during the call. Thus a typical 900 service might be 'Dial up Weather Forecast' or 'Dial Up Sport News'.

The value added information is provided by a service independent of the PTO who pays for the provision of 900 service facilities but receives revenue from the PTO for each call made.

The role of the SCP in the 900 service is to ensure translation of dialled 900 numbers and to record call attempts for later settlement of account between PTO and service provider.

In other countries the service may be known under different names; the UK equivalent for example is the '0898' service.

Centrex Service and virtual private network

Many companies run their own automatic telephone and data networks on their own premises, using automatic 'private branch exchanges (PBXs)' and private packet switches etc. Some of these companies also lease transmission capacity from public telecommunications operators (PTOs) in order to connect together a number of geographically widespread sites into a single, company-wide network. These private networks are always tailored to the company's particular needs, often supporting service facilities which are to available from the public network. For example, on a company's own telephone network, the allocation of extension numbers may be set according to departmental, or company whim. In addition, other special features may be made available, such as 'ring back when free', 'conference calls' and special call barring facilities (to prevent some extensions from dialling trunk or international calls).

The decreasing cost of private networks equipment, coupled with the all-too-restricted service facilities of public networks, has recently stimulated a rapid growth of private networks. If allowed to continue by the public telecommunication operators (PTOs), this could pose a threat to revenue income, since less income is available from

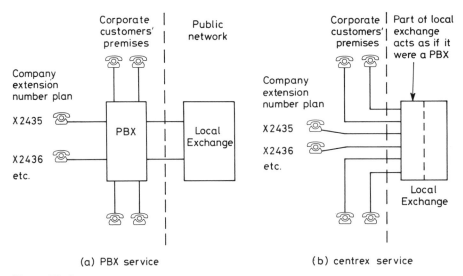

Figure 26.4
Company extension number plan using PBX centrex service.

leased circuits than from the equivalent public network service. From their point of view it will be worse still in countries whose governments allow the resale of private network services. Faced with this, a number of public telecommunications operators and main exchange manufacturers have been developing new services to protect their market shares. Centrex and virtual private network (VPN) services are both products of the counter-reaction.

The centrex service provides facilities similar to that of a PBX, but from the public network's 'local exchange'. This gives the customer benefits equivalent to owning an on-site PBX but without the 'up-front' capital investment, and without the ongoing need for expertise and accommodation to maintain it. All of the customer's 'on-site' telephones are connected directly to the public network's local exchange, which acts as if it were a PBX. For example, the customer may determine the extension numbering plan. In addition, features like 'call interrupt', or 'ring back when free' etc. may be made available between extension numbers. Furthermore, just as in a PBX, only the 'extension number' need be dialled in order to call other on-site company 'extensions'. Figure 26.4 compares a centrex service with the comparable service provided using a PBX.

To the user of extension number 2435, on either the centrex (Figure 26.4(b)), or the PBX network (Figure 26.4(a)) it is not apparent which type of network is being used since the network and special service capabilities are identical. This makes it feasible for a small company to consider first subscribing to centrex service from the public telecommunications operator, and later installing an on-site PBX, when it is cost justified.

A major advantage of the centrex service is that it allows companies which are spread thinly over a number of different buildings within a locality to have their own PBX. If any of the locations are altered, there is no need for PBX upheaval; the PTO can adjust the centrex service to match.

A limiting factor of some PTO's centrex services is that the company's 'on-site' network must all lie within a single local exchange area. Where the company network spans a large area, spread over a number of local exchange catchment areas, then centrex service is inadequate, and an 'enhanced' or 'networked centrex', or a 'virtual private network (VPN)' service is needed instead. The distinction between the two is not clear-cut; both use public network resources to provide for networking needs of geographically-diverse companies spanning several local exchange services. In this chapter we shall not labour the fine distinctions other than to say that while 'enhanced centrex' assumes that the customer has no PBXs at all, VPN assumes that PBXs are in use at some of the sites.

As its name suggests, a 'virtual private network' provides features and services similar to a multiple-site 'private network', while making use of the public network's resources to do so. The economies of scale available from a public network, both in switching and transmission, allow virtual private networks to be cost competitive when compared with private networks.

In Figure 26.5(a) a company's 'private network', comprising two PBXs and a 'leased line' interconnecton between them, has allowed different extensions to get private network style services, using the company's 4-digit extension numbering plan. Extension numbers 2434-7 are illustrated: the user of extension 2436 need only dial the digits '2435' to call the user of that extension. By contrast, in Figure 26.5(b) the same services have been provided using a virtual private network. The leased line is not required, the connections between sites being made by the virtual private network (VPN) embedded in the public network. Nevertheless the user regards the network as a 'private network', and the user on extension 2436 need only dial '2435' to call that extension.

A feature of VPN service illustrated in Figure 26.5(b) is that a PBX or a centrex service may be used as the interface between the extension lines and the local exchanges of the public network. Thus extensions 2435 and 2437 are connected via a PBX (on site 1) while at the customer's second site, extensions 2434 and 2436 are centrex subscribers.

Both centrex and virtual private network services rely heavily upon network intelligence. As in freephone service, the number dialled by the calling customer requires number translation, before it can be routed through the network. A charge for use may have to be recorded, in line with the PTO's overall tariff structure.

Line information database (LIDB)

An early application of network intelligence in North America, was the 'line information database' (or LIDB). In this application, the SCP contains a wealthy store of information about *each* line, and reveals what services are subscribed to by that customer. This enables the network to determine, at call set-up, whether a call attempt to a particular service should be allowed or barred. An example of the use of LIDB in North America, is its use by local telephone companies to record customers' preferred toll (or long distance) telephone company. (It has been a requirement, since deregulation of telephone networks in the United States, for customers to inform their

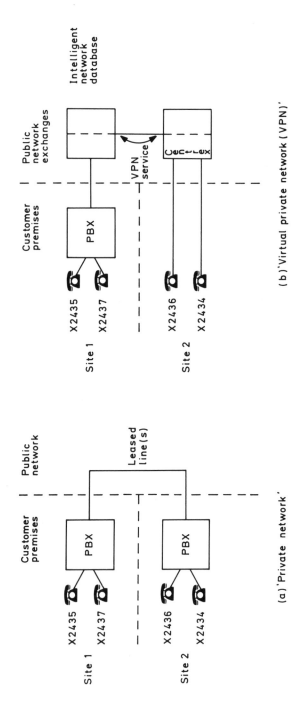

Figure 26.5

Company extension number plan extended to several sites using virtual private network.

local telephone company which toll carriers they wish to pre-subscribe to. The requirement is laid out in the regulations of so-called 'equal access' (1986). Few telephone exchanges at the time had the ability to record and react to the pre-subscription, hence the development of LIDB.)

Another use of LIDB is to hold the current status of a particular piece of information, for instance whether a customer is currently 'at home', or instead wishes incoming calls to be redirected to an alternative number. If redirection of calls is required, the customer may programme the LIDB with the alternative number using a special dialling procedure. On receipt of each incoming call, interrogation of the LIDB by the customer's local exchange secures the redirection.

Televoting

An increasing practice on television programmes is to take an instant poll of viewers' opinions. Viewers are asked to dial one of a set of different directory numbers according to their answer to, or opinion on, a question issue. The following might be an example, on a TV sports programme:

Question to TV viewers

Which footballer, who has played in the World Cup, do you believe most warrants the accolade 'Best Footballer Ever'?

Alternative answers

For Pele	ring 0898 222001
For Diego Maradona	ring 0898 222002
For Bobby Charlton	ring 0898 222003
For Johan Cruyff	ring 0898 222004
For George Best	ring 0898 222005

Viewers ring one of the numbers to cast their vote, and a poll can be taken. This can be done manually by answering each telephone call in person, or automatically by using an intelligent network to record and count up all the votes, even going so far as to give 'votes-so-far' before the voting period ends. The service is called 'televoting'. It is implemented by instantaneous or 'real-time' call counting at the SCP.

Cellular Radio Telephone Service

A 'cellular radio telephone' handset allows its user to roam around a wide geographic area, making and receiving calls anywhere within that area. Outgoing calls from the handset are made via a nearby radio base station in response to the user dialling the public telephone number he wants. Incoming calls may be received by the handset, but herein lies a problem: how do we know where the mobile subscriber is, so that the call can be sent to the radio base station nearest to him? Well, in the early days of mobile telephones, callers were expected to know the geographic location of the mobile user before making the call, and to dial an appropriate area code.

In modern 'cellular radio' networks, intelligence embedded within the network 'keeps an eye' on the location of the mobile user by a continual polling (or 'registration') process. This makes it possible to use the same area code and directory number for calls made to the mobile user, wherever he may be.

Cellular radio networks are split up into a number of cell areas, each served by its own transmitting 'base station' which provides for radio contact between the 'fixed telephone network' and mobile telephone users within the cell. When a user migrates to another cell, radio contact must be transferred to the base station in the new cell (hand-off). The transfer is initiated by a mobile switching centre (MSC), an exchange which controls a number of base stations. While the mobile user stays within any of the cells of a mobile switching centre area, the mobile may be 'polled' to receive its incoming call. Should it enter the area corresponding to a different MSC then the user must be 're-registered' in the new area. In essence, re-registration is an act of updating an 'intelligent network' database, called the 'home location register (HLR)', which records each user's up-to-date location. Any MSC wishing to know the location of a user (in order to complete a call) asks the HLR for the information and then resumes call set-up. (*Note*: Actually the HLR concept is defined by CEPT's GSM digital cellular radio scheme. TACS and AMPS work in a similar way, although the HLR and VLR are not specifically named as such.)

Figure 26.6 shows the registration procedure in outline; the subject is discussed further in Chapter 27.

The mobile user in the car shown on Figure 26.6 has migrated from the cell of a base station in MSC1's catchment area into the catchment area of MSC2. All incoming and outgoing calls must be set up via MSC2, and this is recorded by the database associated with MSC2 in the mobile's 'home location register' (at MSC3). Note how, in this example, the SCP function has actually been 'distributed' around the various MSCs. Although only one copy of each customer's data exists, not all the data resides

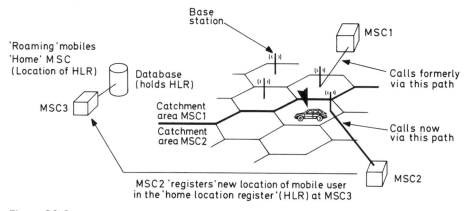

Figure 26.6
Cellular radio registration procedure.

in one SCP location. In some instances there is a clear benefit in distributing the SCP function in this way, since it allows for more traffic to be handled. As traffic growth outstrips the capacity of a single SCP to handle it, the 'service logic' and customer data can be shared out between a number of exchanges so that many more call attempts can be handled simultaneously. (In essence some or all of the SCP's functions are relocated in the exchanges. The SSP part of an exchange still needs to make the referral to the SCP part to take over advanced call control, but this function is no longer located in a specialized and distant SCP computer.) The SMS updating systems works in exactly the same way—but now direct to the SSP/SCP combined unit.

26.5 NETWORK INTELLIGENCE AND PBXs

By adding intelligence' to a PBX, a wealth of new services becomes possible. Three examples are:

(i)   *Automatic 'wake-up' call* for hotel customers. The request for the wake-up call is made by the hotel customer by dialling a service selection code, plus the time. The 'intelligence' stores this information and initiates the PBX to make the wake-up call at the right time.

(ii) *Selection of cheapest public network carrier.* In a country where the public telephone service has been deregulated, it sometimes happens that a PBX is connected to the networks of two or more competing public network carriers. At a particular time of day, one of the carriers may be the 'preferred carrier' on grounds of cost or network congestion but at a different time circumstances may favour the other carrier. By giving the PBX some appropriate 'intelligence', the 'preferred' public network carrier can be adjusted according to time of day.

(iii) *Computer prompting on incoming calls.* In company telephone bureaux receiving large numbers of incoming calls, and where the operators key information into a computer during those calls, it is valuable for the computer and

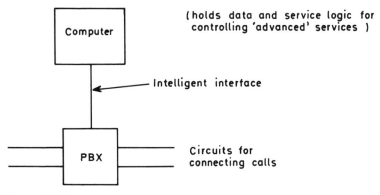

Figure 26.7
An intelligent PBX architecture.

telephone equipment to be linked. The effect can generate a fresh computer 'form' automatically for each new caller. In addition, the computer can signal to the telephone equipment when the last form has been completed—making it ready to receive the next caller.

Figure 26.7 shows an intelligent network based upon a small computer and a PBX. In the example shown, switch and signalling developments have been necessary on both the PBX and the computer. But unlike the public network case, CCITT 7 signalling has not been used. Instead the intelligent network signalling is of a special type, called HCI or 'host computer interface', but there is no overriding standard as yet. All the 'intelligent' interfaces in the PBX market are 'proprietary' to particular manufacturers and several different versions exist.

26.6 THE FUTURE OF INTELLIGENT NETWORKS

Intelligent networks are a sophisticated but complicated service-enabling architecture, and so their achievement on a network-wide basis will take an extensive programme of signalling and switch enhancements over several years. Definition of appropriate standards will be essential to the interworking of intelligent networks around the world and even between different vendors' equipment within the same network. Only in this way will the network providers be able to offer effective and cost-efficient services meeting the bulk of customer needs. The enhancements, however, will mean many more powerful and flexible services for the network users of tomorrow.

BIBLIOGRAPHY

Intelligent Network Access Point (INAP) Framework. Bellcore TA-TSY-000922. September 1988.
Intelligent Peripheral (IP) Framework. Bellcore TA-TSY-000923.
Service Control Point (SCP) Node/Service Management System (SMS) Generic Interface Specification. Bellcore TA-TSY-000365. Issue 2. July 1988.
Service Logic Interpreter 1 and Framework. Bellcore TA-TSY-000924. September 1988.
Service Management Systems/Line Information Data Base (SMS/LIDB)—Interface Specification. Bellcore TA-TSY-000446.
Service Switching Point 1 and Framework. Bellcore TA-TSY-000921. September 1988.
Service Switching Points (SSPs) Generic requirements. Bellcore TR-TSY-000024. October 1985.

27

MOBILE COMMUNICATIONS

Being away from a telephone, a telex or a facsimile machine has become unacceptable for many individuals, not only because they cannot be contacted, but also because they may be deprived of the opportunity to refer to others for advice or information. For these individuals, the advent of mobile communications promises a new era, one in which there will never be an excuse for being 'out-of-touch'. This chapter discusses some of the most recent mobile radio communication technologies, covering the principles of 'radiopaging', 'radiophone', and 'cordless telephones' as well as describing telephone communication with ships, aircraft and trains.

27.1 RADIOPAGING

'Radiopaging' is a method of alerting an individual in a remote or unknown location to the fact that someone wishes to converse with them by phone.

In order to be paged, an individual needs to carry a special radio receiver, called a 'radiopager'. The unit is about the size of a cigarette box, and is designed to be worn on a belt, or clipped inside a pocket. The person carrying the pager may roam freely and can be 'paged' provided they are within the radiopaging service area. The service may provide a full nationwide coverage. Figure 27.1 illustrates a typical radiopaging receiver.

The radiopager is allocated a normal telephone number as if it were a standard telephone. 'Paging' is achieved by dialling this number, as if making a normal telephone call. Instead of being connected through to the radiopager the caller either speaks to a radiopaging service operator, or hears a recorded message confirming that the radiopager has been 'paged'. 'Paging' is done by sending a radio signal to the radiopager, causing it to emit an audible 'bleeping signal' to alert its wearer.

The simplest types of radiopager provide no further information to the wearer than the bleep. Having been alerted, the wearer must find a nearby telephone and ring a prearranged telephone number (say the radiopaging operator, or the wearer's own secretary) to be given the message or the telephone number of the caller who wishes to speak with him. Thus for a caller to 'page' an individual they must first inform the intermediary office (the radiopaging operator or the 'roaming individual's' secretary).

Figure 27.1
Message display radiopager. A relatively sophisticated radiopager, allowing not only bleeping facility, but also the conveyance of a short textual message. (*Courtesy of British Telecom.*)

The caller leaves either a message for the paged person, or a telephone number to be called. Figure 27.2 gives a general schematic view of a radiopaging network.

The key elements of a radiopaging system are the paging access control equipment (PACE), the paging transmitter and the paging receiver. The PACE contains the electronics necessary for the overall control of the radiopaging network. It is the PACE which codes up the necessary signal to alert only the appropriate receiver. This signal is distributed to all the radio transmitters serving the whole of the geographic area covered by the radiopaging service. On receiving its individual alerting signal the receiver bleeps.

A special code is used between the PACE and all the paging receivers. It enables each receiver to be distinguished and alerted. The earliest codes used a discrete signal tone, modulated on to a radiofrequency to identify each receiver. Such systems, available in the late 1950s could address a small number of receivers. Two-tone systems rapidly followed in the 1960s, as the popularity of radiopaging grew. In the two-tone system, up to around seventy tones are used, any two of which are sent in consecutive short bursts, allowing up to $70 \times 70 = 4900$ receivers to be alerted individually. Two-tone systems were used for on-site paging applications, such as summoning medical staff in a large hospital.

Two-tone systems were too small in capacity to be considered for use over a wide area covering large cities or a nation. Hence followed the development of systems using more bursts of tone. A number of proprietary five-tone systems were developed typically using a repertoire of ten different tone frequencies, and allowing up to 25 bursts of tone per second. This amounts to a system capacity of $10 \times 10 \times 10 \times 10 \times 10 = 100\,000$ receivers, and a 'calling' or 'paging' rate of $25/5$ calls per second. However,

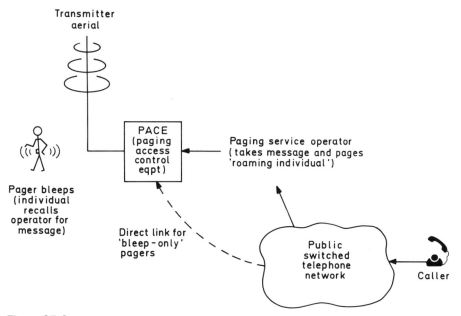

Figure 27.2
Radio paging a 'roaming individual'

even five-tone systems were unable to cope with the explosion in demand that many of the radiopaging operators saw in the late 1970s and new digital coding systems for paging became necessary.

The digital codes were not as sensitive as their predecessors, but they had enhanced performance capabilities in terms of overall calling rate and capacity. Further, they offered the scope for short alphanumeric messages to be paged to the receiver and promised lower overall unit costs, both of the PACE and of the individual receivers. A number of digital codes were developed in the late 1970s and early 1980s, amongst them the Swedish MBS code (1978), the American GSC code (1973), and the Japanese NIT code (1978). The most important code, now common throughout the world, is that stimulated by the British Post Office. Known as the POCSAG code, after the advisory group that developed it (the Post Office Code Standardization Advisory Group), it was developed over the period 1975–1981 and was accepted by the CCIR (International Radiocommunication Consultative Committee) as the first international radiopaging standard. It has a capacity of two million pagers (per zone) and a paging rate of up to 15 calls per second. Further, it has the capability for transmitting short alphanumeric messages to the paging receiver. It works by transmitting a constant digital bit pattern of 512 bits per second. The bit pattern is segregated into 'batches', with each batch sub-divided into eight 'frames'. A particular pager will be identified by a 21-bit 'radio identity code', transmitted within one (and always the same one) of the eight 'frames'. It is this code, when recognized by the paging receiver, that results in the alterting bleep.

An extra feature of the POCSAG code is that an extra 2-bit code can be used to provide four different bleeping cadences in each pager. These can be assigned with different telephone numbers for paging and correspond to four different recall telephone numbers. This may be useful for a user who for most of the time is contacted by a small number of different people, since it potentially removes the need for the intermediary. Figure 27.3 shows how each caller uses a different telephone number to page the 'roaming individual', and produce a distinctive 'bleeping cadence'.

Text messages (consisting of alphanumeric characters) have in the past been conveyed by the radiopaging operator. The most advanced receivers when used in a suitably equipped radiopaging network, are capable of messages up to 80 characters long.

The pager itself is a small, cheap and reliable device. Most are battery operated, but if the pager were to be on all of the time, the battery life would be very short, so a technique of battery conservation has become standard. We have already described how the 'radio identity code' is always transmitted in the same frame of an eight-frame batch to a particular receiver. This means that receivers need 'look' only for their own identity code in one particular frame, and can be 'switched off' for seven-eighths of the time. This prolongs battery life.

Paging receivers include a small wire loop aerial, and because of the low battery power can only detect strong radio signals. This fact needs to be taken into account by the radiopaging system operator when establishing transmitter locations and determining transmitted power requirements, and by the user when expecting important calls. The 'radio fade' near large buildings can be a major contributor to low probability of paging success.

The paging access control equipment (PACE) stores the database of information for

Caller no.	Caller calls telephone number	Roaming individual hears bleep cadence	Recall telephone number
1	A	· · · · · · · · ·	62111 (corresponds to caller 1)
2	B	· — · — · — ·	72372
3	C	— — — —	84923
4	D	· · — · · —	53224

Figure 27.3
Different paging cadence identifies appropriate recall number.

determining which zones the customer has paid for, and for converting the telephone numbers dialled by callers into the code necessary to alert the pagers, and in addition it performs the coding of textual alphanumeric messages. The PACE also has the capability to queue up calls if the incoming calling rate is greater than that possible for alerting receivers over the radio link. Further, the PACE prepares records of customer usage, for later billing and overall network monitoring.

27.2 RADIO TELEPHONE SERVICE

The first radio telephone services were manually operated in the high-frequency radio band. These supported international telephone services, as well as communication with ships and aircraft. Immediately after the Second World War, a new breed of VHF (very high frequency) radio transmitters and receivers (transceivers) was developed. First used for applications such as the police, the fire service, and in taxis, later development led to their use as full 'radio telephones', connected to the public-switched telephone network (PSTN) for the receipt and generation of ordinary telephone calls.

The technology used a radio mast located on a hill and equipped with a powerful multichannel radio transceiver. The mobile stations were weighty but were none the less popular for commercial 'car telephone' service, which grew rapidly in popularity in the mid 1970s.

For calls made to or from the radio telephone user, the public telephone network is connected via 'mobile switching centres (MSCs)' to a number of transmitter 'base stations' which emit and receive radio signals from the mobile telephones. Two radio channels (using different frequencies) are needed to connect each telephone during conversation, one radio channel for each direction of conversation. Figure 27.4 illustrates a typical automatic radio telephone system of the late 1970s.

Figure 27.4 illustrates a single mobile switching centre and a single transmitter base station, serving 3600 mobile customers within an area of 1200 km^2, within a radius of about 20 km from the base station. Calls are made or received by the mobile telephone station by monitoring a particular radio channel, called the 'signalling', or 'call up' channel. If making a call, the mobile station signals a message to the base station

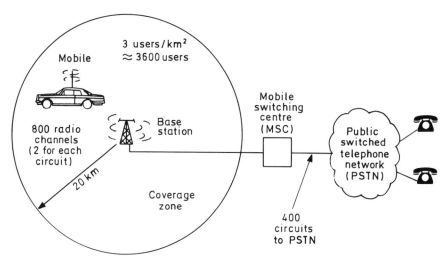

Figure 27.4
A simple radio telephone system.

in much the same way as an ordinary telephone sends an 'off-hook' signal and a dialled digit train (Chapter 7 refers). The base station next allocates a free pair of radio channels to be used between the base station and the mobile for the subsequent period of conversation, and the two switch over to the allocated channels simultaneously.

At the end of the call the radio channels are released for use on another call. Incoming calls to the radio telephone connect in the same way. An ordinary telephone number is dialled by an ordinary customer connected to the public-switched telephone network (PSTN). A particular 'area code' within the normal telephone numbering plan is usually allocated to identify the mobile network, and all calls with this dialled area code are routed to the mobile switching centre and via the appropriate base station transmitter to reach the desired mobile receiver.

A hurdle to be overcome in the design of radio telephone systems is the need for incoming calls to be routed only to the nearest base station. Failure to do so causes the failure of incoming calls, since the mobile cannot guarantee to be within the coverage area of a particular transmitting base station. Early radio telephone systems overcame the problem by requiring the caller to know the whereabouts of the mobile that he wished to call. The same customer number would be used on all incoming calls, but a different area code would have to be used, depending on which base station the caller wished the call to be routed to (i.e. according to which zone the caller thought the mobile was in). Figure 27.5 illustrates this principle.

We see here that if the mobile customer's number is '12345', then when a caller believes that the mobile is in zone 1 the number 0331 12345 should be dialled. Similarly for zones 2 and 3, the numbers 0332 12345 and 0333 12345 are appropriate. Some more advanced radio telephone systems were developed with the ability to 'remember' the last location from which a call was made and route the call there. Others used 'trial-and-error' methods, but these systems were all superseded by the advent of 'cellular radio', which eliminated the market for old-style radio telephones.

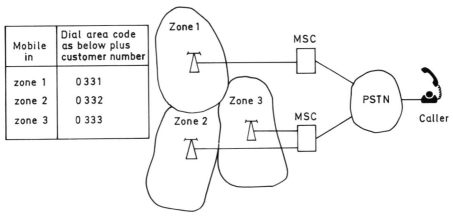

Figure 27.5
Area code routing for incoming radio telephone calls.

Another drawback of the early radio telephone systems was their low user capacity, and their 'unfriendly' usage characteristics. Not only did the automatic systems rely on the caller to know where the mobile was, but some demanded the 'press-to-speak' action of the mobile user. The systems allowed the mobile user to continue to move about during the call, but if he or she moved out of the coverage area of the base station, the call would be terminated. There were no facilities for transferring to other base stations during the course of a call. Together, these drawbacks led to the demise of radiotelephony in its original guise and to its replacement with 'cellular radio', discussed next.

27.3 CELLULAR RADIO

A difficulty with early radio telephone systems was the small capacity in number of users. This arose because the early systems were designed to have very wide area coverage, but had a limited number of radio channels available within each zone. Furthermore, the 're-use' of radio channels in other zones was precluded by the risk of interference except where base stations were separated by large distances. Because the 1970s saw a boom in radio telephone demand, and because radio channel availability was (and still is) a limited resource, there was pressure for development of revised methods, more efficiently using the radio bandwidth—thereby greatly increasing the available user capacity. The system that evolved has become known as 'cellular radio'. The basic components of cellular radio networks are mobile switching centres (MSCs), base stations, and mobile units, as Figure 27.6 shows.

Cellular radio networks make efficient use of the radio spectrum, 're-using' the same radio channel frequencies in a large number of base stations or 'cells'. Oriented in a 'honeycomb' fashion, each 'cell' is kept small, so that the radio transmitting power required at the transmitting 'base station' can be kept low. This limits the area over which the radio signal is effective, and so reduces the area over which radio signal interference can occur. Outside the 'interfering zone' of the transmitter, i.e. in a non-adjacent cell at a sufficient distance away from the first, the same radio channel

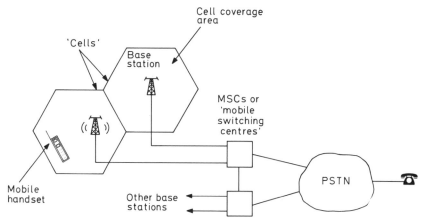

Figure 27.6
The basic components of a cellular radio network.

Figure 27.7
Cellular radio carphone. A cellular radio mounted, and in use in a car. The user base of such services is expanding rapidly throughout the world. The year 1991 will see the commencement of Pan-European service, eventually permitting users to roam anywhere in Europe and still use the same handset and telephone number. (*Courtesy of British Telecom.*)

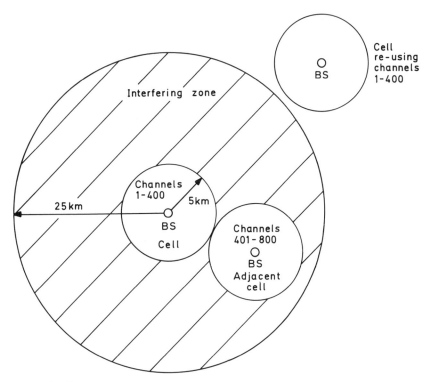

Figure 27.8
Cellular radio channel interference and re-use. BS = Base Station.

frequencies may be 're-used'. Figure 27.8 illustrates the interfering zone of a given cell base station, and shows another cell using the same radio channel frequencies.

A whole honeycomb of cells is established, re-using radio channels between the various cells according to a predetermined plan. A seven-cell 're-use pattern' is shown in Figure 27.9. Seven different radio channel frequency schemes are repeated over each cluster of seven hexagonal cells, each cell using a different set of frequencies. By such planning the same radio frequency can be used for different conversations two or three cells away.

Figure 27.9 shows a seven-cell 're-use' pattern. Other patterns, some involving as few as three-cells, and some more than 30 cells can also be used. Large repeat patterns are necessary to cater for heavy traffic demand in built-up areas where small non-adjacent cells may still interfere with one another.

Each cell is served initially by a single base station at its centre and complemented as traffic grows with directional antennas and more radio channels. Using a directional antenna helps to overcome radio wave 'shadows'. For example, to locate three directional antennas, one at each alternate corner of the cell, helps to overcome the 'shadow' effects that might otherwise occur near tall buildings, by giving an alternative transmission path, as Figure 27.10 shows.

A feature of cellular radio networks is their ability to cope with an increasing level of demand first by using more radio channels and more antennas in the cell, and then by reducing the size of cells, splitting the old cells to form a multiplicity of new ones. Only a limited number of radio channels can be made available in a cell at the same time, and this limits the number of simultaneous telephone conversations. By increasing the number of cells, the overall call capacity can be increased. The number of channels needed in a given cell is determined by the normal Erlang formula (Chapter 10).

The total call demand during the busy hour of the day depends upon the number of callers within the cell at the time. Reducing the size of the cell has the effect of

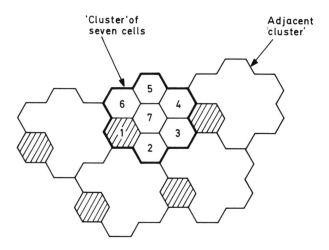

[///] Cells in other clusters, using the same radio channels as cell number

Figure 27.9
Cellular radio re-use pattern.

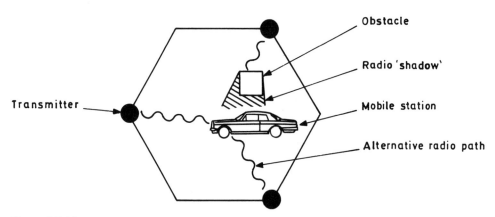

Figure 27.10
Location of multiple base stations.

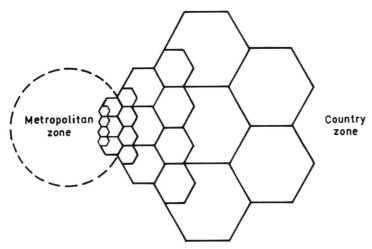

Figure 27.11
Cell splitting to increase call capacity.

reducing the number of mobile stations that are likely to be in it at any time, and so relieves congestion. Figure 27.11 shows a simple splitting of cells and a gradual reduction in cell size in the transitional region between a low traffic (country) area and the high traffic region surrounding a major metropolitan zone. When splitting cells in this manner, due care needs to be taken when allocating radio frequencies to the new cells, and a new frequency re-use plan may be necessary to prevent inter-cell interference.

27.4 MAKING CELLULAR RADIO CALLS

One or more 'control' or 'paging' radio channels are used to make or receive calls between the mobile station. If a mobile user wants to make a call, the mobile handset scans the predetermined channels to determine the strongest control channel, and monitors it to receive network status and availability information. When the telephone number of the destination has been dialled by the customer and the 'send' key has been pressed, the mobile handset finds a free control channel and broadcasts a request for a 'user' (i.e. 'radio telephone') channel. All the base stations use the same control channels, and monitor them for call requests. On the receipt of such a request by any base station, a message is sent to the nearest mobile switching centre (MSC) indicating both the desire of the mobile station to place a call and the strength of the radio signal received from the mobile. The MSC determines which base station has received the greatest signal strength, and on this decides which cell the mobile is in. It then requests the mobile handset to identify itself with an authorization number that can be used for call charging. The authorization procedure eliminates any scope for fraud.

Following authorization of an outgoing call a free radio channel is allocated in the appropriate cell for the carriage of the call itself, and the call is extended to its destination on the public-switched telephone network. Incidentally, the appropriate cell need

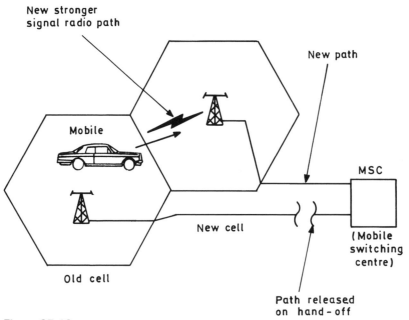

Figure 27.12
'Hand-off' during a call.

not necessarily imply the nearest base station—more sophisticated cellular networks might also choose to use adjacent base stations if this will help to alleviate channel congestion. At the end of the call, the mobile station generates on 'on-hook' or 'end of call' signal, which causes release of the radio channel, and reverts the handset back to monitoring the control and paging channel.

Each mobile switching centre controls a number of radio 'base stations'. And if, during the course of a call, the mobile station moves from one cell to another (as is highly likely, since the cells are small), then the MSC is able to transfer the call to route via a different base station, appropriate to the new position. The process of change-over to the new cell occurs without disturbance to the call, and is known as *'hand-off'*. It is initiated either by the active base station, or by the mobile station, depending upon the system type. The relative strengths of signals received at all the nearest base stations are compared with one another continuously and when the current station signal strength falls below a predetermined threshold, or is surpassed by the signal strength available via an adjacent base station, then hand-off is initiated. The mobile switching centre establishes a duplicate radio and telephone channel in the new cell, and once established, the call is transferred to the new radio path by a control message to the mobile handset. When confirmed on the new channel and base station, the original connection is cleared, as Figure 27.12 illustrates.

27.5 TRACING CELLULAR RADIO HANDSETS

Whenever the mobile handset is switched on, and at regular intervals thereafter, it uses the control channel to register its presence to the nearest mobile switching centre. This enables the local mobile switching centre at least to have some idea of the location of the mobile user. If away from its 'home MSC', the local MSC undertakes a 'registration procedure', in which it interrogates the 'home MSC' (or the intelligent network database associated with it) for details of the mobile, including the authorization number and other information. The information is held by the home MSC in a database called the 'home location register', or HLR. It contains the mapping information necessary for completing calls to the mobile user from the PSTN (its network identity, authorization and billing information). The local MSC duplicates some of this information in a temporary 'visitor location register' or VLR, until the caller leaves the MSC area. Once the visiting location register has been established in the local MSC, outgoing calls may be made by the mobile user.

The 'registration procedure' is a crucial part of the mechanism used for tracing the whereabouts of mobile users, so that incoming calls can be delivered. Incoming calls are first routed via the nearest mobile switching centre (MSC) to the point of origin. This MSC interrogates the 'home location register' for the last known location of the mobile user (this is known as a result of the most recent mobile registration). The call can then be forwarded to the mobile switching centre where the mobile was 'last heard', whereupon a paging mechanism, using the base station control channels, can locate the exact cell in which the mobile is currently located. A suitable free radio channel may then be selected for completion of the call.

Both the handsets and the network infrastructure needed to support cellular radio are complex and expensive, although increase in user demand is reducing the cost. The mobile handsets come in a number of different forms, from the traditional car-mounted telephone, to the pocket versions. The latter are expensive not only because of the feats of electronics miniaturization that have been necessary but also because the trend towards pocket telephones has necessitated advanced battery technology and the use of battery conservation methods also used in radiopaging receivers. The mobile switching centres (MSCs) rely on advanced computers capable of storing and updating large volumes of customer information, and also of rapidly interrogating other MSCs for the location of out-of-area, or 'roaming' mobiles. The interrogation relies on the use of the 'mobile application part (MAP)' and the 'transaction capability (TCAP)' user parts of CCITT 7 signalling (which we shall describe in Chapter 28).

27.6 CELLULAR RADIO STANDARDS

A number of different cellular radio standards have evolved, with the result that hand portables purchased for use on one system are unsuitable for use on another. The most common of the systems currently in use worldwide are:

AMPS (Advanced Mobile Telephone System) was first introduced in Chicago in 1977 by Illinois Bell, at that time one of the operating companies of AT&T in the United States. Developed by AT&T's Bell Laboratories, this has become the most commonly used system in North America. It operates in the 800 MHz and 900 MHz radio bands, at a channel spacing of 30 kHz.

NMT (Nordic Mobile Telephone Service) commenced service in the Nordic countries in 1981 and is used in many countries in Europe. The original system was 450 MHz based but has been gradually extended and to some extent replaced since 1986 with the later NMT-900 version.

C (Network C). Network C is a development of Network B, a digitally controlled mobile radio system introduced in the Federal Republic of Germany in 1971. Network B was 450 MHz based. Its replacement, Network C can be either 450 MHz or 900 MHz based (hence C-450 and C-900); it is used in Germany and Portugal.

TACS (Total Access Communication System) is a derivative of the American AMPS standard, modified to operate in the 900 MHz band, with a more efficient 25 kHz radio channel spacing. TACs was the system introduced into the UK and Ireland during 1985. ETACs, or extended TACs is a compatible derivative of TACs which gives greater channel availability—particularly in very congested areas, like the Metropolitan London area. Other mobile telephone standards include NAMTS (Nippon Automatic Mobile Telephone System—used in Japan), Radiocom 2000 (used in France), RTMS (second generation Mobile Telephone—used in Italy), UNITAX (used in China and Hong Kong) and Comvik (used in Sweden).

The trend of the above systems to migrate to 900 MHz follows the allocation of frequencies in the band by the World Administrative Radio Council (WARC) in 1979. Subsequently there has been pressure to digitalize the radio speech channels, as well as the control channel, to improve radio spectrum usage. Digital channels allow closer channel spacing, due to reduced interference between channels.

In 1982, CEPT (the European Conference of Posts and Telecommunications) decided to start work on specifying new technical standards for the support of a Pan-European mobile telephone service, based entirely on digital radio transmission. Called GSM (after the group that invented it; i.e. Groupe Spéciale-Mobiles), it is an ambitious programme which aims to put in place by 1991 a network spanning the whole of Europe, allowing mobile handset users to roam anywhere on the European land mass and still be able to make and receive calls. By 1991 it seems unlikely that a large interconnected set of networks will be in place, but isolated networks conforming to the GSM standard will exist and it seems likely that international roaming will be possible—and will increase in scope over time. This initiative will certainly advance the capabilities of mobile telephony.

27.7 CORDLESS TELEPHONES

Cordless telephony is the term used to describe telephone sets connected to the ordinary ('fixed') telephone network, but in which the handset communicates with the network by a radio transmission link insisted of wires. The cradle part of a cordless telephone terminal acts as a radio transceiver or 'base station' to connect a radio path to the handset, which also acts as a radio transceiver. The base station is connected to the public telephone network in the normal way. Figure 27.13 illustrates a typical cordless telephone configuration. The maximum range of these systems is typically 50 metres.

Cordless telephones were popular for some time in North America and Japan before

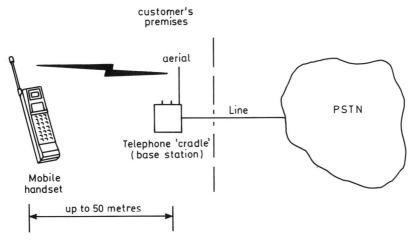

Figure 27.13
A cordless telephone.

they took off in Europe, because the European (CEPT) design specifications were more complex, making the products comparatively expensive. The exception was West Germany, where cordless phones were rented out by the Bundespost at little more than the rental cost of ordinary telephones.

Cordless telephones are very simple in comparison with cellular radio telephones, comprising a (duplex) two-way conversational radio channel, with a relatively simple signalling system. A major hurdle in the design of cordless telephones is ensuring that telephones in adjacent customers' premises do not interfere with one another. A customer is not prepared to pay for the next-door-neighbour's calls, made on the wrong base station. This may happen if a handset 'interferes' with the base station next door, and it is the main reason for the very strict CEPT specifications.

The advantage of cordless telephones is the freedom to carry them about the house, down the garden, around the workshop, so saving users from being away from the phone and not hearing the phone ring. Simple cordless telephones can be used only within range of their own base station. They are thus useless away from 'home', but make the customer more mobile about his own premises.

27.8 'TELEPOINT' OR 'CORDLESS TELEPHONE (CT2)'

An extension of the idea of cordless telephones is the concept of 'telepoint', or 'wide area cordless telephone'. In telepoint, a new type of digital cordless telephone is used with a number of base stations. Besides the base station in his house, the customer has access to public 'telepoint' base stations situated in well populated locations, such as airports, stations, and street corners (much as public payphones are located today). Standing within 50–200 m of a telepoint, a caller with a telepoint handset is able to make outgoing calls into the public-switched telephone network in a similar way to a cellular radio customer making an outgoing call—except that he may not move from one base station to another during the call. Incoming calls, however, are not possible. Telepoint hardware includes a mobile handset and a number of base stations, each connected directly to the public-switched telephone network, as Figure 27.14 illustrates.

To make a call, the handset sends a signal, including a special handset identity code, over a control channel to the base station, which confirms the identity and authorization of the user, and then allocates a radio channel in a way similar to that used in cellular radio. Onward connection of the call is made directly via the PSTN—apply dial tone, collecting dialled digits etc.—while the base station records call details for later billing of the customer. (There is one exception to this—and that is when the customer has installed a private base station in his own premises. In this case the customer pays for public network calls in the normal way—as recorded by the PSTN operator.)

As we have seen, telepoint or 'second generation cordless telephone ('CT2') as it is also known, is capable only of making outgoing calls. The technological problems of tracking the mobile handset location have not been solved, so that incoming calls are not possible, but work has already started on future generations of equipment. Some CT2 handsets may have a built-in radiopaging receiver, so that the user may be 'paged' with a displayed telephone number to dial when next he is near a telepoint base station.

A further development of the paging idea is the 'personal communicator'. In this device the handset automatically redials the original caller when in range of a base station—using the number delivered by the pager. In this way the effect of receiving incoming calls is achieved—at much lower cost than the equivalent cellular radio technology. But the new solution still does not provide for a 'hand-off' procedure for moving between base station zones—so users are forced to stay in range of a single base station for the duration of each call. Thus a car needs to be parked in a 'telepoint' car park—it cannot be on the move.

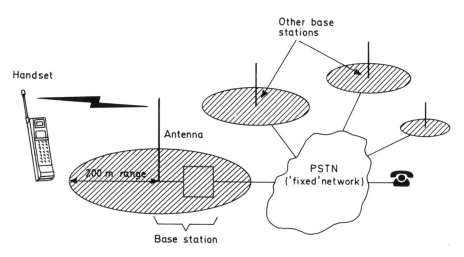

Figure 27.14
Telepoint service.

Recognizing the significance of this new technology, the Europeans have once again joined forces to develop a digital and pan-European standard system (called Digital European Cordless Telephone or DECT). While it is being prepared, some network operators are attempting to place an embargo on today's *de-facto* standard for the 'handset-to-base-station' interface called the 'common air interface' (CAI). CAI is the system introduced for commercial service in the United Kingdom in 1989 and 1990. All UK telepoint network operators and equipment providers are mandated to conform to the CAI standard by 1990. In the period before this, however, some commercial offerings used earlier variants of the standard.

Competition between this 'hybrid' technology and cellular radio is bound to be intense in some market segments. Given the relative cheapness of telepoint technology to the user it is no wonder that some people have described it as the 'poor man's cellular radio'! But an initial problem will be the capital investment needed to establish a mass market.

27.9 PERSONAL MOBILE COMMUNICATIONS (PMC) AND PERSONAL COMMUNICATIONS NETWORKS (PCN)

Other areas of growing interest are those of 'personal mobile communications' (PMC) and personal communication networks, but these systems have no rigid technical standards to date. The concept of PMC is that some mobile attributes can be achieved even within the fixed network. So, for example, telephone numbers could be allocated to individuals rather than telephone handsets, and would follow them around. Contacting a given individual would then be easy for all callers, simply by dialling the same number on all occasions. The difficult job (of knowing where the individual is) is catered for by the network, possibly using a registration procedure similar to that already used in cellular radio network.

The similar short-term spin-off from the work on PMC is the appearance of the personal communication networks (PCN). This is a concept developed by the Europeans for a mass market in mobile communications. At the time of writing it looks likely to adopt a hybrid standard similar to both the GSM (Pan European cellular radio) and DECT (digital European cordless telephone).

27.10 AERONAUTICAL AND MARITIME MOBILE COMMUNICATIONS SERVICES

High-frequency radio communications have long been used in aeronautical and maritime applications for navigational purposes and distress calls. Further, many public telephone network operators have for many years offered ordinary telephone customers the opportunity to place or receive HF radio telephone calls to and from ships in coastal waters. Unfortunately high-frequency radio systems, when used for communication over long distances at sea, suffer from poor signal propagation, atmospheric disturbance, interference, and signal fading. As a result, the ships' telephone radio service is of poor quality, and unreliable.

As a means of getting over this problem, a number of organizations, during the late

1960s and early 1970s, were studying the use of satellites for aeronautical and maritime communications. These studies led in 1976 to the launch of MARISAT, the world's first commercial maritime satellite system. MARISAT was conceived by a consortium of United States companies and comprised three satellites in geostationary orbit, providing communications services for military and civilian use. At the outset heavy use of the system was made by the US Navy.

Further worldwide interest, and an initiative of the Inter-Governmental Maritime Consultative Organization (IMCO) in 1973 led to the signing of an agreement by 26 countries in 1976, and the establishment in 1979 of the International Maritime Satellite Organization (INMARSAT).

INMARSAT's satellites have revolutionized the world of communications with ships. Geostationary satellites now placed over the Indian, Atlantic and Pacific Oceans make possible communication to any suitably equipped ship, no matter where it is located around the globe. Automatic telephone calls, dialled directly by customers of some countries' public-switched networks can be made to people aboard ships—the only requirement is that the caller knows which ocean the ship is in, so that the appropriate ocean country code number (part of the international telephone number) can be dialled, directing the call via the appropriate ocean's satellite. Similarly, persons aboard ships may make direct-dialled calls to other customers connected to the international public telephone network.

Figure 27.15 illustrates the main components of the maritime satellite network, comprising earth station equipments both on shore and aboard ship, together with a sophisticated 'access control and signalling equipment'. The ACSE performs similar functions to those of cellular radio network 'base stations' and 'mobile switching centres' (MSCs). It allocates radio channels for individual calls and carries out billing, accounting, and other administrative functions.

A cheap form of satellite communication, allowing 'telex-like' communication to

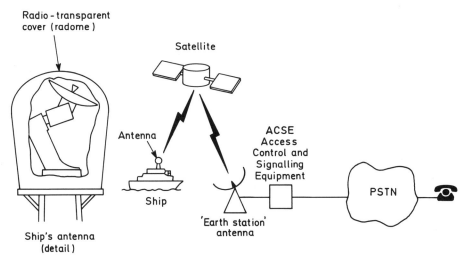

Figure 27.15
Maritime satellite communications.

Figure 27.16
Inmarsat ship's antenna. The radome cover protecting a ship's INMARSAT telecommunications satellite antenna. (*Courtesy of British Telecom.*)

and from terminals on board smaller craft, vehicles and lorries is the INMARSAT Standard 'C' service. This allows two-way communication using an omnidirectional antenna about the size of a saucepan.

By the end of 1987, the number of INMARSAT signatory countries had increased to 50. INMARSAT's original aim was to provide maritime communications, thereby improving the safety and management of ships. In 1985 the charter was expanded to include aeronautical satellite communications.

A number of initiatives are now in hand to provide aeronautical satellite services. These are expected to bring the public telephone network to passengers and crew aboard commercial airliners in 1990. This is not to say that telephone calls cannot already be made from aircraft (indeed on some domestic flights within the United States for example, passengers are already able to make calls via radio direct to the ground). But satellite communications will bring a much broader coverage area, not restricted to flight paths above land masses, but covering the whole globe. With it comes a new scale in complexity: spare a thought for the aircraft antenna designer who has to create a directional device always 'pointing' at the satellite!

27.11   RADIODETERMINATION SATELLITE SERVICES (RDSS)

In July 1985 the Federal Communications Commission (FCC) of the USA authorized the use of a new satellite radio service to be called Radiodetermination Satellite Service (RDSS). It will have widespread benefit in the field of navigation, and personal communication technology. The technical and operational standards adopted by the FCC became the basis for worldwide standards agreement under the auspices of ITU. Other services will be available from the Locstar Corporation, who will start a European service in the early 1990s, and the German 'PRODAT-System' (Prozessdaten—Erfassungs und Auswertesystem).

RDSS is a set of techniques combining radio and computer capabilities which is capable of determining precise geographical locations of points on the ground and relaying this information for the use of other equipment.

Early forms of RDSS are already in use in the USA and Europe. It is used for applications such as tracking vehicle or shipping fleets. Its potential includes the scope for keeping track of individuals—in support of global cellular radio services. Indeed, some claim that RDSS offers a more efficient means of tracking cellular handsets than the current method of continual updating. One company active in this field, Geostar Corporation of the United States, offers its precise location service for only a few cents per transmission.

An RDSS system consists of a set of geostationary satellites, a control centre and a number of user terminals, called 'transceivers'. The control centre repeatedly sends an interrogation signal via the satellite to all the user terminals. The signal is interpreted by each terminal and, if relevant, a response message is generated to:

● alert the control centre of current position (if requested), or

● reply to or request some other information from the control centre.

The relative position of the user terminal from one satellite is computed by the control centre from the round-trip alert-and-response signal time scaled by the velocity of light. The relative range from three different geostationary satellites enables the control centre to compute exactly the position of the terminal in three-dimensional coordinates of latitude, longitude and altitude. This information is then associated with the terminal identity (ID) code and can either be passed to a third party or transmitted back to the user terminal. Thus a fleet operator could trace a lorry on the road or a shipowner could determine a ship's position at sea. Shipping companies will be familiar with the 'Navstar' system.

In parallel with RDSS, a number of simpler systems have also evolved, which while similar are not RDSS systems in the true sense, since they do not determine user locations precisely. Examples include satellite-based radiopaging services. Companies active in this field in the United States include 'Omninet', 'Mobilesat' and 'Sky link'.

27.12 MOBILE DATA COMMUNICATIONS

Mobile radio is an awkward medium for carrying data. Interference, fading, screening by obstacles, the voice channel signalling, and the hand-off procedure all conspire to increase errors, so while the digital fixed telephone network may expect to achieve error rates no greater than 1 in 10^5 bits, the error rate over cellular radio can be as high as 1 in 50 bits.

Very basic systems with slow transmission speeds (say 300 bit/s) have been used. At these rates little data is lost and connections that are lost can be re-established manually. However, for more ambitious applications error correcting procedures must be used—normally a technique employing 'forward error correction and automatic request retransmission'. In this technique sufficient redundant information is sent for data errors to be detected and the original data reconstructed even if individual bits are corrupted during transmission. Typical speeds achieved are 2.4–4.8 kbit/s.

As a particular example, Racal Vodafone in the United Kingdom has developed a special 'cellular data link control' (CDLC) for sending data over its cellular telephone network. Based on the OSI layer 2, HDLC (high level data link control), it is extended with additional error correction capabilities but provides a connection for a standard V24 modem (though of course extra software is needed for the CDLC).

27.13 MAKING USE OF MOBILE RADIO TECHNOLOGY

The companies that propose to use mobile communications must give careful attention to the co-ordination and management of the system. Proper control must be established over:

- equipment purchasing;
- air time subscription;
- cost management;
- user-justification rules.

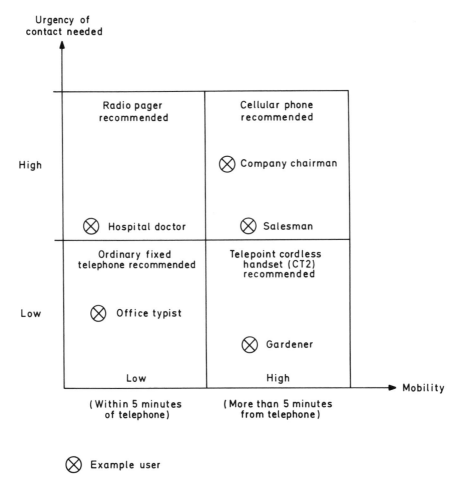

Figure 27.17
'Qualification' chart for mobile communication users.

By dealing with a single supplier for carphone and hand-held units, and also for 'air-time' supply, the administration and control of costs are made considerably easier, but also 'buying-power' leverage over the supplier is increased. The decision to purchase cellular or other mobile telephones for use by individual departments needs to be tempered by a central co-ordinating unit. One of the problems is that the falling cost of handsets is bringing many of them within the financial authorization of the lowest levels of management.

To get around the problem of integrating new and old generation equipment in this fast-changing environment one tactic could be to rent the handsets. However, because much of the cost involved is made up of usage charges—subscription and call charges—users should not be tempted to attach too much importance to the secondary issue of handset cost. Usage costs of cellular radio technology in the United Kingdom are typically £25 per month per handset (1989 subscription) plus a per minute rate for calls up to three times the ordinary telephony charges. Telepoint technology is slightly cheaper (£9 per month subscription plus call charges similar to ordinary fixed network charges) but none the less can represent a substantial extra cost over normal telephones or radiopagers. For this reason, companies are recommended to employ some means of assessing their users' need for service. A simple technique is shown diagrammatically in Figure 27.17.

Figure 27.17 classifies users into four board categories according to their 'mobility' (how long they are likely to be away from the telephone) and their 'need for urgent communication' (i.e. whether they are likely to need to contact or be contacted by others urgently). Examples of typical users in each of the four categories of mobile communications are shown. A hospital doctor with low mobility but high urgency of contact is recommended to have a radiopager, while for a company chairman or salesman with high mobility and high contact urgency, a cellular telephone is recommended. But if this service is heavily congested, as in the London area of the UK, CT2 might provide some degree of alternative. For an office typist we recommend no mobile telephone at all, while our recommendation for the CT2 handset is the humble gardener, assuming of course he does not intend to 'lie low'.

27.14 SUMMARY

In step with the general expectation of individuals constantly to remain 'in-touch', the world of mobile communications has developed rapidly in the last twenty years to the point where almost no part of the globe is inaccessible.

BIBLIOGRAPHY

Calhoun, G., *Digital Cellular Radio*. Artech House, 1988.
Gibson, S. W., *Cellular Mobile Radio Telephones*. Prentice-Hall, 1987.
Gosling, W., *Radio Receivers*. Peter Peregrinus (for IEE), 1986.
Law, P. E. Jr, *Shipboard Antennas*. 2nd edn, Artech House, 1986.
Lee, W. C. Y., *Mobile Cellular Telecommunications Systems*. McGraw-Hill, 1989.
Mobile Communications. Proceedings of the Cellular and Mobile Communications conference held in London, December 1987. Online Publications, London.
Mobile Communications—Developments and Regulations. Proceedings of the 1984 International Conference. Online Publications, 1984.
Pan-European Mobile Communications. IBC Technical Services, Summer 1989.
Rothblatt, M. A., *Radiodetermination Satellite Services and Standards*. Artech House, 1988.
The Applications, Benefits and Experience of Cellular Radio. Conference Transcript, Oct. 1983. Oyez Scientific and Technical Services Ltd.

28

THE CCITT 7 SIGNALLING SYSTEM

Stored program control (SPC) of exchanges and the capabilities of common channel signalling are contributing to the speed of connection and the range of services available from telecommunications networks.

The CCITT SS Number 7, CCITT 7, C7 or 'number seven' signalling system is the most recently developed of CCITT's signalling systems, and it will be used in future as a 'cornerstone' of ISDNs and intelligent networks. It is a complex, common channel signalling system, which enables the controlling processors of two digital exchanges or databases to communicate directly and interact with one another in a manner optimized for digital transmission media. CCITT 7 has also formed the basis of a number of further-developed regional signalling systems. In the United States for example, 'signalling system 7' or 'SS7' is a derivative, while the UK national version is 'C7/BT'. This chapter describes the overall structure and capabilities of CCITT 7.

28.1 CCITT 7 SIGNALLING BETWEEN EXCHANGES

The CCITT 7 signalling system is described in the Q700 series of CCITT recommendations. A common channel system, optimized for digital networks, it allows direct transfer of call information between exchange processors. Comprising a number of layered and modular parts, each with a different function, it is a powerful general purpose signalling system capable of supporting a range of applications and administrative functions. including:

- ISDN;

- 'Intelligent networks';

- Mobile services (e.g. cellular radio);

- Network administration, operation and management.

In addition, its modular nature lends itself to the development of new 'user parts' which may be designed to support almost any new service that can be conceived. The 'user' parts of the system that have been developed so far are:

MTP Message transfer part;
SCCP signalling connection and control part;

TUP telephone user part;
DUP data user part;
ISUP ISDN services user part;

TC   transaction capabilities (used by 'intelligent' networks);
OMAP operation and maintenance application part;
MSP mobile application part.

The MTP and SCCP form the 'foundations' of the system, providing for carriage of messages. The TUP, DUP, and ISUP use the MTP and/or SCCP to convey messages relating to call control, for telephone, data, and ISDN networks respectively. The TC, OMAP and MAP are other 'application parts' for intelligent network database interaction, operation and maintenance interaction, and mobile network control, respectively.

Initially the CCITT 7 system was designed so that the MTP could be used in association with any or all of the telephone, data and ISDN user parts. However, following the emergence of the OSI model, the SCCP was developed as an adjunct to the MTP—the two in combination provide the functions of the OSI network service (layers 1–3).

CCITT 7 signalling can be installed between two exchanges provided that the necessary signalling functions are available in both exchanges. The functions reside in a unit termed a 'signalling point'. This may be a separate piece of hardware to the exchange, but usually it is a software function in the exchange central processor. CCITT 7 signalling points (SPs)—basically exchanges—intercommunicate via 'signalling links' and are said to share a 'signalling relation'.

A single CCITT 7 'signalling link' enables information to be passed directly between two exchange processors, allowing the set-up, control, and release of not just one, but a large number of traffic-carrying circuits between the exchanges. Messages over the unit take the form 'connect circuit number 37 to the called customer number 01-234 5678'. The term 'common channel' signalling aptly describes this method of operation, distinguishing it from the 'channel-associated' signalling method, wherein call set-up signals pertinent to a particular circuit are sent down that circuit. CCITT 7 is not the first common channel signalling system to be developed; CCITT 6 was also a common channel system, but CCITT 6 was less flexible than CCITT 7 and not so suitable for digital network use.

Having a 'common channel' for conveyance of signalling messages saves equipment at both exchanges, since only one 'sender' and one 'receiver' is required at each end of the link, as against the one per circuit required with channel associated systems. The combination of a CCITT 7 sender and receiver is normally referred to as a 'signalling terminal'. In practice signalling terminals are a combination of a software function in the exchange central processor and some hardware to terminate the line and undertake the basic bit transfer function (OSI layer 2—datalink).

A label attached to each message as it passes over the signalling link enables the receiving signalling point to know which of the many circuits it relates to. Figure 28.1 illustrates the network configuration of a simple CCITT 7 signalling link. It shows

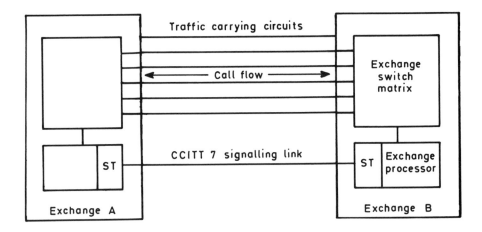

ST = signalling terminal

Figure 28.1
Linking two exchanges using CCITT 7 signalling.

calls flowing over a large number of traffic-carrying circuits which are connected to the switch matrix part of the exchange. Meanwhile all these circuits are controlled according to the information passed directly between the exchange processors. The 'signalling terminal' (ST) function is shown residing within the exchange processor.

28.2 CCITT 7 SIGNALLING NETWORKS

Networks employing CCITT 7 signalling comprise two separate subnetworks. One subnetwork is the network of traffic-carrying circuits interconnecting the exchanges. The second subnetwork is that of the signalling links. In Figure 28.1 we saw this separation of traffic-carrying circuits from the signalling link as it would apply on a single connection between two exchanges. Figure 28.2 now shows a more complicated example to illustrate another powerful feature of CCITT 7: the fact that signalling networks and traffic-carrying networks may be designed and implemented almost in isolation from one another. Just because there are direct traffic-carrying circuits between two exchanges (they have a direct 'traffic-carrying' relation) it does not follow that the signalling information (or 'signalling traffic') has to travel over direct signalling links, though clearly a 'signalling relation' of some sort is needed.

Figure 28.2 shows the traffic-carrying networks and signalling networks interconnecting four exchanges, A, B, C and D. The traffic circuits directly connect A–C, A–B, B–C and B–D. All traffic to or from exchange D passes via exchange B and all traffic to or from exchange A passes either via B or C, and so on. The signalling network, however, is different. Signalling links only exist between A–B, B–C and B–D, so that 'signalling traffic' has to be routed differently from the actual traffic. In the case of the actual traffic from A to B, there exist both direct traffic circuits and a direct signalling link. In effect, this is the same as Figure 28.1, so that both signalling messages and

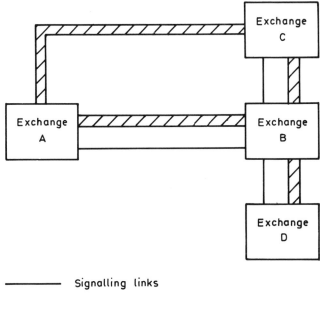

Signalling links

Traffic-carrying circuits

Figure 28.2
Traffic-carrying and signalling networks in CCITT 7.

traffic can be passed directly between the two. Similarly exchange B may pass signalling messages and traffic directly either to exchange C or exchange D, and may also act as a normal 'transit exchange' for two-link routing of traffic from exchange A to either of exchanges C or D. These are all examples of 'associated mode signalling', in which signalling links and traffic circuits have a similar configuration, and signalling messages and traffic both route in the same manner. In short, there is a signalling link associated with each link of direct traffic-carrying circuits.

By contrast, although exchange A is directly connected to exchange C by traffic-carrying circuits, there is no direct signalling link. Signalling information for these circuits must be passed on another route via exchange B. This is known as the 'quasi-associated mode' of signalling, and the signalling point (SP) in exchange B, is said in this instance to perform the function of a 'signal transfer point', or 'STP', as illustrated in Figure 28.3.

Signalling information is passed over CCITT 7 signalling links in short bursts—indeed, a CCITT 7 signalling network is like a powerful packet-switched data network. In order to identify each of the signalling points for the purpose of signalling message delivery around the network, each is assigned a numerical identifier, called a 'signalling point code'. This code enables an SP to determine whether received messages are intended for it, or whether they are to be transferred (in STP mode) to another SP. The codes are allocated on a network by network basis. Thus the code is only unique within, say, national network A, national network B or the international network.

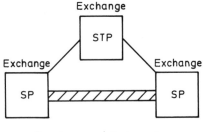

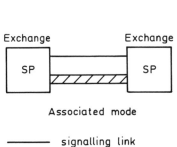

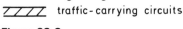

Figure 28.3
Modes of CCITT 7 signalling.

28.3 THE STRUCTURE OF CCITT 7 SIGNALLING

Thanks to the modular manner in which the CCITT 7 system has been designed, it encourages the development of new modules in support of future telecommunications services and functions. Figure 28.4 illustrates the functional structure of the CCITT 7 system, relative to the layers of the open systems interconnection (OSI) model (see Chapter 9).

In the same way as the OSI model has a number of functional layers, each an important foundation for the layers above it, so CCITT 7 signalling is designed in a number of functional 'levels'. Notice in Figure 28.4, that the component 'levels' and 'parts' of CCITT 7 do not align with the OSI layered model. The lack of alignment of signalling 'levels' with OSI 'layers' is unfortunate and it arises from the fact that the two models were developed concurrently but for different purposes. The lack of alignment of 'levels' with 'layers' means that not all higher layer OSI protocols are currently suitable for use in conjunction with the lower 'levels' of CCITT 7 signalling. The various standards development bodies are trying to rationalize the component parts of CCITT 7 to conform with the OSI model. Already the signalling connection and control part (SCCP) is being adjusted to deliver the OSI network service (OSI layer 3 service), so that a communications system can used the SCCP (and MTP below it) to support layers 4–7 OSI based protocols. The levels in CCITT 7 signalling provide a convenient separation of signalling functions, and in the remainder of the chapter the signalling level model is used in explanation.

28.4 THE MESSAGE TRANSFER PART (MTP)

The foundation level of the CCITT 7 signalling system is the message transfer part defined by CCITT Recommendations Q701–Q707. The message transfer part takes care of the conveyance of messages, fulfilling signalling level functions 1 to 3 as follows:

Level 1 (data-link functions). The first level defines the physical, electrical, and func-

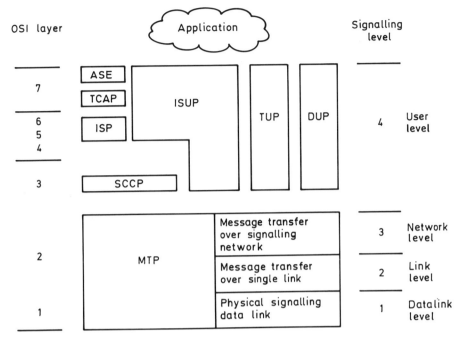

Figure 28.4
The structure of CCITT 7 signalling. ASE, Application service element; TCAP, Transaction capability; ISP, Intermediate service part; ISUP, ISDN services user part; TUP, Telephone user part; DUP, Data user part; SCCP, Signalling connection and control part; MTP, Message transfer part.

tional requirements of the signalling data link. The level 1 function is attuned to the particular transmission medium as laid down by CCITT G series recommendations. The level 1 function allows for an unstructured bit stream to be passed between SPs over an isolated signalling data link.

Level 2 (signalling link functions). This level defines the functions and procedures relating to the structure and transfer of a signal. Message flow control, and error detection and correction are included. (Flow control prevents the over-spill and consequent loss of messages that result if a message is sent when the receiving end is not ready to receive it: error detection and correction procedures eliminate message errors introduced on the link.)

Level 3 (signalling network functions). This level defines the functions and procedures for conveying signalling messages around an entire signalling network. It provides for the routing of messages around the signalling network. In this role it has a number of 'signalling network management' capabilities including 'load sharing' of signalling traffic between different signalling links and routes (illustrated in Figure 28.5) and re-routing around signalling link failures. Link sharing on the same route between signalling points (SPs) guards against lineplant failure (Figure 28.5(a)). Route sharing may additionally provide protection against failure of STPs. Thus in Figure 28.5 the signalling traffic from SP A to SPs B and C is shared over the two STPs, D

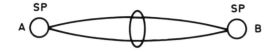

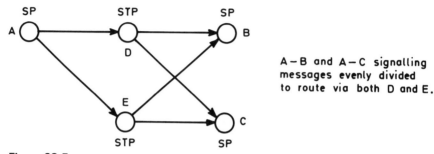

A—B and A—C signalling
messages evenly divided
to route via both D and E.

Figure 28.5
Load sharing over signalling. A—B and A—C signalling messages evenly divided to route via both D and E.

and E. In the event of a failure of any of the routes shown, signalling messages could be re-routed.

MTP is useless on its own for setting up telephone or other connections. To perform these functions MTP needs to be used in association with one of the CCITT 7 'user parts' which are *level 4 or user functions*. Examples are the telephone user part (TUP) and the Integrated Services Digital Network User Part (ISDN-UP or ISUP). These define the content and interpretation of the message, and they provide for connection control.

The structure of an MTP message is shown in Figure 28.6. It comprises four parts, transmitted in the following order:

Flag. The 'flag' is the first pattern of bits sent. This is an unmistakable pattern to distinguish the beginning of each message, and delimiting it from the previous message. It is comparable with the synchronization (SYN) byte in data communications (Chapter 9).

MTP information. The flag is followed by a number of 'fields' of information, which together ensure the correct message transfer. These fields include (i) the message sequence numbers that keep the messages in the correct order on receipt, and allow lost messages to be re-sent; (ii) information about the type and length of the information held in the main 'signalling information field'—it might say which 'user part' is in operation and record the length of the message.

Signalling information field. This is the main information field or the 'substance' of the message. The information is inserted by one of the user parts, as appropriate for

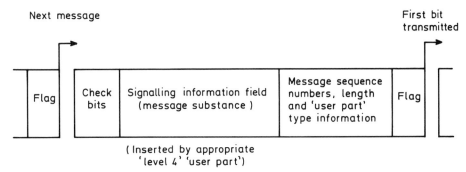

Figure 28.6
CCITT 7 MTP message structure.

the particular application (e.g. telephone user part (TUP), or integrated services user part (ISUP)). The structure of this 'field' depends on which 'user part' is in use.

Check bits. Finally, each MTP message is concluded with a 'check bit' field. This is the data needed to perform the error detection and correction mechanism of the MTP level 2. The check bits are followed by the flag at the start of the next message.

28.5 THE USER PARTS OF CCITT 7

The various user parts of CCITT 7 are alternative functions meeting the requirements of level 4 of the signalling level model. The user parts may be used in isolation, or sometimes may be used together. Thus the telephone user part (TUP) and the MTP together are sufficient to provide telephone signalling between exchanges. The data user part (DUP), ISDN user part (ISUP) and other user parts need not be built into a pure telephone exchange. An example where more than one user part is employed is the combination of SCCP (signalling connection and control part), ISP (intermediate service part), and TCAP (transaction capability application part). (These are all necessary for the support of the 'intelligent networks' described in Chapter 26.) The remainder of the chapter describes the capabilities of each of the level 4 user parts of CCITT 7.

The telephone user part (TUP)

The telephone user part comprises all the signalling messages needed in a telephone network for setting up telephone calls. (We described the sequence of call set-up in Chapter 7.) Thus an exchange using the CCITT 7 signalling system carries out the normal process of digit analysis and route selection, 'seizes' the outgoing circuit and sends the dialled digit train on to the next exchange in the connection by using the CCITT 7 signalling link, conveying TUP encoded messages using the MTP. Crudely put, an example of a TUP message might be 'connect the call on circuit number 56 to the destination directory number 071-234 5678'. Backward messages such as 'destination busy' are also included in the telephone user part.

TUP message

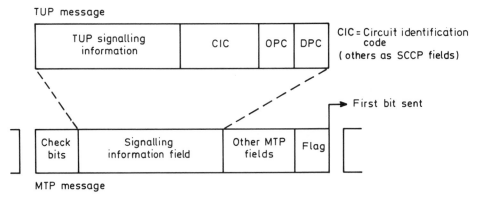

Figure 28.7
TUP message structure and relation to MTP.

The structure of TUP messages is shown Figure 28.7. TUP messages occupy the 'signalling information field' of the underlying MTP message. The messages comprise a *TUP signalling information field* which is used to convey 'dialled digits', 'line busy', 'answer' signals, and other circuit related information, together with four administrative fields as follows:

Destination point code (*DPC*). This code identifies the signalling point to which the signalling message is to be delivered by the MTP. (The destination of a signalling message is not necessarily the same as the final destination of the call.) The signalling point is in the exchange that forms the next link of the connection (for 'forward' messages) or in the previous exchange (for 'backward' messages).

Originating point code (*OPC*). This code identifies the signalling point which originated the message (again not necessarily the origination point of the call).

Circuit identification code (*CIC*). This is a number that indicates to the exchange at the receiving end of the signalling link, which traffic circuit each message relates to.

The telephony user part is defined in CCITT Recommendations Q721–Q725.

The data user part (DUP)

The data user part is similar to the telephone user part, but it is optimized for use on circuit-switched data networks. The message structure of the DUP is very similar to that of the TUP, illustrated in Figure 28.7. The DUP is defined by CCITT recommendation Q741 but is hardly ever used. It has been largely superseded by the ISUP.

The integrated services user part (ISUP)

Used in conjunction with the MTP, the CCITT 7 Integrated Services Digital Network User Part, ISDN-UP or ISUP, is the signalling system designed for use in ISDNs. In effect it is a combination of capabilities similar to TUP and DUP, which allow voice and data switched series to be integrated within a single network. The message structure is similar to that of TUP and DUP, but the messages used are incompatible with both of the other systems. ISUP is defined by CCITT Recommendations Q761–Q764.

The ISDN user part (ISUP) interacts as necessary with the ISDN D-channel signalling (Q931) to convey end-to-end information between ISDN user terminals. Such information includes the 'terminal compatibility' checking procedure which ensures that a compatible receiving terminal is available at the location dialled by the caller. As we learned in Chapter 24, the procedure prevents the connection of a group 4 facsimile machine to a videoconference at the receiving end.

The enhanced telephone user part (TUP +)

The TUP + is an enhanced version of the TUP, though incompatible with it. It was developed by CEPT as Recommendation T/SPS 43-02 for use as an interim ISDN-like signalling system supporting an early pan-European ISDN. It will be used mostly widely in Europe in the early 1990s before being superseded by ISUP.

The signalling connection control part (SCCP)

The SCCP is used to convey 'non-circuit related' information between exchanges or databases or between an exchange and a database. By 'non-circuit related' we mean that although a signalling relation is established between an exchange and a database, no traffic circuit is intended to be set up. In essence the SCCP (in conjunction with the TC and relevant 'application part') provides a means for 'querying' a reference store of information, as is necessary during call set-up on 'intelligent networks'. It is an ideal data transfer mechanism for:

- interrogation of a central database (Chapter 26);
- updating cellular radio 'location registers' (Chapter 27);
- remote activation and control of services or exchanges;
- data transfer between network management or network administration and control centres.

The SCCP is a recently developed 'user part', which in conjunction with the MTP allows a CCITT 7 signalling network to conform to the OSI network service (OSI layers 1 to 3), and to support protocols designed according to layers 4–7 of the OSI model. Most importantly, this allows new user parts to conform with the OSI model.

SCCP controls the type of connection made available between the two signalling points in the exchange and the database. In effect it establishes the 'signalling relation' in preparation for one of the higher level 'application parts'. Four classes of 'transfer service' can be made available. These can be grouped into 'connectionless' and 'connection-oriented' types as shown in Table 28.1.

Table 28.1
The classes of SCCP.

Class	Type
Class 0	Connectionless Message sequence not guaranteed.
Class 1	Connectionless Message sequence guaranteed
Class 2	Basic connection-oriented Message segmentation and reassembly
Class 3	Connection oriented Message segmentation and reassembly. Flow control. Detection of message loss and mis-sequence

In the context of the SCCP classes shown in Table 28.1, 'connection-oriented' data transfer (classes 2 and 3) is that in which an association is established between sender and receiver before the data is sent. In reality this means the establishment of a 'virtual' or 'packet-switched' connection between the ends, and not a circuit-switched connection. Thus before the application part signalling commences operation over the CCITT 7 signalling link, a 'virtual connection' is created by the SCCP. In the alternative 'connectionless' mode of data transfer (SCCP classes 0 and 1), messages are despatched on to the CCITT 7 signalling network without first ensuring that the recipient is ready to receive them. Connection-oriented procedures are useful when a large amount of data needs to be transferred. The connectionless mode is better suited to small and short messages, since it avoids the burden of extra messages to establish the 'connection'. By their nature, connectionless messages always include address information.

The structure of SCCP messages is similar to that of the TUP, as we can see from Figure 28.8, except that the SCCP includes no 'circuit identification code' (CIC). The CIC is superfluous since no circuit will be established.

Figure 28.9 shows an example of the use of SCCP and TC for a database query during the call set-up phase of a Freephone (800) call. The caller (who happens to be an ISDN subscriber) dials the 0800 number, which is conveyed to the ISDN exchange by the D-channel signalling protocol (Q931). Following analysis of the number, the ISDN exchange realizes that it must refer to the intelligent network databases for a number translation. It does so using the SCCP and TC. Meanwhile the circuit set-up is suspended. When the database interaction is over and the ISDN exchange has the appropriate information, the circuit can be connected using the standard ISUP signalling.

The SCCP is defined by CCITT Recommendations Q711–Q714.

Transaction capabilities (TC)

Building on the foundation of MTP and SCCP the transaction capabilities (TC) are that part of the CCITT 7 signalling system which conveys non-circuit-related infor-

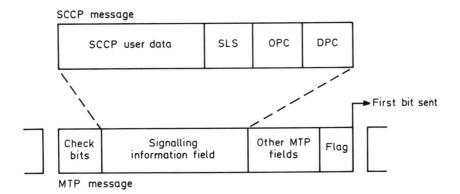

SCCP message

SLS = signalling link selection
OPC = originating point code
DPC = destination point code

Figure 28.8
SCCP message structure and relation to MTP. SLS, signalling link selection; OPC, originating point code; DPC, destination point code.

mation. Its development has been intertwined with the development of 'intelligent networks' (Chapter 26). The transaction capability is ideally suited for supervising short 'ping-pong' style dialogue between signalling points, typically between an exchange and an intelligent network database. Transaction capabilities (TC) break down into three component parts, and undertake the functions of OSI layers 4–7. The

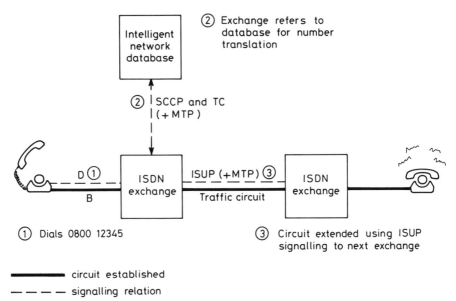

Figure 28.9
A database query using SCCP and TC.

underlying foundation (OSI layers 1–3) is the SCCP and MTP. The Intermediate Service Part (ISP) carries out the functions of OSI layers 4–6, while the 'component sublayer' and 'transaction sublayer' exist within OSI layer 7 and together form the 'transaction capabilities application part (TCAP)'.

Transaction capabilities exist to serve a 'TC user', normally called an 'application entity' (AE). An AE contains the necessary functions to serve a particular application. In addition, every application entity also contains the transaction capabilities application part (TCAP). TCAP is, in essence, a copy of the 'rules' which enable the messages to be interpreted. (For example an AE may support VPN service. Another might support 'freephone'.) Figure 28.10 illustrates the architecture of TC.

The ISP is very poorly defined at present (1990). It is required only when large amounts of data are to be transferred, using one of the SCCP connection-oriented classes. The reason the ISP is so poorly defined is that the 'intelligent network' applications so far conceived for TC involve only relatively short messages, requiring fast dialogue, in order that the calling customer may not perceive a long call set-up time.

The TCAP controls the dialogue between the exchange and the database, overseeing

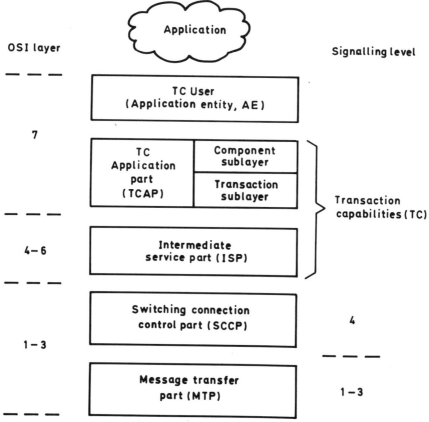

Figure 28.10
The transaction capabilities.

requests (questions or instructions) and making sure that corresponding responses are generated. The content of the request (the question itself) is prepared by the TC user (i.e. the 'application entity (AE)').

An application entity is the logical set of questions, responses and instructions which constitute the dialogue necessary to support an application. An AE comprises one or a number of simple functions, called application service elements (ASEs). The idea is that a small set of multi-purpose ASEs can be combined together in different permutations to serve different applications. Bellcore's intelligent network IN/2 (Chapter 26), for example, defines a set of primitive network actions which it calls 'functional components'. These are examples of ASEs. Figure 28.11 illustrates the concept of application service elements.

By the end of 1989, two application entities had been defined by CCITT. One was called the 'mobile application part' (MAP) and is used to support 'roaming' mobile telephone services. The other is the 'operation and maintenance application part' (OMAP), used for control and maintenance of remote equipment and exchanges. MAP comprises one complex ASE, OMAP comprises two—MRVT and SRVT (the MTP and SCCP routing verification tests).

Returning to the TCAP itself, let us briefly describe the functions of the 'transaction' and 'component' sublayers. The 'transaction sublayer' is responsible for initiating, maintaining, and closing the dialogue between the signalling points. Classification of messages into one of the four types listed below help the two end signalling point devices to relate each message back to the previous dialogue. Thus each message also has a transaction identity code.

- 'Begin' (dialogue or 'transaction');

- 'Continue';

- 'End';

- 'Abort'.

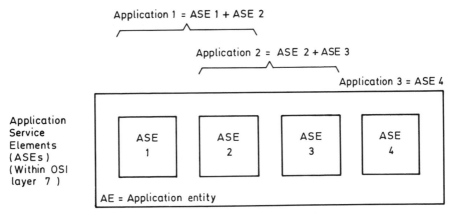

Figure 28.11
Permutation of ASEs to serve different applications.

The 'component sublayer' provides machine discipline to the dialogue, controlling the 'invocation' of requests and making sure that they receive proper responses. The ASE information within the requests and responses is thus classified into one of five types:

- to invoke an action;

- to return the final response (to a sequence);

- to return an intermediate (but not final) response;

- to return a message to signal an error;

- to reject a message (if a request is not understood, or is out of sequence).

Although the information content of the requests and responses is not known by the component layer (it is understood only by the ASE or ASEs), the component sublayer is able to make sure that commands are undertaken and responses are given.

The transaction capabilities are defined in CCITT Recommendations Q771–Q775.

The mobile application part (MAP)

The mobile application part is an example of an application entity of CCITT 7 signalling, developed to serve a particular application. It is used between a 'mobile' telephone network exchange and an intelligent network database, called a 'home' (HLR) or 'visitor location register' (VLR). The database is kept informed of the current location of the mobile telephone handset. Thus the mobile telephone customer's incoming and outgoing calls can be handled at any time. Chapter 27 on mobile services describes this application more fully.

Operation and maintenance application part (OMAP)

OMAP is another 'application entity' of CCITT 7 signalling. It provides for network maintenance as well as other network operations and management functions of remote exchanges and equipment. OMAP contains two ASEs—the MTP Routing Verification Test (MVRT) and the SCCP Routing Verification Test (SRVT). These are procedures designed to enable the network operator to test the integrity of the signalling networks and identify faults.

28.6 THE USE AND EVOLUTION OF CCITT 7 SIGNALLING

CCITT 7 is an adaptable and continuously evolving signalling system that has been designed to meet the challenging and ever-changing service needs of public and major private network exchanges making up the ISDN and 'intelligent network'. Network operators may choose to implement the subset of user parts which best matches their needs, adopting new user parts as they become available.

Depending upon their particular circumstance, some network operators may choose to implement an adapted version of some of the CCITT 7 standards, taking up some of the permitted signalling 'options'. The options allow the operator to 'tailor' the system to particular national requirements (e.g. C7/BT is the UK national version of TUP/ISUP and T1-ISUP is the version of ISUP in the United States). Because of the considerable capital investment already committed to CCITT 7, there is a pressure for use of a common system, and there is pressure also from established CCITT 7 users to ensure that new developments are 'backward compatible' with previous versions of the system.

One of the ways in which backward compatibility is ensured is by building into all C7 implementations a mechanism for handling unrecognized information. Such information is bound to get sent occasionally from a more advanced exchange to an exchange with an older version of CCITT 7. The common methods for dealing with this information are:

- to discard it;

- to ignore it;

- to assume that some other 'expected' response (or 'default') was actually received;

- to reply with a message of 'confusion';

- to terminate the call and reset;

- to raise an alarm to a human.

Such a 'backward compatibility' mechanism obviates any need to stop the development of CCITT 7, or the extension of its services. If backward compatibility is not built into any new signalling standard then joint discussions between the operators of inter-connected networks using different versions will be necessary to agree amendments allowing compatible operation.

28.7 SIGNALLING NETWORK PLANNING AND TESTING

The signalling links of a CCITT 7 'signalling network' need careful planning and implementation just like any other packet-switched network. The links need to be sufficient in number to handle the overall signalling traffic demand, and to be oriented in a topology that gives good resilience to network failures.

Two possible topologies are illustrated in Figure 28.12. Figure 28.12(a) shows a meshed network in which exchanges are capable of both SP and STP functions and rely upon one another for the resilience of their signalling relations. In contrast, Figure 28.12(b) shows a topology commonly used in North America, where dedicated and duplicated computers perform the STP function alone; the exchanges are not capable of the STP function.

Because of the very complex nature of CCITT 7 signalling, and because of the heavy network reliance upon it, it is normal to undertake a comprehensive validation testing programme prior to the introduction of each new link and exchange. Exchanges built by different manufacturers can sometimes be incompatible at first, and a testing

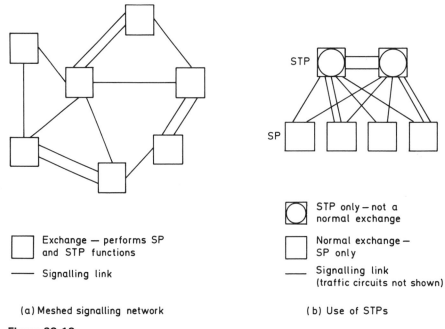

STP

SP

STP only — not a
normal exchange

Normal exchange —
SP only

Signalling link
(traffic circuits not shown)

Exchange — performs SP
and STP functions

Signalling link

(a) Meshed signalling network

(b) Use of STPs

Figure 28.12
Typical CCITT 7 signalling networks.

programme is invaluable in ensuring that their problems are ironed out before being
brought into service.

28.8 INTERCONNECTION OF CCITT 7 NETWORKS

Within a network owned and operated by a single network operator, the CCITT 7
signalling system (or a variant of it) will make for close control and monitoring of
the network and highly efficient call routing. However, when networks belonging to
different network operators are connected together, neither operator is likely to want
the other to have full control of his network. For example:

(i)   In an international network, an operator in one country is unlikely to let
      the operator in another control his network management and routing
      rearrangements.

(ii) Competing network operators in a single country may have their networks
      interconnected but each guards the monitoring and control of his own network
      fiercely.

(iii) Public network operators, while they may allow direct CCITT 7 signalling from
      company private exchanges, are unlikely to give up control of the network.

For these reasons, a number of options are permitted within the signalling system specifications. These allow operators by mutual agreement to restrict the capability of the signalling system when it is used for network interconnection. Thus a period of negotiation is necessary prior to the interconnection of networks. A mutual testing period confirms that the options have been selected correctly.

BIBLIOGRAPHY

CCITT Recommendations Q701–Q795.
CEPT Telephony User Part (enhanced) (TUP +), CEPT Recommendation T/SPS 43–02.
Data User Part (DUP), CCITT Recommendation Q741.
Integrated Services User Part (ISUP), CCITT Recommendation Q760–9.
Message Transfer Part (MTP), CCITT Recommendation Q701–709.
Mobile Application Part (MAP), CCITT Recommendation Q.
Operations and Maintenance Application Part (OMAP), CCITT Recommendation Q795.
Signalling Connection and Control Part (SCCP), CCITT Recommendation Q711–4.
Telephony User Part (TUP), CCITT Recommendation Q721–725.
Transaction capabilities (TC), CCITT Recommendation Q.

29

BROADBAND COMMUNICATION AND MULTI-SERVICE NETWORKS

The advantages of multi-service networks providing a wide range of telecommunications services using shared resources and common interfaces have been clear to network operators and developers for many years. Since the first modem introduced data traffic to the telephone network, researchers have sought ways of using common networks to provide services.

We saw in Chapter 24 how the Integrated Services Digital Network (ISDN) is one step in the development chain, and how it may replace the conventional telephone and data networks in the next few years. But now we hear that ISDN in its current narrowband form falls a long way short of the ultimate needs of a multi-service network, because of its relatively restricted bit rate (bandwidth) and because of the lack of service flexibility arising from the fact that it evolved out of telephone network principles. The maximum rate currently possible on ISDN is $n \times 64$ kbit/s, up to 2 Mbit/s. Such rates are not enough for very high speed file transfer between mainframe computers, or for interactive switching of broad bandwidth signals such as high definition television (HDTV) and video.

Increasing the unit bit rate (64 kbit/s) of ISDN would increase bit rates and make 'broadband' services possible but it would be at the expense of gross inefficiency in network resources for more 'run-of-the-mill' services. So to suit the needs of 'broadband services' (high bit rate services) new types of technology are emerging. Capitalizing on all the best aspects of LANs, packet switching and ISDN, these new technologies from a new generation of multi-service networks under the name integrated broadband communication (IBC). Particular types of IBC include 'broadband ISDN' or 'B-ISDN'. This short chapter discusses the hurdles to be overcome and describes some of the IBC and B-ISDN techniques developed so far.

29.1 THE EVOLUTION OF BROADBAND NETWORKS

The rate at which information can be transported over a telecommunications connection is limited by the bandwidth. The greater the bandwidth, the greater the maximum rate of information transfer. But in addition, the type of services that can be carried will be influenced by the manner in which the network operates. The broadband network will cater for a wide range of digital bit rates and a variety of different service demands. The technology to support the broadband service is still in the course of development. Initial use is expected to include video, television, and high-speed data networks, particularly when very high definition pictures or rapid data response times are required.

Like the ISDN, IBC and B-ISDN (as shown in Figure 29.1) will be a technology of integrated digital switching and transmission, and many of the network control functions may be carried over from ISDN, but the method of information transfer will be radically different and more flexible in order to cope with the wide and varying bandwidths and services which are envisaged. The concept of integrated service engineering (ISE) will separate the intelligence, conveyance and other capabilities of the network into modular units, which combined in different ways will allow the needs of a wide range of individual and diverse services to be met in an optimum way.

Multi-media applications will become commonplace. For example, word processor terminals might be combined with high definition video—so that diagrams in the text of a report could actually be short video films!

Ultimately, IBC will be of such unprecedented complexity, that the evolution will have to be slow, and phased. A number of characteristics can be expected in even the earliest switched broadband networks:

- high bandwidth carrying capacity;

- support of many different services;

- common user-to-network interface for all applications;

- simultaneous support of various applications;

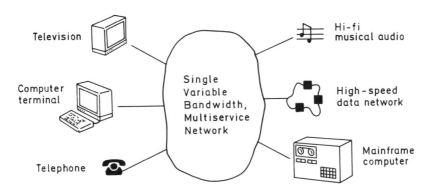

Figure 29.1
The concept of integrated broadband communication (IBC).

- communications isolation between different services (i.e. no interferences of services);

- low unit cost (at least in the longer term), and modular design;

- low initial investment;

- interconnection with established, dedicated networks, and gradual absorption of these services and capabilities.

Early networks not meeting these criteria are likely to be superseded before full implementation.

Ambitious field trials, including the EEC's RACE (Research into Advanced Communications in Europe), will contribute to the introduction of IBC, progressing to European community-wide services by 1995.

The evolution will be complex and sometimes frustrating, as are all long-term Research and Development (R&D) projects. There will be the difficulties of getting services to share a common network, and others in contriving a smooth transition from current network technology. There will be a need to reappraise results, keeping in mind all the investment sunk in the networks, and the need for continuing support of established services. The technology and the cost of its continued development will also need constant reappraisal in the light of other emerging technologies. While a number of parallel developments are currently progressing, ultimately only a small number will prevail.

29.2 THE DEMANDS ON MULTI-SERVICE BROADBAND NETWORKS

A number of existing network techniques, including packet-switching, ISDN, and local area networks (LANs), are under study as the basis of a future broadband and multi-service network. A common failing of existing techniques is their restriction to low bit rate *data* conveyance or fixed bandwidth circuit switching. Without development, this makes them unsuitable for use in broadband networks. The major difficulty is that it is difficult to optimize a network to carry simultaneously both low and very high bit rate services, and to meet simultaneously the exacting demands of rapid signal conveyance necessary for speech and equivalent signals.

Of today's technologies, packet networks and LANs offer the best prospect of developing into the variable bandwidth carrying mechanisms which are needed for IBC or B-ISDN. However, today's networks put too much emphasis on assured and error-free delivery of data, at the expense of relatively prolonged and—worse than that—uneven times for message propagation and delivery. Slight delay in data networks is unimportant and may be imperceptible to the end user; messages are queued up until transmission capacity becomes available. But long or variable delays are quite unacceptable in services such as speech and video signals, because they result in irregular pauses giving a staccato, snowy or clipping effect to voice or picture.

The advantage of packet-switching as the basis for broadband networks is that any number of connections of differing bandwidth can be established at the same time. The different bandwidths are accommodated by using a large number of packets per second to give a high bandwidth, or a smaller number of packets for modest bit rates.

Thus the overall bit rate can be shared between a number of users. All quite acceptable—that is, unless the network gets too close to its maximum capacity. If too many connections are established (i.e. there is a period of overload) all users end up in the data queue, and experience slow throughput and response. Data users will at least be grateful that no one is cut off and no information is lost, but for voice and video users it would have been better to refuse to connect any new callers.

Figure 29.2 illustrates the degradation of a packetized speech channel, sharing a packet network with other speech and data users. The network is subject to congestion, so that every now and then packets have to be held up in the buffer to await the availability of transmission capacity. Thus occasional speech samples are delayed in the network, and the result is a rather disjointed received signal. Unlike computers, we cannot ask a chatting couple to speak at the precise moments when we would like!

A second important limitation of today's packet networks, when used for voice and video, is their relative inefficiency. When sending single packets of a small number of bits (eight for a single speech 'sample'), the network is kept heavily loaded just carrying and processing all those packet headers! Furthermore the error detection and correction procedures which ensure that each packet value is received correctly are largely irrelevant for speech and video signals, since the ear or eye tends to fill in any slight variance in the signal anyway.

In summary, the important limitations of present day LAN and packet-switching techniques for the carriage of voice and video over broadband networks are as follows:

- excessive delay and irregular pause lengths;

- inability to offer guaranteed bandwidth and assured call quality;

- during network overload the available bandwidth is divided between the calls with the effect of degrading all of them, even those previously set up;

- poor network overload control;

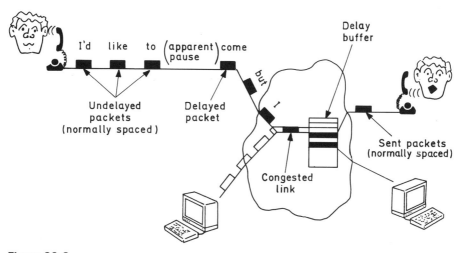

Figure 29.2
Degradation of speech on normal packet networks.

- inefficient employment of bandwidth when used for voice and video applications, arising from unnecessary error correction and other data administration techniques.

We now discuss four maturing techniques for bradband networks, explaining how the above limitations can variously be overcome. The four techniques are called:

- Metropolitan Area Networks (MANs);

- Asynchronous Transfer Mode (ATM) switching;

- Fibre Distributed Data Interface (FDDI);

- Broadband Passive Optical Network (BPON).

29.3 METROPOLITAN AREA NETWORKS (MANs)

MAN refers to a new, extended breed of local area network (LAN) technology, but one in which high-speed data transfer is not possible within the confines of a single office or building but across whole cities or metropolitan areas. Ultimately they may evolve into a globally-linked infrastructure supporting multi-services. At the 140 Mbit/s speed of some prototype MANs, a very large computer file (say 1 Mbyte) could be retrieved from a distant data source across the city in well under a second. In addition, MANs will make possible sophisticated video and graphics—integrated computer workstations combining word processors, electronic mail, high definition video and videotelephones are already appearing.

Various different MAN technologies are appearing. They are LAN-type rings, buses and access protocols to make the broad-bandwidth and high bit rates possible. A number of variants exist, and these loosely include ATM and FDDI which we discuss shortly. However, the technology for which the term is most often applied is defined under IEEE Standard 802.6.

Networks conforming to IEEE 802.6 use a protocol called distributed queue dual bus (DQDB). This was codeveloped by Telecom Australia, the University of Western Australia, and their joint company, QPSX Communications Limited. It is designed to give a possible migration path to IBC and B-ISDN. IEEE 802.6 is the standard Bellcore of the United States specified to provide the initial subscriber interconnection to the 'switched multimegabit digital service (SMDS)', a proposed public data communications service which is due to start service in 1991.

In Europe, the asynchronous transfer mode (ATM) is appearing. Similar in nature to the 'fast packet switching' techniques of IEEE 802.6, the European version also includes some techniques derived from TDM—and by so doing gives very good performance when carrying services which demand the high speeds and other characteristics previously only found in circuit-switched networks.

Finally, there is FDDI, a high bit rate optical fibre backbone technology allowing the interconnection of LANs over a wide area spanning up to 100 km. A key question concerns the relationship of MANs (IEEE 802.6) and FDDI (IEEE 802.8)—the degree to which they evolve to duplicate and/or complement one another. For now, we can only hope to understand them as they are today and accept the pundits' view that FDDI is intended currently only for high-speed data—and is considered to be a 'private network solution' (e.g. for a university campus) rather than a public metropolitan service (as MANs).

29.4 ASYNCHRONOUS TRANSFER MODE (ATM)

In CCITT's 1988 recommendations, the asynchronous transfer mode (also called 'fast packet switching') was provisionally accepted as the basis for future B-ISDNs. A number of ATM research projects are in progress. These seek to improve the throughput of (asynchronous) packet-switching networks and give them attributes of assured connection today found only in circuit-switched networks. All the projects seek to provide:

- simultaneous carriage of mixed telecommunications services;

- some means of overload control;

- a limit to the delay experienced by video, voice and other delay-sensitive signals;

- relatively high efficiency in the use of bandwidth.

The information conveyed over an ATM network must first be encoded into 'packets'. Complete data is already held in binary form, so that it merely needs to be sorted into packets, each packet consisting of an information field (in which the user data is held and conveyed) and a packet 'header' which 'tells' the network where to deliver the packet and provides a sequence number so that packets can be re-assembled in the correct order at the receiving end.

Typical packet lengths used in broadband ATM networks are 128 bits, but longer packets may be available for supporting very high bit-rates. Shorter packet lengths tend to be avoided as far as possible, because of the increased load they inflict on the network in terms of the total demand for bit carriage and the increased workload in interpreting the headers.

When speech signals are conveyed over broadband ATM networks, they must be packet encoded. The inefficiency in use of the overall bit rate and the overhead associated with processing packet headers militate against sending packets with only eight bits of information, corresponding to individual speech samples. Instead a larger single packet is used to convey a number of consecutive speech samples of the same conversation. A packet size of 128 bits allows 16 samples to be sent simultaneously, corresponding to a period of speech of 2 milliseconds duration. Sending simultaneous samples like this inflicts a delay in the propagation path but since it is not large enough to cause difficulties for most listeners, it is a small price to pay for a transmission network efficiency in excess of 75 per cent. Incidentally this same technique is applied in TDMA satellites (Chapter 13), where several samples are sent in each burst of transmission.

We next consider an example of ATM in order to illustrate how the various constraints of current packet technology are circumvented. The example we use is the *Orwell slotted ring*, developed by British Telecom.

British Telecom's 'Orwell ring'

The switches in an Orwell multi-service network are structured in a ring, with links to other exchanges radiating outwards in a star pattern, as shown in Figure 29.3.

The flow of data packets on to the ring from the transmission links is controlled by the 'Orwell protocol'. This is similar to the 'token-ring' passing method used on some local area networks. When controllers at the intersection of transmission lines and the ring are given the 'token', they may send packets around the ring in any free information carrying 'slots'. Meanwhile, all the other controllers continuously scan the headers of the packets passing around the ring, removing those addressed to themselves.

An ingenious feature of the Orwell protocol is the way it maximizes the use of network bandwidth by the use of an efficient bandwidth release mechanism, emptying the 'slots' at their destination. The positive acknowledgement procedures which confirm the receipt of data by returning the token to the originating node, can consume up to half the available bandwidth. By performing error correction and data administration in a different way, the Orwell protocol conserves this bandwidth.

'Tokens' are allocated from the overall control processor of the switch to each 'controller' in the ring. They allow a controller to send a limited volume of data (or number of packets) around the ring. Having sent this volume, the controller must stop and wait for a new token. This prevents one controller from hogging the ring at the expense of others, and is one of the key aspects of the Orwell technique that minimizes the delay for voice and video packets. The amount of data that each controller is allowed to send for each token is determined by the switch control mechanism, and may be varied from instant to instant, so giving a sensitive means of overload control which helps to reduce packet delivery delays and guarantee bandwidth for the entire duration of established calls.

Rather than allow a queue to accumulate, with an increase in delay or the loss of an individual packet, the switch measures the load on the network to ensure that sufficient capacity is available to carry each new call. Depending on intended application, other broadband ATM techniques do not necessarily have this feature.

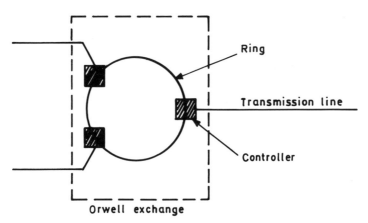

Figure 29.3
Architecture of an Orwell network.

29.5 FIBRE DISTRIBUTED DATA INTERFACE

The Fibre Distributed Data Interface (FDDI) is a 100 Mbit/s LAN token ring under development to become an American National and International Standard. It will interconnect LANs over an area spanning up to 100 km, allowing high-speed data transfer. It will be known as IEEE 802.8 or ISO 8802.8. A second generation version, FDDI-2 will include a capability similar to circuit-switching to allow voice and video to be carried reliably in addition to packet data. First proposed as a high-speed link for the needs of broadband terminal devices, FDDI is now perceived as the optimum backbone transmission system for campus-wide wiring schemes, especially where network management and fault recovery are required.

Computer manufacturers are rapidly incorporating FDDI into their architectures and product ranges, and initial networks are appearing for such uses as providing backbone interconnection networks for existing IEEE 802 LANS (e.g. Ethernet and token ring).

The standard is defined in four parts:

- Media access control (MAC), like IEEE 802.3 and 802.5 (see Chapter 9), defines the rules for token passing and packet framing;

- Physical layer protocol (PHY) defines the data encoding and decoding;

- Physical media dependent (PMD) defines drivers for the fibre optic components;

- Station management (SMT) defines a multilayered network management scheme which controls MAC, PHY and PMD.

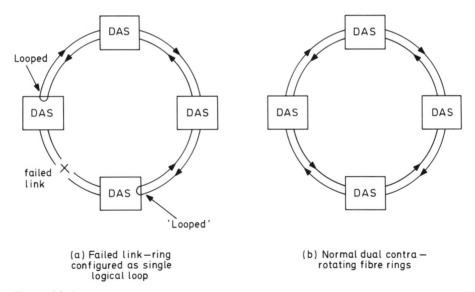

(a) Failed link—ring configured as single logical loop

(b) Normal dual contra—rotating fibre rings

Figure 29.4
The fibre distributed data interface (FDDI) fault recovery mechanism for double attached stations. DAS, double attached station.

Campus

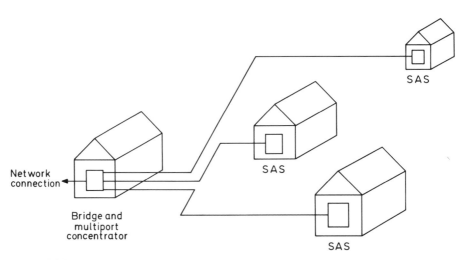

Network connection

Bridge and multiport concentrator

SAS

SAS

SAS

Figure 29.5
Star configuration of FDDI. SAS, single attached station.

The ring of an FDDI is composed of dual optical fibres interconnecting all stations. The dual ring allows for fault recovery even if a link is broken by reversion to a single ring, as Figure 29.4(a) shows. The fault need only be recognized by the CMTs (connection management mechanisms) of the station immediately on either side of the break. To all other stations the ring will appear still to be in its normal contra-rotating state (Figure 29.4(b)).

When configured as a ring, each of the stations is said to be in dual-attached connection. Alternatively, a fibre star connection can be formed using single-attached stations with a 'multiport concentrator' at the hub (Figure 29.5). Single-attached stations (SASs) do not share the same capability for fault recovery as double-attached stations (DASs) on a dual ring.

Like token ring and Ethernet, FDDI is essentially only physical layer (OSI layer 1) and data-link layer (OSI layer 2) standard. At layers 3 and above, protocols such as X25, TCP/IP may be used.

FDDI-2, the second generation of FDDI (Figure 29.6) has a maximum ring length of 100 km and a capability to support around 500 stations including telephone and packet data terminals. Because of this, it holds the prospect of supporting entire company telecommunications requirements, and has blurred further the distinctions between it and other types of metropolitan area network (MAN).

The FDDI-2 ring is controlled by one of the stations, called the cycle master. The cycle master maintains a rigid structure of cycles (which are like packets or data 'slots') on the ring. Within each cycle a certain bandwidth is reserved for circuit-switched traffic (e.g. voice and data). This guarantees bandwidth for established connections and ensures adequate delay performance. Remaining bandwidth within the cycle is available for packet data use.

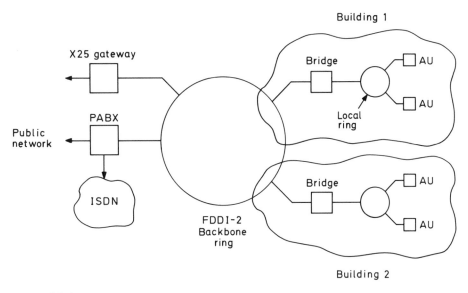

Figure 29.6
The fibre distributed data interface-2 (FDDI-2). AU, access unit.

An important application of FDDI-2 is its interworking with the new integrated voice data (IVD) LAN standard which is defined in IEEE 802.9.

29.6 BROADBAND PASSIVE OPTICAL NETWORK

Broadband Passive Optical Network, or BPON, is a simple approach to broadband networking with a very clearly focused commercial application. It is a technology proposed and developed by British Telecom and intended to bring fibre to the home. Basically, it is a network composed of monomode fibres (either in a ring or star topology) connecting telephone exchanges and people's homes, but allowing not only basic telephony, but also the 'broadcasting' of cable television and video programmes.

The foundation of BPON is a technique known as TPON (telephony passive optical network). This is the method by which telephone services are provided in residential homes by means of fibre connected back to the exchange (again in either a ring or star topology). Optical couplers enable the various fibre distribution joints to be made without active electronic components (Figure 29.7). Individual calls from customers are time division multiplexed (TDM) at the exchange and selectively demultiplexed by the appropriate subscriber's receiver.

By using TDM and a technique known as wavelength division multiplex (WDM—basically the use of another laser of a different wavelength light), other broadband signals may be carried over the same fibre network. Thus the broadcast of cable television and video services is possible simultaneously with the telephone operation. This is the principle of BPON.

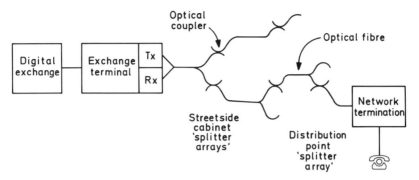

Figure 29.7
Telephony passive optical network (TPON).

29.7 OTHER BROADBAND TECHNIQUES

A broadband switching technique under study by CCITT and worthy of mention, but not with the same scope for support of multi-service networks, is the *associated packet mode bearer service* (*APMBS*). APMBS is an efficient method of broadband packet data transfer being developed for the ISDN, D-channel and the CCITT 7 signalling network. It is intended to complement the narrowband and dedicated packet service networks already available. It achieves the broadband capacity by relaying whole frames of data ('frame relaying') rather than conveying small individual packets of data. The associated packet mode bearer service is valuable for connection between computers when intermittent rapid response times are required, without the expense of a permanent high-speed connection.

Another evolving broadband technique discussed here only briefly is the millimetric microwave videodistribution system, MMMVDS, or M^3VDS. This is a radio-based technique for local distribution of video and broadband services.

29.8 NETWORK EVOLUTION TO BROADBAND

It will be many years before we see a fully integrated variable bit rate, broadband, and multi-service network. Leading up to this time we will see the emergence of small-scale and specialized broadband networks, optimized for particular applications. But as technology develops and demand grows, a gradual migration of services from existing networks is inevitable. Cynics might say 'what demand?', and 'how can we possibly wish to convey so much information?' But cynics have been confounded before. It is only a matter of time!

BIBLIOGRAPHY

IEEE 802.6, Metropolitan Area Networks (MANs).
IEEE 802.7, Broadband LANs.
IEEE 802.8, FDDI/Fibre optic LANs.

IEEE 802.9, Integrated voice and data (IVD) LAN.

Kim, G. Y. *Broadband LAN Technology*. Artech House, 1988.

Wideband Communications. Proceedings of the International Conference held in London (Oct. 1986). Online Publications.

Wilson, R. G. and Squibb, N., *Broadband Data Communications and Local Area Networks*. Collins., 1986.

Part 5

RUNNING A BUSINESS BASED ON TELECOMMUNICATIONS

30

MEETING BUSINESS NEEDS AND CREATING COMPETITIVE EDGE

Computer and telecommunications systems cannot be treated any longer as operations merely for the backroom. Instead electronic mail and messaging systems, customer databases, computer-aided design (CAD) systems, management information systems, voice and image systems, together with the telecommunications networks that make them possible, are revolutionizing the way we do business, with profound effects on company success and profitability.

Companies nowadays must plan and manage their information technology (IT) resources with as much concentrated effort and imagination as they apply to any other resource or activity. The potential benefits are enormous—costs and lead times can be reduced, along with the manpower and other resource needs, while quality and serviceability can be improved.

It follows that a company's IT development cannot be left entirely to the data processing men and women, the computer centre staff and the telecommunications manager. IT has become the concern of the company's top people, and they must build it solidly into the company's central strategy. Too many companies have caught a cold over the lack of a proper and coherent IT strategy, and many now face the long and costly process either of integrating a variety of small computer or telecommunications systems that have been installed in penny numbers or clawing back the competitive ground they have lost.

30.1 CONTENT OF AN IT STRATEGY

Effective IT strategy combines an understanding of company needs, mission, and working practices, with practical understanding of what is technologically possible. Computer and telecommunications equipment should be planned and selected to serve the company's overall interests. The strategy will combine idealistic targets for handling and processing the company's 'information', with a recognition of the strengths and weaknesses of any existing company and IT structure, and will evolve this in a

552

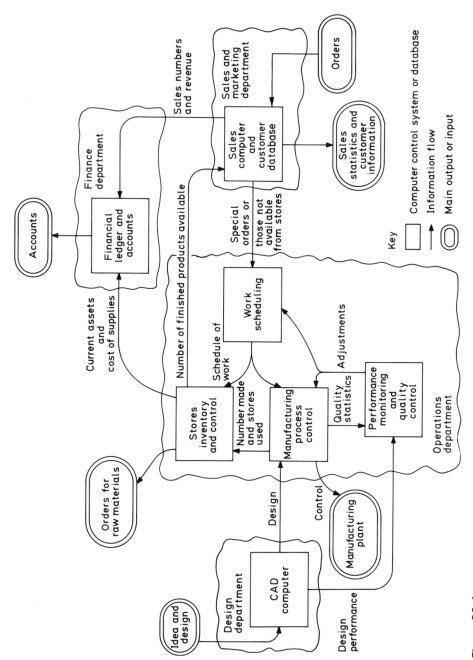

Figure 30.1
IT network for a manufacturing company.

manageable way. Priorities set for development will make allowance for time pressures and short-term goals, but while working towards the long-term objective.

For some large companies, with the time, money, and expertise, it may be right to consider immediate full interconnection of all existing computer and telecommunications systems, and by so doing streamline the operation of previously diverse functions such as financial ledger, warehouse stock control, customer orders and enquiries. Such interconnection could speed up the flow of information between various company departments, with consequential improvement in information detail and accuracy, better customer service, and faster preparation of accounts. However, the systems can only be interconnected where individual systems have been designed and built according to the plan of a previous integration strategy.

Figure 30.1 illustrates a simple network of computers and telecommunications being used to support a manufacturing operation. In total, seven compatible computer systems are in use. Each has its own function and has been designed to carry it out in an optimum way, but each is able to communicate information with the other six systems. While the computer-aided design (CAD) computer relies on working hand-in-hand with the human designer for creative input, the computer controlling the manufacturing plant will instead have to run for long periods without human interference—and at high levels of accuracy and reliability. The two systems are very different, but need to communicate the design information.

Overall, the computer and telecommunications systems of Figure 30.1 support an optimum company organizational structure and the necessary 'key information flows'. This does not mean that all the systems need to be implemented at the same time—in one expensive and traumatic step. Instead the individual systems can be rolled out in an order dictated by company priorities and resources.

For smaller companies using computing and advanced telecommunications technology for the first time it is right to produce a long-term 'information system strategy', but to invest in the first place in only one of the component parts. This allows the company to adapt more gradually to the different working practices demanded by the computer environment.

The study of information flows should not be restricted to those which can be computerized. The strategy also needs to take cognizance of business information, flowing either by word-of-mouth or on paper deliveries. Today's technology allows considerable integration and streamlining of all these modes of information flow, as we shall see.

30.2 THE STUDY OF INFORMATION FLOWS

Many companies have recognized the benefits of computerization and are at various stages in the automation of their main 'supply chain'. Computers control the taking of orders, the design of products, the manufacture of goods and the scheduling of deliveries. Very few companies, however, have appreciated the need for complete information flow analysis. This is borne out by the huge expenditure on telephone bills, site visits, and mailing of documents.

Let us return to our model of information flow methods presented in Chapter 1. The model is shown again in Figure 30.2.

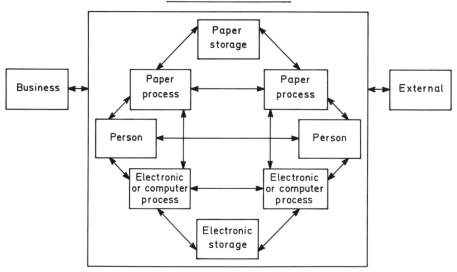

Figure 30.2
Modes of information flow.

The model simplifies the issues of information flow, helping us to analyse business processes and to perform value and cost-benefit analyses. The model shows three basic modes of information transfer and two methods of storage, namely:

● paper flow;

● person-to-person flow;

● inter-computer (electronic) flow;

● paper storage;

● electronic storage.

Far-sighted companies have probably already set themselves a target to achieve widespread computerization and automation of information flows, but how many have truly achieved it? A couple of examples bring our model to life.

Figure 30.3 shows an example of near achievement of the goal for entire computerization. It is drawn from the motor-car industry. Note how all the basic information flows are held, processed, or conveyed by computer. There is still interaction with humans for design input and tooling, but there is no human 'bottleneck' or breach in the main flow of information—from right to left of the diagram—the key information can always be retrieved from the computers.

Next we show a far less well organized company common in many industries. Figure 30.4 illustrates a company rife with manual processes, paper, and person-to-person information flows which are strangling the business—even though a considerable amount of computing equipment has been installed. The example is of a simple

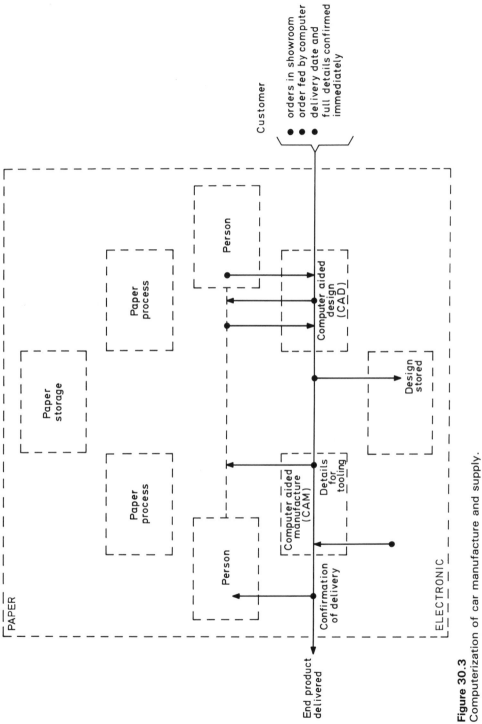

Figure 30.3
Computerization of car manufacture and supply.

distribution company. The company takes orders from retail and wholesale outlets and delivers goods by lorry from its warehouse. The orders are taken by 'telesales' operators who make telephone calls out to customers according to a scheduled call-out list. These orders are then typed into a computer system, which records customers' invoices, and 'picks' stock from the warehouse. Next, the 'order picking list' is printed out on pages in the warehouse, and a delivery schedule is produced for the lorry driver. These enable the goods to be loaded and delivered.

The problems and missed opportunities are manifold. For a start, by calling the customer over the telephone we miss the opportunity to allow him to order electronically—from a PC or some other keyboard device. His order has to be made when we ring him—rather than when he wants. And if our competitor were to ring him up first, we might lose the order. If instead we supplied the customer with a PC and a 24-hour communication path then we might be able to eliminate our competitors. A further benefit of this approach would be greater accuracy in processing orders—and greater responsiveness to changed orders.

Changed orders, traffic jams and other problems snarl up the ideal schedule. Stock may go missing, be damaged, or broken. The result is that the stock available may not meet that demanded by the picking list. Thus what is loaded on to the lorry may differ from what ought to be. Some is broken on route. On an unacceptably high number of occasions, when the goods are actually delivered, they neither match the original order nor the pre-prepared delivery note. So the delivery man amends the delivery note by hand, gets the customer to sign it to confirm delivery, and then takes it back to a clerk at the depot, who keys the actual delivery details back into the central computer.

If the invoice is produced at this point, then with luck the customer will not dispute it, but if, as in some companies, the invoice was produced at the time of the original order, then further correction will be necessary to sort out the inevitable customer query. The result is a spaghetti of information flows—messages and enquiries—on paper and by word-of-mouth, trying to sort out the mess. Local 'entrepreneurs' introduce their own isolated personal computer to help with their own departmental problem, and at least the job finally gets done, but not without a heavy penalty of cost, delayed payment and customer dissatisfaction. Customer enquiries are a nightmare because the information is so difficult to trace—word of mouth? computer? paper records?

Much simpler to identify the important information flows, get them in the correct sequence, and support them with information technology at the right place. In the example of Figure 30.4 a computer terminal in the cab of the delivery truck could enable the delivery note and even the invoice to be produced at the point of delivery. This would mean both would always be correct and delivered in a timely manner to the customer, leading to fewer queries, reduced customer frustration, and timely customer payment.

The lesson is simple and obvious—information flow studies need to address all information flows. These include not only the obvious and ideal flows but also the more 'out-of-sight' ones—a weather eye on the rising phone bill or the cost of mailed complaints gives a telling warning of things amiss.

I hope the value of Figure 30.2 in analysing business processes is apparent already, but its benefits do not end there. The model also helps us to recognize and compare

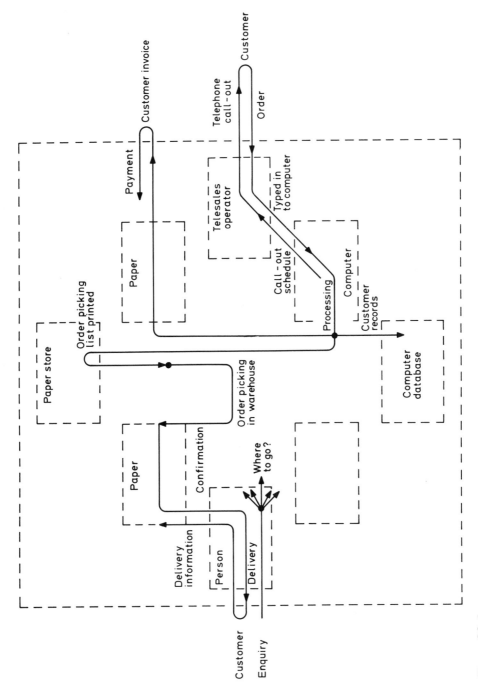

Figure 30.4
Inefficient company information flows.

similar methods of conveying the same information. For example, our analysis of information flows may conclude that customer enquiries and helpline support are best carried out in a person-to-person manner. We therefore next have to choose the best method available, there are of course many:

- site visit;

- telephone call;

- video or 'videoconference';

- voice messaging (e.g. answerphone);

- facsimile;

- telex;

- electronic mail;

- cellular telephone.

Each method suits different needs, since each has it own characteristics of speed, cost and reliability. The challenge for users is to recognize the differences and make the appropriate choice from the portfolio. Do not be tempted, however, to use a person-to-person technique when either an inter-computer flow is more appropriate or when a paper flow is paramount (e.g. for the recording of a signature).

Many companies could learn from this analysis, perhaps the most significant benefit being the conversion of routine processes (e.g. order-taking) into an electronic form, thereby freeing people to provide personal services which today receive too little priority (e.g. customer account development, and assistance with problems).

As an example of a pertinent 'portfolio' study of inter-personal mail communications, consider the differences between mail, telex, facsimile and electronic mail. All convey a page of written text from one person to another, and have little other value. Mail is probably the cheapest, but the time for a response is several days. Telex is a bit quicker but requires an experienced telex terminal operator. Facsimile requires only a minimum of skills to operate, can transmit even hand-written text, and is often cheaper and quicker than sending a telex. Of all the electronic methods, electronic mail is cheaper on a word-for-word basis for large scale and regular information transfer, it is more reliable and has the benefit of underlying integration with other computer processes.

30.3   THE TACTICAL DEVELOPMENT PLAN

Having decided upon a strategy and long-term goal for use of information technology, a tactical development plan needs to be laid out, setting down the order in which various specific functions will be carried out, and the overall structure, technical standards, design, and supply guidelines to be followed.

30.4 BUSINESS APPLICATIONS OF IT

The business applications of IT are wide and diverse. In the remainder of the chapter we discuss some examples of how computer and telecommunications technology has been combined harmoniously to serve business needs in a few important market sectors.

Manufacturing Computer-aided design (CAD), computer-aided manufacturing (CAM) and computer-integrated manufacturing (CIM) have enabled faster turnaround times to be achieved and much closer quality control. A number of the major car manufacturers, for example, are now entirely dependent upon computer-aided design and manufacture. Telecommunications networks link the various design organizations to speed the process of design, and to aid in the rapid programming of robotic production line equipment.

An important OSI initiative in the manufacturing area, undertaken by the United States Society of Manufacturing Engineers, is the development of the MAP/TOP OSI profile (Manufacturing Application Procedure/Technical Office Protocol). This is a particular set of OSI protocols supplemented with extra application layer protocols to meet manufacturing industry and office networking needs in particular.

Distribution and transport—EDI (electronic data interchange) allows for paperless trading, promising the elimination of huge volumes of invoices, orders and delivery confirmations in favour of much faster electronic methods.

EDI requires the use of common telecommunications methods between trading partners (e.g. customers/suppliers), but in addition it requires agreement on the exact format and information content of messages that will be sent. The underlying public telecommunications networks needed for EDI are provided by dial-up lines, leased data lines, X25 packet networks, and by the emerging X400 message handling system (MHS—see Chapter 25), but over and above these standards, agreement is also necessary on the format, syntax and information content of messages.

In the United Kingdom the interim standard for EDI is the TRADACOMS standard developed by the UK's Article Numbering Association (ANA). ANA is the association set up by the retailing and distribution industry to administer 'bar coding' numbers on retail goods. The TRADACOMS standards are based on the United Nations Trade Data Interchange (UNITDI) syntax and design method.

TRADACOMS defines formats for transaction files, orders, stock adjustment, delivery notes, invoices, delivery confirmations, credit notes, pricing lists, statements, availability reports, general text and stock snapshots, as well as the product price and customer information master files.

Over 1000 companies across the complete spectrum of retail and service industries in the UK use the standards—the largest EDI community in Europe.

The standards are available from:

Article Numbering Association (UK) Ltd
6 Catherine Street
London WC2B 5JJ
United Kingdom

Tel: 071-836 2460

Another important interim EDI standard is the American National Standard Institute's ANSI X.12. This standard is in widespread use in North America. In the longer term EDIFACT (Electronic Data Interchange for Administration, Commerce and Transport) syntax and message format will be the international standard. The standards will comprise two parts, syntax and message design. The first part, EDIFACT EDI syntax has already been designed and accepted as a full international standard, ISO 9735 but a number of project teams are still working under the direction of the EDIFACT board to develop all the UNSMs (United Nations standard messages) that will also be needed.

EDI offers the potential for much reduced paperwork and business bureaucracy; and much improved competitive advantage through locking-in customers using technology. But the greatest, and as yet unfulfilled potential, is that for dramatic change in the fundamental processes of business.

Faster and more accurate communication may remove the need for parts of the business process which previously we thought indispensable. Electronic orders will speed suppliers' knowledge of their customers needs—eliminating the need for brokers, distributors and middlemen—and enabling manufacturers to make-to-order rather than manufacture and store according to a forecast. Meanwhile, electronic-goods-delivered notices could trigger electronic payments—thereby eliminating the need for invoices and credit control departments. The real potential has yet to be imagined—and will not be realised without the application of creative minds to challenging previous preconceptions of the way to conduct business. The greatest impact and challenge of EDI is therefore not technical at all, but rather the opportunity to revolutionise business. The technology is ready, relatively simple, and waiting!

Retail trade—EPOS (Electronic Point of Sale) terminals are the intelligent 'tills' that are replacing the simple cash register in many 'chain' retail outlets. The EPOS terminal is a computer which collects detailed information about each sale. By 'polling' each shop every night, the central headquarters of a major retailing chain can get extremely up-to-date and accurate information about the market, its fast and slow selling brands. Figure 30.5 illustrates a simple EPOS arrangement.

EFTPOS (Electronic Funds Transfer at the Point of Sale), sometimes also known simply as 'EFT', terminals allow retailers to validate credit cards by 'swiping' them

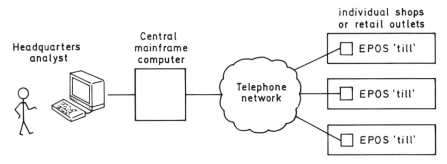

Figure 30.5
Electronic point of sale (EPOS).

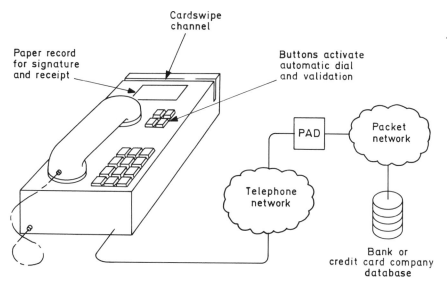

Figure 30.6
Electronic funds transfer/point of sale (EFTPOS).

through the slot of a specially adapted telephone (Figure 30.6). Card users will be familiar with such machines in shops.

The EFT terminals look like adapted telephones, as Figure 30.6 shows. They automatically make a connection to the bank or credit card company's computer—either directly via the telephone network or via the public packet data network. Any necessary confirmation or response to the retailer is given by a liquid crystal display on the terminal.

Videotext has become popular among UK travel agents. They can get up-to-the-minute information on available holidays, and interaction allows them to book holidays and reserve airline seats instantly. Figure 30.7 illustrates a general schematic of the service. In the diagram, a terminal in the travel agent's office is used to dial up the videotext computer over the telephone network. This computer holds the information and administers any necessary charging scheme for 'viewing' pages of information.

Various proprietary names are used for Videotext service. Examples are British Telecom's 'Prestel' service, France Telecom's 'Minitel' and the Deutsche Bundespost's 'Bildschirmtext'.

Satellites are ideally suited to broadcasting sound, video or textual (data) information from a single origin out to a large number of outstations, since they allow simultaneous reception by each of the outstations. Thus for large chains of retail stores they are an ideal medium for sending out to all stores price update information, new catalogue information, training information, new product videos and so on. Within the United Kingdom, a consortium of betting shop chains combined forces in the early 1980s to set up a satellite company called 'Satellite Information Services (SIS)'. The prime business of this company started in televising horse and dog racing meetings and

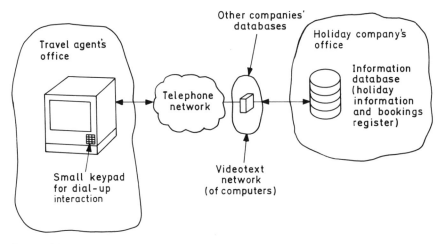

Figure 30.7
The videotext (or 'Prestel') service.

publishing up-to-minute betting odds and race results. The output is broadcast nation-wide by satellite to more than 10 000 betting shops in the UK.

Telesales and telemarketing are the names given to large-scale telephone interaction with customers. Bureaux are often staffed by several thousands of operators equipped with computer terminals and capable of receiving or making telephone calls to customers. Either in the wake of an advertising campaign or on a schedule of 'cold calling', conversations are established for the purpose of promotion, taking customers' orders, etc.

Financial services Dealer-board systems comprise a large console and advanced computer and telecommunications components. At the centre of the console, a 'touch-sensitive' screen displays the latest market prices, which are updated by a direct data-

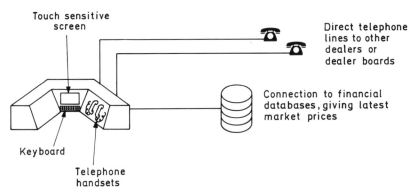

Figure 30.8
A financial 'dealer board' system.

link connection to an appropriate database. The touch sensitive nature of the screen allows the dealer to select topics and get more information quickly about a particular item, or may record the details of a deal. Dealer boards also usually comprise a number of telephone sets so that telephone calls can be set up quickly to other dealers.

Commerce   In some large organizations, an 'electronic office' environment has been created, in which each member of staff has a desktop personal computer, word processor or other terminal which is linked to all the others using either a local area network, or by means of an 'electronic mail' 'store-and-forward' mailbox system. More sophisticated office networks even allow the individual users to connect their desktop personal computers to a range of other devices, including the office mainframe computer, the telex machine, the office user printer, the public packet network, a 'library' database or just other PC users. Figure 30.9 gives a schematic of a simple electronic office network.

Media and advertising   The media world has demands for fast transmission of high quality pictures and graphical material. Such demands can be met by using the latest video cameras and monitors and by the high-speed switched data capabilities available with ISDN and broadband networks. New video-telephones are available for transmitting a moving colour picture and its sound signal down a single ISDN telephone line.

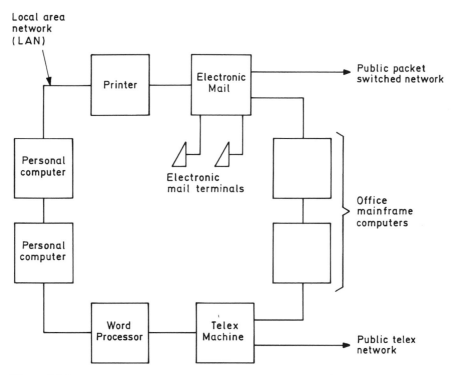

Figure 30.9
An electronic office.

The same device can send a high quality colour picture in a matter of seconds. Given that the receiver (say an advertising agency's client) has a graphics computer available to accept the colour copy, then it could be 'edited' in a few minutes and returned to the advertising agency for copy-printing and publication.

Another new application that eases the transport of high quality paper documents is 'group 4 facsimile'. A group 4 facsimile machine can transmit a photocopier-grade A4 page to a similar machine across an ISDN network in 8–15 seconds.

30.5 SUMMARY

Correctly configured, telecommunications and computer technology have already stimulated a revolution in the business practices and competitive nature of many industries. Further advances in technology mean that systems with almost human capabilities are no longer inconceivable, but the equipment and techniques are complex, and require careful planning to achieve maximum business advantage.

BIBLIOGRAPHY

Camrass, R. and Smith, K. *Wiring Up the Workplace—A Practical Guide for Management.* IBC Technical Services, 1986.

Franco, G. L., *World Communications—Ways and Means to Global Integration.* Le Monde Economique, 1987.

Fritz, J. S., Kaldenbach, C. F. and Progar, L. M., *Local Area Networks—Selection Guidelines.* Prentice-Hall, 1985.

Gandoff, M., *Manager's Guide to Telecommunications.* Heinemann. New Tech. 1987.

Gifkins, M. and Hitchcock, D., *The EDI Handbook: Trading in the 1990s.* Blenheim Online Publications. 1988.

Harper, J. M., *Telecomms and Computing—The Uncompleted Revolution.* Comm. Ed. Books., 1986.

Hollingum, J., *The MAP report.* IFS (Publications) Ltd/Springer-Verlag, 1986.

Hunt, V D., *Computer-Integrated Manufacturing Handbook.* Chapman & Hall, 1989.

Jones, V. C., *MAP/TOP Networking.* McGraw-Hill Manufacturing & Systems Engineering Series, 1988.

Keen, P. G. W., *Competing In Time—Using Telecommunications for Competitive Advantage.* Ballinger Publishing, 1986.

MAP—Manufacturing Automation Protocol Specification Version 3.0. Society of Manufacturing Engineers (USA), July 1987.

Miller, B., *Telecomms for Business—A Manager's Guide.* Comm. Ed. Books, 1986.

Naughton, M., Hall, N. and Wilson, C., *Viewdata—The Business Applications.* Comm. Ed. Books, 1988.

Paterson, A., *Office Systems—Planning, Procurement and Implementation.* Ellis Horwood, 1985.

Pye, C., *What is MAP?* NCC Publications, 1988.

Rizzardi, V. A., *Understanding Map.* Society of Manufacturing Engineers (USA), 1988.

Sommerville, I., Information Unlimited—The Applications and Implications of Information Technology. Addison-Wesley/Small Computer Series. 1983.

The Catalogue—The Definitive Guide for Networking: Data Translation—Networking. Data Translation Limited, The Mulberry Business Park, Wokingham.

The Computer User's Year Book 1989. VNU Business Publications, London.

The Software User's Year Book 1989. VNU Business Publications, London.

TOP—Technical and Office Protocols Specification Version 3.0. Society of Manufacturing Engineers (USA), July 1987.

Townsend, C., *Networking with the IBM Token Ring.* Tab Books Inc., 1987.

Vignault, W. M., *Worldwide Telecommunications Guide for the Business Manager.* Wiley, 1987.

31

COMPANY TELECOMMUNICATIONS MANAGEMENT

While large multinational companies can afford to employ specialist 'telecommunications managers', smaller companies and individuals are less privileged. For these latter groups, telecommunications is often just another subsidiary responsibility for either a computer services manager or even the general business manager. None the less all telecommunications managers can affect the way in which their companies operate. In this chapter we shall discuss the following topics:

- telecommunications performance and management;

- cabling—for telephones and computers;

- office computer networking (LANs etc.);

- private network design and operation.

31.1 TELECOMMUNICATIONS MANAGEMENT

The telecommunications manager should *manage* all aspects of telecommunications. He needs to take *ownership* of the full cost and functional performance of all information flows—by whatever means—throughout his company. The job should not consist only of designing and operating the company's private networks. This loses sight of the 'strategic horizon'—the profound effect that effective all-round communication can have on a company's operations.

Telecommunications management involves the day-to-day operation and administration of resources, and it should include creative and strategic questioning:

- What is our long-term business goal?

- What are the most important 'information flows' supporting it?

- Why is 'key' information being carried on a particular medium?

- What is the capability of new technology—how can it affect the core business process to advantage?

- Why should I maintain private network rather than use public services? Which gives better 'added value'?

- Why are we spending so much money on paper mail?

- Why are we using facsimile rather than electronic mail?

- How can we gain competitive advantage?

- How can we lock-in our customers using telecommunications?

- Which are the right standards for me to use?

- How will legal regulation affect me?

- *Above all*, are our board and senior management properly clued up about telecommunications? If not, is that not my fault?

Good telecommunications management keeps a company using the right technologies, the right suppliers, the right control procedures and the right monitoring measures— maintaining optimum performance at all points in time.

From the suppliers there is a greater need for products which encompass whole business solutions—tailored packages of services, products and support from a single supplier to meet the need of a business process neatly and completely. No longer has the telecommunications manager got the time to bind together a rag-bag of bits— essentially having to design the solution for himself. Neither can he afford the time needed to resolve any subsequent problems or take on any necessary enhancements. What he has to do is to identify the need, understand it completely, and find suitable suppliers. Thus he seeks a communications solution for 'order collection', 'customer payment', customer account management' etc.

Having found solutions to the business communication needs, a framework is needed for performance and overall cost management. We covered this in some detail in Chapter 16. Finally, the manager may turn to the technology—and to the design and operation of networks. The remainder of the chapter discusses some of the important current issues facing telecommunications managers in this particular area of responsibility:

- building cabling;

- computer networking;

- private network design and operation.

31.2 PREMISES CABLING SCHEMES

Until the explosion in number of office computer workstations in the early 1980s, buildings were designed and pre-wired with the needs only of electricity and telephone cabling in mind. However, the large number of personal computers, word processors, facsimile machines, mainframe computer terminals, printers and electronic mail terminals now in use mean that the needs of data terminal cabling now generally exceeds that of simple electricity and telephone.

At first, companies installed dedicated wires for each terminal as it was needed—installation work and the costs associated with it mounted up on a 'pay-as-you-go' basis. In time the wiring became a complicated spaghetti of leads strewn across false ceilings, knotted on cross-connection frames, and squeezed into congested conduits. Every time a new cable was needed, the ceiling tiles would come down again, the conduits would be opened up, and somehow or another a path would be found. The costs began to grow out of proportion. Each new cable was harder to install than the previous one—and each new installation led to faults being inflicted on existing wiring as cables were drawn into already congested ducts.

Most companies now favour formally 'structured wiring schemes' which are planned and installed at the time of initial building development or during refurbishment. Sufficient conduits, equipment cabinets, and cables are provided to cater for wiring in excess of all foreseen needs.

Figure 31.1 illustrates a typical structured wiring scheme. It comprises cables running out from a main trunking room or path frame. First a number of high-grade 'backbone' cables run out (in building conduits—often called 'risers') to 'patch panels' located on a one-per-floor basis. From these secondary locations, a large number of lower grade cables run out to all possible locations within the office floor.

Typically four 4-pair cables (i.e. 8-wire) are run from the floor patch panel to each single occupant office. These allow two telephone extensions and two data terminals per office. Standard telephone gauge 'unshielded twisted pair' is suitable for most LANs, data and telephone needs. For higher performance needs, shielded twisted pair, coaxial cable or fibre optic cable may be needed (see Chapter 8).

The floor patch panel provides all the connections on that floor and for later reconfiguration simply by changing the patch wire. Local area networks of logical 'star', 'bus' and 'ring' topologies can be created and reconfigured at will. In addition the floor patch panel is connected via 'riser' cables to patch panels on another floor and back to the main patch frame. These allow for connection to public telecommunications networks and the mainframe computer installations on other floors. The riser cables are usually either shielded twisted pair or fibre. Coaxial cable can also be used.

Though structured cabling schemes require greater capital outlay at first, they usually reward their owners with much lower running costs, very fast and cheap terminal wiring lead times and much reduced frequency of faults. The largest single cause of wiring errors—tampering with the cables in the conduits—is eliminated by a structured scheme. Unstructured (ad-hoc) wiring schemes should thus only be considered where very low terminal populations exist (less than about 20 terminals).

Structured wiring schemes generally require very little maintenance and operate virtually trouble free, but good management practice demands that a proper inventory be kept of the use of each wire, and that each be properly labelled.

A number of regulatory stipulations have to be considered during the planning and installation phases of cabling. First it is not good practice to run cables in air-conditioning ducts because of the smoke and fire risk arising from the combustion of the plastic insulation. (Actually, US law permits this but it is illegal in the UK.) Separate conduits for electrical and telecommunications wiring are recommended to

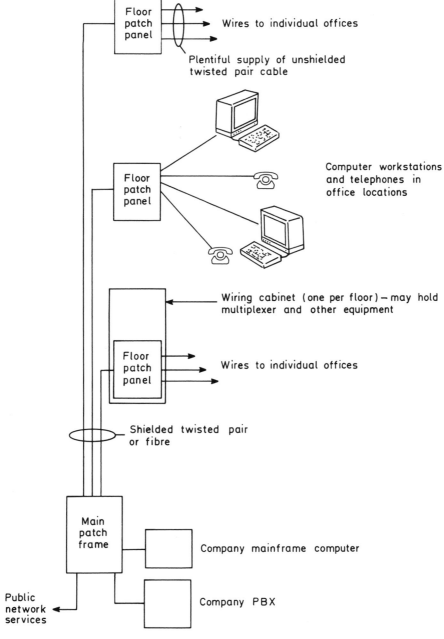

Figure 31.1
A structured wiring scheme.

protect telecommunications users and equipment alike from the dangers of a mains electrical 'shock'.

One further regulatory restriction applies in the United Kingdom which may cause companies to consider separate wiring schemes for data and telephone uses. The restriction is that all cables and patch frames carrying telephone wiring which may possibly be connected to the public telephone network (via the PBX) must be maintained by the registered PBX maintainer. Further, any data or other equipment connected to wires in the same *cable* (irrespective of whether these *wires* are connected to the PBX) must either be 'green dot' approved for connection to the telephone network, or they must be isolated by means of a barrier box (intended to prevent high voltages leaking on to the public telephone network). These restrictions are easy to avoid by installing separate structured cabling schemes for telephone and data wiring. Thus cables and patching equipment are duplicated, but conduits and cabinets may be shared.

31.3 OFFICE COMPUTER NETWORKING

Issues in the design and management of office computer networks are:

- meeting capacity and response time needs;

- flexibility for reconfiguration;

- compatibility of 'application' and 'network management' software.

We discussed network dimensioning—to meet data network capacity and user response needs—in Chapter 10. Typical response times needed from a real computer network are shown in Table 31.1. Achieving such targets takes a considerable amount of skill—there are far too many detailed factors to be considered which we have not the time for here, but many excellent books, consultancy organizations and suppliers are available to assist with complex data networking problems. Basic guidance is easier—'keep it as simple as possible'. Careful design of relevant software, together with the use of the formulae given in Chapter 10 will go a long way!

Table 31.1
Typical response time expectations.

Function	'Maximum' response time
Computer activation	3.0 s
Feedback of error	2–4 s
Respond to identity code	2.0 s
Keyboard entry	0.1 s
Respond to simple enquiry	2.0 s
Request for next page	0.5 s
Respond to 'execute problem'	15.0 s

The flexibility for reconfiguring data networks or for meeting a number of needs simultaneously derives from the use of a number of tools and techniques; e.g.:

- structured cabling;

- local area networks (see Chapter 9);

- open computer architectures (e.g. 'SNA'—see Chapter 9).

The use of LANs or other computer network architectures (e.g. SNA), needs to be tempered by careful preparation. None of the tools are as flexible as they may seem, and lack of appreciation of this fact can lead to equipment incompatibilities.

Local area networks should be designed and run using rigorous 'mini-data centre' procedures. Out of hand they can be very difficult to manage and problems can be hard to diagnose.

Selection of a particular LAN (e.g. 'Ethernet' or 'token ring') or a particular computer architecture (e.g. 'OSI' or 'SNA') needs to take cognizance of the applications to which it will be put (e.g. 'file transfer' or 'printer sharing' or 'mainframe interaction') and also of the computer hardware and software (e.g. PC 'operating system software': 'DOS', 'OS/2', 'UNIX', 'Windows') which it is expected to serve. Equipment from a particular computer manufacturer tends to have been developed with one architecture in mind. For example, the IBM architecture is SNA and the preferred LAN technology is token ring. Meanwhile Digital Electric Corporation's proprietary architecture is called 'DECNET' and the preferred LAN technology is Ethernet. In the longer term, open system interconnection (OSI) standards allow greater interchangeability of computer equipment and network components.

31.4 PRIVATE NETWORKS

Public telecommunications networks are usually large and complex requiring massive capital and manpower resources. As a result, network changes are usually conducted in a careful and steady manner since the cost of mistakes is magnified by the vast scale. Also, because public telecommunications operators (PTOs) are normally required by terms of licence to provide services of uniform availability over wide geographic areas, the development and rollout of innovative and technologically advanced services can be held up, and prices to main business customers can be prejudiced by the subsidization of uneconomic rural areas. Because of this, and driven by the desire to adopt the latest techniques, some companies have developed extensive 'private networks'. The advantage to be gained depends upon the service offerings and tariffs of the public telecommunications operator(s) and also upon the legal constraints placed locally on such a network.

A typical private network is shown in Figure 31.2. The example shows the three private branch exchanges (PBXs) in different office buildings, interconnected by circuits between the buildings. The PBXs and other 'customer premises equipment' (CPE) may be owned by the private company, or leased from the PTO. The wiring in each office normally belongs to the private company. However, the circuits interconnecting the different PBXs in different buildings almost certainly are leased from the PTO.

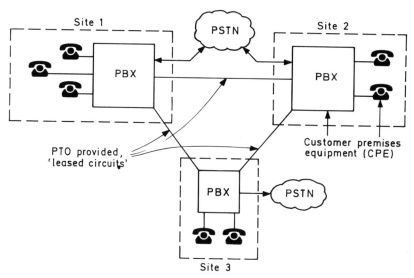

Figure 31.2
A private telephone network.

Thus a private network usually comprises 'customer premises equipment' (or 'customer apparatus') and 'leased circuits' which interconnect the different locations.

The private network illustrated in Figure 31.2 is capable of switching telephone calls between any two telephones in any of the three different offices. If desired, a single numbering plan could cover all three offices, each telephone having a unique 3- or 4-digit 'extension number'. In this way users benefit from a short dialling procedure, and the company can save money by transporting inter-office calls over the leased circuits rather than having to pay the full public telephone tariff for each call. Provided the telephone 'traffic' between different offices exceeds a given threshold value (the exact value of which depends on the relative leased circuit and PSTN call tariffs) it will almost certainly be cheaper to use direct leased circuits in this manner, as the calculation in Table 31.2 shows. The cost benefit of the private network (compared with using the public network) is prone to changes in PTO tariffs. As an example, in the United Kingdom the balance of public service tariffs and leased line rental charges has changed so radically since the early 1980s that many large companies are now

Table 31.2
Threshold for leased circuit consideration.

Leased circuit tariff $= \pounds L$ per quarter
Public telephone tariff per minute $= \pounds p$
Telephone usage minutes per quarter $= m$

Public telephone bill per quarter $= mp$
therefore if $mp > L$
or minutes per quarter $m > L/p$
a leased circuit is worth consideration

---

considering the abolition of the private networks they established at that time. Another factor encouraging companies to give up their private networks is the desire to reduce the in-house manpower needed for the day-to-day responsibilities of network operation. So the pressure is on for the public network operators to assume this greater responsibility!

The calculations shown in Table 31.2 are oversimplified. The full analysis should take account of traffic profile effects (Chapters 10 and 11 refer). Only the point-to-point traffic should be considered, and the number of leased circuits costed needs to take account of the grade of service (following the Erlang formula).

Private networks are common in medium and large companies where the economics of large scale helps them to 'play-in', particularly for intra-company communication between main offices. For communication with suppliers or customers, or even to staff in small remote offices, private networks are often connected to the public switched telephone network (PSTN). In Figure 31.2, each of the PBXs is illustrated with a direct link to the PSTN.

Apart from cost or capabilities of new technology, another motive for a company to establish a 'private network' could be the network security afforded. Take a small private packet-switched, or other data network which connects a computer holding sensitive data to a number of remote workstations. Unauthorized users with workstations which are not connected to the private network cannot gain access to the sensitive information.

When it is not possible or economic to establish a private network to ensure the security of information, 'scramblers' or 'encryption' devices may be employed in conjunction with the public network. 'Scramblers' work in pairs—at each end of a connection—converting speech into a coded, but intelligible signal for transmission on the public line. Equivalent devices when used to 'scramble' sensitive data are usually called 'encryption' devices.

Architecture of private networks

'Public' and 'private' networks are based on similar technical principles, but differ in scale and complexity. Private networks cater for much lighter traffic, but more specialized service needs. An example of the difference in technical standards is illustrated by making a comparison between the signalling systems used for conveying telephone calls between PBXs over a private network with those used between the exchanges of a public network. Inter-PBX signalling systems often only carry short digit strings (equivalent to a 4-digit extension number), but in addition they have to carry information necessary for invoking special features such as 'ring back when free' etc. As a result, the speed at which individual digits of the dialled number are sent between the PBXs may not be the critical factor, because the small number of digits constrain the call set-up time anyway, but the signal vocabulary has to be wide in order to support all the various special functions. By contrast, inter-exchange signalling systems need to carry longer digit strings. A full international number and its prefix could be up to 18 digits long, and must pass quickly between exchanges so that the set-up delay is reasonable. Exchanges in a public network may need to transfer information on charging or routing the call, as well as any customer-requested supplementary services

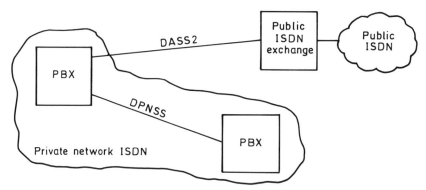

Figure 31.3
PBX-to exchange and PBX-to-PBX ISDN signalling in the UK.

(e.g. 'calling line identity' or 'ring back when free'). Finally, the interexchange signalling system may be used to convey information for network administration, including network management information and control signals, or for remote operation and maintenance commands.

Public network standards tend to be robust, designed to withstand and control the severe and diverse demands of the wide range of uses to which public networks are exposed. They need to be resistant both to failure and fraud. Private network standards are nearly always derivatives of public network standards, and are often developed from the interface used to connect customer premises equipment to public network exchanges. Figure 31.3 illustrates an example pertinent in the UK, where the primary rate ISDN PBX-to-network interface, called DASS2 (Digital Access Signalling System No. 2) is similar to the 'Digital Private Network Signalling System' (DPNSS) which may be used on private networks between PBXs to support ISDN-like services. (CCITT is also developing similar versions of these signalling systems in its Q93X series of recommendations.) The advantage of using similar standards for PBX-to-exchange and PBX-to-PBX interfaces is that it minimizes the complexity of PBXs which are required to operate in both modes. Similar benefits can be gained in other types of networks (e.g. packet networks etc.) by aligning the technical standards used in private and public network variants. Minor differences are inevitable, have always existed and are bound to persist. They are the results of the contrasting network circumstances and demands on public and private networks. Thus a private packet-switched network is likely to be based on CCITT's X25 standard, but it may include a number of customized software features. Likewise the message handling system, as we saw in Chapter 25, recognizes separate public and private 'management domains'.

Planning private networks

The smaller traffic scale of private networks, coupled with their relative freedom from licence constraints, means that their structure can differ from that of public ones. Indeed, optimum topology is not only independent of the real cost of equipment, but can be highly distorted by the PTO's relative tariffs for 'leased circuits' and public network services.

Besides the economy measures of Chapter 20, three particular network routing techniques which offer considerable scope for cost reduction to private network planners are:

- Establishment of telecommunications hubs.

- Public network overflow.

- Public network bypass.

Telecommunications hubs

The small traffic scale of private networks tends to encourage the use of a 'star network' topology in which one, or a small number of exchanges at the 'hub' of the network provide a transit switching point between all other exchanges. The configuration is illustrated in Figure 31.4, where all calls between any pair of exchanges are switched via the 'hub' exchange, A. By employing such a topology a handful of circuits may be sufficient to carry all traffic (let us say three circuits from each outlying exchange to exchange A, a total of 21 circuits), as against a more 'meshed' or fully interconnected network of only one circuit between each pair of exchanges which would require 29 circuits (as shown in the inset of Figure 31.4).

The star topology shown in Figure 31.4 is similar to the hierarchical network structure described for larger scale (public) networks in Chapter 12. The smaller traffic on private networks means that the use of the star topology (or 'hierarchical structure') is nearly always more cost effective than a more meshed network of inter-exchange connections. It not only minimizes the total number of circuits required, but also tends to minimize the total number of circuit miles. Minimizing the number of circuit miles

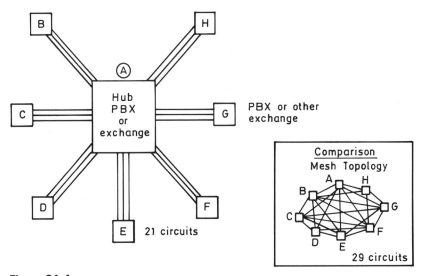

Figure 31.4
'Hub' or 'star network' topology for private networks.

is most important, since it is this value that relates to the 'leased circuit' charges levied by the public telecommunications operator (PTO). Careful siting of the hub exchange is also important. In international private networks, the choice of the hub site may be especially important, since the price of leased circuits may not be the same in each of the countries. Let us return to Figure 31.4. It might be the case that exchange A is the geographical hub of the eight sites, that the prices of leased circuits into and out of the country in which exchange F is situated are cheaper. In such an instance, it may be more economic to use exchange F as the network hub. Similar anomalies of leased circuit prices can arise even within a single country, caused either by a banded (rather than directly distance proportioned) price structure or because of the differential price structure of competing PTOs.

Circuit numbers are calculated in the normal way for the type of network and the circuit lengths are measured as radial distances. The costs can then be worked out accordingly. Repeating the procedure for different optional networks based on the various available hub sites helps to determine the cheapest configuration.

Public network overflow

The traffic between any two exchanges of a network has a 'peaked' profile. But while public networks are mandated to carry this peak of demand, there is no similar mandate for private networks. Especially there is no need for the private network to be able to carry the peak demand using its own resources alone. Instead it is permissible, and may be far more economic, to 'overflow' the peak demand to the public network. Figure 31.5 illustrates this technique. Two PBXs (or other type of private exchange) called A and B, and located at different offices of the same company, are connected together as a 'private network'. Most of the day the traffic demand between exchanges A and B is between 6 and 8 erlangs, but between three and four in the afternoon it peaks at 12 erlangs.

The private network planner has done some cost calculations. He has established a network of eight direct 'leased circuits' on the 'private network' between exchanges A and B, but has elected to overflow the excess 7 erlangs of peak traffic via the public network. The total cost of the configuration is difficult to calculate, since the cost of calls overflowed via the public network must be measured or estimated from the complicated teletraffic formulae presented in Chapters 10 and 12. For each hour of the day the traffic that is carried on the eight circuits direct route, and the traffic overflowed from it, is determined using the Erlang formula, by inputting the traffic demand for that hour. Thus, until 5.00 am there is no traffic and no overflow. From 9.00 am to 3.00 pm the traffic is 7 erlangs, and the overflow from eight circuits is 1.25 erlangs so 75 call minutes per hour are sent via the public network. Similarly during the hour of peak activity (3.00 pm to 4.00 pm) the offered traffic is 12 erlangs, the overflow from the eight circuits is 5.07 erlangs, and 304 minutes route via the public network during this period. Over the whole day (adding also the overflowed minutes in the 4.00 pm to 6.00 pm) a total of 904 minutes overflow daily via the public network. At a cost of £P per minute and £L per quarter tariff for each leased circuit, this configuration gives an approximate overall quarterly charge of:

$$\text{Estimated quarterly charge} = 3 \text{ months } (L + 22 \text{ days}^* \times 904 \times P)$$

(*Approximately 22 business days per month)

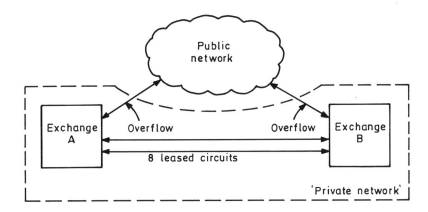

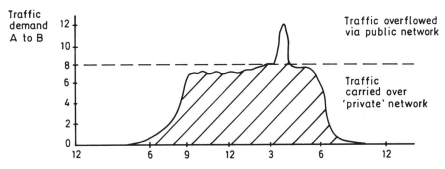

Figure 31.5
Public network overflow.

As in the last example, the optimum number of leased circuits (for minimum cost) is determined in a trial-and-error fashion by calculating two costs for various configurations.

Public network bypass

Where permitted by law, public network bypass may be an additional method of deriving additional telecommunications savings. A cost saving is made on long-distance calls over the public network by routing the call as far as possible within the private network before handing it over to the public network for completion. By so dong, the public network is bypassed as far as possible, the shorter geographical length of the public network connection reducing the chargeable tariff. Figure 31.6 illustrates the method of bypass. In the example shown, a company private network is established in the United Kingdom, linking private exchanges in London and Edinburgh. A caller connected to the private exchange in London wishes to make a call to an office in Edinburgh. Unfortunately the desired destination is not connected directly to the desired network. However, by routing the call over the private network from London to Edinburgh, and then into the public network for the final short hop, the public network (and its distance dependent tariff) has been bypassed for the greater part of

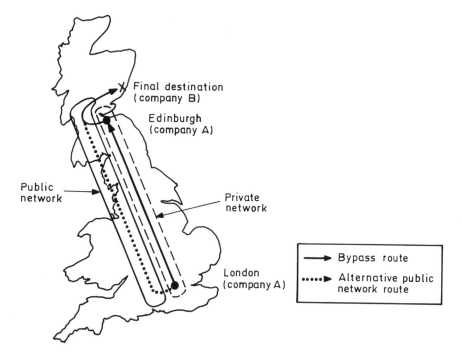

Figure 31.6
Public network bypass.

the connection. Only the tariff of a local call is payable, rather than the much higher tariff that would have been due if the call had been routed over the public network as a 'trunk call' for the entire distance between London and Edinburgh. Whether or not the overall economics favour such a bypass may change from day to day, depending on public network tariffs and on the existing topology of the private network. You should not assume that the example used is economic in all cases—specific evaluation is necessary on each occasion. Further, the transmission quality of the end-to-end connection may preclude certain configurations. Particularly in the data networking case, the speed of data propagation is often critical.

Another means of achieving some degree of public network 'bypass' is by the use of 'out-of-area exchange lines'. Thus our Edinburgh office could have purchased London out-of-area exchange lines from the PTO. In effect, this makes the office appear to the public telephone user as if it were in London. It will have a London telephone number and call charges will assume London is the endpoint. Thus calls from the Edinburgh to London Office, or vice-versa, count as 'local calls'. There is a rental charge, however, for the out of area line—similar to a leased circuit charge for a London/Edinburgh circuit.

31.5 A WORD OF WARNING

Having described a number of clever routing methodologies for reducing the costs of private networks, it is worth recalling—before we end the discussion—that the end-to-end connection must be properly serviceable. The transmission planning and routing 'rules', set out in Chapters 13 and 14, are applicable here. For example, it is not worth creating a super-cheap end-to-end telephone connection which comprises so many component links that speech is unintelligible over it. However, the high performance of digital switching and transmission makes this much less likely.

If the private network is to be connected to the public network there may be legal restrictions on the type and technical characteristics of the signal that may be conveyed. In some countries telephone network bypass is illegal. In nearly all countries there will be signal power and technical line limitations.

31.6 PTO LEASED CIRCUIT OFFERINGS

Generally, PTOs offer a comprehensive range of different 'leased' or 'private circuit' types, each attuned to one or a number of particular types, such as data, telephone, packet switching. With care, the planning of private networks can be relatively straightforward.

The four most common types of leased circuit conform to CCITT Recommendation M1020, or CCITT Recommendation M1040, their national equivalents, or they are digital leased circuits.

Recommendation M1020 defines the line conditioning (i.e. line-up limits—see Chapter 18) of a high grade 4-wire analogue circuit, of 3.1 kHz bandwidth, suitable for high-speed data or speech use. A circuit lined up according to M1020 has a very flat frequency response (i.e. shows very little attenuation distortion), very good group delay response, and good circuit stability. Such a circuit is suitable for most of the analogue data modems described in Chapter 9 and can be used for point-to-point or networked (e.g. packet networks) applications.

Recommendation M1040 defines the line conditioning of a 4-wire leased circuit, intended primarily for voice conversation. Its performance is not as good as an M1020 circuit, but since less time and equipment is needed for line conditioning, the cost to the customer is lower.

Digital leased circuits are usually defined in terms of their bit rate (e.g. 64 kbit/s, 2 Mbit/s, 1.5 Mbit/s), their bit error rate (BER) (typically 10^5 or 10^9) and their overall availability (usually in excess of 99.5 per cent). Provided these parameters suit the given application, the circuit may be used for any type of service (e.g. telephone, data).

Other types of leased circuits that are available are wideband analogue (e.g. FDM group, supergroup) or high-speed digital circuits. In addition some PTOs are willing to provide just the physical medium, and allow the user to configure the bandwidth in the manner desired. Examples include the 'wholesale' of optical fibre capacity and the leasing and sale of satellite dishes for direct customer access to very high bandwidth satellite circuits (e.g. International Business Service (IBS), which is a satellite service of INTELSAT allowing companies direct access for rates up to 2 Mbit/s). Finally, other specialized leased circuit services may be available from particular PTOs.

BIBLIOGRAPHY

Bell, R., *Private Telecommunications Networks—Design and Implementation*. Comm. Ed. Books, 1988.

CCITT Recommendation M1020, 'Characteristics of Special Quality International Leased Circuits with Special Bandwidth Conditioning'.

CCITT Recommendation M1040, 'Characteristics of Ordinary Quality International Leased Circuits'.

Eosys Cabling Guide for Building Professionals. Eosys Ltd., Clove House, The Broadway, Farnham Common, Slough, 1986.

Ettinger, J. E., *Communication Networks—Private Networks within the Public Domain*. Pergamon Infotech Ltd., 1985.

Lane, J. E., *Corporate Communications Networks*. NCC Publications, 1984.

Scott Currie, W., *The LAN Jungle Book—A Survey of Local Area Networks*. Edinburgh University/Kinesis Computing, 1988.

Scott, P. R. D., *Reviewing Your Data Transmission Network*. NCC Publications, 1983.

Vignault, W. M., *Worldwide Telecommunications Guide for the Business Manager*. Wiley, 1987.

32

NETWORK REGULATION AND DEREGULATION

During the 1980s, governments in hot pursuit of general economic development, made dramatic changes in the regulation of their industry. Monopolistic public utilities were privatized, and some of the markets historically dominated by those utilities were deregulated. This chapter discusses the motivations behind telecommunications deregulation, and explains some of the new regulatory measures which protect customers' interests by ensuring that quality services are available at a fair price.

32.1 REASONS FOR DEREGULATION

The word 'deregulation' is something of a misnomer, no country having completely rescinded all the telecommunications laws and regulations. Rather, the regulations are being changed to a new framework (a 'liberalized' one) to encourage competition between companies for the provision of telecommunication services. In fact, the new regulations will be more voluminous than the old ones, and will require more policing to ensure that companies are conforming with their obligations.

As with any change in the legal system, a government wants to convince itself of the benefit of the new regulation; not only its direct effect but also the secondary effect on other elements of the social or industrial strata. The government will wish to ensure maximum efficiency of overall resources, but it will also be interested in the fairness of a new system. In addition, from an economic sense, the impact on the country's balance of trade will be important.

The EEC noted in 1986 that:

> the strengthening of European telecommunications has become one of the major conditions for promoting a harmonious development of economic activities and a competitive market throughout the [European] community and for achieving the completion of the community-wide market for [all] goods and services by 1992.

As a result, the European Commission (the 'government' of the EEC) has set about establishing a European-wide network infrastructure comprising common network standards and a more open competitive market for telecommunications terminals and services.

The pressure for reform has been building in countries worldwide for many years, but different governments are responding at different speeds and in different ways.

Some consider the United States furthest ahead with its process of deregulation. As early as 1967, a small company called 'Microwave Communication Inc' (MCI) lodged with the US government's 'Federal Communication Commission', an application to run a 'common carrier' (transmission circuit) service between Chicago and St Louis. The proposed quality of the service was lower than that of the available leased circuit service using the established large (and licensed) carriers, a market dominated by American Telephone and Telegraph (AT&T) and the Bell Operating Companies (BOCs). The 'back-up' for the MCI service was to be minimal, and no subscriber terminal equipment was to be provided (allowing the customer to connect almost anything, as desired), but the benefit was a significantly lower price, being only half the cost of the established service.

Encouraged by the significantly different nature of the service, and hopeful that competition would stimulate new activity from previously untapped markets, the FCC gave approval for the service in 1969, and went on, a year later, virtually to force the established 'common carriers' to provide interconnection facilities. A flood of other aspirant carriers followed in MCI's footsteps, and the regulatory framework for telecommunications within the United States has been changing ever since.

A recent milestone was the establishment, in 1986, of 'equal access' for all trunk (or toll) telephone carriers. As the phrase suggests, the laws of equal access are designed to promote fair and equal competition between the long-distance (correctly called 'toll' or 'trunk') telephone carriers. They demand that local telephone companies connect in an equal manner with all licensed carriers who request interconnection. The effect is that local telephone subscribers have a 'free choice' of their long-distance carrier.

Outside the United States, the tide of change had been slower until the 1980s, at which time a number of factors brought pressure on many of the world's governments to consider deregulation. The factors are:

- The convergence of computing technology and telecommunications (the 'information technology' explosion), and their contribution to national wealth and economic growth.

- The reduced costs and much increased service capabilities made possible by technological progress.

- The high customer expectations, and in particular the demands of business customers.

- Growing customer dissatisfaction with the high prices charged and limited services made available by public telecommunications operators (PTOs).

- The willingness of private companies to invest 'venture' capital into developing new telecommunications business.

Together these factors are strongly challenging the existing framework of telecommunications regulation, and rapid changes are in progress around the world.

32.2 THE DILEMMA OF DEREGULATION

Deregulation poses a dilemma for governments considering it. Any government will wish to decide the optimum allocation of the country's overall resources and produce a framework of laws and regulations to encourage both public and private companies to use those resources accordingly. A substantial proportion of a country's gross national product (GNP) may depend upon telecommunications, either directly from the sale of telecommunications services, or as a result of business conducted on the telephone. The decisions on the regulatory framework governing telecommunications are therefore crucial.

The main dilemma facing governments is that shown pictorially in Figure 32.1. The question is whether a public service need is best served by a small number of monopolized but strongly regulated public utility companies which have an advantage in their homogeneity of service offerings and in the prices offered over a large geographic area, but tend to be inefficient and slow to respond to change; or whether the public need is better served by a more deregulated market, with a large number of more diverse companies and services, competing for business. The smaller, competing companies will tend to respond more quickly to technological changes and fluctuations in the demands of the market. They will also tend to be cheaper to the customer, but may be so because of a compromise in quality. Further, viewed from a government perspective, the companies of a competitive environment may not appear to use the country's resources efficiently, and may not provide for the needs of some crucial but unprofitable markets (e.g. remote rural telephone service).

The historic argument has been in favour of a small number of 'natural monopolies' as we know. Candidate services for a 'natural monopoly' are those in which:

(i) There is a clearly established public need for a service of a uniform nature and covering a widespread geographic area.

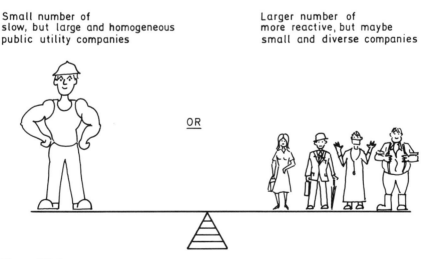

Small number of
slow, but large and homogeneous
public utility companies

Larger number of
more reactive, but maybe
small and diverse companies

OR

Figure 32.1
Deregulation and the monopoly dilemma.

(ii)   There is economy in large scale operation.

(iii) It is cheaper for one firm to serve the market than for two or more.

(iv)   The market power of a large company in competition can enable it to control prices and make entry unattractive for rivals.

(v)   The lack of information available to new market entrants and the limited resources of small companies leave them at significant disadvantage.

The services which have been protected by governments as 'natural monopolies', are electricity supply, gas supply, water supply, sewerage, refuse and telecommunications. All have to some extent benefited from the large scale of the operation that has resulted from monopoly. In telecommunications networks the addition of each new customer to the early networks sometimes at disproportionate expense, brought benefit to the new customer and to established users, for whom the available extent of connectivity had been increased.

Under a 'natural monopoly' it is usual for a government to confer obligations and rights upon the companies operating public utility services. These obligations and rights are summarized in Table 32.1.

The rights and obligations of public utility companies have historically been closely policed by governments keen to dictate these services which may be offered, the prices which may be charged, and the profits which may be made. Laudable though this may seem it slows up the development of the public utility companies and their services and only adds to the inefficiencies, since more bureaucracy is involved in each decision made.

The aim of improved efficiency has been a prime motivator for 'deregulation', the aim being to 'shake up' the established monopoly carrier by introducing competition. But most governments' actions have fallen short of removing all regulations and legal entry barriers, so that some financial protection exists for the established carriers to go on providing unprofitable, but necessary services. The challenge is to create a new regulatory regime that is simple enough to promote competition, increase efficiency, and reduce prices, but without introducing the plethora of legal and bureaucratic barriers that seem necessary to protect unprofitable services. Ideally, the regulatory framework should encourage a greater variety of services, a greater responsiveness to

Table 32.1
The regulation of large public utility companies.

Obligations
1   To provide service to all who apply, even if unprofitable
2   To provide safe, reliable and adequate service
3   To provide fair terms and prices for service provision.

Rights
1   To make money and a fair return on investment
2   The right to withdraw services under given conditions
3   In a competitive environment, the freedom and scope to use initiative to create a 'market edge' for their services, and be fairly financed for unprofitable services

changing customer demands, better quality, and lower costs; all made possible by more efficient use of resources and by new private investment. Next we consider some of the methods of regulation being reviewed by a number of the world's governments.

32.3 OPTIONAL METHODS OF REGULATION

There are two main methods by which a government may impose regulation on industry. The government can control the *structure* of the industry, dictating either what companies may produce or what services may be operated. Monopolies are often structure controlled. Alternatively, the 'deregulated' approach is to control the *conduct* of companies laying out the actions and behaviours which are permissible.

Most regimes comprise elements of both structure and conduct regulation, and real measures are not always easily classified into one of the two types.

Structure regulation controls which companies may operate in an industry, typically restricting entry by licensing operators, and by having a strict monopolies and company merger policy.

Conduct regulation controls include those governing restrictive practices and those prohibiting 'anti-competitive' acts (such as financial cross-subsidization between two different company products).

Conduct regulation requires a 'policing' body to ensure that companies conform with their obligations. Under 'structure' regulation, such a body is not necessary, due to the smaller number of active companies, and also because the public utility company is often state-owned anyway.

In countries undergoing telecommunications deregulation, the shift in the emphasis of regulation tends to be away from *structural* control, towards *conduct* control. Many countries are establishing 'regulatory bodies' to oversee the conduct of companies in the industry. In addition, it is common for the government to sell off part or all of their ownership of the established monopoly company. The main reason for this is to put the established carrier on an equal business and commercial footing with other new telecommunications market entrants. It affords the company more scope to manage itself, and allows the government to handle competing companies more equally.

32.4 TYPES OF REGULATORY BODIES

The regulatory body needed for policing 'conduct' regulations can take one of three forms:

- a government department;

- an independent body, financed by the government or by the industry;

- a self-regulating body, composed of a council of delegates from companies active in the industry.

Central to the decision as to which form of regulatory body is appropriate, is the question of how much information is available to the regulators. A regulator cannot perform the task if unable either to collect or to comprehend information easily.

Government and independent organizations have the benefit of neutrality and their desire to see 'fair play', but have less information and expertise available to them than would be the case for a body formed of delegates from active companies. On the other hand, self-regulating bodies (such as the UK's 'Law Society') have more information and expertise available, but only work fairly where the body has an interest in preserving a good reputation. Customers might not respect the rulings of a self-regulatory body, set up to control conduct in the second-hand car market.

Countries may choose different types of regulatory body to oversee telecommunications, but in the United Kingdom and the United States the regulatory bodies are linked to the government. OFTEL (the Office of Telecommunications) is linked to the UK Department of Trade and Industry (DTI), while in the United States, the Federal Communications Commission (FCC) reports directly to the US Congress.

32.5 DESIGNATION OF CUSTOMER PREMISES EQUIPMENT (CPE)

In nearly all countries, the first step in deregulation has been the opening of the customer apparatus (or 'customer premises equipment' (CPE)) market to free competition. The public telecommunication operator (PTO) is mandated to provide a line from the telephone exchange as far as a standard socket in the customer's premises and is deprived of the exclusive right to provide the first telephone handset. The customer is free to purchase terminal apparatus (i.e. a telephone) from a high street store or some other supplier. He may connect it to the network, provided it is suitably approved (in the UK marked with a green dot). In countries with open customer apparatus markets, fierce competition prevails, promoting a wide diversity of available equipment and low prices. Only a minimum of regulation is necessary in this market. In general, governments have found it adequate to set up 'network approvals' bodies, to verify that new terminal types conform to the network interface and safety standards. This is done at the design and prototype stage of terminal product development, and each manufactured unit then bears an 'approved' marking, although it is not individually conformance tested.

32.6 DEREGULATION OF VALUE-ADDED SERVICES

It is not feasible or desirable to open all parts of the market to competition, since certain parts of it can benefit from being a 'natural monopoly'. On the other hand, regulation must include steps to control the market power of the large established company and its ability to prevent new companies entering the market.

In telecommunications the problem has been tackled by attempting to separate the natural monopoly segments of the market (e.g. the transmission network) from the parts which might benefit from competition (the 'value-added network services' and 'customer premises equipment' (CPE) markets).

The definition of 'value added services' or 'value-added network services' (VANS) varies from country to country, but it is generally worded to cover any service falling outside the definition of a 'basic transmission service'. This definition allows entrepreneurial companies to establish services at a premium price based upon network and transmission equipment leased from the PTO. Examples of value-added services include:

- Entire 'managed networks' (these are networks in which a VANs supplier contracts to provide services akin to those on a private network, in essence taking over the role of the customer telecommunications manager). The added value is the 'management' provided.

- Recorded information services (e.g. 'speaking clock' or share price news).

- Store and forward message service.

- Telephone conference service.

- Voicebank service (message recording service). etc.

Valued added service companies may or may not be required to apply for an operating licence. In the United Kingdom companies providing value added services operate under the 1989—revised 'Branch Systems General Licence' (BSGL). The licence lays down certain codes of conduct, fair trading, technical, and pricing conditions, to ensure fair play. The Office of Telecommunications (OfTel) 'polices' the market.

In nearly all deregulated countries, providers of value added service are not usually permitted to provide the transmission equipment between premises, though sometimes they are permitted to resell the 'basic services' (e.g. the simple telephone service), at a straight mark-up. In some countries, basic services are protected as the sole province of the public telecommunications organizations (PTOs). The situation may change in due course. In the United States many of the new toll telephone carriers established themselves in the mid 1980s by reselling telephone service based on their own purchase of freephone and leased line facilities. They used the bulk traffic discount available on the latter services as a way of creating a margin for undercutting the normal telephone tariffs. Unfortunately, the quality of such services was sometimes seriously impaired. Other observing governments have steered their 'deregulation paths' to avoid this, but resale is bound to be permitted when the markets are more mature. In the UK, the government gave the PTOs, British Telecom and Mercury Communications, five years respite from the 'simple resale' of telephone service when they were first licensed in 1984, but on 1 July 1989 decided that this protection was no longer necessary, since both operators had been given plenty of time for the greater competition demanded by major users.

32.7 COMPETITION IN BASIC SERVICES

In the United States, the United Kingdom and Japan a few small financially-strong companies have been licensed to set up in competition with the established PTO. The financial strength of a new entrant to the field is a prerequisite factor, since heavy capital investment is required to establish a viable network, and the government is keen to ensure protection of both investor and customer interests. For this reason, in the United Kingdom only one major PTO has been licensed to compete with the previously established operator, British Telecom. The new PTO is Mercury Communications Limited. It has a similar licence to British Telecom, enabling it to provide transmission plant and 'basic' telecommunication services. It is backed by the might and telecommunications expertise of its parent, Cable and Wireless, but when originally set up, it also had the financial backing of Barclays Bank and British Petroleum (BP). More companies may be licensed in the future, but the British government will need to be convinced of the benefits that further competition and market diversity might bring to the market.

In the United States and Japan scores of 'toll' (or 'trunk') carriers are now operating, as well as a small number (3–10) of international carriers. Other countries are following suit.

32.8 THE INSTRUMENTS OF PTO REGULATION

Historically, PTOs tended to separate in all areas of the market; not just in basic service provision, but in the value-added service domain and in the provision of customer apparatus. Special regulatory measures may therefore be necessary in order that companies newly entering the market are not disadvantaged because they are small, or because they only compete in one domain of the PTO's business. The regulations ensure that new companies can gain a fair foothold in the market. Two forms of conduct regulation can have this desired effect:

- Price control: to prevent cross-subsidy of PTO services;

- Enforced separation of PTO's business units into 'basic network services', 'value added services', and 'provision of CPE': to prevent unfair advantage because of the greater availability of market information.

An effective, but flexible method of *price control* used in the UK is based on the formula shown in Table 32.2.

Table 32.2
PTO price control.

Price rise limited to: $RPI - X + Y$
RPI—Retail Price Index (Government index of inflation)
X—expected potential for PTO cost reduction from its own improved efficiency
Y—reflects the increase expected in uncontrollable PTO costs. (This factor is more pertinent to other public services, e.g. electricity generation, where the cost of oil/gas is outside the generating agency's control.)

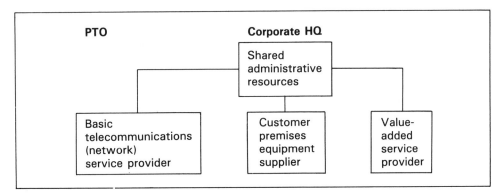

Figure 32.2
Enforced separation of PTO business units.

By applying the formula not just to the price of a single service, but to the prices of a 'basket' of basic telecommunications services, a degree of flexibility is permitted to the PTO, to temper price rises between a number of similar services. Thus the formula may apply to the *overall* price rise of a number of services. One service may increase in price at a rate higher than the RPI provided some other service has little or no price rise in compensation.

Enforced separation of the PTO's business units into sub-organizations dealing with 'basic network services', 'value added services', and 'provision of customer premises equipment' (CPE) using 'chinese walls', helps to reduce the overwhelming and unfair market power which the PTOs would otherwise have over new entrants in each of the sub-markets. A regulated code of conduct prevents the PTO from linked sales, from unfair cross-subsidization of services and from unfair use of information. A linked sale would include a practice such as 'preferential network charges provided you buy your CPE from us'. Cross-subsidization is where the price of one service is kept artificially low by the profits from another service. Unfair information flow includes 'tip-offs' from one part of the business to another. A 'tip-off' from the part of a PTO which supplies exchange lines to the part selling customer apparatus, would enable the latter to corner the market in CPE—largely preventing potential competitors from bidding.

Figure 32.2 illustrates the enforced separation of a PTO's business units. Each is forced to operate in a manner independent of the others, with cross-subsidization and unfair information flow being prohibited. In practice such clean breaks are difficult to maintain, and we can expect some arguments—not least over the funding of standards development work (this will affect service and CPE prices) and the ownership of new technology ideas.

32.9 EUROPEAN AND UK TELECOMMUNICATIONS DEREGULATION

As part of the EEC programme for generating a single European market for all products and services by 1992, removing all trade barriers between the 12 member states, the European Commission issued a green paper in 1986 laying out plans for a European-wide network infrastructure for telecommunications services. An ambitious programme over the period up to 1992 is intended to bring about a community-wide market for customer apparatus and 'portability' of equipment, by setting up a common set of network interface standards and introducing a single approvals mechanism for CPE. In addition, an ongoing period of long-term R&D and EEC government investment will ensure harmonization of the constituent national networks.

The Green Paper of 1986 laid out a framework of recommendations to the governments of member states. The objectives were to:

- Encourage the development of new services and customer apparatus, so stimulating general economic activity in and between member states.

- Encourage market conditions favouring innovation.

- Stimulate the growth of value-added and information services, since these were forecast to have a major impact on the future speed and tradeability of other industrial services, as well as upon the potential geographic locations for economic growth.

- Give wide customer choice.

- Provide employment and economic growth.

- Conform with social obligations both within and between states.

The framework laid out by the EEC recommended:

- The protection from competition of a small number of licensed public telecommunications operators (PTOs) to run the basic services and network infrastructure.

- Regulated competition in value-added and information services.

- European-wide network interface standards and common approvals procedure for network attachments.

- Strict guidelines concerning conduct and policing of these guidelines.

- An open market for customer apparatus (network attachments).

- Separation of PTO activities by 'chinese walls' between those parts of the business acting as basic service provider, value added service provider, customer apparatus provider, and any other part acting on behalf of the regulatory authority. (Some PTOs have been assigned the job of performing network attachment approvals.)

- Continuous review of PTO's conduct and performance on basic service provision, to ensure that the monopoly position is not abused.

- Continuous conduct review of value added providers.

The objective of a single market for telecommunications in Europe was endorsed by the council of ministers on 30 June 1988. European governments are responding at various speeds. Out in front, the United Kingdom currently has the most deregulated of the European markets, but Germany, Spain, France and the other EEC member countries are not far behind. German proposals for deregulation were accepted in principle by the Bundestag in April 1989, and detailed measures are to follow. In support of their objective the European government has elected to issue directives covering the regulation of:

- network framework,

- terminal equipment, and

- services.

In parallel with these objectives a new code of rules known as ONP (open network provision) will define the conditions under which the basic public services in each member state will be opened up to rival private service providers.

For the regulation of terminal equipment, a number of directives will be issued with the following objectives:

- The creation of an entirely open market for connection and maintenance.

- The loss of the PTO's exclusive right to provide CPE by end 1990.

- The establishment of user access to public network terminal points.

- The establishment within member states of independent bodies to oversee technical specification, licensing, type approval etc.

- The right of customers to dissolve long-term contracts concluded during the monopoly era.

The first terminal equipment directive was 86/361/EEC of July 1986, establishing the mutual recognition by member states of equipment type approval.

In the services field, a number of additional directives will:

- remove virtually all restrictions on competitive service provision, and

- review by January 1992 the need for continuing the exclusive rights of PTOs to provide the basic network infrastructure and the basic telephone service.

The ONP framework is the responsibility of a subgroup of the European government, called GAP (*Groupe d'Analyse et Provision*—in English the Group of Analysis and forecasting). The goals of ONP (open network provision) are:

- to harmonize access to the public telecommunications networks of Europe, and

- to create a common framework for open use of the networks.

The first draft directive (AS ART 100A) covering ONP was agreed in principle by the European council of ministers on 6 February 1990 and the full directive came into force in mid 1990. Three main areas are proposed for harmonized conditions:

- Technical interfaces.

- Usage conditions (e.g. provision time, contract period, quality, maintenance, fault reporting, resale).

- Tariff principles.

ONP will apply to all PTOs and eventually to all services, but initial priority will be given to leased lines, data networks and ISDN. ONP will demand that PTOs offering these services do so under published conditions which are objective, equal on both parties and non-discriminatory. Limitations in the contract may relate only to essential requirements such as safety and security. We shall discuss the United Kingdom as a single European case study.

32.10 INSTRUMENTS OF UNITED KINGDOM REGULATION

Legislation regulating the operation of telecommunications networks and services is changing too rapidly to make a worldwide review of much value in this book. However, one example of a country in an advanced state of deregulation, and one conforming to the EEC model, is the United Kingdom. Within the UK the EEC objectives have been met by the following legislation.

(i) The 'British Telecommunications Act, 1981', and the 'Telecommunications Act, 1984' provide the main legal instruments in the UK. The 1981 act set up British Telecom (the former monopoly network operator) as a private company separate from the post office. The 1984 act introduced the framework for competition. Under section 7 of the 1984 act, which allows the Secretary of State and the Director General of Telecommunications to licence PTOs, British Telecom was granted the scope to act as a public telecommunications operator for provision of basic services. Under its licence British Telecom (BT) is required to continue to provide service throughout the UK, to provide rural service, directory enquiry service, public call-boxes, and emergency services (999—Police, Fire Service and Ambulance). BT must publish standard terms for provision of service, must wire customer premises in such a way that it does not constrain the choice of CPE, and BT is not allowed to undertake unfair cross-subsidization of services. In addition the BT licence gives the regulator a number of powers for undertaking quality checks and ensuring fair trading.

(ii) A number of other carriers were also licensed as public telecommunications operators (PTOs) under section 7 of the 1984 act. Mercury Communications Ltd (MCL) has an equivalent licence to that of British Telecom and will provide the main competition for basic services, including long-distance calling and international network services. A number of other, but more restricted PTO licences have also been issued to:

- Telephone Department of the city of Kingston-upon-Hull (who run the telephone service in that city);

- a number of cable television operators;

- the 'cellular radio' telephone companies, 'Vodafone' and 'Cellnet';

- the emerging telepoint (CT2) public cordless telephone operators.

(iii) The terms of each licence constrain the PTO's operation of services and professional conduct.

(iv) All telecommunications networks and systems other than those owned by PTOs are, by definition of the Telecommunications Act, 'private telecommunications systems'. As such they need to be licensed and must conform with the provisions of the relevant licence. In order to obviate the need for cumbersome licensing of each individual telecommunication system, a number of 'class licences' were initially issued. The two most noteworthy issued so far, are the 'Branch Systems General Licence' (BSGL) and the 'Value Added and Data Services' (VADS) licence. However, revision in November 1989 of the BSGL has meant that the old VADS licence is now superfluous, and will be withdrawn.

(v) The Branch Systems General Licence (BSGL) is the general class licence allowing any user to connect telephones, private branch exchanges (PBXs), or any other 'branch system' equipment to the network. It lays out where the PTO's network ends, what equipment may be attached to the network, and what branch network wiring and network connections may be made. It is a catch-all licence which allows every ordinary telephone user to use the phone legally. The provisions of the 'class licence' are such that most users need not register. Instead, the granting of a licence may be assumed unless there has been a special revocation. The BSGL applies to the vast majority of telecommunications systems in the UK, i.e. those run in a single building occupied by the licensee or in premises within 200 m of one another. The licence states that private circuits leased from the PTO must be used to connect premises more than 200 m apart—and a number of constraints apply to the use of such circuits. The 'technical conditions' pertaining to the BSGL are laid out in another Oftel document, called the 'Network Code of Practice', or NCOP. The NCOP lays strict technical demands on private networks which are connected to the public network. These ensure adequate quality of connections—with the interests of both public and private network operators in mind. The technical conditions of NCOP cover:

- safety and installation practice;

- transmission performance;

- transmission levels;

- transmission delay, echo loss and stability;

- quantizing distortion (what we in this book have referred to as 'quantization distortion';

- jitter, wander and network synchronisation;

- other transmission impairments;

- signalling and call control;

- numbering and addressing;

- traffic handling capacity.

The changes to the BSGL made in November 1989 relaxed the previous prohibition on simple resale of basic services and gave greater scope for running value added services under the auspices of BSGL.

Both the BSGL and the NCOP are available, together with simple explanatory guides direct from Oftel at the address listed in Chapter 21.

(vi) The Value Added and Data Services (VADS) licence was the original class licence for value added service providers, and covered any networks falling outside the domain of the original BSGL. It laid down the conditions to be met in running the initial value added networks, the apparatus which could be connected to a VADS network and the circumstances under which the licence could be revoked. It specifically excluded straight resale of basic services and laid down a number of regulations of conduct intended to ensure fair competition between service providers.

An example of a service made possible by the VADS licence is the 'managed data network service' (MDNS) operated in a variety of forms by computer network operators. The 'basic' element of MDNS is data network conveyance, usually over a packet network; the 'value-added' element is the data format conversions which are available, the store-and-forward capabilities, and the procurement from PTOs of necessary lineplant into all relevant customer premises. Overall, the service takes the 'hassle' out of data networking.

An interesting feature of the VADS licence was the idea of a 'major service provider'. In the VADS licence a major service provider was a company whose revenue from telecommunications business exceeded £1m per annum or whose total *group* turnover (whether from telecommunications or not) exceeded £50m per annum. Such a company was considered to be unduly threatening to the viability of the PTO and so was obliged to register with OFTEL and was subject to a number of extra trading conditions. Effectively these conditions allowed Oftel closely to control their activities. The 'major service provider' concept has been incorporated into the revised BSGL, but the definition has changed to one of being any company whose turnover from telecommunications business exceeds £5m per annum. The change reflects OFTEL's view that BT and Mercury no longer need the previous level of protection from competition.

(vii) Under both the BSGL (new and revised forms) and the VADs licence, only approved apparatus may be connected to the network. Unapproved equipment may be connected by means of an approved 'barrier box' which provides for protection of the network against spurious mains voltages generated in error by the CPE. The technical standards for network interfaces are lodged by PTOs with the British Approvals Board for Telecommunications (BABT), who must approve equipment for network attachment.

(viii) Other than the need for network attachment approval and for conformance with normal quality and safety standards there is little constraint on the market in customer premises equipment (CPE).

(ix) Overall, the whole telecommunications market is regulated by the Office of Telecommunications (OFTEL), which is part of the UK government's Department of Trade and Industry (DTI).

32.11 UNITED STATES TELECOMMUNICATIONS REGULATION

Telecommunications regulation as all other United States law is complicated by the two-tier federal (interstate) and state governing bodies. Thus while federal laws and regulations apply these may be supplemented in each state with further local laws.

The FCC (Federal Communications Commission) is the independent United States government agency, responsible directly to Congress, charged with regulating interstate and international communication by radio, television, wire, satellite and cable. The FCC chairman is designated by the President. The jurisdiction of the FCC covers the 50 states plus the Islands. Its responsibilities and rulings may be categorized into seven broad areas:

- Radio regulations for commercial and amateur radio.
- Broadcast services and broadcast stations.
- Other radio services (e.g. Aviation, Marine, Public Safety).
- 'Common carriers' (i.e. telephone and similar service companies).
- Radio licence restrictions.
- Call signs.
- International matters.

Prime instruments of federal regulation are:

- the Communications Act of 1934 and its amendments;
- FCC reports and decisions;
- FCC annual reports;
- FCC registers (of Notices of proposed rulemaking decisions and policy statements);
- Code of Federal Regulations (CFR).

Copies of all these documents are available from the US Government Printing Office

(GPO) at the address below:

US Government Printing Office (GPO)
Superintendent of Documents
Washington DC 20402
United States of America

Tel: (1)–202–783–3238

In particular, title 47 of the code of federal regulations (CFR) details the FCC Rules and Regulations as shown in Table 32.3.

The Rules and Regulations lay down the operating constraints and procedures to be followed by companies providing inter-state and foreign communications services and by customers using them. Applications for construction of new facilities, licence of new services, discontinuance of old services, tariff amendments for new or existing services must be made direct to the FCC, accompanied by the relevant fee. Policies and rulemaking reflect the applications made.

The address of the FCC is given in Chapter 21.

The federal regulations governing telecommunications in the United States have largely come about as the result of a number of major judicial rulings, each plaintiff generally seeking a relaxation in the rules to add to that achieved by a predecessor.

Until the early 1950s, the telephone companies in the United States, dominated by the Bell Telephone System, dictated the services that were available to customers, and laid down strict conditions about their use. In general, these were quite restrictive, for example even going as far as prohibiting subscribers from attaching subscriber-provided equipment to a telephone company circuit. But in 1956, a US circuit court of appeals overturned an FCC ruling that had permitted the Bell Telephone System to prevent a subscriber from attaching a non-Bell device to his telephone. The device was the 'Hush-a-Phone', a mechanical device that clipped on to the handset. This commenced a period of relatively rapid regulatory change.

In June 1968, a new ruling of the FCC prohibited the Bell Telephone System from preventing subscribers using an acoustically-coupled (i.e. sound-interconnected) device known as the 'Carterfone'. The Carterfone was designed to interconnect private mobile radio units to the fixed telephone network without special and expensive equip-

Table 32.3
The FCC Rules and Regulations (Title 47 of the Code of Federal Regulations (CFR)).

Vol. I	Parts 0–19	Commission organization, practice and procedure. Commercial radio service and frequency allocations
Vol. II	Parts 20–39	Common carrier services
Vol. III	Parts 40–69	Common carrier services
Vol. IV	Parts 70–79	Mass media services
Vol. V	Parts 80–end	Private radio and direct broadcast satellite services

ment. The decision opened the way for data communication over the telephone network, and in 1969 the FCC allowed independently manufactured (non-Bell) modems to be used on the public network.

By 1976 the FCC had introduced their 'equipment registration program'. This enables devices produced by non-telephone company manufacturers to be used on the telephone network provided they meet certain stipulated specifications. The specifications are intended to ensure that terminal equipment does not cause physical or electrical damage to the network, does not injure telephone company employees and does not impair the service of other users.

The first real telephone network competition for the Bell Telephone System came in 1969 with the MCI private wire service. But by the early 1980s competition was building rapidly in the resale of long-distance telephone service, and there was growing pressure to reduce the monopoly privileges of the AT&T/Bell Company. The pressure led to the drawn out AT&T 'anti-trust suit', and finally to Judge Greene's 'Modified Final Judgement' which concluded it. The modified final judgement was the vehicle for AT&T (Bell Company) divestiture, breaking the company up into AT&T (as a long-distance carrier and equipment manufacturer) and seven regional Bell operating companies. The seven RBOCs (Ameritech, Bell Atlantic, Bell South, NYNEX, Pacific Telesis, Southwestern Bell, US West) were of roughly equal asset size and were established to take over the local call transport services of the former AT&T/Bell Company. The RBOCs, however, were prohibited from providing long-distance services and from manufacturing equipment. Meanwhile AT&T was prohibited from local call transport services and from activity as a 'yellow pages' operator.

As a toll provider, AT&T competes with a number of other licensed toll telephone service providers, principally MCI, US Sprint, Allnet and Cable & Wireless North America, and to ensure fair competition between them the FCC laid down equal access regulations governing the access arrangements with the local telephone companies. These came into full force in 1986, but paradoxically the framework is not entirely equal, since AT&T as the established and 'dominant' carrier is necessarily subject to greater regulation to curb its *de-facto* monopoly strength. Ongoing congressional debate over time is likely to equalize the rules.

The domains of 'toll traffic' and 'local call transport' are defined according to whether calls pass between or remain within a small geographical area known as a LATA. About 160 LATAs (Local Access and Transport Areas) make up the United States. Toll calls (inter-LATA) are those which cross LATA boundaries and are the sole province of toll carriers. Meanwhile, local (intra-LATA) calls are the domain of local telephone companies. Toll carriers come under federal jurisdiction, while local telephone companies are largely governed by the regulations of the relevant state public utilities commission.

Individual state Rules and Regulations need to be interpreted separately as they affect inter- and intra-LATA telecommunications services. Application is suggested direct to the relevant state agency.

32.12 OTHER COUNTRIES

Other countries are also deregulating and most follow a common philosophy to that of the United States and EEC, but each country learns from what it considers were

Table 32.4
Review of regulations, December 1989.

Country	Legislation governing telecommunications	Regulatory Agency	Is there competition for the PTO on basic services?	Is there a specific VAN licence or registration procedure?
Australia	General policy framework (1988)	Austel	N	—
Austria	—	—	N	—
Belgium	Wise Men Commission report (1986) Law in preparation (1990)	Belgian Institute for Telecommunications	N	Y
Canada	Railway Act (1906) Competitive basis for National Common Carriers (NCCS) (1987)	—	N	—
Denmark	Telephone and Telegraph Act (1987)	—	N	N
EEC (1992 objective)	EC Green Paper (1987) and subsequent directives	European Commission DG IV/DGXIII	N	—
Finland	Telecommunications Act (1987)	TAC	Y	Y
France	Decree on VANS (1987) new law 1 Jan 1991	Direction de la Reglementation Generale	N	Y
FR Germany	1989 law	Ministry of Posts and Telecommunications	N	N
Greece	New law in preparation (1990)	—	N	Y
Iceland	Telecommunications Law (1984)	—	—	—
Ireland	—	—	N	N
Italy	New law in preparation (1990)	Ministero delle Poste & Telecommunicazioni (MPT)	N	Y
ITU	World Administrative Telegraph and Telephone Conference (1988)	—	—	—
Japan	Telecommunications Business Law (1985) NTT Law (1985)	MPT (Ministry of Posts and Telecommunications)	Y	Y
Netherlands	Telecoms structure reform (1989)	—	N	N
New Zealand	Law from 1 April 1989 removed—now subject only to normal industrial trading regulation	Normal trading regulators	Y	N
Norway	—	Ministry for Transport	N	Y
Portugal	—	—	N	Y
Spain	Law Organising Telecommunications (1987, 1990)	Direccion General de Telecomunicaciones	N	Y
Sweden	Orientation of Telecoms Policy (1989)	—	N	N
Switzerland	New law in preparation (1990)	—	N	Y

Table 32.4 (*Continued*)

Country	Legislation governing telecommunications	Regulatory Agency	Is there competition for the PTO on basic services?	Is there a specific VAN licence or registration procedure?
UK	British Telecommunications Act (1981) Telecommunications Act (1984)	Oftel	Y	Y
United States	Code of Federal Regulations Title 47 (published annually) Modified Final Judgement (1984) Computer III enquiry (1986)	FCC, PUCs Judge Greene		

Key N = No; Y = Yes; — = No Information Available

the mistakes of those countries who have already deregulated. Thus while Japan and Australia followed the European model to some degree, so the European model was previously based on the experience of United States deregulation. All the regulations are far from stable. They continue to evolve at a quickening pace, but as a brief review, Table 32.4 gives an indication of the apparent state of regulation in a number of leading countries at end December 1989. Unfortunately, the table is incomplete owing to the difficulty in compiling ever-changing information. Where a new and known framework is known to have been established, or where a new regulatory body has recently been set up, this is noted.

The scope of value-added networks (VANs) and services permitted in most countries remains restricted in some degree, and the definition of them varies widely. Table 32.4 merely attempts to indicate those countries in which a formal licensing or registration procedure has been established. The table should be used only as a guide. It should not be considered to be a full, accurate or comprehensive statement of today's regulations.

BIBLIOGRAPHY

Beesley, M. and Laidlaw, B., *The Future of Telecommunications*. Institute of Economic Affairs (IEA).

Blackman, C. (ed.) *Telecommunications Policy*. Butterworth Scientific. Continuously updated.

Bruce, R. R., Cunard, J. P. and Director, M. D., *From Telecommunications to Electronic Services—A Global Spectrum of Definitions, Boundary Lines and Structures*. Butterworths, 1986.

Forum 87—5th World Telecommunications Forum. Conference Proceedings. Part 3—The Legal Symposium. ITU, Geneva, 1987.

Forum 87—5th World Telecommunications Forum—Conference Proceedings. Part 1—The Executive Telecommunication Policy Symposium. ITU, Geneva, 1987.

Franco, G. L. *World Communications—Ways and Means to Global Integration*. Le Monde Economique, 1987.

Gassman, H-P., *Information, Computer and Communications Policies for the 80's*. An OECD report. North-Holland, 1981.

Hay, A. R. and Roberts, S., *International Directory of Telecommunications.*, 3rd edn. 0582-01890-0. May 1988.

Leeson, K. W., *International Communications—Blue Print for Policy.* North-Holland, 1984.

Macpherson, S., *What are VADS?* NCC Publications, 1987.

Telegraph Regulations/Telephone Regulations—Final Acts of the World Administrative Telegraph and Telephone Conference. ITU Geneva 1973—Final Protocol/Resolutions/Recommendations/Opinions.

The Telecommunications Approvals Handbook. BT Telephone/Frank Lawson. Comm. Ed. Service Ltd., 1989.

The Telecommunications Approvals Manual (UK). Interconnect Communications Limited/Oftel, Comm. Ed. Publishing Limited, 1989.

US Code of Federal Regulations Title 47. Published annually.

White, R. L. and White, H. M. Jr, The Law and Regulation of International Space Communication. Artech House, 1988.

33

MARKETING AND BUSINESS STRATEGY FOR PUBLIC TELECOMMUNICATIONS OPERATORS

Historically, telecommunications networks have been run by state-owned organizations, which have operated as legally protected monopolies. These organizations tended to run telecommunications in a manner similar to other public utility services (electricity, gas, water etc.), providing only a few very basic services, but in a reliable and well-engineered fashion, and available at almost any geographic location to anyone who cared to apply. Recently, the trend has been to 'privatize' the established 'public telecommunications operator' (PTO) and introduce competition. The objectives in so doing are to improve the quality and diversity of available services, while simultaneously reducing unit costs and attracting private capital investment—in other words there is an attempt to make the PTOs more commercially responsive. Where competition is now a reality, the public telecommunications operators (PTOs) are forced to minimize running costs in order to work at maximum efficiency and so keep their prices competitive. In addition, PTOs are forced to adopt a much more commercial attitude towards marketing and business strategy. This chapter reviews briefly how some of the basic principles of marketing may be applied to a telecommunications marketplace and in addition outlines some of the PTOs' options for business development.

33.1 COMPANY 'MISSION'

Like any good business, a PTO needs to define its business objectives clearly, including its overall mission and its strategy for achieving it.

A typical 'mission' of a PTO might be:

- To provide a sufficiently diverse range of telecommunications networks and services to meet business (and/or residential) customers' needs.

- To provide services of high quality at a fair price and to operate in an efficient manner, developing the business in order to sustain growth in company earnings.

- (Possibly) to accept an obligation to serve the community at large, providing telecommunications for remote areas, and other specialized services for minority groups, including the handicapped.

- To establish and maintain a certain market share.

The 'mission' needs to be sufficiently clear to motivate company staff appropriately, but sufficiently flexible to allow for change in business circumstances.

Realizing the company mission and determining day-to-day strategy is the task of company management. Each of the various functions and organizational departments need to be kept working in concert—serving their customers. It is no good having a brilliantly well-designed product or service, if the customers for whom it is intended are so unaware of it that they do not buy it. Similarly if it is far too expensive, not well enough made, simply too difficult to purchase or the customer service is poor, then customers will not want to buy it.

33.2 IDENTIFYING AND ADDRESSING THE PTO'S MARKET

The first part of a PTO's marketing programme must be to identify a customer requirement. A good deal of insight, imagination, and research is required in order to analyse potential ways in which a customer's business can be improved by a product or service, in return for payment. The product or service may make the customer's activity easier, more effective, cheaper, more enjoyable, or may be attractive for some other reason.

A large PTO will be wise to maintain a balanced portfolio of products and services. Dependence on too few products leaves a company prone to rapid loss of business whenever there is a sudden and significant change in market demand, or when a rival company rapidly introduces a much better product. None the less, each product must be clearly defined in terms of the customer's need that it serves. In classical marketing terms this means getting the four P's right—product, price, place and promotion. In PTO service terms we interpret these as follows.

The product Telecommunications products (or services) should not be specified only in terms of their technical characteristics; the quality and customer service aspects of the product should be defined as well. The 'quality' is not only the service or manufacturing quality, but also the target in-service reliability and back-up measures (if any). The customer service arrangements should include any necessary enquiry service, and the standard and expected speed of repair should be defined.

The price Price is an important consideration in the customer's final decision whether or not to purchase. The customer may be unable to afford too high a price, no matter how 'fair' the price may be for a very high quality product. Paradoxically, however, too low a price may conjure customer perceptions of a low quality product.

As a first estimate of a suitable price, it is valuable to undertake a 'cost-plus' analysis. Figure 33.1 illustrates the costs and revenue associated with a given product. Costs include 'fixed costs' and 'variable costs'. Fixed costs are those costs that are not related to sales volume. For the telephone service, the fixed costs are the costs of the telephones and the exchanges. These items are needed by all customers, irrespective of how many calls they make.

The variable costs include those associated with the lines interconnecting different exchanges. The number of circuits provided (and so the cost of them) can be adjusted to match the actual inter-exchange call demand. Other variable costs include those associated with staff costs (service level could be adjusted), administration etc.

The revenue from the telephone service is the total money collected, and equal to the total number of call minutes times the price charged per-minute. The relationship of costs and revenues is as shown in Figure 33.1.

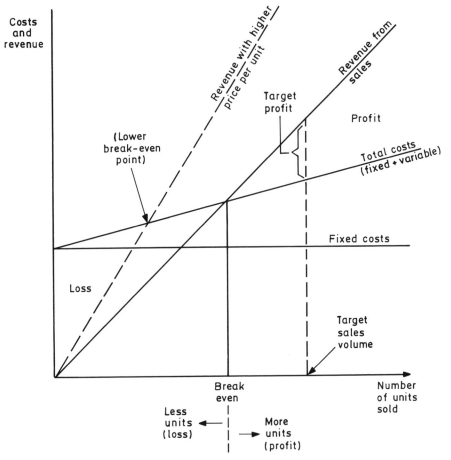

Figure 33.1
Determining product price.

The 'break even' point shown in Figure 33.1, is the point at which the revenue just equals the costs. At this 'point', sufficient products (i.e. calls or whatever) have been made for the company to 'break even'. Selling fewer units than this will result in a loss. Selling more will generate a profit.

The number of units that have to be sold to 'break-even' can be adjusted, by adjusting the price charged for each unit. The higher the price charged to customers for each 'unit', the smaller the number of units that need to be sold in order to break even. This is also illustrated in Figure 33.1.

The degree to which the number of sales is sensitive to price is known as the 'price elasticity'. Normally, one expects a lower price to attract more sales and a higher price to deter sales; but different products have different priced elasticities, so that with some products a higher price has little effect on the number sold, while with other 'commodity' products, the only important factor in the buyer's decision may be which competitor's product is the cheapest.

The place   The place at which a product or service is available is also important. For some products, customers' loyalty will make them travel long distances to purchase a given 'brand' of product. For other products (for example, foodstuffs) the customer is likely to choose from the brands available on the nearest supplier's shelf (e.g. supermarket).

'The place' element of telecommunications marketing plan needs to lay out the geographic availability of the service (for example 'lines are available nationwide' or 'lines are available in major cities'). The plan also needs to lay out the location of sales and customer support staff, and the method by which customers may 'buy' the service (e.g. by going to the local telephone office, or by calling a freephone number etc.).

Promotion   Promotion is the means by which the intended customers of a product or service get to hear about it. Including 'advertising' and 'publicity', the reader will be familiar with many techniques used in this field.

33.3   PTO PRODUCT DEVELOPMENT

During product development, it is helpful to develop a set of documentation and procedures detailing the product—its launch and subsequent support. One possible means of doing this is by means of a 'product manual'. Suggested topics to be covered in such a manual are given below.

Product manual

Product description   A full description of the product (or service), describing its customer applications and its technical attributes.

Product positioning   The competing products (both internal products and those of competing companies). The expected product lifetime, and the changeover strategy for products which it is planned to replace.

Availability The geographic availability of the product and any relevant service availability or 'upgrade' dates.

Quality and performance The expected reliability of the product, and the methodology for pre-service and in-service 'quality assurance' monitoring checks.

Product price and conditions The price of the product and any discounts. The warranty, service, maintenance and regulatory conditions.

Regulatory approval Safety and technical standards may have to be declared. Customer premises equipment (CPE) may have to be type-approved for connection to the network.

Promotion Is the product to be sold by 'direct sales' or will an 'agency' outlet be used? What advertising and promotional methods will be used?

Management of product stock Is the product to be sold and transported to the customer's premises (e.g. an item of CPE)? Or does it need specialized installation, as in the case of a network access wire to the customer's premises (for a network service)? Products and back-up spares should either be freely available to meet the demand of customers or be deliverable within the target lead-time.

Installation or provision Ordering lead times and installation duration times need to be determined. Resources need to be allocated, and training given to company staff. An ordering procedure should to be established.

Sales Sales targets and incentives need to be set. An exhibition suite is needed to demonstrate complex services, and shops are needed for retail sales. A full price list must be available.

Maintenance and after sales Records of customer names, and of the applications to which products will be put, can be invaluable. For example, the knowledge that a telephone line is used predominantly in conjunction with a facsimile machine may help maintenance and network design staff.

Customer support A telephone number is needed for customer enquiries and fault reporting.

Training Adequate training needs to be given to all staff, and reference information made available.

Post-launch monitoring Means and procedures are needed to measure the ongoing performance of the product—not just in technical, quality and reliability terms, but also from the marketing and sales perspective, and in financial viability terms.

Updates A proper procedure should be established for updating the product manual.

33.4 PTO BUSINESS DEVELOPMENT

Finally, we consider the directions in which the world's major public telecommunications companies are likely to evolve and develop their business over the next few years. As a result of their historic backgrounds, most PTOs are today restricted to providing telecommunications network services within a single country, with international extension of network services made available as a result of 'bilateral agreement' between two or more PTOs.

Figure 33.2 illustrates a typical spread of a PTO's current activities, with the predomination of business in the home country. Most business is in the 'basic interconnection market', though there is growing business resulting from international traffic interconnection, and there may be a considerable source of income from CPE sales and value added services.

From this position, each PTO must develop its business strategy taking into consideration a number of factors, including size, legal constraints, the degree of deregulation in its home market, the sophistication of its customers, the degree of its network development, and its capital resources.

One PTO may decide that its appropriate short-term goal is simply to make its established business grow in financial terms, without a significant change of emphasis in its sphere of activity. In this case, the activity and resources of the PTO can be concentrated upon developing the sophistication of its established network, upon increasing its coverage area (perhaps by increasing the proportion of households with a telephone), and maybe upon diversifying the number of basic services offered (for example, by introducing a packet-switched data network). In support of this strategy,

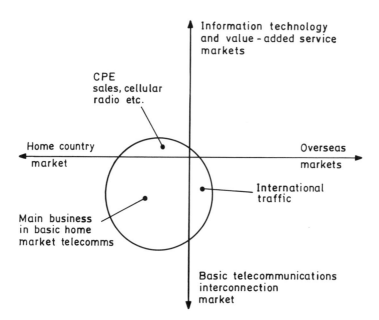

Figure 33.2
Typical PTO business (late 1980s).

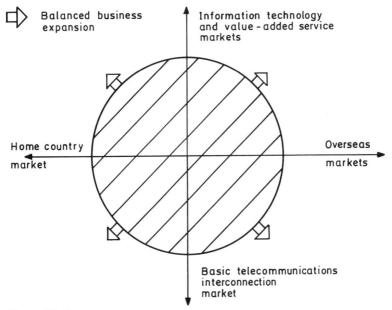

Figure 33.3
Business sphere of major international PTOs (mid to late 1990s).

finance is needed for network developments, modernization, research and development (R&D) and product innovation in the home market, and revenue return accrues from the expanded home customer base, until the point of home market saturation is reached.

In contrast, some of the world's large PTOs, already operating in deregulated and well developed home 'interconnection' markets, may wish to consider diversification of their business into other spheres of activity. Opportunities for diversification include the specialist international service markets, and the whole field of value-added and information services. It seems possible that by the mid to late 1990s a small number of major PTOs will have grown into large international conglomerates with spheres of business activity much wider than their previous operations. Figure 33.3 illustrates an example in which a PTO's business is diversified over a large range of markets throughout the world, with steady and balanced growth in each of the markets.

But how will these companies achieve this level of expansion? Well, their first step will be the generation of spare cash from their established business. Then, diversification into the new markets will come about by financial venture in those countries where new deregulation permits it, by joint venture, or by simple company acquisition. We can therefore expect the establishment or acquisition of a number of PTO subsidiary companies. Largely autonomous at first, these companies may be kept small and dynamic to cope with the initial market entry, but will be able to call on the massive resources of the parent. Later on, fuller integration of the subsidiary organization into the main company structure may help further development of the company's 'core' business. Each PTO's prime business and its prime competitors will constantly be changing. We may see major alliances of PTOs with telecommunications manufacturers, computer hardware and software suppliers, and may even see enormous PTO conglomerates, operating in a number of regions of the world. Aspirations to become a 'global' PTO have already been demonstrated by a number of PTOs, which have sought variously to infiltrate overseas CPE markets or to run foreign networks on a franchise basis.

Some PTOs are already in close joint co-operation, specifically aiming to meet the telecommunications and value added service needs of the world's multinational corporations. This is a crucial market, since the top 60 corporations have annual turnover of between $10 billion and $100 billion, and valuable telecommunications business. PTOs will ignore this business at their peril.

In Europe we can expect a fight for the domination of the Single European Market (SEM) of 320 million customers from 1992. The fight for the North American domestic market is already raging and the gloves are off for the world market battle! A major business conflict is in prospect, and we can expect some of the world's major PTOs, as well as some computer and telecommunications equipment manufacturers, to get bruised on the way! But let the fittest company survive!

BIBLIOGRAPHY

Bruce, R. E., Cunard, J. P. and Director, M. D., *From Telecommunications to Electronic Services—A Global Spectrum of Definitions, Boundary Lines and Structures.* Butterworths, 1986.

Forum 87—5th World Telecommunications Forum. Conference Proceedings. Part 5—Symposium on Regional Network Development and Cooperation. ITU, Geneva, 1987.

Forum 87—5th World Telecommunications Forum. Conference Proceedings. Part 1—The Executive Telecommunication Policy Symposium. ITU, Geneva, 1987.

Forum 87–5th World Telecommunications Forum. Conference Proceedings. Part 4—Symposium on Economic and Financial Issues of Telecommunications. ITU, Geneva, 1987.

Franco, G. L., World Telecommunications—ways and means to global integration. Le Monde Economique, 1987.

Fuld, L. M., *Competitor Intelligence—How to Get It; How to Use It.* John Wiley & Sons, 1985.

Hunter, J. M., Lawrie, N. and Peterson, M., *Tariffs, Traffic and Performance—the Management of Cost Effective Telecommunications Services.* Comm. Ed. Publishing, 1988.

Lancaster, K. L., *International Telecommunications—User Requirements and Supplier Strategies.* Lexington Books, 1982.

Leeson, K. W., *International Communications—Blue print for Policy.* North-Holland, 1984.

Majoro, S., *International Marketing—A Strategic Approach to World Markets.* George Allen & Unwin, 1977.

Sciberras, E. and Payne, B. D., *Telecommunications Industry—Technical Change and International Competitiveness.* Longman, 1986.

GLOSSARY

adaptive differential pulse code modulation (ADPCM) A highly efficient, low bit speed method of digital encoding of voice signals.

amplification The boosting of a faint signal—counteracting the effects of attenuation in order to ensure good comprehension by the receiver.

amplitude The relative strength or volume of a signal.

amplitude modulation (AM) A technique used to enable a low bit speed or comparatively low bandwidth formation signal to be carried on a higher frequency 'carrier' signal—by adjustment of the latter's amplitude to match the former's wave shape.

analogue transmission A method of signal transmission in which the shape of the electrical current waveform on the telecommunications transmission line is analogous to the air pressure waves of the sound or music signal it represents.

application layer The uppermost layer (layer 7) of the open systems interconnection (OSI) model. This 'layer' is a software function providing a program in one computer with a communication service to a 'co-operating' computer.

ASCII The American Standard Code for Information Interchange. A widely-used 7-bit binary code used to represent alphabetic and numeric characters in computer understandable form.

asynchronous data transfer A method of data transfer in which each alphabetic or numeric character (represented by 7 or 8 bits) is preceded by 'start' and 'stop' bits to delineate the 7/8 bit pattern from the idle bit pattern which otherwise occupies the (digital) transmission medium.

attenuation The effect of signal dwindling experienced with accumulating line length or distance of radio transmission.

attenuation distortion The various frequency components of a complex sound or data signal are often attenuated by the line to different extents. The tonal balance of the original signal is thus distorted—attenuation distortion.

audio circuits Copper or aluminium-wired circuits being used to carry a single signal—typically a speech signal in the audio bandwidth.

bandwidth   Measured in Hertz or cycles per second, the range of component frequencies making up a complex speech or data circuit. Thus speech comprises frequencies between 300 Hz and 3400 Hz, a total bandwidth of 3.1 kHz.

basic rate interface (BRI)   The simplest form of network access available on the ISDN (integrated services digital network). The BRI comprises 2B + D channels for carriage of signalling and user information.

bearer service   Relating in particular to ISDN, various bit rates or bandwidths (called bearer services) may be made available between the two ends of the connection. Examples include 'speech' bearer service, 64 kbit/s data service.

binary code   A two-state (on/off) code easily interpreted by electronic circuits. Seven and 8-Binary digIT (bit) codes are usually used in computers to represent alphabetic and numeric characters.

bit error rate (BER)   A measure of the quality of a digital transmission line, either quoted as a percentage, or more usually as a ratio, typically 1 error in 10^5 or 10^9 bits carried. The lower the number of errors, the better quality the line.

bit speed.   The 'bandwidth' of a digital line or signal—governing the rate at which individual alphabetic or numeric characters may be carried by the line.

busy-hour traffic   A measure of the maximum user demand for network throughput.

carrier frequency   A high frequency signal used in FDM or radio transmission. The information is 'modulated' on to the carrier prior to transmission.

cellular radio   A mobile, call-in and call-out telephone service in which radio base stations are arranged in individual cellular coverage areas. Variants include TACS, AMPS, NMT, GSM, network-C, Radiocom-2000.

channel associated signalling   A method for interexchange signalling in which dialled digit and other call set-up and control information is carried by the associated circuit.

circuit multiplication equipment (CME)   A device enabling line economy by processing signals (e.g. speech) in such a way that it appears that more than one circuit is available, hence circuit multiplying. Often CMEs rely upon the fact that the individual signals rarely use all their available circuit bandwidths. The term is usually applied to multiplexing equipment.

circuit-switching   A type of exchange or network designed to be capable of providing a direct connection between originating and terminating points of a communication for the entire duration of a call.

coaxial cable   A cable used for high frequency and high bandwidth analogue and digital signal transmission. The cable comprises a single inner conductor plus a conducting outer sleeve—in cross-section a dot in a circle.

codec (coder/decoder)   A device for interfacing a 4-wire digital transmission line to a 4-wire analogue. The coder codes the received 2-wire digital signal in a form suitable for 2-wire analogue transmission. The decoder meanwhile decodes the 2-wire analogue signal received in the other direction.

common channel signalling A type of interexchange signalling in which the dialled digit, call set-up and other circuit control information for a large number of circuits (typically 200) is conveyed directly between the exchange processors on 'common channel signalling links'.

cordless telephone A telephone handset connected to the ordinary telephone network by means of a short radio path (typically up to 150 metres).

crosstalk The interference of telephone and other signals resulting from the induction of an unwanted signal from an adjacent telephone line. (The crossed-wire.)

CT2 The second generation type of cordless telephone, in which handsets may operate in conjunction with public radio base stations—distributed around the streets much like public phone kiosks.

customer premises equipment (CPE) The telephone or data equipment on a customer's premises—that using the public telephone or data network for connection to a remote location.

data Data is the term used to describe alphabetic and numeric information stored in and processed by computers.

data compression Any of various techniques, of both reversible and irreversible types, designed to reduce the bit speed needed for a particular type of communication (e.g. facsimile).

data circuit terminating equipment (DCE) The equipment terminating and controlling the transmission line, and often marking the endpoint of the public data network. Data terminal equipments (DTEs) such as computers are connected directly to DCE.

datalink layer Layer 2 of the open systems interconnection (OSI) model. This software function operates on an individual link within the connection to ensure that individual bits are conveyed without error. The best known layer 2 protocol is HDLC (high level data link control).

data terminating equipment (DTE) The jargon used to describe any type of computer or other equipment, when connected to a data communications network.

decibel (dB) A measure of signal strength, volume, amplification or signal loss.

digital transmission A modern method of signal transmission in which the speech or data information is represented as a string of binary digits.

digit analysis The process of analysis of dialled digits carried out by an exchange in order to determine the destination of a call and therefore enable the selection of an appropriate outgoing route.

distortion The corruption of a signal brought about by electrical disturbance or non-uniform line properties, particularly causing different frequency components of a complex signal to be unequally affected.

duplex Transmission means allowing simultaneous two-way signal transmission—in both received and transmit directions without interference of the signals.

EBCDIC Extended Binary Coded Decimal Interchange Code. An 8-bit computer code for representing alphabetic and numeric characters.

echo The effect resulting from a delayed reflection of a signal. Echo in a telephone circuit manifests to the talker as a slightly delayed repeat of his own voice—returned from the distant end—just like a sound echo.

electronic ticketing The process of recording call details at the time of call-set-up and conversation in order that charges can be calculated and billed to the calling customer.

en-block signalling A method of interexchange signalling in which the entire dialled number is transferred from one exchange to the next exchange in the connection as a continuous string of digits. Digits are not conveyed to any subsequent exchange in the connection until all the dialled string has been received. Compare with overlap signalling.

equalization The process carried out on long lines in order to counteract the ill-effects of attenuation distortion—rebalancing the signal strengths of various component frequencies to be equal to that of the original signal.

erlang The measure of traffic intensity on circuit-switched networks. One erlang of traffic represents an average demand of one call in progress continuously throughout the 'busy hour' of a given route or exchange.

Erlang lost-call formula developed in 1917 by A.K. Erlang, this formula is used to calculate circuit requirements of a given interexchange route given the traffic demand in erlangs and the required 'grade of service'.

Erlang waiting-call formula A derivative of the lost-call formula, the Erlang waiting-call formula is used to determine circuit requirements, queue length requirements and the server numbers needed in a 'waiting system' (e.g. an operator switchroom).

error-free seconds (EFS) A performance measure used to specify the quality of data lines. Higher specification lines require a higher proportion of error-free seconds.

framing The structuring framework required in digital line systems in order that a number of different users or channels may share the line by time division means.

frequency distortion A signal impairment caused by transmission line phenomena (e.g. attenuation) which affect the various component frequencies of a complex signal unequally. The tonal balance of the original complex signal is lost—distorted.

frequency division multiplex (FDM) A technique allowing a number of different audio signals to share the same line without impact upon one another. A high bandwidth transmission line is shared out by dividing into a number of smaller bandwidths—each used to convey a different end-user signal.

frequency modulation (FM) A technique used to enable a low bit speed or low bandwidth information signal to be carried on a much higher frequency 'carrier' signal—by adjustment of the latter's frequency by a small amount corresponding to the amplitude of the former.

frequency shift distortion An impairment caused by slight, but different shifts in frequency of the component frequencies of a signal.

frequency shift keying (FSK) When a carrier signal is frequency modulated with a binary data signal, the effect is merely to alternate the carrier signal between two or more fixed values. This is known as frequency shift keying.

full duplex In contrast to half-duplex devices, full duplex ones allow permanent, simultaneous two-way transmission of information, without interaction or interference of receive and transmit signals.

gain hits An impairment causing errors on data transmission lines. Gain hits are intermittent but only moderate and relatively short disturbances in the amplitude of a received signal.

geostationary satellite A type of communications satellite in orbit above the earth's equator but moving at the same speed as the rotation of the earth. From the ground, the satellite therefore appears to be geographically stationary.

grade of service (GOS) The proportion of calls in a circuit-switched network which are lost, owing to network congestion. Since it is uneconomic to provide networks of enormous uncongested proportion, it is usual to design networks to a given grade of service—typically to be incapable of completing around 1–3 per cent of calls.

group delay A signal impairment caused by different speeds of propagation of the component frequencies of a complex signal.

half duplex A telecommunications device allowing two-way transmission of signals or other information, but only in one direction at a time. Thus a half-duplex device cannot simultaneously transmit and receive, though interspersed bursts in each direction are possible.

harmonic disturbance Harmonic disturbance is an impairment resulting from the intermodulation of two different signal frequencies F_1 and F_2 when passed through non-linear transmission devices. Stray signals, $F_1 + F_2$, $F_1 - F_2$, $F_1 + 2F_2$ etc. are produced.

high-level data link control (HDLC) A protocol conforming with the open systems interconnection (OSI) model, layer 2, allowing for the error-free conveyance of bits over a single link of a connection.

holding time The total time for which a circuit or exchange is in use on a call, commencing from the moment when the calling customer picks up the phone, or when the circuit or exchange is 'seized', until the time when the connection is cleared.

hybrid A hybrid or hybrid transformer provides 2-wire to 4-wire transmission line conversion. Often used at each end of long transmission lines to covert the local customer's line (if analogue and 2-wire) to the 4-wire central trunk section (usually 4-wire to allow easier intermediate signal amplification).

IA2 alphabet International alphabet number 2. A 5-bit mark/space code used between telex machines to represent the alphabetic, numeric and punctuation characters used in text.

IA5 alphabet International alphabet number 5. A 7-bit computer code used to represent alphabetic, numeric and control characters. Developed from the ASCII code.

impulse noise Impulse noise is characterized by large 'spikey' waveforms and arises from unsuppressed power surges or mechanical switching noise. To the listener it manifests as a loud chatter or cracking sound.

inband range The term inband is usually applied to signal frequencies lying within the bandwidth of a normal speech channel, i.e. between 300 Hz and 3400 Hz.

integrated services digital network (ISDN) A recent development of the telephone and circuit-switched data network principle, in which a common, integrated and digital network is provided to cater for both voice and data applications.

interference A signal impairment caused by the interaction of an unwanted adjacent signal. Interference particularly affects FDM and radio systems.

intermodulation noise Intermodulation noise is the name given to the undesirable signals which can result, of frequencies $F_1 + F_2$, $F_1 - F_2$, $F_1 + 2F_2$, etc., when two different component signal frequencies F_1 and F_2 interact with one another.

International Telegraph and Telephone Consultative Committee (CCITT) An influential subcommittee of the International Telecommunications Union (ITU)—a leading body in technical and operational standards for international telephone networks.

I420 interface An alternative expression to define the ISDN basic rate, or $2B + D$ interface, I420 is the number of the CCITT recommendation in which the interface is defined.

jitter Jitter or 'phase jitter' arises when the timing of the pulses of a digital signal varies slightly, so that the pulse pattern is not quite regular. The effects of jitter can accumulate over a number of regenerated links and result in received bit errors.

junction A circuit interconnecting two telephone or other circuit-switched exchanges. In particular, the word junction usually relates to the circuits directly interconnecting local exchanges in localized zones.

local area network (LAN) A type of data network allowing the interconnection of computer devices within a relatively restricted area, typically lying along a spine or circle of length less than about 200 metres.

line code The technique used when actually transmitting a digital signal on to a transmission line. The line code helps to ensure the synchronization and correct receipt of data.

loudness rating (LR) The loudness rating of a telephone circuit gives a measure of the received signal volume (in the listener's ear) compared with that spoken by the talker. A transmission plan allocates how much loss is acceptable overall and apportions design limits on the individual sections of the circuit.

mark A binary digit of value '1'.

message handling system (MHS) A conceptual model specified by CCITT Recommendation X400, describing the way in which message devices (e.g. facsimile and telex machines, computers, word processors etc.) intercommunicate. By laying down a standard method, the MHS seeks to make the interconnection of different types of equipment easier.

microwave radio   Radio systems operating with a carrier frequency in the range 1 GHz–30 GHz. In telecommunications networks, microwave radio is used to span long distances and difficult terrain. Microwave towers with their dish and horn-type antennas are usually found on hill-tops.

modem   A term derived from a shortening of MOdulator/DEModulator. A modem provides for the carriage of digital data information over analogue transmission lines.

modulation   Modulation is a means of encoding information on to a carrier signal in order that it may be transmitted over a radio link or transmission line.

monomode fibre   An optical fibre with a narrow central core of a different refractive index from the cladding which surrounds it. The narrowness of the core allows only very few ray paths to exist—so that dispersion is minimized—a single mode of transmission prevails—hence 'monomode'.

multiplexing   Multiplexing is a technique for combining a number of full duplex channels together to share the same transmission line or radio medium. Unlike circuit multiplication, multiplexing does not affect the full duplex nature of the channels. Forms in common usage are time division multiplexing (TDM) and frequency division multiplexing (FDM).

network layer   The network layer is layer 3 of the open systems interconnection (OSI) model. This layer sets up an end-to-end connection across a real network, determining which permutation of individual links to be used. Thus the network layer performs overall routing functions.

noise   Stray, unwanted and random signals interfering with the desired signal. Common forms of noise manifest as a low level crackling noise.

notched noise   Notched noise arises in data circuits—caused by the way in which the signal is processed over the course of the connection. Its name arises from the way it is measured—a pure frequency tone is applied at one end and removed with a notch filter at the other. Remaining noise is the notched noise.

NT1 (network terminating equipment 1)   The line terminating device providing customers' basic rate interface (2B + D) to the ISDN.

off-hook signal   The signal generated by a telephone on lifting the handset, indicating to the exchange that the customer wishes to make a call.

open systems interconnection (OSI)   A conceptual model specified by CCITT recommendations in the X200 series. The model describes the 7-layer process of communication between 'co-operating' computers. The model provides a standard for the development of communication protocols allowing for computers of different manufacturers to be interconnected.

optical fibre   A glass fibre of only a hair's breadth dimension, capable of carrying very high digital bit rates with very little signal attenuation.

OSI network service   The service provided by a protocol conforming to the OSI network layer (layer 3). The network service provides an end-to-end connection for the use of the OSI transport layer. An X25 network could provide a network service to the front-end processor of a mainframe computer.

out-of-band range   The term used to describe the frequency range within an allocated 4 kHz circuit bandwidth but outside of the filtered speech range of 300 Hz–3400 Hz. The out-of-band range is used by some interexchange signalling systems in order that the signalling tones neither interfere with the speech nor can be mimicked by telephone fraudsters.

overlap signalling   A method of interexchange signalling in which dialled number digits are transferred from one exchange to the next exchange in the connection one-by-one. As soon as received, digits may also be onpassed to any subsequent exchange, even if the entire dialled number string has not yet been received. Two links may thus be signalling simultaneously (overlapping). The advantage of overlap signalling over *en-bloc* signalling is that subsequent exchanges may start their digit analysis earlier, so that the overall call set-up time can be minimized.

packet assembler/disassembler (PAD)   A device enabling a character (i.e. asynchronous) terminal to be connected to a packet network.

packet-switching   A type of exchange or network which conveys a string of information from origin to destination by cutting it up into a number of packets and carrying each independently. A packet-switched effect could be achieved by sending individual pages of a book through the post separately. The receiving device reassembles the message. Thus a direct connection between origin and destination does not exist at any point.

parallel data transmission   Data transmission between computer devices using multiple lead wires or ribbon cables. Whole bytes of information (8 bits) are sent simultaneously by using a separate wire for each of the individual bits of the byte.

phase hits   Phase hits are impairments affecting data circuits. A phase hit is an intermittent but moderate and relatively short disturbance in the phase of the received signal. Where these hits accumulate over a multilink connection they can affect the correct interpretation of the data, resulting in errors.

phase jitter   Phase jitter or simply jitter arises when the timing of the pulses on an incoming data signal varies slightly, so that the pulse pattern is not quite regular. Where these effects accumulate over a multilink connection, received bit errors can result.

phase modulation   A technique used to enable a low bit speed or comparatively low bandwidth information signal to be carried on a higher frequency 'carrier' signal—by adjustment of the latter's phase to match the former's wave shape.

phase shift keying (PSK)   When a binary data signal is phase modulated on to a carrier the effect is to create a transmitted carrier signal of one of a number of fixed phases. Jumping between these phases according to the incoming bit stream is known as keying, hence phase shift keying.

physical layer   Layer 1 of the open systems interconnection (OSI) model. The physical layer protocol is the hardware and software in the line terminating device which converts the databits needed by the datalink layer into the electrical pulses, modem tones, optical signals or other means which will transmit the data.

presentation layer Layer 6 of the open systems interconnection (OSI) model. The presentation layer defines the manner in which the data is encoded, e.g. binary, ASCII, IA5, IA2, EBCDIC, facsimile etc.

primary multiplex (PMUX) A device performing the first stage of time division multiplexing. In the UK and Europe, a primary multiplexor converts 32 analogue signals into a 2.048 Mbit/s time division multiplexed signal. In the US equivalent, instead 24 channels are multiplexed into a 1.544 Mbit/s stream.

primary rate interface (PRI) The 23B + D or 30B + D ISDN interface typically used to connect ISDN PBXs to the public ISDN.

proceed-to-send (PTS) A backwards interexchange signal used to indicate the exchange's readiness, following the prompting of the forward seizure signal, to receive dialled digits for analysis, switching and onward call routing.

propagation delay The delay encountered by a signal during the course of transmission from origin to destination.

protocol A protocol is a rule of procedure by which computer devices intercommunicate. Thus a protocol is the equivalent of a human language, with punctuation and grammatical rules.

PTO (public telecommunications operator) A term of legal significance in the United Kingdom, applying to companies licensed to provide public network services, including telephone and data network services.

pulse code modulation (PCM) The method of conversion used to enable speech or some other analogue audio signal to be carried over a digital transmission path.

pulse metering The historical method of monitoring telephone call duration for later customer invoicing. Pulses are generated at regular intervals during the call and recorded on the customer's cyclic meter. At the end of a one or three month period the customer is billed for the number of units used.

quadrature amplitude modulation (QAM) A complex method of modulation used in modern modems to allow very high data rates to be carried reliably and relatively error-free. QAM is a combination of phase and amplitude modulation.

quantization A process fundamental to pulse code modulation, when converting an analogue signal into a digital equivalent. The analogue signal is sampled at regular periods of time, its amplitude being estimated by the nearest match to a number of predetermined or 'quantum' values. Each quantum value is numbered. This value is transmitted on the line and allows the original signal to be reproduced.

quantization noise Quantization noise is heard when an analogue signal is reproduced at the receiving end of a pulse code modulated transmission link. It arises from the slight discrepancy between the reproduced 'quantum' amplitude level and the actual amplitude of the original signal.

quantization distortion unit (qdu) When PCM encoded signals encounter non-linear processing during their course of transmission (e.g. numerous analogue/digital conversions, or an intermediate ADPCM link) then further quantization noise can result on final reception. The measure of overall error is measured in terms relative to the amplitude difference of adjacent quantum levels. This unit is called the qdu.

quantizing   An alternative term for quantization.

radiopaging   A radio-based method for alerting field based staff. Staff wear a cigarette box-sized unit which either bleeps or displays a short alphanumeric message.

regenerator   A device inserted at an intermediate stage of a digital transmission line to counteract the effects of signal deterioration which occur on long lines. As far as possible the device regenerates the original bit stream.

repeater   A device inserted at an intermediate stage of an analogue transmission line to counteract the effects of signal deterioration (particularly attenuation) which occur on long lines.

ringing current   The current applied to a destination telephone to ring the bell and thereby alert its owner to answering it.

ringing tone   The tone, quite separate from ringing current, but applied at the same time, to inform the caller that the destination telephone is ringing.

routing table   A table held in the exchange which relates dialled digit information to the appropriate outgoing route to be selected.

RS–232C   A standard interface, using a 25-pin D-type plug and connector by which computing devices may be physically interconnected. The standard also defines the layer 1 protocol for data transfer.

satellite   In telecommunications terms, a radio repeater station operating in the microwave frequency range, which has been placed in space to orbit the earth. Most telecommunications satellites nowadays are geostationary (described elsewhere).

SDLC   Synchronous data link control. IBM's equivalent of HDLC, forming the layer 2 of IBM's Systems Network Architecture (SNA).

segment of data   A measure of data quantity used to monitor data network (particularly packet data network) usage. A segment is equal to 512 bytes.

seizure signal   The interexchange signal sent forwards along the connection to alert the subsequent exchange of a desire to establish a new connection for a call currently being set up.

serial data transmission   Data transmission between computer devices using only a single circuit path. Whole bytes of information (8 bits) are sent in a sequential pattern. Compare with parallel transmission. Parallel transmission is often used internally within computing devices because of the higher processing speeds which are possible, but for long-distance telecommunication, serial transmission is more economic in terms of lineplant.

session layer   Layer 5 of the open systems interconnection (OSI) model. When established for a session of communication, the two devices at each end of the communication medium must conduct their 'conversation' in an orderly manner. They must listen when spoken to, repeat as necessary, and answer questions properly. The session layer provides for this discipline.

sidetone The term used to describe the talker's own voice when heard immediately in his own earpiece (if there is any delay then the effect is instead called echo). With no sidetone at all most telephone users are caused to think that the telephone is dead, but with too much sidetone, the talker is distracted and can be tempted to talk too softly for good listener comprehension.

simplex A transmission means allowing only one direction of transmission. (For example public broadcast radio.)

single sideband A method of bandwidth and power economy used in radio and FDM systems.

space A term used to describe a bit of binary value '0'.

speech interpolation A method of achieving circuit multiplication, by interspersing other clippets of other conversations in the gaps between individual words, sentences, and in the period while the distant user is talking.

S/T interface An alternative term used to describe the ISDN basic rate interface.

statistical multiplexing A method of data lineplant economy similar in operation to speech interpolation devices. Various data communication conversations share the same line by making use of each other's idle periods.

supplementary service Supplementary services exist in ISDN. On their own they have no value, but serve to add a supplementary value to a bearer service such as speech or 64 kbit/s data. An example of a supplementary service is 'calling line identity' or 'ring back when free'.

synchronization The method by which the bit patterns appearing on digital line systems may be properly 'clocked' and interpreted—allowing the beginning of particular patterns and frame formats to be correctly identified.

synchronous data transfer Data transfer employing a strictly regular pattern, rather than (as in synchronous operation) using start and stop bits to distinguish character patterns from idle line co-operation. In synchronous operation, certain regularly spaced bits within the overall string should always be interpreted to be the first bit of a byte pattern.

systems network architecture (SNA) A standard framework of communications methods used in particular with IBM and compatible computer devices.

tandem A non-direct connection between two exchanges. The term can also be used to describe the intermediate exchange in such a connection.

telepoint The generic name used in the United Kingdom to describe cordless telephones of the second generation (CT2). These allow portable handsets to be used in public locations, provided the user is within about 150–200 metres of an appropriate telepoint radio base station. Calls are onward routed via the ordinary telephone network.

teleservice A term used to describe a particular type of ISDN service, a teleservice comprises all elements necessary to support an end user's particular purpose or application. Examples of teleservices include 'telephony', 'facsimile', 'videotelephone' etc.

time division multiplex (TDM) A technique allowing a number of different digital signals to share the same line without impact upon one another. A high bit rate digital transmission line is shared out by dividing it into a number of smaller component bit rates, interleaved in time with one another and each used to convey a different end-user signal.

traffic intensity The measure of instantaneous demand on a network, quantified in terms of the number of users using the network simultaneously.

transmission bridge The circuitry provided in telephone exchanges to power individual customers' lines but retain isolation between them so that they do not overhear one another's conversations.

transmultiplexor (TMUX) A device enabling an FDM group or supergroup to be carried on digital line plant.

transponder A term derived from TRANSmatter/resPONDER used to describe the radio repeater device associated with telecommunications satellites.

transport layer Layer 4 of the open systems interconnection (OSI) model. The transport layer provides for end-to-end data relaying service across any type of data network (e.g. packet network or circuit-switched network).

tropospheric scatter A form of radio wave reflection from the earth's troposphere, allowing over-the-horizon communication.

trunk A circuit interconnecting telephone exchanges. Particularly a long-distance circuit.

twisted pairs Metallic conductor wires, twisted in pairs corresponding to the legs of a 2-wire circuit.

virtual connection A means of conveying data from origin to destination without there ever being a complete direct path between the two. An example is a packet-switched connection. Imagine sending a book through the post in a number of envelopes, each containing only one numbered page. Receiver gets all the information, but never talks directly to sender.

weighted noise Weighted noise is a measure of noise in the middle of a channel bandwidth (in other words the noise which is hardest to remove by simple filtering).

X25 network/recommendation X25 is CCITT's recommendation for the protocol to be used between a data terminal device when connected to a packet network. In less strict usage packet networks are sometimes referred to as X25 networks.

X400 recommendation The CCITT recommendation defining the message handling system (described earlier).

X500 recommendation The CCITT recommendation defining the directory service, used in conjunction with X400 message handling systems.

2B + D interface An alternative term used to describe the ISDN basic data interface.

2-wire communication A communication path between end points using only two wires. Its difficulties and limitations arise from the shared use by transmit and receive signals of the same path.

4-wire communication A communication path between end points using four wires, two each for each direction of transmission, transmit and receive. Four-wire communication paths have better isolation of transmit and receive signals, and are easier to regenerate or repeat on long sections.

INDEX